T0305503

Homotopy-Based Methods in Water Engineering

Most complex physical phenomena can be described by nonlinear equations, specifically, differential equations. In water engineering, nonlinear differential equations play a vital role in modeling physical processes. Analytical solutions to strong nonlinear problems are not easily tractable, and existing techniques are problem-specific and applicable for specific types of equations. Exploring the concept of homotopy from topology, different kinds of homotopy-based methods have been proposed for analytically solving nonlinear differential equations, given by approximate series solutions. *Homotopy-Based Methods in Water Engineering* attempts to present the wide applicability of these methods to water engineering problems. It solves all kinds of nonlinear equations, namely algebraic/transcendental equations, ordinary differential equations (ODEs), systems of ODEs, partial differential equations (PDEs), systems of PDEs, and integro-differential equations using the homotopy-based methods. The content of the book deals with some selected problems of hydraulics of open-channel flow (with or without sediment transport), groundwater hydrology, surface-water hydrology, general Burger's equation, and water quality.

Features:

- Provides analytical treatments to some key problems in water engineering
- Describes the applicability of homotopy-based methods for solving nonlinear equations, particularly differential equations
- Compares different approaches in dealing with issues of nonlinearity

Homotopy-Based Methods in Water Engineering

Manotosh Kumbhakar
and Vijay P. Singh

CRC Press
Taylor & Francis Group
Boca Raton London New York

CRC Press is an imprint of the
Taylor & Francis Group, an **informa** business

Contents

PART II   Algebraic/Transcendental Equations

PART III   Ordinary Differential Equations (Single and System)

PART IV   *Partial Differential Equations (Single and System)*

PART V *Integro-Differential Equations*

Preface

Most of the complex physical phenomena are described by the nonlinear equation, specifically the differential equation. It is difficult to achieve analytical approximations to strong nonlinear problems. Basically, the techniques for the nonlinear equations are problem-specific and often break down as the nonlinearity becomes stronger. The concept of the perturbation theory showed a new direction to obtain an approximate solution to a nonlinear equation. However, its applicability is dependent on the presence of a small physical parameter in the governing equation and/or boundary or initial conditions. Moreover, the perturbation solutions are valid for a restricted region of the domain, making the rate and region of convergence limited. In the 1990s, Shijun Liao proposed a general analytical tool, known as the *homotopy analysis method* (HAM) that does not depend on a small physical parameter, unlike the perturbation method, and is valid for strongly nonlinear equations. The core idea of the method is based on *homotopy* – a concept of topology in abstract mathematics. The HAM provides an efficient way to control the rate and region of convergence of series solutions. Further, it is shown that many existing analytical techniques such as Adomian's decomposition method, Lyapunov's artificial small parameter method, δ-expansion method, etc., are special cases of HAM, which make it a unified approach. Almost at the same time, the *homotopy perturbation method* (HPM) was proposed by He, which follows the same concept as HAM but with certain restrictions. The other variant of HAM was proposed in 2008 by Marinca and Herisanu, known as the *optimal homotopy asymptotic method* (OHAM). In the past three decades, a wide variety of nonlinear problems appearing in fluid and solid mechanics, chemical engineering, biology, finance, theoretical physics, aerospace engineering, etc., have been solved by means of these homotopy-based methods.

In the area of water engineering, nonlinear equations (specifically, differential equations) play a vital role in modeling the physical processes. Several pieces of literature exist on analytical techniques applied to water engineering problems. However, there seems to be no book available on homotopy-based methods for water engineering readership. Therefore, looking into the successful applications of these methods to different branches of science and engineering, there exists a need for a book that introduces the homotopy-based methods and shows their applications to several nonlinear equations appearing in water engineering. This book attempts to serve this need. To show the wide applicability of these methods, all kinds of nonlinear equations, namely algebraic/transcendental equations, ordinary differential equations (ODEs), systems of ODEs, partial differential equations (PDEs), systems of PDEs, and integro-differential equations, are considered. In addition, the book puts more attention on HAM as it is fundamental among them all.

The subject matter of the book is divided into 17 chapters organized into five parts. Part I comprises two chapters, which deal with the introduction and preliminaries. Chapter 1 introduces the homotopy-based methods. In Chapter 2, first, the basic concepts of the methods are described in detail from a general point of view. Then, a specific example is considered to show the comparison between several analytical methods and the homotopy-based methods. Apart from that, the basic properties of some important functions and techniques are presented. Related codes are attached as supplements.

Part II contains Chapter 3, which shows the application of the methods to nonlinear algebraic/transcendental equations. Several examples are considered to show the effectiveness of the Newton-like methods based on HPM and HAM. It then goes on to consider the Colebrook equation, which is a nonlinear transcendental equation, and applies the homotopy-based Newton-like methods to find the roots of the equation numerically considering several test cases.

Part III consists of eight chapters dealing with the ordinary differential equation (ODE)–based models for both single equations and systems of equations. Chapter 4 solves the governing ODE using HAM, HPM, and OHAM, representing the velocity distribution in smooth open-channel flows. The governing equation is a first-order linear ODE with variable coefficients. In Chapter 5, sediment concentration distribution models are solved analytically using three methods. Two types of models are considered: one is for the linear shear stress profile, and the other is based on a refined shear stress equation. The exact solution of the latter is not available, and the homotopy-based methods provide an approximate series solution for that model. Chapter 6 applies HAM, HPM, and OHAM to obtain series solutions for the Richards equation under gravity-driven infiltration and constant rainfall intensity. The solutions are derived for the soil hydraulic conductivity function based on Toricelli's law and Brooks and Corey's formulation. In Chapter 7, the error equation for unsteady uniform flow is solved analytically using the homotopy-based methods. In addition to standard HAM, it also proposes a modified HAM approach, which first converts the domain into several subdomains and then applies HAM to each of those subdomains. The approach shows the importance of handling the nonlinear equations having a larger domain. Chapter 8 revisits the spatially varied flow equations when the flow is subcritical, with the objective of finding improved and efficient approximate analytical solutions in comparison to the existing perturbation-based solutions. Both the frictionless and frictional cases are taken into consideration to obtain solutions using HAM, HPM, and OHAM. In Chapter 9, the nonlinear differential equation with an arbitrary exponent that arises for a nonlinear reservoir used for rainfall runoff modeling is solved using the homotopy-based methods. It shows how to tackle the equations with noninteger power nonlinearity using the Padé approximant technique. Chapter 10 considers the frequently used flood routing approach, namely the Muskingum method, which is represented by a first-order ODE. Based on the storage-discharge relationship, the nonlinear equation having noninteger power nonlinearity is considered and solved using HAM, HPM, and OHAM to achieve efficient series solutions. In Chapter 11, a system of coupled nonlinear ODEs is considered that represents the vertical distribution of velocity and sediment concentration in open-channel flow. The system is solved using the homotopy-based method.

The application of the homotopy-based methods to PDEs is the subject of Part IV, which contains five chapters. Chapter 12 solves the unsteady confined radial groundwater flow equation using HAM, HPM, and OHAM. An equivalent boundary value problem is considered after converting the partial differential equation (PDE) into an ODE using similarity transformation. It also shows the comparison of the derived solutions with the well-known *Theis solution*. In Chapter 13, the homotopy-based methods are applied to solve Burger's equation. First, the PDE is converted into a second-order nonlinear ODE using a group-theoretic approach. Chapter 14 solves the PDE of the diffusive wave equation with later inflow. It directly solves the PDE using HAM, HPM, and OHAM. In Chapter 15, the kinematic wave equation is solved using the homotopy-based methods. To deal with the noninteger power nonlinearity, the original PDE is first converted into a system of ODEs by discretizing it in one direction. Chapter 16 considers a system of PDEs representing the multispecies convection-dispersion transport equation with variable parameters. It solves the PDE using HAM, HPM, and OHAM considering different test cases.

Part V contains Chapter 17 that considers an ordinary integro-differential equation representing the absorption equation in unsaturated soils. The homotopy-based methods are applied to obtain the approximate series solutions. In addition, the derived solutions are compared with the classical *Philip's solution*.

Manotosh Kumbhakar
Vijay P. Singh
College Station, Texas

About the Authors

Manotosh Kumbhakar is a postdoctoral researcher at National Taiwan University, Taiwan. Previously, he was a postdoctoral research associate at Texas A&M University, USA, from 2020 to 2021. He specializes in entropy theory, mechanics of sediment transport, and semi-analytical methods. Dr. Manotosh has published several papers in reputed international journals.

Vijay P. Singh, PhD, DSc, D Eng (Hon), PhD (Hon), PE, PH, Hon D WRE, DistM ASCE, NAE, is a distinguished professor, a Regents professor, and Caroline & William N. Lehrer Distinguished Chair in Water Engineering in the Department of Biological and Agricultural Engineering and Zachry Department of Civil & Environmental Engineering at Texas A&M University. He specializes in surface-water hydrology, groundwater hydrology, hydraulics, irrigation engineering, environmental and water resources engineering, entropy theory, and copula theory. Professor Singh has published extensively and has received 110 national and international awards, including three honorary doctorates. He is a member of the National Academy of Engineering and a fellow or member of 12 international engineering/science academies.

Part I

Introduction

1 Introduction

Natural phenomena are often described by differential equations. Depending on the linearity or nonlinearity of these phenomena, the differential equations are also either linear or nonlinear. Linear differential equations are relatively easy to solve, and a multitude of theories are available. However, most of the natural phenomena are nonlinear and are described by nonlinear differential equations. For solving a nonlinear differential equation, both analytical and numerical methods are available. Due to the development of high-performance computers, several numerical methods have been developed for solving nonlinear differential equations in complex domains. However, numerical methods may become unstable and are not easily amenable to accommodate complex boundaries and may sometimes lack accuracy, and the error needs to be controlled to obtain a satisfactory approximation to the *true* solution. Further, numerical methods are problem-specific and provide values at discontinuous points, and hence it is costly and time-consuming to get the complete solution. Also, it is difficult to numerically solve nonlinear equations with singularity or multiple solutions.

An analytical solution is preferable, for it gives insights into the general properties of the solution. However, traditional analytical methods are mostly equation-specific and are for simple equations. Therefore, one often seeks an approximate analytical solution to the nonlinear differential equation because most of the time exact analytical solutions are difficult to find. To sum up, both numerical and analytical solutions have their own advantages and disadvantages, but it is desirable to have an analytical solution.

One of the widely applied methods for finding an approximate solution to a nonlinear differential equation is the perturbation technique, which is essentially based on the existence of a small parameter known as the *perturbation quantity*. The perturbation method converts a nonlinear differential equation into an infinite number of linear subdifferentials with the aid of the perturbation parameter, and the final solution is obtained by considering the first few terms of approximation. This method has proved to be of great success and is used in a variety of areas, including the study of planetary motion (Nayfeh 2000). However, the existence of a small parameter in the governing equation and/or boundary conditions and the breakdown of the solution for strongly nonlinear equations limit the applicability of the method. In 1892, Lyapunov introduced an artificial small-parameter technique that can avoid the dependence on the small parameter (Lyapunov 1892). This method was further generalized, which led to the development of the δ-expansion method (Karmishin et al. 1990). Both the Lyapunov method and the generalized method showed an improvement over the perturbation method for the differential equations considered. However, a fundamental way to place the artificial parameter at an appropriate place in the equation is still lacking.

One of the well-known non-perturbation methods was developed by George Adomian and is commonly known as the Adomian decomposition method (ADM). The ADM is a powerful analytical tool for solving strongly nonlinear differential

DOI: 10.1201/9781003368984-2

equations and does not require the presence of a small/large parameter (Adomian 1994). This method decomposes the original nonlinear equation into linear and nonlinear parts and provides an approximate solution, which often contains polynomials. However, the power series solution has a small region of convergence, which restricts its applicability. Another non-perturbation method, known as the homotopy perturbation method (HPM), is rather a general way of analytically solving nonlinear equations (He 1999). This method uses the fundamental concept of *homotopy* from topology to gain series solutions, although many times the solutions may not converge due to the lack of freedom for the choice of base function, restricted region, and rate of convergence.

Despite the advances made in developing methods for analytically solving nonlinear equations as mentioned earlier in the text, there is a need for a unified method having three aspects: (i) development of a unified method that is valid for strongly nonlinear equations and is not confined to the presence of a small parameter, (ii) finding a convenient way to monitor the region and rate of convergence of the series solution, and (iii) freedom to choose various kinds of base functions to approximate a nonlinear differential equation. In 1992, Shijun Liao developed a powerful analytical method, popularly known as the homotopy analysis method (HAM), for solving nonlinear equations efficiently, even if they contain strong nonlinearity (Liao 1992).

For the last two decades, HAM has been applied successfully to different branches of science and engineering (Liao 2003). With the use of this method, it has been possible to obtain some groundbreaking solutions that had not been obtained earlier even by numerical methods. HAM has been employed to solve some complicated nonlinear partial differential equations (PDEs) and enrich and deepen our physical understanding. For example, the wave-resonance criterion for an arbitrary number of traveling waves with large amplitudes was found by Liao in 2011, for the first time, by means of HAM, which logically contains the famous Phillips' criterion for four waves with small amplitudes (Liao 2011). Although HAM has proved to be a powerful analytical tool for nonlinear equations, one often needs to compute a large number of iterations to obtain an accurate solution. This disobeys the definition of the so-called *classical* analytical solution, which can be written explicitly in a single or few terms. With this aim, Marinca and Herisanu proposed a method known as the optimal homotopy asymptotic method (OHAM) by completely modifying HAM (Marinca and Herisanu 2012). This method can provide an approximate analytical solution to a nonlinear problem by considering only a few terms of an infinite series. OHAM has been successfully applied to a variety of nonlinear equations, from heat transfer to Navier-Stokes equations. Because of the enormous success of HAM, efforts are continuing to either modify it to overcome its weaknesses or apply it to new problems. For example, in 2016, Liao proposed the method of directly defining inverse mapping (MDDiM), which does not require one to compute higher-order terms using the inverse linear operator (Liao and Zhao 2016). For computers, calculation of the inverse linear operator is a tiresome task and hence needs a lot of central processing unit (CPU) time. Using MDDiM, one can compute higher-order terms without much difficulty.

Although the application of non-perturbation methods (specifically, HAM) to different branches of science and engineering is a well-established area of study, these methods are not much known in the water engineering field. It is therefore desirable to apply the HAM and its variants to a variety of problems in water engineering. This constituted the motivation for preparing this book. To appreciate the usefulness of the method, the book is written in a simple and understandable way illustrated with figures and tables. This book attempts to solve all kinds of nonlinear equations, namely algebraic/transcendental equations, ordinary differential equations (single/system), partial differential equations (single/system), and integro-differential equations arising in water engineering using homotopy-based methods. The content of the book is organized as follows: Chapters 3–5, 8, and 11 deal with the problems of hydraulics of open-channel flow (with or without sediment transport); Chapters 6–7, 12, and 17 solve equations from groundwater hydrology; Chapters 9, 10, and 15–16 focus on the problems of surface-water hydrology; and Chapters 13 and 14 deal with the general Burger's equation and a problem of water quality, respectively.

REFERENCES

Adomian, G. (1994). *Solving frontier problems of physics: the decomposition method*. Kluwer Academic Publishers, Boston.

He, J. H. (1999). Homotopy perturbation technique. *Computer Methods in Applied Mechanics and Engineering*, *178*(3–4), 257–262.

Karmishin, A. V., Zhukov, A. I., and Kolosov, V. G. (1990). *Methods of dynamics calculation and testing for thin-walled structures*. Mashinostroyenie, Moscow, 137–149.

Liao, S. J. (2003). *Beyond perturbation: introduction to the homotopy analysis method*. Chapman & Hall/CRC Press, Boca Raton.

Liao, S. J. (2011). On the homotopy multiple-variable method and its applications in the interactions of nonlinear gravity waves. *Communications in Nonlinear Science and Numerical Simulation*, *16*(3), 1274–1303.

Liao, S. J. (1992). *The proposed homotopy analysis technique for the solution of nonlinear problems* (Doctoral dissertation, PhD thesis, Shanghai Jiao Tong University).

Liao, S. J., and Zhao, Y. (2016). On the method of directly defining inverse mapping for nonlinear differential equations. *Numerical Algorithms*, *72*(4), 989–1020.

Lyapunov, A. M. (1892). The general problem of motion stability. *Annals of Mathematics Studies*, *17*, 1892.

Marinca, V., and Herisanu, N. (2012). *Nonlinear dynamical systems in engineering: some approximate approaches*. Springer–Verlag, Berlin, Heidelberg.

Nayfeh, A. H. (2000). *Perturbation methods*. John Wiley & Sons, New York.

2 Basic Concepts

2.1 DEFINITION OF HOMOTOPY

The core idea of the homotopy analysis method (and also the homotopy perturbation method and optimal homotopy asymptotic method) is based on the concept of *homotopy* from topology, which is a study concerned with the geometric properties of a mathematical object that remains unaffected under continuous deformation (Armstrong 1983). Homotopy defines a connection between two mathematical objects. Specifically, two objects are homotopic if one can be *continuously deformed* into the other. For example, in two dimensions, a circle can be continuously deformed into an ellipse or a square; similarly, in three dimensions, a doughnut and a coffee cup are homotopic. However, a coffee cup cannot be deformed continuously into a football.

If we consider two continuous functions $f(t)$ and $g(t)$ of a dimension t, say time or space, a homotopy can always be defined as $\mathcal{H}(t;q) = (1-q)f(t) + qg(t)$, where $q \in [0,1]$ is an embedding parameter. A homotopy $\mathcal{H}(t;q)$ between these two functions is itself a continuous function, defined as $\mathcal{H}: T \times [0,1] \to U$ and satisfies $\mathcal{H}(t;q) = f(t)$ when $q = 0$ and $\mathcal{H}(t;q) = g(t)$ when $q = 1$, where T and U are the topological spaces. This shows that as q goes from 0 to 1, $\mathcal{H}(t;q)$ varies from $f(t)$ to $g(t)$. The functions $f(t)$ and $g(t)$ are called homotopic. To develop an intuitive idea, we consider two examples where (i) $f(t) = \sin(t)$ and $g(t) = \cos(t)$ over the interval $\left[0, \dfrac{\pi}{2}\right]$ and (ii) $f(t) = t^2$ and $g(t) = -t^2$ over the interval $[-5,5]$. We can define the homotopies as (i) $\mathcal{H}(t;q) = (1-q)\sin(t) + q\cos(t)$ and (ii) $\mathcal{H}(t;q) = (1-q)t^2 + q\left(-t^2\right)$ for $q \in [0,1]$. Figure 2.1 shows that $\mathcal{H}(t;q)$ varies *continuously* from $f(t)$ to $g(t)$ as q goes from 0 to 1. Since algebraic or differential equations represent curves (or functions), the concept of homotopy can be employed for solving nonlinear differential or algebraic equations.

2.2 HOMOTOPY PERTURBATION METHOD

Applying the concept of homotopy, a new perturbation technique that does not depend on the presence of small parameters was formulated by He (1999). The technique is known as the homotopy perturbation method (HPM). We describe the technique for a nonlinear differential equation in what follows. For illustration, we consider the following nonlinear differential equation:

$$\mathcal{A}[v(\mathbf{s},t)] = f(\mathbf{s},t) \tag{2.1}$$

subject to the boundary conditions:

$$\mathcal{B}\left(v, \frac{\partial v}{\partial n}\right) = 0 \tag{2.2}$$

DOI: 10.1201/9781003368984-3

7

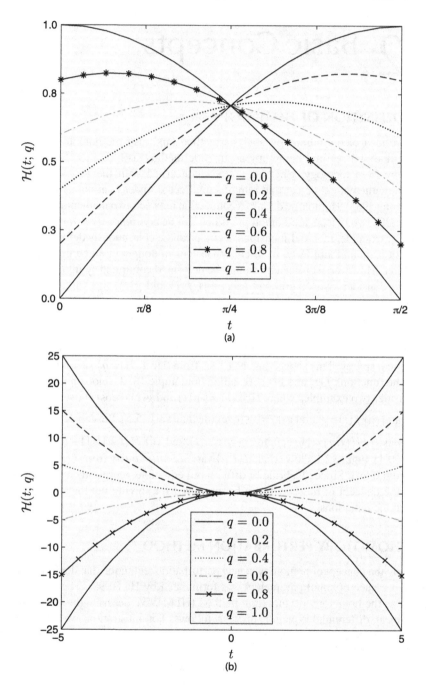

FIGURE 2.1   Plot of homotopy $\mathcal{H}(t;q)$ between functions (a) $\sin(t)$ and $\cos(t)$ over the interval $\left[0, \dfrac{\pi}{2}\right]$ and (b) t^2 and $-t^2$ over the interval $[-5,5]$ for $q = 0$, 0.2, 0.4, 0.6, 0.8, and 1.

where $\mathcal{A}$ and $\mathcal{B}$ are the general differential operator and boundary operator, respectively; s and t are the spatial and temporal (independent) variables, respectively; v is the unknown (dependent) variable; and f is the right side of the differential equation involving independent variables only.

In general, the differential operator $\mathcal{A}$ can be split into two parts, say $\mathcal{L}$ and $\mathcal{N}$, where $\mathcal{L}$ is the linear operator and $\mathcal{N}$ is the nonlinear operator. For example, we consider two ordinary differential equations (ODEs) and one partial differential equation (PDE) as (i) $\dfrac{dv}{dt}+v^2=1$, (ii) $\dfrac{d^2v}{dt^2}+v\dfrac{dv}{dt}+exp(v)=0$, and (iii) $\dfrac{\partial^2 v}{\partial x^2}+v\dfrac{\partial v}{\partial x}+\dfrac{\partial^2 v}{\partial y^2}+\dfrac{\partial^2 v}{\partial t^2}+\dfrac{\partial v}{\partial y}\dfrac{\partial v}{\partial t}+v^2-xyt^2=0$. In relation to Eq. (2.1), the operators for these equations can be given as (i) $\mathcal{A}\big[v(t)\big]=\dfrac{dv}{dt}+v^2$, $\mathcal{L}\big[v(t)\big]=\dfrac{dv}{dt}$, $\mathcal{N}\big[v(t)\big]=v^2$, and $f(t)=1$;(ii) $\mathcal{A}\big[v(t)\big]=\dfrac{d^2v}{dt^2}+v\dfrac{dv}{dt}+exp(v),\mathcal{L}\big[v(t)\big]=\dfrac{d^2v}{dt^2},\mathcal{N}\big[v(t)\big]=v\dfrac{dv}{dt}+exp(v)$, and $f(t)=0$; and (iii) $\mathcal{A}\big[v(x,y,t)\big]=\dfrac{\partial^2 v}{\partial x^2}+v\dfrac{\partial v}{\partial x}+\dfrac{\partial^2 v}{\partial y^2}+\dfrac{\partial^2 v}{\partial t^2}+\dfrac{\partial v}{\partial y}\dfrac{\partial v}{\partial t}+v^2,\mathcal{L}\big[v(t)\big]=\dfrac{\partial^2 v}{\partial x^2}+\dfrac{\partial^2 v}{\partial y^2}+\dfrac{\partial^2 v}{\partial t^2}$, $\mathcal{N}\big[v(x,y,t)\big]=v\dfrac{\partial v}{\partial x}+\dfrac{\partial v}{\partial y}\dfrac{\partial v}{\partial t}+v^2$, and $f(x,y,t)=xyt^2$. For the other methods, namely homotopy asymptotic method (HAM) and optimal homotopy asymptotic method (OHAM), the same approach should be followed in identifying the operators of the differential equations. Therefore, Eq. (2.1) can be rewritten as follows:

$$\mathcal{L}\big[v(s,t)\big]+\mathcal{N}\big[v(s,t)\big]-f(s,t)=0 \tag{2.3}$$

Using the concept of homotopy, one can construct a homotopy as follows:

$$\mathcal{H}\big(\Phi(s,t);q\big)=(1-q)\Big[\mathcal{L}\big[\Phi(s,t;q)\big]-\mathcal{L}\big[v_0(s,t)\big]\Big]$$
$$+q\Big[\mathcal{A}\big[\Phi(s,t;q)\big]-f(s,t)\Big]=0 \tag{2.4}$$

Using Eq. (2.3), Eq. (2.4) can be equivalently written as:

$$\mathcal{H}\big(\Phi(s,t);q\big)=\mathcal{L}\big[\Phi(s,t;q)\big]-\mathcal{L}\big[v_0(s,t)\big]+q\mathcal{L}\big[v_0(s,t)\big]$$
$$+q\Big[\mathcal{A}\big[\Phi(s,t;q)\big]-\mathcal{L}\big[\Phi(s,t;q)\big]-f(s,t)\Big]=\mathcal{L}\big[\Phi(s,t;q)\big]$$
$$-\mathcal{L}\big[v_0(s,t)\big]+q\mathcal{L}\big[v_0(s,t)\big]+q\Big[\mathcal{N}\big[\Phi(s,t;q)\big]-f(s,t)\Big]=0 \tag{2.5}$$

where $q\in[0,1]$ is an embedding parameter, and $v_0(s,t)$ is an initial approximation to the solution that satisfies the boundary conditions. Now, at $q=0$ and $q=1$, we have from Eq. (2.4):

$$\mathcal{H}\big(\Phi(s,t);0\big)=\mathcal{L}\big[\Phi(s,t;0)\big]-\mathcal{L}\big[v_0(s,t)\big]=0 \tag{2.6}$$

$$\mathcal{H}\big(\Phi(s,t);1\big)=\mathcal{A}\big[\Phi(s,t;1)\big]-f(s,t)=0 \tag{2.7}$$

This shows that the process of changing q from 0 to 1 corresponds to the change from the initial approximation $v_0(s,t)$ to the final solution $v(s,t)$. It can be seen that the expressions $\mathcal{L}\left[\Phi(s,t;q)\right]-\mathcal{L}\left[v_0(s,t)\right]$ and $\mathcal{A}\left[\Phi(s,t;q)\right]-f(s,t)$ are homotopic.

The embedding parameter q can be considered a 'small' quantity as $0\leq q\leq1$. Therefore, expanding the solution in a power series of q, one obtains:

$$\Phi(s,t;q)=\Phi_0(s,t)+q\Phi_1(s,t)+q^2\Phi_2(s,t)+\cdots \tag{2.8}$$

where $\Phi_m(s,t)$, $m\geq1$ are the higher-order terms of the series. Finally, the solution can be obtained as:

$$\begin{aligned}v(s,t)&=\lim_{q\to1}\left[\Phi_0(s,t)+q\Phi_1(s,t)+q^2\Phi_2(s,t)+\cdots\right]\\&=\Phi_0(s,t)+\Phi_1(s,t)+\Phi_2(s,t)+\cdots\end{aligned} \tag{2.9}$$

As mentioned earlier, $v_0(s,t)$ is an initial approximation to the solution. One can choose a form for $v_0(s,t)$ subject to the condition that it satisfies the given boundary conditions. On the other hand, the terms $\Phi_i(s,t)$ for $i\geq1$ can be obtained by substituting Eq. (2.8) into Eq. (2.5) and then equating the like powers of q. The steps concerning the methodology will be provided in a flowchart diagram in later subsections considering a particular example.

2.3   HOMOTOPY ANALYSIS METHOD

The concept of homotopy was applied by Liao (1992) to solve nonlinear differential equations, giving rise to the well-known HAM. Here, the methodology is described for a (single) general nonlinear differential equation. The methodology can be employed in a similar manner to a system of ODEs and PDEs. The sequence of steps constituting the method will be given in a flowchart diagram in a later subsection when a particular example is considered. For illustration, we consider a nonlinear differential equation:

$$\mathcal{N}\left[v(s,t)\right]=0 \tag{2.10}$$

subject to the boundary conditions:

$$\mathcal{B}\left(v,\frac{\partial v}{\partial n}\right)=0 \tag{2.11}$$

where $\mathcal{N}$ is a nonlinear operator or the operator of the original equation; $v(s,t)$ is the unknown (or dependent) variable; and s and t are the spatial and temporal (independent) variables, respectively. The aim is to find an approximate solution to Eq. (2.10). Now, constructing the homotopy $\mathcal{H}$ and setting it to zero, we get:

$$\begin{aligned}\mathcal{H}\left(\Phi(s,t;q);q,\hbar,\;H(s,t)\right)&=(1-q)\mathcal{L}\left[\Phi(s,t;q)-v_0(s,t)\right]\\&\quad-q\hbar H(s,t)\mathcal{N}\left[\Phi(s,t;q)\right]=0\end{aligned} \tag{2.12}$$

where $q \in [0,1]$ is the embedding-parameter; $\hbar$ is a non-zero auxiliary parameter; $H(s,t)$ is a non-zero auxiliary function; $\Phi(s,t;q)$ is an unknown function; $v_0(s,t)$ is an initial approximation to the solution $v(s,t)$; and $\mathcal{L}$ is a linear operator with the property:

$$\mathcal{L}[g(s,t)] = 0 \text{ when } g(s,t) = 0, \text{ } g \text{ being any function} \qquad (2.13)$$

It can be seen that the solution of Eq. (2.12) varies (or deforms) continuously with respect to q. The core idea of HAM is that a continuous mapping is employed to relate the solution $v(s,t)$ and the unknown function $\Phi(s,t;q)$ with the aid of the embedding parameter q. Mathematically, when $q = 0$, Eq. (2.12) becomes:

$$\mathcal{L}[\Phi(s,t;0) - v_0(s,t)] = 0 \qquad (2.14)$$

Using the property of Eq. (2.13), one obtains from Eq. (2.14):

$$\Phi(s,t;0) = v_0(s,t) \qquad (2.15)$$

Now, at $q = 1$, Eq. (2.12) takes on the form:

$$\hbar H(s,t) \mathcal{N}[\Phi(s,t;1)] = 0 \qquad (2.16)$$

Since $\hbar \neq 0$, $H(s,t) \neq 0$, we get:

$$\mathcal{N}[\Phi(s,t;1)] = 0 \qquad (2.17)$$

which shows that any function $\Phi(s,t;1)$ satisfies the original nonlinear Eq. (2.10). Therefore, as q goes from 0 to 1, the unknown function $\Phi(s,t;q)$ varies from the initial approximation $v_0(s,t)$ to the final solution $v(s,t)$. In other words, the function $\Phi(s,t;q)$ agrees with the initial approximation at $q = 0$ and with a solution of the nonlinear equation when $q = 1$. To this end, Eq. (2.12) is known as the *zeroth-order deformation equation*.

Initially, in the formulation of HAM, the auxiliary parameter $\hbar$ and the auxiliary function $H(s,t)$ were not included, or, equivalently, $\hbar = -1$ was assumed. The essence of the auxiliary parameter in the framework of HAM will be discussed later. Also, it can be noted that unlike perturbation methods (or other similar analytical methods), HAM is not directly dependent on a small parameter. Here, the homotopy parameter q plays the role of a small parameter as $0 \leq q \leq 1$, for which the solution is expressed in a power series of q, and hence the method does not depend on the nature of the physical parameters present in the equation.

We now define

$$v_m(s,t) = \frac{1}{m!} \frac{\partial^m \Phi(s,t;q)}{\partial q^m}\bigg|_{q=0} \qquad (2.18)$$

where $\dfrac{\partial^m \Phi(s,t;q)}{\partial q^m}\bigg|_{q=0}$ is called the *m-th−order deformation derivative*. Eq. (2.18) provides the higher-order terms v_m in terms of derivatives of the unknown function

$\Phi(s,t;q)$ that deforms continuously with respect to q. Using Maclaurin series expansion, one can expand $\Phi(s,t;q)$ in a power series with respect to the embedding parameter q as follows:

$$\Phi(s,t;q) = \Phi(s,t;0) + \sum_{m=1}^{\infty} \frac{1}{m!} \frac{\partial^m \Phi(s,t;q)}{\partial q^m}\bigg|_{q=0} q^m = v_0(s,t) + \sum_{m=1}^{\infty} v_m(s,t) q^m \quad (2.19)$$

Eq. (2.19) shows that the solution can be written as a combination of the initial approximation $v_0(s,t)$ and the higher-order terms $v_m(s,t)$ for $m \geq 1$. It will be seen shortly that $v_m(s,t)$ is governed by a system of linear differential equations, depending upon the terms $\mathcal{L}, H(s,t), v_0(s,t)$, and the parameter $\hbar$. To that end, it is assumed that $\mathcal{L}, H(s,t), v_0(s,t)$, and $\hbar$ are properly chosen such that the series Eq. (2.19) converges at $q = 1$. Then, at $q = 1$, the series becomes:

$$v(s,t) = v_0(s,t) + \sum_{m=1}^{\infty} v_m(s,t) \quad (2.20)$$

Eq. (2.20) provides an explicit relationship between the initial approximation $v_0(s,t)$ and the final solution $v(s,t)$. However, to obtain the full solution, the higher-order approximations $v_m(s,t)$ for $m \geq 1$ need to be determined. To that end, differentiating the zeroth-order deformation in Eq. (2.12) with respect to q, we have:

$$(1-q)\mathcal{L}\left[\frac{\partial \Phi(s,t;q)}{\partial q}\right] - \mathcal{L}\left[\Phi(s,t;q) - v_0(s,t)\right] = \hbar H(s,t)\mathcal{N}\left[\Phi(s,t;q)\right]$$

$$+ q\hbar H(s,t)\frac{\partial \mathcal{N}\left[\Phi(s,t;q)\right]}{\partial q} \quad (2.21)$$

It may be noted that $\dfrac{\partial}{\partial q}\left[\mathcal{L}(\blacksquare)\right] = \mathcal{L}\left(\dfrac{\partial}{\partial q}[\blacksquare]\right)$ since $\mathcal{L}$ is a linear operator. Now, using $\Phi(s,t;0) = v_0(s,t)$ and Eq. (2.18) and then setting $q = 0$ in Eq. (2.21), one obtains:

$$\mathcal{L}\left[v_1(s,t)\right] = \hbar H(s,t)\mathcal{N}\left[v_0(s,t)\right] \quad (2.22)$$

Again, differentiating Eq. (2.12) two times, or equivalently differentiating Eq. (2.21), with respect to q, we have:

$$(1-q)\mathcal{L}\left[\frac{\partial^2 \Phi(s,t;q)}{\partial q^2}\right] - \mathcal{L}\left[\frac{\partial \Phi(s,t;q)}{\partial q}\right] - \mathcal{L}\left[\frac{\partial \Phi(s,t;q)}{\partial q}\right]$$

$$= \hbar H(s,t)\frac{\partial \mathcal{N}\left[\Phi(s,t;q)\right]}{\partial q} + \hbar H(s,t)\frac{\partial \mathcal{N}\left[\Phi(s,t;q)\right]}{\partial q}$$

$$+ q\hbar H(s,t)\frac{\partial^2 \mathcal{N}\left[\Phi(s,t;q)\right]}{\partial q^2} \Rightarrow (1-q)\mathcal{L}\left[\frac{\partial^2 \Phi(s,t;q)}{\partial q^2}\right] - 2\mathcal{L}\left[\frac{\partial \Phi(s,t;q)}{\partial q}\right]$$

$$= 2\hbar H(s,t)\frac{\partial \mathcal{N}\left[\Phi(s,t;q)\right]}{\partial q} + q\hbar H(s,t)\frac{\partial^2 \mathcal{N}\left[\Phi(s,t;q)\right]}{\partial q^2} \quad (2.23)$$

Now, using Eq. (2.18), setting $q = 0$, and then dividing by 2, we have from Eq. (2.23):

$$2\mathcal{L}\big[v_2(s,t)\big] - 2\mathcal{L}\big[v_1(s,t)\big] = 2\hbar H(s,t)\frac{\partial \mathcal{N}\big[\Phi(s,t;q)\big]}{\partial q}\bigg|_{q=0}$$

$$\Rightarrow \mathcal{L}\big[v_2(s,t) - v_1(s,t)\big] = \hbar H(s,t)\frac{\partial \mathcal{N}\big[\Phi(s,t;q)\big]}{\partial q}\bigg|_{q=0} \qquad (2.24)$$

Differentiating Eq. (2.12) three times, or equivalently differentiating Eq. (2.23), with respect to q, we have:

$$(1-q)\mathcal{L}\left[\frac{\partial^3 \Phi(s,t;q)}{\partial q^3}\right] - 3\mathcal{L}\left[\frac{\partial^2 \Phi(s,t;q)}{\partial q^2}\right] = 3\hbar H(s,t)\frac{\partial^2 \mathcal{N}\big[\Phi(s,t;q)\big]}{\partial q^2}$$

$$+ q\hbar H(s,t)\frac{\partial^3 \mathcal{N}\big[\Phi(s,t;q)\big]}{\partial q^3} \qquad (2.25)$$

Now, using Eq. (2.18), setting $q = 0$, and then dividing by 6, we have from Eq. (2.25):

$$6\mathcal{L}\big[v_3(s,t)\big] - 6\mathcal{L}\big[v_2(s,t)\big] = 3\hbar H(s,t)\frac{\partial^2 \mathcal{N}\big[\Phi(s,t;q)\big]}{\partial q^2}\bigg|_{q=0}$$

$$\Rightarrow \mathcal{L}\big[v_3(s,t) - v_2(s,t)\big] = \frac{1}{2}\hbar H(s,t)\frac{\partial \mathcal{N}\big[\Phi(s,t;q)\big]}{\partial q}\bigg|_{q=0} \qquad (2.26)$$

Proceeding in a similar way, we can arrive at a recurrence relation for the terms $v_m(s,t)$. In other words, differentiating the zeroth-order deformation Eq. (2.12) m times with respect to q, setting $q = 0$, and then dividing by $m!$, one obtains:

$$\mathcal{L}\big[v_m(s,t) - \chi_m v_{m-1}(s,t)\big] = \hbar H(s,t)R_m(\vec{v}_{m-1}), \; m = 1,2,3,\ldots \qquad (2.27)$$

where

$$\chi_m = \begin{cases} 0 & \text{when } m = 1, \\ 1 & \text{otherwise} \end{cases} \qquad (2.28)$$

and

$$R_m(\vec{v}_{m-1}) = \frac{1}{(m-1)!}\frac{\partial^{m-1} \mathcal{N}\big[\Phi(s,t;q)\big]}{\partial q^{m-1}}\bigg|_{q=0} \qquad (2.29)$$

It can be seen from Eq. (2.27) that the HAM converts the original nonlinear equation into an infinite system of linear equations that are easier to solve, subject to the choice of the linear operator $\mathcal{L}$. The convergence of the series solution, based on HAM, depends on the choice of the initial approximation $v_0(s,t)$, linear operator $\mathcal{L}$, and auxiliary function $H(s,t)$. To choose these three components appropriately,

Liao (2003) proposed three generalized rules: (1) *rule of solution expressions*, (2) *rule of coefficient ergodicity*, and (3) *rule of solution existence*. These rules will be discussed in detail in a later section.

2.4 OPTIMAL HOMOTOPY ASYMPTOTIC METHOD

The homotopy method has been hugely successful in solving strong nonlinear equations arising in science and engineering. However, most of the time, we may need to consider a large number of terms to find an accurate approximation to the true solution, which can be time-consuming and is not very elegant. It is also true that with the use of high-performing symbolic computational software, such as MATLAB, Mathematica, etc., it is easy to implement HAM (MATLAB 2020, Wolfram Research 2014). Intending to develop an analytical technique that needs only a few terms to produce a sufficiently accurate solution of a nonlinear differential equation, Marinca and Herisanu (2008) proposed a somewhat different HAM, known as the OHAM, which is described in what follows. The steps concerning the method will be given in a flowchart diagram in a later subsection where a particular example is considered.

We consider a nonlinear ODE:

$$\mathcal{L}[v(s,t)] + \mathcal{N}[v(s,t)] + h(s,t) = 0 \tag{2.30}$$

subject to the initial/boundary conditions:

$$B\left[v(s,t), \frac{\partial v(s,t)}{\partial n}\right] = 0 \tag{2.31}$$

where s and t are the independent variables, $v(s,t)$ is the dependent variable, $\mathcal{L}$ is the linear operator, $\mathcal{N}$ is the nonlinear operator, $h(s,t)$ is a known function, and B is the boundary operator. Following HAM, one can construct a family of equations as follows:

$$(1-q)\left[\mathcal{L}\big(\Phi(s,t,C_i;q)\big) + h(s,t)\right]$$
$$= H(s,t,C_i;q)\left[\mathcal{L}\big(\Phi(s,t,C_i;q)\big) + h(s,t) + \mathcal{N}\big(\Phi(s,t,C_i;q)\big)\right] \tag{2.32}$$

where $q \in [0, 1]$ is the embedding parameter, $\Phi(s,t,C_i;q)$ is the unknown function, C_i are the unknown parameters, and $H(s,t,C_i;q)$ is the auxiliary function defined as:

$$H(s,t,C_i;q) = 0 \text{ for } q = 0$$

$$\neq 0 \text{ for } q \in (0,1] \tag{2.33}$$

When $q = 0$ and $q = 1$, we get from Eq. (2.32):

$$\Phi(s,t,C_i;0) = v_0(s,t) \tag{2.34}$$

$$\Phi(s,t,C_i;1) = v(s,t,C_i) \tag{2.35}$$

It is seen from Eqs. (2.34) and (2.35) that as q goes from 0 to 1, the unknown $\Phi(s,t,C_i;q)$ varies from $v_0(s,t)$ to the final solution $v(s,t,C_i)$. The initial approximation $v_0(s,t)$ can be found by solving the following equation, which is obtained after substituting $q = 0$ in Eq. (2.32):

$$\mathcal{L}\left(v_0(s,t)\right) + h(s,t) = 0 \tag{2.36}$$

subject to the boundary conditions:

$$B\left[v_0(s,t), \frac{\partial v_0(s,t)}{\partial n}\right] = 0 \tag{2.37}$$

The auxiliary function $H(s,t,C_i;q)$ can be expressed as follows:

$$H(s,t,C_i;q) = qH_1(s,t,C_i) + q^2 H_2(s,t,C_i) + q^3 H_3(s,t,C_i) + \cdots \tag{2.38}$$

where $H_i(s,t,C_i)$ are the auxiliary functions that depend on the independent variable t and parameters C_i. It can be noted that the series Eq. (2.38) is in accordance with the property Eq. (2.33).

The solution of Eq. (2.30) can be assumed to be of the form:

$$\Phi(s,t,C_i;q) = v_0(s,t) + \sum_{j=1}^{\infty} v_j(s,t,C_i) q^j \tag{2.39}$$

Substituting Eq. (2.39) into Eq. (2.32) and equating the like powers of q, the following equations are obtained [q^0 corresponds to Eqs. (2.36) and (2.37)]:

$$\mathcal{L}\left(v_1(s,t,C_i)\right) = H_1(s,t,C_i)\mathcal{N}_0\left(v_0(s,t)\right) \tag{2.40}$$

subject to the boundary conditions:

$$B\left[v_1(s,t), \frac{\partial v_1(s,t)}{\partial n}\right] = 0 \tag{2.41}$$

For $k = 2,3,4,\ldots,$

$$\mathcal{L}\left[v_k(s,t,C_i) - v_{k-1}(s,t,C_i)\right] = H_k(s,t,C_i)\mathcal{N}_0\left(v_0(s,t)\right)$$

$$+ \sum_{j=1}^{k-1} H_j(s,t,C_i)\left[\mathcal{L}\left[v_{k-j}(s,t,C_i)\right] + \mathcal{N}_{k-j}\left[v_0(s,t), v_1(s,t,C_i),\ldots,v_{k-j}(s,t,C_i)\right]\right] \tag{2.42}$$

subject to the boundary conditions:

$$B\left[v_k(s,t), \frac{\partial v_k(s,t)}{\partial n}\right] = 0 \tag{2.43}$$

where the term $\mathcal{N}_{k-j}\left[v_0(s,t), v(s,t,C_i),\ldots,v(s,t,C_i)\right]$ is the coefficient of q^m, which is obtained by expanding $\mathcal{N}\left(\Phi(s,t,C_i;q)\right)$ as follows:

$$\mathcal{N}\left(\Phi\left(\mathbf{s},\mathrm{t},C_i;q\right)\right)=\mathcal{N}_0\left(v_0\left(\mathbf{s},t\right)\right)+q\mathcal{N}_1\left(v_0\left(\mathbf{s},t\right),v_1\left(\mathbf{s},t,C_i\right)\right)$$
$$+q^2\mathcal{N}_2\left(v_0\left(\mathbf{s},t\right),v_1\left(\mathbf{s},t,C_i\right),v_2\left(\mathbf{s},t,C_i\right)\right)+\cdots \quad (2.44)$$

It can be observed from Eqs. (2.40) and (2.42) that similar to HAM, OHAM also converts the original nonlinear equation into an infinite set of linear sub-equations. Further, the method does not depend on the presence of a small parameter in the governing equation. The convergence of the series in Eq. (2.39) depends on the choice of $H_j\left(\mathbf{s},t,C_i\right)$, and there exist many ways to choose it. According to Marinca and Herisanu (2015), one can choose $H_j\left(\mathbf{s},t,C_i\right)$ in such a way that the product $H_j\left(\mathbf{s},t,C_i\right)\left[\mathcal{L}\left[v_{k-j}\left(\mathbf{s},t,C_i\right)\right]+\mathcal{N}_{k-j}\left[v_0\left(\mathbf{s},t\right),v_1\left(\mathbf{s},t,C_i\right),\dots,v_{k-j}\left(\mathbf{s},t,C_i\right)\right]\right]$ of Eq. (2.42) is of the same form as that of $H_j\left(\mathbf{s},t,C_i\right)$. The considered functions can be of any type, such as polynomial, exponential, trigonometric, etc.

Now, based on the choice of auxiliary function, if the series Eq. (2.39) converges at $q=1$, then

$$v\left(\mathbf{s},\mathrm{t},C_i\right)=v_0\left(\mathbf{s},t\right)+\sum_{j=1}^{\infty}v_j\left(\mathbf{s},\mathrm{t},C_i\right) \quad (2.45)$$

Finally, the approximate solution can be obtained after considering up to a finite number of terms, say, m, as follows:

$$v\left(\mathbf{s},\mathrm{t},C_i\right)\approx v_0\left(\mathbf{s},t\right)+\sum_{j=1}^{m}v_j\left(\mathbf{s},\mathrm{t},C_i\right) \quad (2.46)$$

The determination of parameters C_i and the choice of auxiliary function will be discussed later in relation to a differential equation.

2.5   AN ILLUSTRATIVE EXAMPLE

Here we consider a simple problem of modeling runoff from a small watershed subject to rainfall occurring for a finite period of time. For simplicity, the watershed is represented by a single nonlinear reservoir (Singh 1988). The runoff response can be modeled using the spatially lumped form of continuity equation and a storage-discharge relation or storage equation. These two equations, when coupled, lead to a nonlinear governing equation. We apply various analytical methods, including HAM, HPM, and OHAM, to solve the governing equation. Consider a nonlinear reservoir, as shown in Figure 2.2.

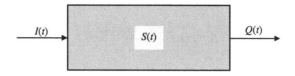

FIGURE 2.2   Storage element.

The equation of continuity produces:

$$\frac{dS}{dt} = I(t) - Q(t) \tag{2.47}$$

where:

$$Q(t) = K[S(t)]^x, \, x > 1, \text{ and } \neq 1 \tag{2.48}$$

Here $S(t)$ is the storage of water per unit area at time t, $Q(t)$ is the runoff or discharge from the plane per unit area represented by the nonlinear reservoir, $I(t)$ is the effective rainfall intensity (depth per unit time), K is the storage coefficient, and x is an exponent. For simplicity, $I(t)$ can be assumed as a constant (as a rectangular pulse):

$$I(t) = \begin{cases} I \text{ (a constant)}, & 0 \le t < T \\ 0, & t \ge T \end{cases} \tag{2.49}$$

where T is the duration of rainfall. Using Eqs. (2.48) and (2.47) can be written as:

$$\frac{dS}{dt} + KS^x = I \tag{2.50}$$

One can also express the governing equation in terms of runoff as:

$$\frac{d}{dt}\left[\frac{Q}{K}\right]^{\frac{1}{x}} + Q = I \tag{2.51}$$

It is assumed that the rainfall event has already satisfied the initial watershed abstractions. Thus, $I(t)$ denotes effective or excess rainfall and $Q(t)$ denotes surface runoff. The solution $Q(t)$ will describe the surface runoff hydrograph, which will have a rising part and receding part. Likewise, the solution $S(t)$ will describe the storage as a function of time or storage hydrograph.

It can be noted that the exact solution to Eq. (2.50) can be achieved for $I(t) = 0$, i.e., when $t \ge T$. Therefore, our focus is to solve the equation using HAM for $0 \le t < T$, even though the complete solution will be given. Before solving the equation, it may be convenient to nondimensionalize the variables as:

$$S_* = \frac{S}{S_\infty}, \, t_* = \frac{t}{t_c}, \, I_* = \frac{I}{I_0}, \text{ and } Q_* = \frac{Q}{I_0} = \frac{Q}{Q_e} \tag{2.52}$$

Here, the quantities with an asterisk represent the corresponding dimensionless forms. S_∞ is the maximum or equilibrium storage, t_c is the characteristic time or time to reach the equilibrium, I_0 is the maximum intensity of inflow, and Q_e is the outflow (peak) at the equilibrium time. Using Eq. (2.52), the governing Eq. (2.50) becomes:

$$\frac{dS_* S_\infty}{dt_* t_c} + K(S_* S_\infty)^x = I_* I_0 \tag{2.53}$$

$$\Rightarrow \frac{dS_*}{dt_*} + K\frac{t_c}{S_\infty}(S_\infty)^x S_*^x = I_* I_0 \frac{t_c}{S_\infty} \tag{2.54}$$

Defining $t_c = \dfrac{1}{KS_\infty^{x-1}}$, we obtain:

$$\frac{dS_*}{dt_*} + S_*^x = I_* \tag{2.55}$$

where $I_0 = KS_\infty^x$, since $I = I_0$, $I_* = 1$. Therefore, Eq. (2.55) reduces to:

$$\frac{dS_*}{dt_*} + S_*^x = 1 \tag{2.56}$$

For simplicity, we can consider a specific value of the exponent, say $x = 2$. The initial condition can be prescribed as $S_*(t_* = 0) = 0$. Then, we need to solve the following governing equation:

$$\frac{dS_*}{dt_*} + S_*^2 = 1 \text{ subject to } S_*(t_* = 0) = 0 \tag{2.57}$$

Equivalently, considering $x = 2$, Eq. (2.57) can be also written in terms of Q_* as follows:

$$\frac{dQ_*^{\frac{1}{2}}}{dt_*} + Q_* = 1 \text{ subject to } Q_*(t_* = 0) = 0 \tag{2.58}$$

For simplicity, henceforth, the asterisk will be dropped from symbols.

2.5.1 Solution Using Various Analytical Methods

Here, we solve Eq. (2.57) using different analytical methods. For practical implementation, flowchart diagrams for each of the methods are provided. Also, the advantages and disadvantages of the methods are discussed in detail.

2.5.1.1 Exact Solution

Eq. (2.57) can be rearranged as follows:

$$\frac{dS}{1 - S^2} = dt \tag{2.59}$$

subject to the initial condition $S(0) = 0$:

$$\Rightarrow \frac{1}{2}\left[\frac{dS}{1 + S} + \frac{dS}{1 - S}\right] = dt \tag{2.60}$$

subject to the initial condition $S(0) = 0$:

The solution of Eq. (2.60) can be obtained as:

$$S = \frac{\exp(2t)-1}{\exp(2t)+1} \tag{2.61}$$

The solution can also be given as:

$$Q = K \left[\frac{\exp(2t)-1}{\exp(2t)+1} \right]^2 \tag{2.62}$$

Eq. (2.61) is the exact solution (rising hydrograph) of the governing Eq. (2.57) for $0 \le t < T$. For $t > T$, the governing equation takes on the form:

$$\frac{dS}{dt} + S^2 = 0 \text{ subject to } S(t = T) = S_T \tag{2.63}$$

It may be noted that the initial condition for this equation is given by the solution obtained in Eq. (2.61), i.e., S_T is the value of S in Eq. (2.61) evaluated at $t = T$. The solution of Eq. (2.63) reads as:

$$S = \frac{S_T}{1+(t-T)S_T} \tag{2.64}$$

Equivalently, for $t > T$, the solution in terms of Q can be obtained as:

$$Q = \left[\frac{S_T}{1+(t-T)S_T} \right]^2 \tag{2.65}$$

2.5.1.2 Perturbation Solution

The perturbation methods need the presence of a small parameter in the governing equation and/or boundary/initial conditions (Nayfeh 1981). We assume the dimensionless time as a small quantity (perturbation parameter). The solution for storage can then be expressed as:

$$S(t) = \mu_0 + \mu_1 t + \mu_2 t^2 + \mu_3 t^3 + \cdots \tag{2.66}$$

where μ_i, $i = 0, 1, 2,\ldots$ are the coefficients. Substituting Eq. (2.66) into the initial condition $S(0) = 0$, we obtain $\mu_0 = 0$. Using Eq. (2.66) in Eq. (2.57), one gets:

$$\frac{d}{dt}\left(\mu_0 + \mu_1 t + \mu_2 t^2 + \mu_3 t^3 + \cdots\right) + \left(\mu_0 + \mu_1 t + \mu_2 t^2 + \mu_3 t^3 + \cdots\right)^2 - 1 = 0$$

$$\Rightarrow \left(\mu_1 + 2\mu_2 t + 3\mu_3 t^2 + \cdots\right) + \left(\mu_0 + \mu_1 t + \mu_2 t^2 + \mu_3 t^3 + \cdots\right)^2 - 1 = 0$$

$$\Rightarrow \left(\mu_1 + \mu_0^2 - 1\right) + \left(2\mu_2 + 2\mu_0\mu_1\right)t + \left(3\mu_3 + \mu_1^2 + 2\mu_0\mu_2\right)t^2$$

$$+ \left(4\mu_4 + 2\mu_0\mu_3 + 2\mu_1\mu_2\right)t^3 + \cdots = 0 \tag{2.67}$$

Equating the like powers of t in Eq. (2.67), we obtain the following equations:

$$\mu_1 + \mu_0^2 - 1 = 0 \Rightarrow \mu_1 = 1 \tag{2.68}$$

$$\mu_2 = -\mu_0\mu_1 \tag{2.69}$$

$$\mu_3 = -\frac{1}{3}\left(2\mu_0\mu_2 + \mu_1^2\right) \tag{2.70}$$

$$\mu_4 = -\frac{1}{4}\left(2\mu_0\mu_3 + 2\mu_1\mu_2\right) \tag{2.71}$$

The closed form of the coefficients can be achieved as:

$$\mu_0 = 0, \ \mu_1 = 1 \tag{2.72}$$

$$\mu_{k+1} = -\frac{1}{k+1}\sum_{j=0}^{k}\mu_j\mu_{k-j}, \ k \geq 1 \tag{2.73}$$

Therefore, the perturbation solution can be obtained explicitly as:

$$S_{pert}(t) = t - \frac{t^3}{3} + \frac{2t^5}{15} - \frac{17t^7}{315} + \cdots \tag{2.74}$$

The perturbation solution in terms of Q can be given as:

$$Q_{pert}(t) = \left(t - \frac{t^3}{3} + \frac{2t^5}{15} - \frac{17t^7}{315} + \cdots\right)^2 \tag{2.75}$$

Also, the solution over the entire time domain can be given as:

$$S = \begin{cases} S_{pert}(t) & \text{for } 0 \leq t < T \\ \dfrac{S_T}{1+(t-T)S_T} & \text{for } t > T \end{cases} \tag{2.76}$$

where $S_{pert}(t)$ is given by Eq. (2.74). The perturbation method can be implemented efficiently in a mathematical solver. For that purpose, a flowchart diagram containing the steps is provided in Figure 2.3.

2.5.1.3 Lyapunov's Artificial Small Parameter Method–Based Solution

Lyapunov's artificial small parameter method (LASPM) introduces an artificial small parameter in the governing equation and then solves the equation similarly like a perturbation method (Lyapunov 1992). However, placing the parameter inside the equation is not unique. Here, we introduce a small parameter, say, ε, and write the governing Eq. (2.57) in the following form:

$$\frac{dS}{dt} + \varepsilon S^2 = 1 \tag{2.77}$$

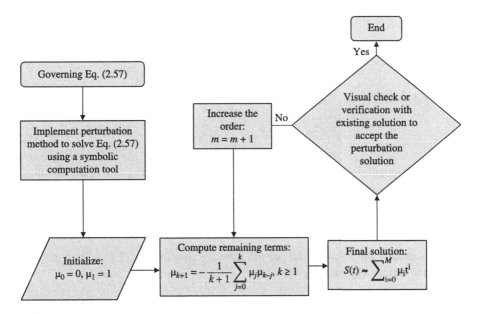

FIGURE 2.3 Flowchart for the perturbation solution in Eq. (2.74).

It can be seen that as $\varepsilon \to 1$, Eq. (2.77) recovers the original Eq. (2.57). Now, expressing the solution in series form using ε, we write:

$$S(t) = S_0(t) + \varepsilon S_1(t) + \varepsilon^2 S_2(t) + \varepsilon^3 S_3(t) + \cdots \qquad (2.78)$$

where $S_i(t)$, $i = 0,1,2,\ldots$ are the coefficients of the series that are functions of t. Substituting Eq. (2.78) into Eq. (2.77), we get:

$$\frac{d}{dt}\left[S_0(t) + \varepsilon S_1(t) + \varepsilon^2 S_2(t) + \varepsilon^3 S_3(t) + \cdots \right]$$
$$+ \varepsilon \left[S_0(t) + \varepsilon S_1(t) + \varepsilon^2 S_2(t) + \varepsilon^3 S_3(t) + \cdots \right]^2 - 1 = 0 \qquad (2.79)$$

Equating the like powers of ε and using Eq. (2.78) in the initial condition, the following system of differential equations is obtained:

$$\frac{dS_0}{dt} - 1 = 0, \; S_0(0) = 0 \qquad (2.80)$$

$$\frac{dS_1}{dt} + S_0{}^2 = 0, \; S_1(0) = 0 \qquad (2.81)$$

$$\frac{dS_2}{dt} + 2S_0 S_1 = 0, \; S_2(0) = 0 \qquad (2.82)$$

Proceeding in a like manner, one can arrive at the following recurrence relation:

$$\frac{dS_i}{dt} + \sum_{j=0}^{i-1} S_j S_{i-1-j} = 0, \; S_i(0) = 0 \; \text{for} \; i = 1, 2, 3, \ldots \tag{2.83}$$

Solving these equations, we get:

$$S_0(t) = t, \; S_1(t) = -\frac{t^3}{3}, \; S_2(t) = \frac{2t^5}{15}, \ldots \tag{2.84}$$

Using these quantities from Eq. (2.84) and taking $\varepsilon \to 1$ in Eq. (2.78), the solution becomes:

$$S_{LASPM}(t) = t - \frac{t^3}{3} + \frac{2t^5}{15} - \cdots \tag{2.85}$$

which is exactly the same as the perturbation solution for Eq. (2.74). The LASPM solution in terms of Q can be given as:

$$Q_{LASPM}(t) = \left(t - \frac{t^3}{3} + \frac{2t^5}{15} - \frac{17t^7}{315} + \cdots \right)^2 \tag{2.86}$$

Also, the solution over the entire time domain can be given as:

$$S = \begin{cases} S_{LASPM}(t) & \text{for } 0 \le t < T \\[2mm] \dfrac{S_T}{1 + (t - T) S_T} & \text{for } t > T \end{cases} \tag{2.87}$$

where $S_{LASPM}(t)$ is given by Eq. (2.84). The LASPM can be implemented efficiently in a mathematical solver. For that purpose, a flowchart diagram containing the steps is provided in Figure 2.4.

2.5.1.4 Adomian Decomposition Method–Based Solution

In the Adomian decomposition method (ADM), the original nonlinear equation is decomposed into linear and nonlinear parts and then solved using an approximation method (Adomian 2013). We consider the original Eq. (2.57) in the following form:

$$\mathcal{L}[S] = R[S] + \mathcal{N}[S] \tag{2.88}$$

where $\mathcal{L}$ is a linear operator, $\mathcal{N}$ is a nonlinear operator, and R is the remainder term of the equation. Here, we consider $\mathcal{L}[S] = \dfrac{dS}{dt}$, $\mathcal{N}[S] = -S^2$, and $R[S] = 1$. One may choose different kinds of linear operators; however, it is convenient to use the single-term linear operator. Applying the inverse operator $\mathcal{L}^{-1}$ to both sides of Eq. (2.88), one obtains:

$$S(t) - S(0) = \mathcal{L}^{-1}(R[S]) + \mathcal{L}^{-1}(\mathcal{N}[S])$$

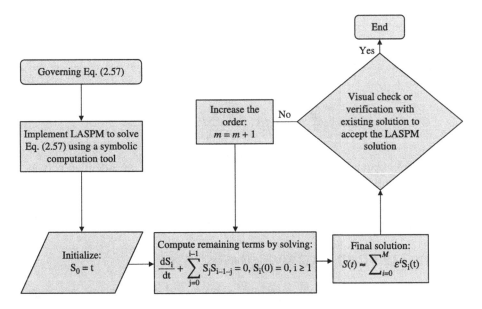

FIGURE 2.4 Flowchart for the LASPM solution in Eq. (2.85).

$$\Rightarrow S(t) = t - \int_0^t S^2(\tau)\,d\tau \tag{2.89}$$

The solution of Eq. (2.89) is given by:

$$S(t) = S_0(t) + \sum_{k=1}^{\infty} S_k(t) \tag{2.90}$$

where:

$$S_0(t) = t \tag{2.91}$$

$$S_k(t) = -\int_0^t A_{k-1}(\tau)\,d\tau, \; k \geq 1 \tag{2.92}$$

Here, A_k are known as the Adomian polynomial, which depends on the terms $S_0(t)$, $S_1(t), \ldots, S_k(t)$. The term A_k is associated with the nonlinear term of the equation as follows:

$$\mathcal{N}[S] = \sum_{k=0}^{\infty} A_k(S_0, S_1, \ldots, S_k) \tag{2.93}$$

For an arbitrary function $f(S)$, A_k are given by (Adomian 1994):

$$A_0 = f(S_0) \tag{2.94}$$

$$A_1 = S_1 \frac{df(S_0)}{dS_0} \tag{2.95}$$

$$A_2 = S_2 \frac{df(S_0)}{dS_0} + \frac{S_1^2}{2!} \frac{d^2 f(S_0)}{dS_0^2} \tag{2.96}$$

$$A_3 = S_3 \frac{df(S_0)}{dS_0} + S_1 S_2 \frac{d^2 f(S_0)}{dS_0^2} + \frac{S_1^3}{3!} \frac{d^3 f(S_0)}{dS_0^3} \tag{2.97}$$

For the present equation, $f(S) = S^2$. Using Eqs. (2.94–2.97), the Adomian polynomial can be written in a general form as:

$$A_k(t) = \sum_{j=0}^{k} S_j(t) S_{k-j}(t) \tag{2.98}$$

Using Eqs. (2.91), (2.92), and (2.98), one obtains:

$$S_1(t) = -\frac{t^3}{3}, \ S_2(t) = \frac{2t^5}{15}, \ S_3(t) = -\frac{17t^7}{315}, \dots \tag{2.99}$$

Therefore, the Adomian decomposition–based solution is given by:

$$S_{ADM}(t) = t - \frac{t^3}{3} + \frac{2t^5}{15} - \frac{17t^7}{315} \cdots \tag{2.100}$$

which is exactly the same as that obtained using the perturbation technique.

The ADM solution in terms of Q can be given as:

$$Q_{ADM}(t) = \left(t - \frac{t^3}{3} + \frac{2t^5}{15} - \frac{17t^7}{315} + \cdots \right)^2 \tag{2.101}$$

Also, the solution over the entire time domain can be given as:

$$S = \begin{cases} S_{ADM}(t) & \text{for } 0 \le t < T \\ \dfrac{S_T}{1+(t-T)S_T} & \text{for } t > T \end{cases} \tag{2.102}$$

where $S_{ADM}(t)$ is given by Eq. (2.100). The ADM can be implemented efficiently in a mathematical solver. For that purpose, a flowchart diagram containing the steps is provided in Figure 2.5.

2.5.1.5 Homotopy Perturbation Method–Based Solution

The fundamental idea of HPM is the same as HAM, which is based on the concept of homotopy from topology (He 2003). To apply HPM, we consider the governing equation in the following form:

$$\mathcal{N}(S) = f(t) \tag{2.103}$$

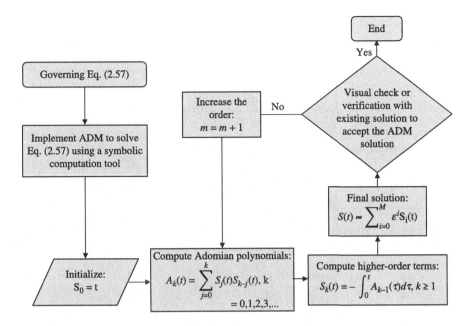

FIGURE 2.5   Flowchart for the ADM solution for Eq. (2.100).

We construct a homotopy $\Phi(t;q)$ which satisfies:

$$(1-q)\Big[\mathcal{L}\big(\Phi(t;q)\big)-\mathcal{L}\big(S_0(t)\big)\Big]+q\Big[\mathcal{N}\big(\Phi(t;q)\big)-f(t)\Big]=0 \qquad (2.104)$$

where q is the embedding parameter that lies in the interval $[0,1]$, $\mathcal{L}$ is the linear operator, and $S_0(t)$ is the initial approximation to the final solution. It can be seen from Eq. (2.104) that when $q=0$:

$$\mathcal{L}\big(\Phi(t;q)\big)-\mathcal{L}\big(S_0(t)\big)=0 \Rightarrow \Phi(t;0)=S_0(t) \qquad (2.105)$$

At $q=1$:

$$\mathcal{N}\big(\Phi(t;1)\big)-f(t)=0 \Rightarrow \mathcal{N}\big(\Phi(t;1)\big)=f(t) \qquad (2.106)$$

It shows that as q goes from 0 to 1, the homotopy $\Phi(t;q)$ varies from the initial approximation $S_0(t)$ to the final solution $S(t)$. Considering parameter q as a small parameter, the following expression can be obtained:

$$\Phi(t;q)=\Phi_0+q\Phi_1+q^2\Phi_2+q^3\Phi_3+\cdots \qquad (2.107)$$

As $q \to 1$, Eq. (2.107) produces the final solution as:

$$S(t)=\lim_{q\to 1}\Phi(t;q)=\sum_{i=0}^{\infty}\Phi_i \qquad (2.108)$$

We consider the following forms for the operators:

$$\mathcal{L}\big(\Phi(t;q)\big) = \frac{\partial \Phi(t;q)}{\partial t} \tag{2.109}$$

$$\mathcal{N}\big(\Phi(t;q)\big) = \frac{\partial \Phi(t;q)}{\partial t} + \Phi^2(t;q) - 1 \tag{2.110}$$

Using these expressions, Eq. (2.104) takes on the form:

$$(1-q)\left[\frac{\partial \Phi(t;q)}{\partial t} - \frac{dS_0}{dt}\right] + q\left[\frac{\partial \Phi(t;q)}{\partial t} + \Phi^2(t;q) - 1\right] = 0 \tag{2.111}$$

Substituting Eq. (2.107) into the initial condition $S(0) = 0$, one can obtain $\Phi_0(0) = 0$, $\Phi_1(0) = 0$, $\Phi_2(0) = 0,\ldots$. Now, using Eq. (2.107) in Eq. (2.111), we get:

$$(1-q)\left[\left(\frac{d\Phi_0}{dt} + q\frac{d\Phi_1}{dt} + q^2\frac{d\Phi_2}{dt} + q^3\frac{d\Phi_3}{dt} + \cdots\right) - \frac{dS_0}{dt}\right]$$
$$+ q\left[\left(\frac{d\Phi_0}{dt} + q\frac{d\Phi_1}{dt} + q^2\frac{d\Phi_2}{dt} + q^3\frac{d\Phi_3}{dt} + \cdots\right) + \left(\Phi_0 + q\Phi_1 + q^2\Phi_2 + q^3\Phi_3 + \cdots\right)^2 - 1\right] = 0 \tag{2.112}$$

Equating the like powers of q in Eq. (2.112), the following system of differential equations is obtained:

$$\frac{d\Phi_0}{dt} - \frac{dS_0}{dt} = 0 \text{ subject to } \Phi_0(0) = 0 \tag{2.113}$$

$$\frac{d\Phi_1}{dt} + \frac{dS_0}{dt} + \Phi_0{}^2 - 1 = 0 \text{ subject to } \Phi_1(0) = 0 \tag{2.114}$$

$$\frac{d\Phi_2}{dt} + 2\Phi_0\Phi_1 = 0 \text{ subject to } \Phi_2(0) = 0 \tag{2.115}$$

Proceeding in a like manner, one can obtain the following recurrence relation:

$$\frac{d\Phi_k}{dt} + \sum_{j=0}^{k-1}\Phi_j\Phi_{j-1-k} = 0 \text{ subject to } \Phi_k(0) = 0 \text{ for } k \geq 2 \tag{2.116}$$

Taking the initial approximation $\Phi_0 = t$ and solving these equations, we obtain:

$$\Phi_0 = t, \ \Phi_1 = -\frac{t^3}{3}, \ \Phi_2 = \frac{2t^5}{15},\ldots \tag{2.117}$$

Therefore, the HPM-based solution is obtained from Eq. (2.108) as:

$$S_{pert}(t) = t - \frac{t^3}{3} + \frac{2t^5}{15} - \cdots \tag{2.118}$$

which is exactly the same as the perturbation solution. The HPM solution in terms of Q can be given as:

$$Q_{HPM}(t) = \left(t - \frac{t^3}{3} + \frac{2t^5}{15} - \frac{17t^7}{315} + \cdots\right)^2 \qquad (2.119)$$

Also, the solution over the entire time domain can be given as:

$$S = \begin{cases} S_{HPM}(t) & \text{for } 0 \leq t < T \\ \dfrac{S_T}{1+(t-T)S_T} & \text{for } t > T \end{cases} \qquad (2.120)$$

where $S_{HPM}(t)$ is given by Eq. (2.118). The HPM can be implemented efficiently in a mathematical solver. For that purpose, a flowchart diagram containing the steps is provided in Figure 2.6.

It is observed that the solutions obtained using the perturbation method, LASPM, ADM, and HPM are the same. Here, we compare the exact solution of the equation with the solution obtained using these methods. In Figure 2.7, we compare both solutions over $t \in [0, 10]$. It is observed that the solution breaks down once t exceeds 1.5, which shows that these methods are valid for a small region of t. That means the series solution obtained using these methods has a small region of convergence. For numerical comparison, we have considered some values of t over $[0, 2]$ and compared both methods in Table 2.1, which shows that the higher the order of approximation,

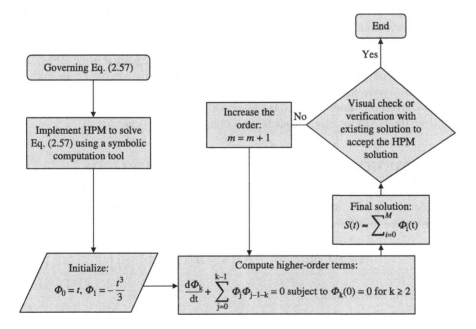

FIGURE 2.6 Flowchart for the HPM solution in Eq. (2.118).

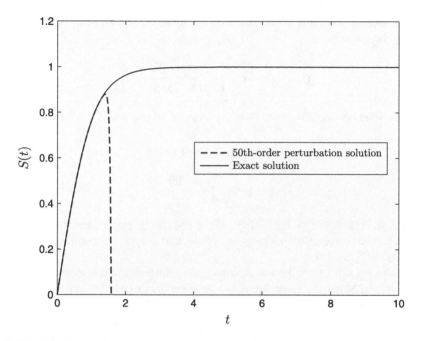

FIGURE 2.7 Comparison of analytical solutions using perturbation, Lyapunov's artificial small parameter, Adomian decomposition, and homotopy perturbation methods with the exact solutions.

the better the accuracy of analytical methods. However, after a certain range, the solution curve breaks down, and it does not provide any feasible solution even for the 50th order of approximation.

Further, it can be noted that while solving the nonlinear equation using HPM, we considered a simple linear operator and an initial approximation. One can indeed consider different kinds of linear operators and initial approximations to obtain a rapidly converging series solution. However, although freedom exists in HPM, it differs from the HAM in several aspects, as will be discussed later.

2.6 HOMOTOPY ANALYSIS METHOD–BASED SOLUTION

The methodology of HAM described in the previous section is followed here. Using the embedding parameter q, the zeroth-order deformation equation for Eq. (2.57) can be constructed as follows:

$$(1-q)\mathcal{L}\big[\Phi(t;q)-S_0(t)\big]=q\hbar H(t)\mathcal{N}\big[\Phi(t;q)\big] \qquad (2.121)$$

subject to the initial condition:

$$\Phi(0;q)=0 \qquad (2.122)$$

TABLE 2.1

Comparison of Analytical Solution Using Perturbation, Lyapunov's Artificial Small Parameter, Adomian Decomposition, and Homotopy Perturbation Method with the Exact Solution

t	Exact Value	5th-Order Approximation	10th-Order Approximation	15th-Order Approximation	20th-Order Approximation	30th-Order Approximation	50th-Order Approximation
0.2	0.19738	0.19738	0.19738	0.19738	0.19738	0.19738	0.19738
0.4	0.37995	0.37995	0.37995	0.37995	0.37995	0.37995	0.37995
0.6	0.53705	0.53708	0.53705	0.53705	0.53705	0.53705	0.53705
0.8	0.66404	0.66464	0.66404	0.66404	0.66404	0.66404	0.66404
1.0	0.76159	0.76790	0.76153	0.76159	0.76159	0.76159	0.76159
1.2	0.83365	0.87524	0.83084	0.83385	0.83364	0.83365	0.83365
1.4	0.88535	1.08538	0.82209	0.90536	0.87902	0.88472	0.88535
1.6	0.92167	1.68693	0.00162	2.02781	-0.40821	-1.00059	-3.09455
1.8	0.94681	3.40940	-8.66747	38.48237	-145.59759	-2232.73749	-518950.7227
2.0	0.96403	7.88924	-76.57718	869.19196	-9720.57313	-1218812.7	-19157642229

It can be seen from Eq. (2.121) that as q varies from 0 to 1, the representation to the solution $\Phi(t;q)$ varies from the initial approximation $S_0(t)$ to the final solution $S(t)$. For Eq. (2.57), the linear operator in general form can be selected as:

$$\mathcal{L}\left[\Phi(t;q)\right] = a_1(t)\frac{\partial\Phi(t;q)}{\partial t} + a_2(t)\Phi(t;q) \tag{2.123}$$

where $a_1(t)$ is a non-zero function. The selection of $a_1(t)$ and $a_2(t)$ will be discussed later. The nonlinear operator is selected as the operator of the original nonlinear equation, i.e.,

$$\mathcal{N}\left[\Phi(t;q)\right] = \frac{\partial\Phi(t;q)}{\partial t} + \Phi^2(t;q) - 1 \tag{2.124}$$

Next, the same steps can be followed, as discussed in the previous section, to obtain the solution using HAM. Here we write down the final solution as follows:

$$S(t) = S_0(t) + \sum_{m=1}^{\infty} S_m(t) \tag{2.125}$$

The higher-order terms can be calculated from the following higher-order deformation equation:

$$\mathcal{L}\left[S_m(t) - \chi_m S_{m-1}(t)\right] = \hbar H(t) R_m\left(\vec{S}_{m-1}\right), \quad m = 1,2,3,\ldots \tag{2.126}$$

subject to the initial condition:

$$S_m(0) = 0$$

where:

$$\chi_m = \begin{cases} 0 & \text{when } m = 1, \\ 1 & \text{otherwise} \end{cases} \tag{2.127}$$

and:

$$R_m\left(\vec{S}_{m-1}\right) = \frac{1}{(m-1)!}\frac{\partial^{m-1}\mathcal{N}\left[\Phi(t;q)\right]}{\partial q^{m-1}}\bigg|_{q=0} \tag{2.128}$$

For the considered equation, $\mathcal{N}[\Phi(t;q)]$ is given by Eq. (2.124). Using that, the terms of $R_m\left(\vec{S}_{m-1}\right)$ can be calculated from Eq. (2.128) as follows:

$$R_1(S_0) = \frac{1}{(0)!}\frac{\partial^0\mathcal{N}\left[\Phi(t;q)\right]}{\partial q^0}\bigg|_{q=0} = \frac{dS_0}{dt} + S_0^2 - 1 \tag{2.129}$$

$$R_2\left(\vec{S}_1\right) = \frac{1}{(1)!} \left.\frac{\partial^1 \mathcal{N}\left[\Phi(t;q)\right]}{\partial q^1}\right|_{q=0} = \frac{dS_1}{dt} + S_0 S_1 + S_1 S_0 \qquad (2.130)$$

$$R_3\left(\vec{S}_2\right) = \frac{1}{(2)!} \left.\frac{\partial^2 \mathcal{N}\left[\Phi(t;q)\right]}{\partial q^2}\right|_{q=0} = \frac{dS_2}{dt} + S_0 S_2 + S_1^2 + S_2 S_0 \qquad (2.131)$$

Proceeding in this way, we can find the recurrence relation as follows:

$$R_m\left(\vec{S}_{m-1}\right) = \frac{dS_{m-1}}{dt} + \sum_{j=0}^{m-1} S_j S_{m-j-1} - \left(1 - \chi_m\right) \qquad (2.132)$$

Using Eq. (2.132), the higher-order terms can be calculated from Eq. (2.126), which will be discussed later in relation to an example. The final solution can be found in approximate form, considering some of the terms of the exact solution, as follows:

$$S(t) = S_0(t) + \sum_{m=1}^{\infty} S_m(t) \approx S_0(t) + \sum_{m=1}^{N} S_m(t) \qquad (2.133)$$

To find the solution using HAM given by Eq. (2.133), one needs to select the linear operator $\mathcal{L}$, the auxiliary function $H(t)$, and the auxiliary parameter $\hbar$. As mentioned in the previous section, Liao (2003) proposed some rules, which are described here.

For a given nonlinear equation, the solution can be expressed in terms of a proper set of base functions. In fact, different kinds of base functions can be considered for a problem to represent its solution. Fortunately, in the framework of HAM, one has great freedom to choose the linear operator, the auxiliary function, and the initial approximation (Liao 2003). In this way, a better base function can always be selected out of many. The selection of a base function for a particular equation sometimes can be made by looking into the physical properties of the problem, such as the nature of the governing equation and/or boundary/initial condition. We suppose that:

$$\left\{e_m(t) \mid m = 1,2,3....\right\} \qquad (2.134)$$

denotes a set of base functions for the problem considered in Eq. (2.57). Therefore, the solution $S(t)$ can be expressed as:

$$S(t) = \sum_{m=1}^{\infty} \alpha_m e_m(t) \qquad (2.135)$$

where α_m are the coefficients. Looking into the nature of the base function Eq. (2.134), the linear operator $\mathcal{L}$, the auxiliary function $H(t)$, and the initial approximation $S_0(t)$ can be chosen – the rule that directs us for that choice is known as the *rule of solution expression*. As mentioned earlier, the choice is not unique, as one can have different kinds of base functions for a particular equation.

The *rule of coefficient ergodicity* directs us to choose the auxiliary function $H(t)$. As per the rule, the function should be selected so that all the coefficients α_m in the solution expression in Eq. (2.136) appear in the solution to guarantee the completeness of the set of base functions. With the help of the *rule of solution expression* and the *rule of coefficient ergodicity,* auxiliary functions should be selected to ensure the existence of the solution of higher-order deformation equations. This provides us with the so-called *rule of solution existence*. For the problem considered by Eq. (2.57), here we explain the HAM by choosing two sets of base functions: first, the polynomial one, and second, the exponential one.

2.6.1 SOLUTION IN TERMS OF A POLYNOMIAL

The choice of a polynomial as a base function is straightforward, as the perturbation method produced the solution of the equation in terms of a polynomial. To that end, the base function can be chosen as:

$$\left\{ t^{2n+1} \mid n = 0, 1, 2, 3... \right\} \tag{2.136}$$

so that:

$$S(t) = \sum_{n=0}^{\infty} a_n t^{2n+1} \tag{2.137}$$

where a_n represent coefficients.

Eq. (2.136) provides the *rule of solution expression* for the present problem. Following this rule, the initial approximation can simply be chosen as:

$$S_0(t) = t \tag{2.138}$$

The general form of the linear operator was given by Eq. (2.123), where the coefficients $a_1(t)$ and $a_2(t)$ can be chosen, based on the form of the base function. According to Eq. (2.136), the following linear operator is selected:

$$\mathcal{L}\left[\Phi(t;q)\right] = \frac{\partial \Phi(t;q)}{\partial t} \tag{2.139}$$

with the property:

$$\mathcal{L}\left[C_1\right] = 0 \tag{2.140}$$

where C_1 is an integral constant. Using the rule of solution expression and the higher-order deformation equation, the auxiliary function can be chosen as:

$$H(t) = t^{2\beta} \tag{2.141}$$

Using the expressions defined earlier, the higher-order deformation in Eq. (2.126) becomes:

$$S_m(t) = \chi_m S_{m-1}(t) + \hbar \int_0^t \tau^{2\beta} R_m\left(\vec{S}_{m-1}(\tau)\right) d\tau + C_1 \qquad (2.142)$$

where constant C_1 needs to be calculated from the given initial condition. Most of the time, the auxiliary function $H(t)$ can be uniquely determined following the *rule of solution expression*, as we will see here.

It can be noticed that if we set $\beta \leq -1$, the term t^{-1} appears in the expression of $S_m(t)$, which certainly disobeys the *rule of solution expression* in Eq. (2.137). On the other hand, when $\beta \geq 1$, the term t^3 always disappears in the expression of $S_m(t)$, which disobeys the *rule of coefficient ergodicity*, which tells us to consider all the terms of solution expression for the completeness of the base function. This leads to the only choice $\beta = 0$, i.e., $H(t) = 1$. Using the initial condition, i.e., $S_m(0) = 0$, we obtain $C_1 = 0$. For calculating the higher-order terms, first, we need to find $R_m\left(\vec{S}_{m-1}\right)$ for the previous iteration. To that end, one obtains successively:

$$R_1(S_0) = \frac{dS_0}{dt} + S_0^2 - 1 = t^2 \qquad (2.143)$$

$$S_1(t) = \chi_1 S_0(t) + \hbar \int_0^t R_1(S_0(\tau)) d\tau = 0 + \hbar \int_0^t \tau^2 d\tau = \frac{\hbar t^3}{3} \qquad (2.144)$$

$$R_2\left(\vec{S}_1\right) = \frac{dS_1}{dt} + S_0 S_1 + S_1 S_0 = \hbar t^2 + \frac{2\hbar t^4}{3} \qquad (2.145)$$

$$S_2(t) = \chi_2 S_1(t) + \hbar \int_0^t R_2\left(\vec{S}_1(\tau)\right) d\tau = \frac{\hbar t^3}{3} + \hbar \int_0^t \left(\hbar \tau^2 + \frac{2\hbar \tau^4}{3}\right) d\tau$$

$$= \frac{\hbar t^3}{3} + \frac{\hbar^2 t^3}{3} + \frac{2\hbar^2 t^5}{15} \qquad (2.146)$$

$$R_3\left(\vec{S}_2\right) = \frac{dS_2}{dt} + S_0 S_2 + S_1^2 + S_2 S_0 = \hbar t^2 + \hbar^2 t^2 + \frac{4\hbar^2 t^4}{3} + \frac{2\hbar t^4}{3} + \frac{17\hbar^2 t^6}{45} \qquad (2.147)$$

$$S_3(t) = \chi_3 S_2(t) + \hbar \int_0^t R_3\left(\vec{S}_2(\tau)\right) d\tau = \frac{\hbar t^3}{3} + \frac{\hbar^2 t^3}{3} + \frac{2\hbar^2 t^5}{15}$$

$$+ \hbar \int_0^t \left(\hbar \tau^2 + \hbar^2 \tau^2 + \frac{4\hbar^2 \tau^4}{3} + \frac{2\hbar \tau^4}{3} + \frac{17\hbar^2 \tau^6}{45}\right) d\tau = \frac{\hbar t^3}{3}$$

$$+ \frac{2\hbar^2 t^3}{3} + \frac{4\hbar^2 t^5}{15} + \frac{\hbar^3 t^3}{3} + \frac{4\hbar^3 t^5}{15} + \frac{17\hbar^3 t^7}{315} \qquad (2.148)$$

Finally, the N-th order HAM-based approximation can be found as:

$$S(t) \approx \sum_{m=0}^{N} S_m(t) \qquad (2.149)$$

where the terms $S_m(t)$ can be calculated as noted earlier. The solution using HAM contains the auxiliary parameter $\hbar$, which produces a family of solution expressions, as can be seen from Eq. (2.137). It is interesting to note that $\hbar = -1$, corresponding to the solution obtained by the perturbation method, LASPM, ADM, and HPM. This shows that HAM is a unified method that logically contains other methods as special cases.

2.6.2 SOLUTION IN TERMS OF EXPONENTIAL FUNCTIONS

Since the HAM provides great freedom to choose different kinds of base functions for a given equation, here we consider the base function as:

$$\{\exp(-nt) \mid n = 0,1,2,3...\} \qquad (2.150)$$

to represent the solution $S(t)$, i.e.,

$$S(t) = \sum_{n=0}^{\infty} a_n exp(-nt) \qquad (2.151)$$

where a_n represents coefficients. As $t \to \infty$, $\exp(-nt) \to 0$ for $n \geq 1$, which means the solution is finite when t approaches infinity. This guides us to choose an exponential base function for the present problem. Eq. (2.151) provides us with the *rule of solution expression*. Following this rule and the initial condition, the initial approximation can simply be chosen as:

$$S_0(t) = 1 - \exp(-t) \qquad (2.152)$$

According to Eq. (2.151), the following linear operator is selected:

$$\mathcal{L}\big[\Phi(t;q)\big] = \frac{\partial \Phi(t;q)}{\partial t} + \Phi(t;q) \qquad (2.153)$$

with the property:

$$\mathcal{L}\big[C_2 \exp(-t)\big] = 0 \qquad (2.154)$$

where C_2 is an integral constant. Using the rule of solution expression and the higher-order deformation equation, the auxiliary function can be chosen as:

$$H(t) = \exp(-\kappa t) \qquad (2.155)$$

Using the expressions defined earlier, the higher-order deformation Eq. (2.126) becomes:

$$S_m(t) = \chi_m S_{m-1}(t) + \hbar\exp(-t)\int_0^t \exp(\tau)\exp(-\kappa\tau)R_m\left(\vec{S}_{m-1}(\tau)\right)d\tau + C_2\exp(-t) \quad (2.156)$$

where constant C_2 can be determined from the given initial condition. It is found that when $\kappa \le 0$, the term $t\exp(-t)$ appears in the expression of $S_m(t)$, which certainly disobeys the *rule of solution expression* in Eq. (2.151). On the other hand, if we set $\kappa \ge 2$, the term $\exp(-2t)$ always disappears in the expression of $S_m(t)$, which disobeys the *rule of coefficient ergodicity*. Therefore, we must set $\kappa = 1$, i.e., $H(t) = \exp(-t)$. Using the initial condition, i.e., $S_m(0) = 0$, we obtain $C_2 = 0$. Using Eq. (2.156), the higher-order terms can be obtained successively as:

$$R_1(S_0) = \exp(-2t) - \exp(-t) \quad (2.157)$$

$$S_1(t) = -\frac{\hbar\exp(-t)}{2} + \hbar\exp(-2t) - \frac{\hbar\exp(-3t)}{2} \quad (2.158)$$

$$R_2\left(\vec{S}_1\right) = \frac{\hbar\exp(-t)}{2} - 2\hbar\exp(-2t) + \frac{3\hbar\exp(-3t)}{2}$$
$$+2\left(1 - \exp(-t)\right)\left(-\frac{\hbar\exp(-t)}{2} + \hbar\exp(-2t) - \frac{\hbar\exp(-3t)}{2}\right) \quad (2.159)$$

$$S_2(t) = -\frac{\hbar\exp(-t)}{2} - \frac{\hbar^2\exp(-t)}{2} + \hbar\exp(-2t) + \frac{\hbar^2\exp(-2t)}{2}$$
$$-\frac{\hbar\exp(-3t)}{2} - \frac{\hbar^2\exp(-3t)}{2} + \frac{\hbar^2\exp(-4t)}{2} - \frac{\hbar^2\exp(-5t)}{4} \quad (2.160)$$

Finally, the N-th-order HAM-based approximation can be found as:

$$S_{HAM}(t) \approx \sum_{m=0}^{N} S_m(t) \quad (2.161)$$

where the terms $S_m(t)$ can be calculated as noted earlier.

The HAM solution in terms of Q can be given as:

$$Q_{HAM}(t) = \left(\sum_{m=0}^{N} S_m(t)\right)^2 \quad (2.162)$$

Also, the solution over the entire time domain can be given as:

$$S = \begin{cases} S_{HAM}(t) & \text{for } 0 \le t < T \\ \dfrac{S_T}{1 + (t-T)S_T} & \text{for } t > T \end{cases} \quad (2.163)$$

where $S_{HAM}(t)$ is given by Eq. (2.161).

TABLE 2.2

A Typical Set of Base Functions for Eq. (2.57)

Name	Base Function $e_i(t)$	Linear Operator $\mathcal{L}$	Initial Approximation $S_0(t)$	Auxiliary Function $H(t)$
Polynomial	$\{t^{2n+1} : n = 0,1,2,...\}$	$\dfrac{\partial \Phi(t;q)}{\partial t}$	t	$t^{2\alpha}$
Exponential	$\{\exp(-nt) \mid n = 0,1,2,3...\}$	$\dfrac{\partial \Phi(t;q)}{\partial t} + \Phi(t;q)$	$1 - \exp(-t)$	$\exp(-\alpha t)$
Fractional	$\{(1+t)^{-n} : n = 0,1,2,...\}$	$(1+t)\dfrac{\partial \Phi(t;q)}{\partial t} + \Phi(t;q)$	$1 - \dfrac{1}{1+t}$	$\dfrac{1}{(1+t)^{\alpha}}$

In the previous text, we considered two kinds of base functions for solving the nonlinear equation using HAM. One may consider other types of base functions also, such as fractional, if feasible (Liao 2003). A typical set of base functions with the corresponding operators and functions for the present equation is given in Table 2.2. Further, it can be noted that the calculation of terms R_m, S_m, etc., becomes difficult to perform by manual calculation. The development of efficient symbolic computation software, such as MATLAB, Mathematica, etc., makes it easy to perform the iterations using HAM. For the convenience of readers, the codes for the equations considered in this chapter are included. One may refer to Liao (2012) or the webpage http://numericaltank.sjtu.edu.cn/ for more details regarding the implementation of HAM in Mathematica.

It is seen that the solution obtained using HAM contains an auxiliary parameter $\hbar$, whose determination is essential for the assessment of the solution. In the framework of HAM, the auxiliary parameter plays one of the most vital roles. A suitable choice for the parameter leads the series solution toward the exact solution over the entire domain. For this, $\hbar$ is commonly termed the convergence-control parameter. Also, unlike the other perturbation and non-perturbation methods, these convergence-control parameters greatly enhance the radius and rate of convergence of the series solution.

Now, we discuss two methods developed over the years for the calculation of $\hbar$. The first method is based on the concept of the so-called $\hbar$-curve. In this case, for a particular order of approximation, we plot the approximate solution $S(t)$ (or its derivatives $S'(t)$, $S''(t)$, etc.) at some point, say t_{dom}, in the domain of the problem. The flatness of the $\hbar$−curve determines a suitable choice for the auxiliary parameter $\hbar$. In Figure 2.8, we plot the $\hbar$-curve for the 20th-order HAM-based approximation using both polynomial and exponential base functions. For the polynomial base function, quantities $S(1)$ and $S'(1)$ are considered for the $\hbar$-curve. From the exact solution, those quantities can be calculated, and it is observed from the figure that the curves exhibit the flat nature for a specific range of $\hbar$. Any choice of $\hbar$ within this range determines an optimal value for which the series solution converges. A similar conclusion can be made for the $\hbar$−curve obtained using the exponential base function. For mathematical understanding about the reason

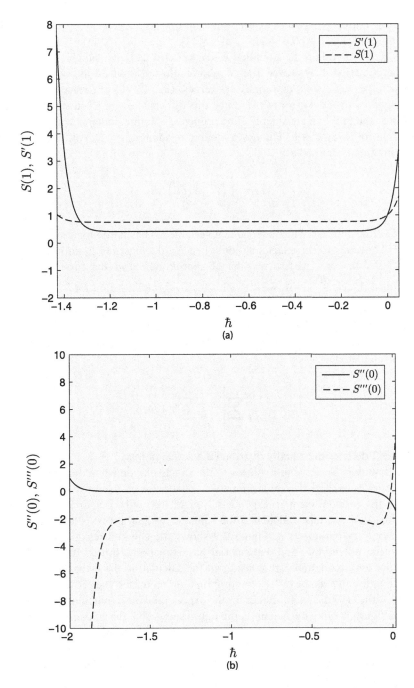

FIGURE 2.8 $\hbar$-Curve for the 20th-order HAM approximation: (a) polynomial base function and (b) exponential base function.

behind the convergence of the series solution for the occurrence of the horizontal line, one can refer to Abbasbandy et al. (2011).

As discussed earlier, the so-called $\hbar$-curve helps us find a suitable value for the auxiliary function $\hbar$. However, such $\hbar$-curves do not produce the best value of $\hbar$ for which the series solution rapidly converges. Liao (2010) introduced the concept of the squared residual error to overcome this drawback, which is discussed here in relation to the problem considered. The proposed solution contains a single convergence-control parameter $\hbar$. The exact squared residual error (Δ_m) at the m-th-order approximation is defined as:

$$\Delta_m = \int\limits_{t\in\Omega} \left(\mathcal{N}\left[S(t)\right]\right)^2 dt \tag{2.164}$$

where Ω is the domain of the independent variable t. The approximate series solution using HAM leads to the exact solution when Δ_m decreases to its minimum value, zero. Therefore, for a particular order of approximation m, the corresponding Δ_m is minimized, i.e., $\dfrac{d\Delta_m}{d\hbar} = 0$, which yields an algebraic equation for $\hbar$. The solution of this equation produces an optimal value of the convergence-control parameter. However, due to the analytical integration, Eq. (2.164) may sometimes create computational difficulty. To avoid this difficulty, the averaged squared residual error, E_m, is often calculated as:

$$E_m = \frac{1}{L+1}\sum_{j=0}^{L}\left[\mathcal{N}\left(\sum_{k=0}^{m}S_k\left(j\Delta t\right)\right)\right]^2 \tag{2.165}$$

where $L+1$ denotes the equally distributed discrete points.

The homotopy series solution leads to the exact solution when the squared residual error tends to zero. Therefore, for the convergence of a solution, it is sufficient to check the residual error in Eqs. (2.164) or (2.165) only. For different orders of approximations, we have calculated the squared residual error for different values of the auxiliary parameter $\hbar$. Figure 2.9 shows the squared residual error curves for both the polynomial and exponential base functions. Fifth-, 10th-, 15th-, and 20th-order approximations are considered for calculating the error. It can be seen from the figure that as the order of approximation increases, the error E_m decreases. Moreover, the error is much smaller for the exponential base function, as can be seen from the figure. Unlike the $\hbar$-curve, the figure shows that the calculation of squared residual error leads to an optimal value for $\hbar$, and thus it is more useful for finding a convergent series solution using HAM. The HAM can be implemented efficiently in a mathematical solver. For that purpose, a flowchart diagram containing the steps is provided in Figure 2.10.

The optimal value of the auxiliary parameter $\hbar$ ensures the convergence of the series solution using HAM. It is seen that the calculation of squared residual error produces an optimal value for the parameter. Using the same technique, we

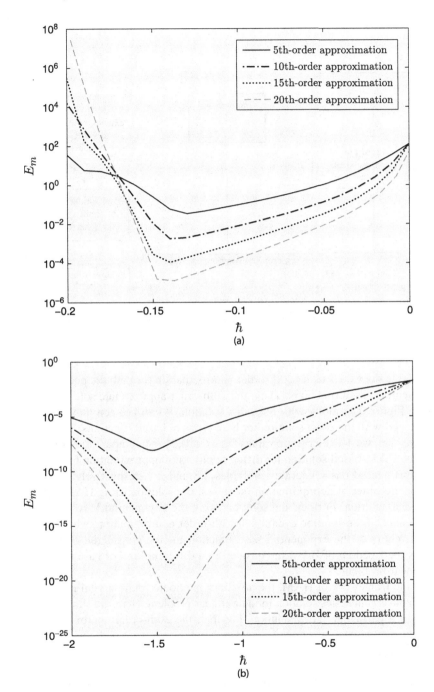

FIGURE 2.9 Squared residual error for a different order of approximations: (a) polynomial base function and (b) exponential base function.

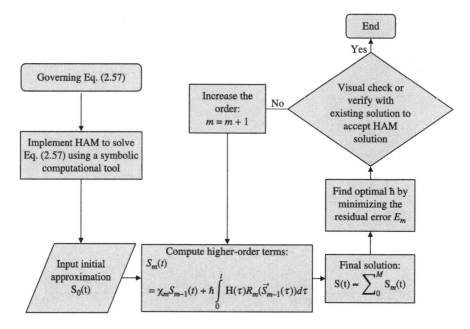

FIGURE 2.10   Flowchart for the HAM solution in Eq. (2.133).

determine the values using 20th-order approximation for both the polynomial- and exponential-based functions. Then, the 20th-order approximate solutions are plotted in Figure 2.11 along with the exact solution. It can be seen that the solutions agree well with the exact solution for both types of based functions. To get a quantitative idea, we consider some values from $t \in [0,5]$ and compare the exact solution and the HAM-based solution for different orders of approximation for polynomial and exponential base functions in Tables 2.3 and 2.4, respectively. One can see that as the order of approximation increases, the solution using HAM is closer to the exact solution. Further, the solution using the exponential base function provides an accurate solution even for the 5th-order approximation, which shows the effectiveness of the exponential base function for the considered equation. This shows the freedom of HAM in considering a different kind of base function for a nonlinear problem.

In the framework of HAM, the auxiliary parameter plays a vital role in obtaining a convergent series solution for a nonlinear problem. Here, we show the essence of $\hbar$ for a particular series solution. For that, the 20th-order approximation using the exponential base function is considered and is compared with the exact solution for different values of the auxiliary parameter $\hbar$ in Figure 2.12. Using the squared residual error method, an optimal value of $\hbar$ is obtained, which is $\hbar = -0.15$. Now, we select some values of $\hbar$ starting from $\hbar = -1$ and move to -0.15. It can be seen from the figure that as we move to $\hbar = -0.15$, the region of convergence of the series solution increases greatly. This shows the significance of auxiliary parameter $\hbar$ and hence of HAM as compared to other analytical methods.

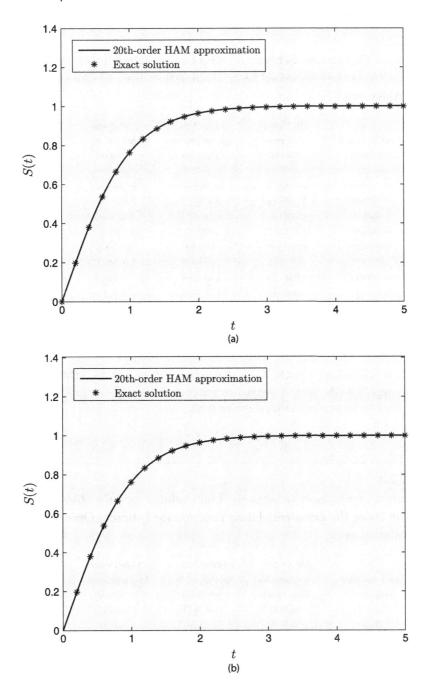

FIGURE 2.11 Comparison between the exact solution and the 20th-order HAM-based approximation: (a) polynomial base function and (b) exponential base function.

TABLE 2.3

Numerical Comparison between the Exact Solution and the HAM-Based Solution Using the Polynomial Base Function for Different Orders of Approximations

t	Exact Solution	5th-Order Approximation	10th-Order Approximation	15th-Order Approximation	20th-Order Approximation
0.0	0.000000	0.000000	0.000000	0.000000	0.000000
0.5	0.462117	0.479523	0.469491	0.464959	0.463147
1.0	0.761594	0.848373	0.790007	0.769824	0.763804
1.5	0.905148	1.049321	0.935734	0.910894	0.906244
2.0	0.964028	1.101148	0.981544	0.967006	0.964820
2.5	0.986614	1.076562	0.997891	0.989829	0.987630
3.0	0.995055	1.049164	1.007790	0.999004	0.996220
3.5	0.998178	1.047845	1.013450	1.002544	0.999380
4.0	0.999330	1.058710	1.016268	1.003803	1.000506
4.5	0.999753	1.066963	1.017440	1.004151	1.000894
5.0	0.999909	1.000677	1.023861	1.002199	1.002249

2.6.3 OPTIMAL HOMOTOPY ASYMPTOTIC METHOD–BASED SOLUTION

The steps concerning OHAM have already been discussed for ODEs. Here, we directly apply it to the nonlinear equation under consideration. For applying OHAM, Eq. (2.57) can be written in the following form:

$$\mathcal{L}\big[S(t)\big] + \mathcal{N}\big[S(t)\big] + h(t) = 0 \tag{2.166}$$

TABLE 2.4

Numerical Comparison between the Exact Solution and the HAM-Based Solution Using the Exponential Base Function for Different Orders of Approximations

t	Exact Solution	5th-Order Approximation	10th-Order Approximation	15th-Order Approximation	20th-Order Approximation
0.0	0.000000	0.000000	0.000000	0.000000	0.000000
0.5	0.462117	0.462117	0.462117	0.462117	0.462117
1.0	0.761594	0.761588	0.761594	0.761594	0.761594
1.5	0.905148	0.905106	0.905148	0.905148	0.905148
2.0	0.964028	0.963970	0.964027	0.964028	0.964028
2.5	0.986614	0.986566	0.986614	0.986614	0.986614
3.0	0.995055	0.995021	0.995055	0.995055	0.995055
3.5	0.998178	0.998156	0.998178	0.998178	0.998178
4.0	0.999330	0.999316	0.999329	0.999329	0.999330
4.5	0.999753	0.999745	0.999753	0.999753	0.999753
5.0	0.999909	0.999904	0.999909	0.999909	0.999909

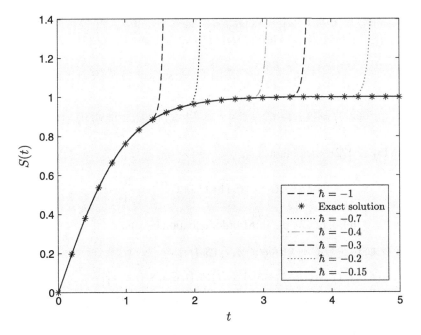

FIGURE 2.12 Comparison between the exact solution and the 20th-order HAM-based approximation for different values of $\hbar$.

where $\mathcal{L}[S(t)] = \dfrac{dS}{dt}$, $\mathcal{N}[S(t)] = S^2(t)$, and $h(t) = -1$. With these considerations, the zeroth-order Eq. (2.36) becomes:

$$\mathcal{L}(S_0(t)) + h(t) = 0 \text{ subject to } S_0(0) = 0 \tag{2.167}$$

Solving Eq. (2.167), one obtains:

$$S_0(t) = t \tag{2.168}$$

The first-order Eq. (2.40) reduces to:

$$\frac{dS_1}{dt} = H_1(t, C_i) S_0^{\,2} \text{ subject to } S_1(0) = 0 \tag{2.169}$$

We simply choose $H_1(t, C_1) = C_1$. Then, solution of Eq. (2.169) produces:

$$S_1(t, C_1) = \frac{C_1}{3} t^3 \tag{2.170}$$

Putting $k = 2$ in Eq. (2.42), the second-order equation becomes:

$$\mathcal{L}[S_2(t, C_i) - S_1(t, C_i)] = H_2(t, C_i) \mathcal{N}_0(S_0(t))$$
$$+ H_1(t, C_i)[\mathcal{L}[S_1(t, C_i)] + \mathcal{N}_1[S_0(t), S_1(t, C_i)]] \text{ subject to } S_2(0) = 0 \tag{2.171}$$

Expanding $\mathcal{N}\big(\Phi(t,C_i;q)\big)$, we obtain the coefficient of q as $\mathcal{N}_1\big[S_0(t),S_1(t,C_i)\big]=2S_0(t)S_1(t,C_i)$. Further, we choose $H_2(t,C_i)=C_2$. Then, Eq. (2.171) becomes:

$$\frac{dS_2}{dt}=(1+C_1)\frac{dS_1}{dt}+C_2S_0{}^2+C_1\mathcal{N}_1 \text{ subject to } S_2(0)=0$$

$$\Rightarrow \frac{dS_2}{dt}=\big[C_1(1+C_1)+C_2\big]t^2+\frac{2C_1{}^2}{3}t^4 \text{ subject to } S_2(0)=0 \qquad (2.172)$$

Solving Eq. (2.172), one obtains:

$$S_2(t,C_1,C_2)=\big[C_1(1+C_1)+C_2\big]\frac{t^3}{3}+\frac{2C_1{}^2}{15}t^5 \qquad (2.173)$$

Putting $k=3$ in Eq. (2.42), the third-order equation becomes:

$$\mathcal{L}\big[S_3(t,C_i)-S_2(t,C_i)\big]=H_3(t,C_i)\mathcal{N}_0\big(S_0(t)\big)$$
$$+H_1(t,C_i)\big[\mathcal{L}\big[S_2(t,C_i)\big]+\mathcal{N}_2\big[S_0(t),S_1(t,C_i),S_2(t,C_i)\big]\big]$$
$$+H_2(t,C_i)\big[\mathcal{L}\big[S_1(t,C_i)\big]+\mathcal{N}_1\big[S_0(t),S_1(t,C_i)\big]\big] \text{ subject to } S_3(0)=0 \qquad (2.174)$$

Expanding $\mathcal{N}\big(\Phi(t,C_i;q)\big)$, we obtain the coefficient of q^2 as $\mathcal{N}_2[S_0(t),S_1(t,C_i),S_2(t,C_i)]=S_1{}^2(t,C_i)+2S_0(t)S_2(t,C_i)$. Further, we choose $H_3(t,C_i)=C_3$. Therefore, Eq. (2.174) becomes:

$$\frac{dS_3}{dt}=(1+C_1)\frac{dS_2}{dt}+C_2\frac{dS_1}{dt}+C_3S_0{}^2+C_1S_1{}^2+2C_1S_0S_2+2C_2S_0S_1 \text{ subject to } S_3(0)$$

$$=0 \Rightarrow \frac{dS_3}{dt}=\big\{(1+C_1)\big[C_1(1+C_1)+C_2\big]+C_3+C_1C_2\big\}t^2$$

$$+\frac{2C_1}{3}\big\{\big[C_1(1+C_1)+C_2\big]+C_2+C_1+C_1{}^2\big\}t^4+\frac{17C_1{}^3}{45}t^6 \text{ subject to } S_3(0)=0 \quad (2.175)$$

Solving Eq. (2.175), we get:

$$S_3(t,C_1,C_2,C_3)=\big\{(1+C_1)\big[C_1(1+C_1)+C_2\big]+C_3+C_1C_2\big\}\frac{t^3}{3}$$

$$+\frac{2C_1}{15}\big\{\big[C_1(1+C_1)+C_2\big]+C_2+C_1+C_1{}^2\big\}t^5+\frac{17C_1{}^3}{315}t^7 \quad (2.176)$$

Therefore, the approximate solution can be found as:

$$S_{OHAM}(t)\approx S_0(t)+S_1(t,C_1)+S_2(t,C_1,C_2)+S_3(t,C_1,C_2,C_3) \qquad (2.177)$$

where the terms are given by Eqs. (2.168), (2.170), (2.173), and (2.176). It is interesting to note that for $C_1=-1$, $C_2=0$, and $C_3=0$, the terms of the solution become:

$$S_0(t)=t,\ S_1(t)=-\frac{t^3}{3},\ S_2(t)=\frac{2t^5}{15},\ S_3(t)=-\frac{17t^7}{315} \qquad (2.178)$$

which are exactly the same as the perturbation solution. The OHAM solution in terms of Q can be given as:

$$Q_{HAM}(t) = \left(\sum_{m=0}^{N} S_m(t) \right)^2 \tag{2.179}$$

Also, the solution over the entire time domain can be given as:

$$S = \begin{cases} S_{OHAM}(t) & \text{for } 0 \le t < T \\ \dfrac{S_T}{1+(t-T)S_T} & \text{for } t > T \end{cases} \tag{2.180}$$

where $S_{OHAM}(t)$ is given by Eq. (2.176).

It can be noted that we considered the auxiliary function $H(t,C_i)$ simply as a constant function. One can have many choices for this, and a better choice can indeed quicken the convergence of the series (Marinca and Herisanu 2015). Now, to assess the solution Eq. (2.177), one needs to calculate the constants C_i. For that purpose, Marinca and Herisanu (2015) followed the same approach as computing the auxiliary parameter in HAM. If $S_{OHAM}(t,C_i)$ is an approximate solution to the equation, then the residual can be calculated as:

$$R(t,C_i) = \mathcal{L}\left[S_{OHAM}(t,C_i) \right] + \mathcal{N}\left[S_{OHAM}(t,C_i) \right] + h(t), \ i = 1,2,\dots,s \tag{2.181}$$

When $R(t,C_i) = 0$, $S_{OHAM}(t,C_i)$ becomes the exact solution to the problem. But for nonlinear equations, one cannot expect $R(t,C_i)$ to be zero. However, we can minimize it to obtain an accurate solution. One of the ways to obtain the optimal C_i for which the solution converges is the minimization of squared residual error, i.e.,

$$J(C_i) = \int_{t \in D} R^2(t,C_i)\,dt, \ i = 1,2,\dots,s \tag{2.182}$$

where D is the domain of the equation. The minimization of Eq. (2.182) leads to a system of equations as follows:

$$\frac{\partial J}{\partial C_1} = \frac{\partial J}{\partial C_2} = \cdots = \frac{\partial J}{\partial C_s} = 0 \tag{2.183}$$

Solving this system, one can obtain the optimal values of the parameters.

Another way to obtain the parameters is to first consider some points, say t_i, in the domain D, and then to solve the equations:

$$R(t_1,C_i) = R(t_2,C_i) = \cdots = R(t_s,C_i) = 0, \ t_i \in D, \ i = 1,2,\dots,s. \tag{2.184}$$

For the equation considered, we calculate the parameters C_1, C_2, and C_3 by minimizing the residual error over $t \in [0,5]$, and they are found as $C_1 = -0.1445$, $C_2 = -0.01007$,

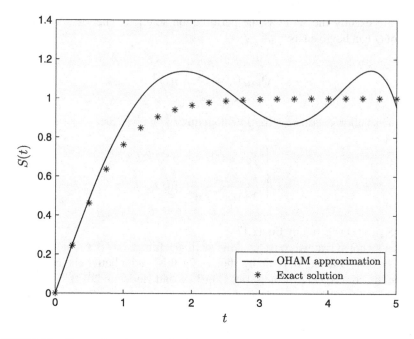

FIGURE 2.13   Comparison between the OHAM-based approximation and the exact solution.

and $C_3 = 0.02588$. With these values, the approximate solution is plotted against the exact solution in Figure 2.13. The numerical comparison for some selected values of t is also shown in Table 2.5. It is seen from the figure and the table that the OHAM-based solution agrees well with the exact solution only for a small region of t. Once t exceeds 1, the solution starts to deviate periodically. Indeed, a better solution can be achieved

TABLE 2.5

Comparison between the OHAM-Based Solution and the Exact Solution

t	Exact Solution	OHAM Approximation
0.0	0.000000	0.000000
0.5	0.462117	0.482885
1.0	0.761594	0.869178
1.5	0.905148	1.091409
2.0	0.964028	1.133942
2.5	0.986614	1.042863
3.0	0.995055	0.916628
3.5	0.998178	0.871080
4.0	0.999330	0.972387
4.5	0.999753	1.131530
5.0	0.999909	0.953900

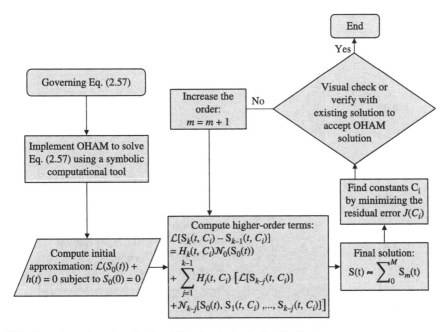

FIGURE 2.14 Flowchart for the OHAM solution in Eq. (2.177).

with the proper selection of the auxiliary functions $H_i(t, C_i)$, which we will see in subsequent chapters. The OHAM can be implemented efficiently in a mathematical solver. For that purpose, a flowchart diagram containing the steps is provided in Figure 2.14.

In this section, we obtained the analytical solution of the governing equation of one-dimensional overland flow using various methods. It is observed that the perturbation method, the LASPM, the ADM, and the HPM yield the same solution, whereas the solution using HAM and OHAM logically contains the other solutions as special cases. Further, HAM allows great freedom in choosing different kinds of base functions for a nonlinear equation, and the auxiliary parameter plays a vital role in enhancing the rate and region of convergence of the series solution. It is true that the solution using HPM and OHAM can be further improved, based on the selection of a better linear operator and auxiliary functions, respectively. However, they still differ from HAM in several aspects, such as the freedom of base function, role of the auxiliary parameter, etc. As an overview of the solutions obtained so far using these methods, in Figure 2.15, we plot different solutions over $t \in [0, 5]$ against the exact solution of the equation. It can be observed that only the solution using HAM agrees with the exact solution over the whole domain.

2.7 HOMOTOPY DERIVATIVE AND ITS PROPERTIES

In the section describing HAM, we observed that the key is to find R_m to obtain the higher-order deformation equations. For convenience of the reader, here we derive and report homotopy derivatives for some of the important specific forms and also for a general form. For more elaborated guidance, the reader may refer to Liao (2012).

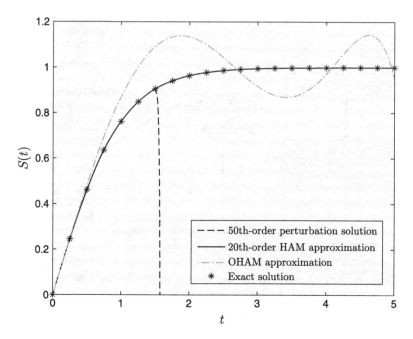

FIGURE 2.15 Comparison between the analytical solutions using different methods and the exact solution.

Let f be a function involving the embedding parameter q. Then D_n, defined by:

$$D_n(f) = \frac{1}{n!} \frac{d^n f}{dq^n}\bigg|_{q=0} \tag{2.185}$$

is called the *n-th-order homotopy derivative* of f. Here, D_n is the operator and $n \geq 0$ is an integer. Based on the form of f, a few specific cases are considered in what follows.

1. Consider $f = \psi$, where the homotopy series of ψ is given as $\psi = \Sigma_{i=0}^{\infty} u_i q^i$. Then the homotopy derivative becomes:

$$D_n(f) = \frac{1}{n!} \frac{d^n \left(\sum_{i=0}^{\infty} u_i q^i\right)}{dq^n}\bigg|_{q=0} = \frac{1}{n!}\left(n! u_n + \text{terms containing q}\right)\big|_{q=0} = u_n$$

$$\tag{2.186}$$

Therefore, $D_n(\psi) = u_n$.

2. Consider $f = \psi_1 \psi_2$, i.e., the product of two functions whose homotopy series are given as $\psi_1 = \Sigma_{i=0}^{\infty} u_i q^i$, $\psi_2 = \Sigma_{i=0}^{\infty} v_i q^i$. Using Leibniz's rule for derivatives of the product of functions (Olver 1986), we have:

$$\frac{d^n\left(\psi_1\psi_2\right)}{dq^n} = \sum_{j=0}^{n} \frac{n!}{j!(n-j)!} \frac{d^j\psi_1}{dq^j} \frac{d^{n-j}\psi_2}{dq^{n-j}} \tag{2.187}$$

Therefore, using Eqs. (2.186) and (2.187), we get:

$$D_n\left(f\right) = \frac{1}{n!} \frac{d^n\left(\psi_1\psi_2\right)}{dq^n}\bigg|_{q=0} = \sum_{j=0}^{n} \left(\frac{1}{j!} \frac{d^j\psi_1}{dq^j}\bigg|_{q=0}\right) \left[\frac{1}{(n-j)!} \frac{d^{n-j}\psi_2}{dq^{n-j}}\bigg|_{q=0}\right]$$

$$= \sum_{j=0}^{n} D_n\left(\psi_1\right) D_{n-j}\left(\psi_2\right) = \sum_{j=0}^{n} u_j v_{n-j} \tag{2.188}$$

Hence, $D_n\left(\psi_1\psi_2\right) = \Sigma_{j=0}^{n}\, u_j v_{n-j}$. In a similar way, one can also obtain $D_n\left(\psi_1\psi_2\right) = \Sigma_{j=0}^{n}\, v_j u_{n-j}$.

3. Consider $f = c_1\psi_1 + c_2\psi_2$, where c_1 and c_2 are independent of q but can be a function of the independent variable. We have:

$$D_n\left(f\right) = \frac{1}{n!} \frac{d^n\left(c_1\psi_1 + c_2\psi_2\right)}{dq^n}\bigg|_{q=0} = c_1 \left[\frac{1}{n!} \frac{d^n\left(\psi_1\right)}{dq^n}\bigg|_{q=0}\right]$$

$$+ c_2 \left[\frac{1}{n!} \frac{d^n\left(\psi_2\right)}{dq^n}\bigg|_{q=0}\right] = c_1 D_n\left(\psi_1\right) + c_2 D_n\left(\psi_2\right) = c_1 u_n + c_2 v_n \tag{2.189}$$

Therefore, $D_n\left(c_1\psi_1 + c_2\psi_2\right) = c_1 u_n + c_2 v_n$.

4. Consider $f = \psi^\gamma$, where $\gamma \geq 2$ is a positive integer. We first consider some specific values of γ. For $\gamma = 2$, taking $\psi_1 = \psi_2 = \psi$ in Eq. (2.188), we have:

$$D_n\left(\psi^2\right) = \sum_{j=0}^{n} u_j u_{n-j} \tag{2.190}$$

For $\gamma = 3$, using Eqs. (2.188) and (2.189), we get:

$$D_n\left(\psi^3\right) = D_n\left(\psi^2 . \psi\right) = \sum_{j=0}^{n} D_{n-j}\left(\psi\right) D_n\left(\psi^2\right)$$

$$= \sum_{j=0}^{n} u_{n-j} D_n\left(\psi^2\right) = \sum_{j=0}^{n} u_{n-j} \sum_{k=0}^{j} u_k u_{j-k} \tag{2.191}$$

In a similar manner, for $\gamma = 4$, one obtains:

$$D_n\left(\psi^4\right) = \sum_{j=0}^{n} D_{n-j}\left(\psi\right) D_n\left(\psi^3\right) = \sum_{j=0}^{n} u_{n-j} \sum_{k=0}^{j} u_{j-k} \sum_{i=0}^{k} u_i u_{k-i} \tag{2.192}$$

Proceeding like this, one can derive:

$$D_n\left(\psi^\gamma\right) = \sum_{\sigma_1=0}^{n} u_{n-\sigma_1} \sum_{\sigma_2=0}^{\sigma_1} u_{\sigma_1-\sigma_2} \sum_{\sigma_3=0}^{\sigma_2} u_{\sigma_2-\sigma_3} \cdots \sum_{\sigma_{\gamma-1}=0}^{\sigma_{\gamma-2}} u_{\sigma_{\gamma-1}} u_{\sigma_{\gamma-2}-\sigma_{\gamma-1}} \quad (2.193)$$

The identity Eq. (2.193) can easily be proved using the principle of mathematical induction. One can also refer to Liao (2012).

5. Consider $f = \exp(\beta\psi)$, where $\beta \neq 0$ is independent of parameter q. We can see that for $n = 0$:

$$D_0\left(\exp(\beta\psi)\right) = \frac{1}{0!} \left.\frac{d^0\left(\exp(\beta\psi)\right)}{dq^0}\right|_{q=0} = \exp(\beta u_0) \qquad (2.194)$$

Before deriving the homotopy derivative for $n \geq 1$, consider:

$$\frac{d^n}{dq^n}\left(\exp(\beta\psi)\right) = \frac{d^{n-1}}{dq^{n-1}}\left(\frac{d\exp(\beta\psi)}{dq}\right) = \beta\frac{d^{n-1}}{dq^{n-1}}\left(\exp(\beta\psi)\frac{d\psi}{dq}\right) \quad (2.195)$$

Using the Leibniz rule given by Eq. (2.187), we obtain:

$$\frac{d^{n-1}}{dq^{n-1}}\left(\exp(\beta\psi)\frac{d\psi}{dq}\right) = \sum_{j=0}^{n-1} \frac{(n-1)!}{j!(n-j-1)!} \frac{d^j\left[\exp(\beta\psi)\right]}{dq^j}\frac{d^{n-j}\psi}{dq^{n-j}} \quad (2.196)$$

From Eqs. (2.195) and (2.196), we have:

$$\frac{d^n}{dq^n}\left(\exp(\beta\psi)\right) = \beta\sum_{j=0}^{n-1} \frac{(n-1)!}{j!(n-j-1)!} \frac{d^j\left[\exp(\beta\psi)\right]}{dq^j}\frac{d^{n-j}\psi}{dq^{n-j}} \quad (2.197)$$

Therefore,

$$
\begin{aligned}
D_n\left(\exp(\beta\psi)\right) &= \frac{1}{n!}\left.\frac{d^n\left(\exp(\beta\psi)\right)}{dq^n}\right|_{q=0} = \frac{\beta}{n}\sum_{j=0}^{n-1}\frac{1}{j!(n-j-1)!}\left.\frac{d^j\left[\exp(\beta\psi)\right]}{dq^j}\frac{d^{n-j}\psi}{dq^{n-j}}\right|_{q=0} \\
&= \beta\sum_{j=0}^{n-1}\frac{(n-j)}{n}\left[\frac{1}{(n-j)!}\frac{d^{n-j}\psi}{dq^{n-j}}\right]\left.\left(\frac{1}{j!}\frac{d^j\left[\exp(\beta\psi)\right]}{dq^j}\right)\right|_{q=0} \\
&= \beta\sum_{j=0}^{n-1}\left(1-\frac{j}{n}\right)D_{n-j}\left(\psi\right)D_j\left(\exp(\beta\psi)\right) \\
&= \beta\sum_{j=0}^{n-1}\left(1-\frac{j}{n}\right)u_{n-j}D_j\left(\exp(\beta\psi)\right) \qquad (2.198)
\end{aligned}
$$

6. Consider $f = \sin(\psi), \cos(\psi)$, then the following holds:

$$D_0\left(\sin(\psi)\right) = \sin(u_0) \qquad (2.199)$$

$$D_0\left(\cos\left(\psi\right)\right)=\cos\left(u_0\right) \tag{2.200}$$

$$D_n\left(\sin\left(\psi\right)\right)=\sum_{j=0}^{n-1}\left(1-\frac{j}{n}\right)u_{n-j}D_j\left(\cos\left(\psi\right)\right),\ n\geq1 \tag{2.201}$$

$$D_n\left(\cos\left(\psi\right)\right)=-\sum_{j=0}^{n-1}\left(1-\frac{j}{n}\right)u_{n-j}D_j\left(\sin\left(\psi\right)\right),\ n\geq1 \tag{2.202}$$

The formulae given by Eqs. (2.201) and (2.202) can be proved without any difficulty using the definition of the homotopy derivative and writing $\sin\left(\psi\right)=\dfrac{exp\left(i\psi\right)-exp\left(-i\psi\right)}{2i}$, $\cos\left(\psi\right)=\dfrac{exp\left(i\psi\right)+exp\left(-i\psi\right)}{2}$ where i is an imaginary number defined as $i^2=-1$.

7. Consider $f=g\left(\psi\right)$, where g is an arbitrary smooth function (a function that possesses continuous derivatives). For $n=0$:

$$D_0\left(g\left(\psi\right)\right)=\frac{1}{0!}\frac{d^0\left(g\left(\psi\right)\right)}{dq^0}\Bigg|_{q=0}=g\left(u_0\right) \tag{2.203}$$

For $n\geq1$, first, we obtain the following result using the Leibniz rule for the derivative of the product of functions:

$$\frac{d^n}{dq^n}\left(g\left(\psi\right)\right)=\frac{d^{n-1}}{dq^{n-1}}\left(\frac{d\psi}{dq}\frac{dg\left(\psi\right)}{d\psi}\right)$$

$$=\sum_{j=0}^{n-1}\frac{(n-1)!}{j!(n-j-1)!}\left[\frac{d^{n-j-1}}{dq^{n-j-1}}\left(\frac{d\psi}{dq}\right)\right]\frac{d^j}{dq^j}\left[\frac{dg\left(\psi\right)}{d\psi}\right] \tag{2.204}$$

Therefore,

$$D_n\left(g\left(\psi\right)\right)=\frac{1}{n!}\frac{d^n\left(g\left(\psi\right)\right)}{dq^n}\Bigg|_{q=0}=\frac{1}{n}\sum_{j=0}^{n-1}\frac{1}{j!(n-j-1)!}\left[\frac{d^{n-j-1}}{dq^{n-j-1}}\left(\frac{d\psi}{dq}\right)\right]\frac{d^j}{dq^j}\left[\frac{dg\left(\psi\right)}{d\psi}\right]\Bigg|_{q=0}$$

$$=\sum_{j=0}^{n-1}\frac{(n-j)}{n}\left[\frac{1}{(n-j)}\frac{d^{n-j}\psi}{dq^{n-j}}\right]\left(\frac{1}{j!}\frac{d^j}{dq^j}\left[\frac{dg\left(\psi\right)}{d\psi}\right]\right)\Bigg|_{q=0}$$

$$=\sum_{j=0}^{n-1}\left(1-\frac{j}{n}\right)D_{n-j}\left(\psi\right)D_j\left(\frac{dg\left(\psi\right)}{d\psi}\right)$$

$$=\sum_{j=0}^{n-1}\left(1-\frac{j}{n}\right)u_{n-j}\frac{d}{du_0}\left(D_j\left(g\left(\psi\right)\right)\right) \tag{2.205}$$

Since $D_j\left(\dfrac{dg\left(\psi\right)}{d\psi}\right)=\dfrac{d}{d\psi}\left(D_j\left[g\left(\psi\right)\right]\right)$ and $\dfrac{d}{d\psi}\cong\dfrac{d}{du_0}$ for $q=0$.

Eq. (2.205) is the homotopy derivative for an arbitrary function. In solving real-life problems, we often come up with differential equations having non-integer power nonlinearity, logarithmic nonlinearity, etc. Eq. (2.205) can be applied to those cases for the formulation of the higher-order deformation equations.

2.8   CONVERGENCE THEOREM OF A HAM-BASED SOLUTION

It is seen that HAM provides the solution to a nonlinear problem in the form of a series. As already stated, the convergence of a series is based on the proper selection of the linear operator, initial approximation, auxiliary function, and auxiliary parameter, which are directed by some fundamental rules. Here, we prove two important theorems in relation to the problem considered in this chapter, which tell us that if the series obtained by HAM is convergent, then it must be the solution of the original nonlinear equation.

Theorem 2.1: *If the homotopy series $\Sigma_{m=0}^{\infty} S_m(t)$ and $\Sigma_{m=0}^{\infty} S_m{}'(t)$ converge, then $R_m\left(\vec{S}_{m-1}\right)$ given by Eq. (2.132) satisfies the relation $\Sigma_{m=1}^{\infty} R_m\left(\vec{S}_{m-1}\right) = 0$. Here, "'" represents the derivative with respect to t.*

Proof: The auxiliary linear operator for Eq. (2.139) was defined as follows:

$$\mathcal{L}[S] = \frac{dS}{dt} \tag{2.206}$$

According to Eq. (2.126), we obtain:

$$\mathcal{L}[S_1] = \hbar R_1\left(\vec{S}_0\right) \tag{2.207}$$

$$\mathcal{L}[S_2 - S_1] = \hbar R_2\left(\vec{S}_1\right) \tag{2.208}$$

$$\mathcal{L}[S_3 - S_2] = \hbar R_3\left(\vec{S}_2\right) \tag{2.209}$$

$$\mathcal{L}[S_m - S_{m-1}] = \hbar R_m\left(\vec{S}_{m-1}\right) \tag{2.210}$$

Adding all the previous terms, we get:

$$\mathcal{L}[S_m] = \hbar \sum_{k=1}^{m} R_k\left(\vec{S}_{k-1}\right) \tag{2.211}$$

As the series $\Sigma_{m=0}^{\infty} S_m(t)$ and $\Sigma_{m=0}^{\infty} S_m'(t)$ are convergent, $\lim_{m \to \infty} S_m(t) = 0$ and $\lim_{m \to \infty} S_m'(t) = 0$. Now, recalling the earlier summand and taking the limit, the required result follows as:

$$\hbar \sum_{k=1}^{\infty} R_k\left(\vec{S}_{k-1}\right) = \hbar \lim_{m \to \infty} \sum_{k=1}^{m} R_k\left(\vec{S}_{k-1}\right) = \lim_{m \to \infty} \mathcal{L}[S_m] = \lim_{m \to \infty} S_m' = 0 \qquad (2.212)$$

Theorem 2.2: *If $\hbar$ is so properly chosen that the series $\Sigma_{m=0}^{\infty} S_m(t)$ and $\Sigma_{m=0}^{\infty} S_m'(t)$ converge absolutely to $S(t)$ and $S'(t)$, respectively, then the homotopy series $\Sigma_{m=0}^{\infty} S_m(t)$ satisfies the original governing Eq. (2.57).*

Proof: We first recall the following definition.

Cauchy product of two infinite series: Let $\Sigma_{i=0}^{\infty} x_i$ and $\Sigma_{j=0}^{\infty} y_j$ be two infinite series of real/complex terms. Then the *Cauchy product* of the two series is defined by the discrete convolution as follows:

$$\left(\sum_{i=0}^{\infty} x_i\right)\left(\sum_{j=0}^{\infty} y_j\right) = \sum_{k=0}^{\infty}\sum_{l=0}^{k} x_l y_{k-l} \qquad (2.213)$$

Therefore, using the previous rule in relation to Eq. (2.132), we get:

$$\sum_{m=1}^{\infty}\sum_{j=0}^{m-1} S_j S_{m-1-j} = \left(\sum_{m=0}^{\infty} S_m\right)^2 \qquad (2.214)$$

Theorem 2.1 shows that if $\Sigma_{m=0}^{\infty} S_m(t)$ and $\Sigma_{m=0}^{\infty} S_m'(t)$ converge then $\Sigma_{m=1}^{\infty} R_m\left(\vec{S}_{m-1}\right) = 0$.

Therefore, substituting the earlier expressions in Eq. (2.132) and simplifying further lead to:

$$\sum_{m=0}^{\infty} S_m' + \left(\sum_{m=0}^{\infty} S_m\right)^2 - 1 = 0 \qquad (2.215)$$

which is basically the original governing equation [Eq. (2.57)]. Furthermore, subject to the initial condition $S_0(0) = 0$ and the conditions for the higher-order deformation equation $S_m(0) = 0$, for $m \geq 1$, we easily obtain $\Sigma_{m=0}^{\infty} S_m(0) = 0$. Hence, the convergence result follows.

2.9 CONVERGENCE THEOREM OF AN OHAM-BASED SOLUTION

Similar to HAM, the OHAM also yields an analytical solution to a nonlinear problem in the form of a series. The convergence of the series depends on the choice of the auxiliary function. Further, the optimal values of the parameters increase the rate of convergence. Here, we prove the convergence of the OHAM-based solution theoretically.

Theorem 2.3: *If the series $S_0(t) + \Sigma_{j=1}^{\infty} S_j(t, C_i)$, $i = 1, 2, \ldots, s$ converges, where $S_j(t, C_i)$ is governed by Eqs. (2.36), (2.40), and (2.42), then Eq. (2.45) is a solution of the original Eq. (2.57).*

Proof: Based on the choice of the auxiliary function, suppose that the series Eq. (2.45) is convergent. Then, we get:

$$\lim_{j \to \infty} S_j(t, C_i) = 0, \ i = 1, 2, \ldots, s \tag{2.216}$$

One can write:

$$\begin{aligned} S_j(t, C_i) &= S_0(t) + \left[S_1(t, C_i) - S_0(t, C_i) \right] \\ &\quad + \left[S_2(t, C_i) - S_1(t, C_i) \right] + \cdots + \left[S_j(t, C_i) - S_{j-1}(t, C_i) \right] \\ &= S_0(t) + \sum_{k=1}^{j} \left[S_k(t, C_i) - S_{k-1}(t, C_i) \right], \ i = 1, 2, \ldots, s \end{aligned} \tag{2.217}$$

Using Eq. (2.216), one can obtain from Eq. (2.15):

$$0 = \lim_{j \to \infty} S_j(t, C_i) = S_0(t) + \sum_{k=1}^{\infty} \left[S_k(t, C_i) - S_{k-1}(t, C_i) \right], \ i = 1, 2, \ldots, s \tag{2.218}$$

Eq. (2.216) can be rearranged as:

$$\begin{aligned} 0 &= S_0(t) + h(t) - h(t) + \left[S_1(t, C_i) - S_0(t, C_i) \right] \\ &\quad + \sum_{k=2}^{\infty} \left[S_k(t, C_i) - S_{k-1}(t, C_i) \right], \ i = 1, 2, \ldots, s \end{aligned} \tag{2.219}$$

Using the property of the linear operator, i.e., $\mathcal{L}[A_1(t) + A_2(t)] = \mathcal{L}[A_1(t)] + \mathcal{L}[A_2(t)]$ and $\mathcal{L}(0) = 0$, and Eqs. (2.36), (2.40), and (2.42), we have from Eq. (2.219):

$$0 = \mathcal{L}(0) = \mathcal{L}\big[S_0(t)\big] + h(t) + \mathcal{L}\big[S_1(t,C_i)\big] - \big(\mathcal{L}\big[S_0(t)\big] + h(t)\big)$$

$$+ \sum_{k=2}^{\infty}\big(\mathcal{L}\big[S_k(t,C_i)\big] - \mathcal{L}\big[S_{k-1}(t,C_i)\big]\big) = H_1(t,C_i)N_0\big[S_0(t)\big]$$

$$+ \sum_{k=2}^{\infty}\Bigg(H_k(t,C_i)N_0\big[S_0(t)\big] + \sum_{l=1}^{k-1}H_l(t,C_i)\big[\mathcal{L}\big[S_{k-l}(t,C_i)\big] + N_{k-l}\big[S_0(t),S_1(t,C_i),...,S_{k-l}(t,C_i)\big]\big]\Bigg)$$

$$= \Bigg[\sum_{k=1}^{\infty}H_k(t,C_i)\Bigg]N_0\big[S_0(t)\big] + \sum_{k=2}^{\infty}\sum_{l=1}^{k-1}H_l(t,C_i)\big[\mathcal{L}\big[S_{k-l}(t,C_i)\big] + N_{k-l}\big[S_0(t),S_1(t,C_i),...,S_{k-l}(t,C_i)\big]\big]$$

$$= H(t,C_i)N_0\big[S_0(t)\big] + \sum_{l=1}^{\infty}\sum_{k=l+1}^{\infty}H_l(t,C_i)\big[\mathcal{L}\big[S_{k-l}(t,C_i)\big] + N_{k-l}\big[S_0(t),S_1(t,C_i),...,S_{k-l}(t,C_i)\big]\big]$$

$$= H(t,C_i)N_0\big[S_0(t)\big] + \sum_{k=1}^{\infty}H_k(t,C_i)\sum_{p=1}^{\infty}\big[\mathcal{L}\big[S_p(t,C_i)\big] + N_p\big[S_0(t),S_1(t,C_i),...,S_p(t,C_i)\big]\big]$$

$$H(t,C_i)N_0\big[S_0(t)\big] + H(t,C_i)\Bigg[\mathcal{L}\Bigg(\sum_{p=1}^{\infty}S_p(t,C_i)\Bigg) + \sum_{p=1}^{\infty}N_p\big[S_0(t),S_1(t,C_i),...,S_p(t,C_i)\big]\Bigg]$$

$$= H(t,C_i)N_0\big[S_0(t)\big] + H(t,C_i)\big[\mathcal{L}(S(t,C_i)) - \mathcal{L}(S_0(t)) + N(S(t,C_i)) - N(S_0(t))\big]$$

$$= H(t,C_i)N_0\big[S_0(t)\big] + H(t,C_i)\big[\mathcal{L}(S(t,C_i)) - [\mathcal{L}(S_0(t)) + h(t)] + h(t) + N(S(t,C_i)) - N(S_0(t))\big]$$

$$= H(t,C_i)\big[\mathcal{L}(S(t,C_i)) + h(t) + N(S(t,C_i))\big], \quad i = 1,2,...,s \tag{2.220}$$

Now, since $H(t,C_i) \neq 0$, from Eq. (2.220) we have:

$$\mathcal{L}(S(t,C_i)) + h(t) + N(S(t,C_i)) = 0, \quad i = 1,2,...,s \tag{2.221}$$

which shows that $S(t,C_i)$ is the exact solution of Eq. (2.57).

2.10 PADÉ APPROXIMANT

A Padé approximant is the approximation of a function by a rational function of a given order (Baker and Grave-Morris 1996). It is considered to be the most accurate approximation of a function. The method often gives a better approximation of a function than by truncating its Taylor series. In the area of nonlinear analysis, as in differential equations, we often have to tackle a strongly nonlinear term. Using this method, one can convert the strongly nonlinear term into a relatively weaker form, and hence it becomes easy to handle the problem. Suppose $f(x)$ is a given function. Then, its $[m,n]$-order Padé approximant, say $f_p(x)$, around the point $x = a$ is given as follows:

$$[f_p(x)]_{m,n} = \frac{\sum_{i=0}^{m}A_i(x-a)^i}{1 + \sum_{j=1}^{n}B_j(x-a)^j} \tag{2.222}$$

where A_i and B_j are constants, which can be uniquely determined (Baker and Moris 1996). The approximant Eq. (2.222) agrees with the function $f(x)$ to the highest possible order (here, $m+n$), i.e.,

$$f(a) = \left[f_p(a)\right]_{m,n}, \; f'(a) = \left[f'_p(a)\right]_{m,n}, f''(a)$$
$$= \left[f''_p(a)\right]_{m,n}, \dots, f^{m+n}(a) = \left[f_p^{m+n}(a)\right]_{m,n} \qquad (2.223)$$

Coefficients A_i and B_j can be obtained by solving the set of Eq. (2.223). The coefficients can also be obtained by expanding $f(x)$ in a Taylor series and then equating the like powers of x in Eq. (2.220). However, the higher-order approximants are not easy to achieve by hand calculation. For that, one can simply use the Mathematica command *PadeApproximant*. Here, we consider some examples and show the command's effectiveness.

Example 2.1

Consider $f(x) = \exp(x)$. Padé approximants of different orders for this function are obtained as follows:

We find out the [1,1]-order Padé approximant for this function. For that purpose, expanding the function $\exp(x)$ in a Taylor series, we have from Eq. (2.222):

$$1 + x + \frac{x^2}{2!} + \dots = \frac{A_0 + A_1 x}{1 + B_1 x} \Rightarrow \left(1 + x + \frac{x^2}{2!} + \dots\right)(1 + B_1 x)$$

$$= A_0 + A_1 x \Rightarrow (1 - A_0) + x(1 + B_1 - A_1) + x^2\left(\frac{1}{2} + B_1\right) + \dots = 0 \quad (2.224)$$

Equating the like powers of x, we obtain the set of equations:

$$1 - A_0 = 0 \qquad (2.225)$$

$$1 + B_1 - A_1 = 0 \qquad (2.226)$$

$$\frac{1}{2} + B_1 = 0 \qquad (2.227)$$

Solving this set, we get $A_0 = 1$, $A_1 = 1/2$, and $B_1 = -1/2$. Therefore, the approximant can be obtained as follows:

$$\left[\exp(x)\right]_{1,1} = \frac{1 + \frac{x}{2}}{1 - \frac{x}{2}} \qquad (2.228)$$

In a similar manner, the higher-order approximants can be obtained. Those are written as follows:

$$\left[\exp(x)\right]_{1,2} = \frac{1 + \frac{x}{3}}{1 - \frac{2x}{3} + \frac{x^2}{6}} \qquad (2.229)$$

$$\left[\exp(x)\right]_{2,1} = \frac{1+\dfrac{2x}{3}+\dfrac{x^2}{6}}{1-\dfrac{x}{3}} \qquad (2.230)$$

$$\left[\exp(x)\right]_{2,2} = \frac{1+\dfrac{x}{2}+\dfrac{x^2}{12}}{1-\dfrac{x}{2}+\dfrac{x^2}{12}} \qquad (2.231)$$

Example 2.2

Consider $f(x) = \cos(x)$. Padé approximants of different orders for this function are obtained as follows:

$$\left[\cos(x)\right]_{1,1} = 1 \qquad (2.232)$$

$$\left[\cos(x)\right]_{1,2} = \frac{1}{1+\dfrac{x^2}{2}} \qquad (2.233)$$

$$\left[\cos(x)\right]_{2,1} = 1-\frac{x^2}{2} \qquad (2.234)$$

$$\left[\cos(x)\right]_{2,2} = \frac{1-\dfrac{5x^2}{12}}{1+\dfrac{x^2}{12}} \qquad (2.235)$$

Example 2.3

Consider $f(x) = a^x$, where a is a positive integer. Padé approximants of different orders for this function are obtained as follows:

$$\left[a^x\right]_{1,1} = \frac{1+\dfrac{1}{2}x\ln a}{1-\dfrac{1}{2}x\ln a} \qquad (2.236)$$

$$\left[a^x\right]_{1,2} = \frac{1+\dfrac{1}{3}x\ln a}{1-\dfrac{2}{3}x\ln a+\dfrac{1}{6}x^2\left(\ln a\right)^2} \qquad (2.237)$$

$$\left[a^x\right]_{2,1} = \frac{1+\dfrac{2}{3}x\ln a+\dfrac{1}{6}x^2\left(\ln a\right)^2}{1-\dfrac{1}{3}x\ln a} \qquad (2.238)$$

$$\left[a^x\right]_{2,2} = \frac{1+\dfrac{1}{2}x\ln a+\dfrac{1}{12}x^2\left(\ln a\right)^2}{1-\dfrac{1}{2}x\ln a+\dfrac{1}{12}x^2\left(\ln a\right)^2} \qquad (2.239)$$

To check the effectiveness of the method, the Padé approximants obtained here are compared with the corresponding functions over the interval $x \in [0,1.5]$. Figures 2.16–2.18 compare the approximants of different orders for the selected functions. It is clear from the figure that the higher the order of approximation, the greater the accuracy. To get a quantitative idea, we report the values of the function and its approximants of different orders for some selected values of x, say $x = 0.2, 0.5, 0.8, 1.2, 1.5$ in Tables 2.6–2.8.

It is observed that the Padé approximant is an efficient tool for approximating a nonlinear function. One can convert a strong nonlinear equation into a relatively weaker nonlinear equation by applying the method to the strong nonlinear term. Further, in the context of application to HAM, this method can be applied to increase the region and rate of convergence of a series. For a given series:

$$\sum_{m=0}^{\infty} a_m x^m,$$

the corresponding $[n,p]$ order Padé approximant is given by:

$$\frac{\sum_{l=0}^{n} b_{n,l} x^l}{\sum_{l=0}^{p} c_{p,l} x^l}$$

where $b_{n,l}$ and $c_{p,l}$ are the coefficients that can be determined by c_i $(i = 0,1,2,\dots,n+p)$.

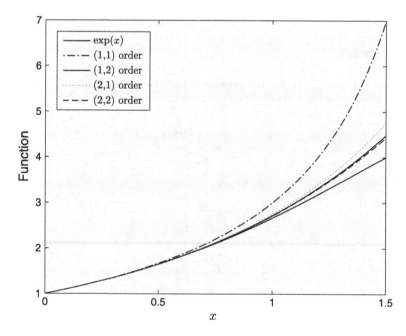

FIGURE 2.16 Comparison of $\exp(x)$ and its Padé approximants of orders (1,1), (1,2), (2,1), and (2,2) for $x \in [0,1.5]$.

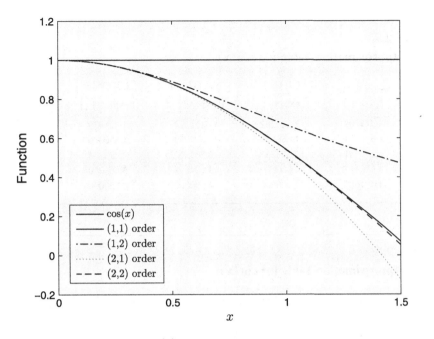

FIGURE 2.17   Comparison of $\cos(x)$ and its Padé approximants of orders (1,1), (1,2), (2,1), and (2,2) for $x \in [0,1.5]$.

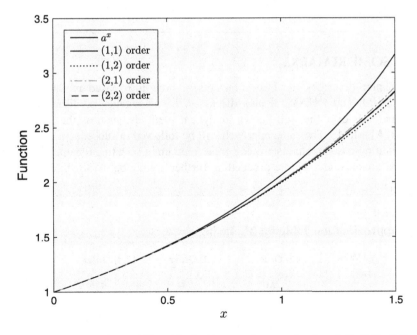

FIGURE 2.18   Comparison of 2^x and its Padé approximants of orders (1,1), (1,2), (2,1), and (2,2) for $x \in [0,1.5]$.

TABLE 2.6
Padé Approximation Table for $\exp(x)$

x	Exact Value $(\exp(x))$	$(1,1)$ Order $\left[\exp(x)\right]_{1,1}$	$(1,2)$ Order $\left[\exp(x)\right]_{1,2}$	$(2,1)$ Order $\left[\exp(x)\right]_{2,1}$	$(2,2)$ Order $\left[\exp(x)\right]_{2,2}$
0.2	1.221403	1.222222	1.221374	1.221429	1.221402
0.5	1.648721	1.666667	1.647059	1.650000	1.648649
0.8	2.225541	2.333333	2.209302	2.236364	2.224490
1.2	3.320117	4.000000	3.181818	3.400000	3.307692
1.5	4.481689	7.000000	4.000000	4.750000	4.428571

TABLE 2.7
Padé Approximation Table for $\cos(x)$

x	Exact Value $(\cos(x))$	$(1,1)$ Order $\left[\cos(x)\right]_{1,1}$	$(1,2)$ Order $\left[\cos(x)\right]_{1,2}$	$(2,1)$ Order $\left[\cos(x)\right]_{2,1}$	$(2,2)$ Order $\left[\cos(x)\right]_{2,2}$
0.2	0.980067	1.000000	0.980392	0.980000	0.980066
0.5	0.877583	1.000000	0.888889	0.8750000	0.877551
0.8	0.696707	1.000000	0.757576	0.680000	0.696202
1.2	0.362358	1.000000	0.581395	0.280000	0.357143
1.5	0.070737	1.000000	0.470588	−0.125000	0.052632

2.11 SOME REMARKS

This chapter describes the basic concepts of homotopy-based methods, namely HPM, HAM, and OHAM. It also discusses the advantages or disadvantages of these methods over the well-known analytical methods, such as the perturbation method, ADM, etc. The same approach will be followed in subsequent chapters. As the homotopy-based methods provide great freedom in selecting the operators, functions, parameters, etc., before proceeding further, it may be worthwhile to mention

TABLE 2.8
Padé Approximation Table for a^x, Taking $a = 2$

x	Exact Value (a^x)	$(1,1)$ Order $\left[a^x\right]_{1,1}$	$(1,2)$ Order $\left[a^x\right]_{1,2}$	$(2,1)$ Order $\left[a^x\right]_{2,1}$	$(2,2)$ Order $\left[a^x\right]_{2,2}$
0.2	1.148698	1.148954	1.148692	1.148704	1.148698
0.5	1.414214	1.419219	1.413900	1.414474	1.414204
0.8	1.741101	1.767242	1.738397	1.743125	1.740972
1.2	2.297397	2.424003	2.277602	2.310407	2.296074
1.5	2.828427	3.165455	2.764909	2.866914	2.823353

some key points here. Also, some of them will be followed in this book. Some remarks are given as follows:

1. The linear and nonlinear operators have their own meanings in mathematics literature, which was discussed in this chapter. However, in the context of homotopy-based methods, these are defined in a general framework in the existing literature. To be more specific, there exist several linear ODE/PDEs that are difficult to solve analytically, and therefore, these methods are equally applicable to those equations (Liao 2012). But in the cases of linear equations, the existing literature follows the same terms such as linear, nonlinear operator, etc., to be consistent with the methodological aspects. While they do not carry proper mathematical meaning, the operators are indeed different and do not violate the steps of the method. The same is followed in this book.

2. The homotopy-based methods provide solutions in the form of a series approximated at a finite step. Therefore, computations are performed with functions instead of numbers. It is true that development of high-performing symbolic computation tools, such as Mathematica, MATLAB, Maple, etc., can reduce the effort of implementing these methods; however, sometimes it might take a lot of time to produce the higher-order terms that depend on the form of the base function, linear operator, etc. It may be noted that one can obtain the higher-order terms numerically to reduce the computation time, though in that case the methods cannot be considered analytical ones.

3. It seems logical to choose a linear operator that does not exceed the order of a differential equation. However, because of the great freedom that the homotopy-based methods provide, it is possible to obtain a convergent series solution considering a higher-order linear operator than the equation itself. Some such examples can be found in Liao (2012).

4. The first step for these methods is to choose a proper set of base functions to represent the series solution. The existing perturbation/analytical solution and physical characteristics of the equation greatly help in choosing the appropriate base functions. However, it is not always possible to have an existing solution for a specific equation. For these cases, there are two ways one can follow: (i) choose different kinds of base functions and see which one works better and (ii) choose a part of the equation that is easier to solve and then consider the base function based on the form of the solution.

5. The OHAM is an extended version of HAM using approximation theory, and it has shown a significant improvement in providing accurate solutions with just a few terms of the series. In some cases, it seems that the OHAM is not able to provide satisfactory solutions, which can be attributed to the selection of the auxiliary functions. A proper choice for auxiliary function can indeed provide convergent solutions for a few terms of the series. Some examples can be found in Marinca and Herisanu (2015).

6. Finally, we emphasize here that the homotopy-based methods provide a great freedom and flexibility in choosing the base function, linear and nonlinear operators, and auxiliary functions. In this chapter and subsequent chapters, we choose some specific cases of these functions and operators. One can indeed find more accurate solutions with other sets of these operators.

SUPPLEMENT TO CHAPTER 2

- Code for perturbation/Lyapunov's artificial parameter/ADM/HPM method-based solution:

```
clc
clear all;

syms t

N=49;

s=sym(zeros(1,N+1));

s(1)=t;

for k=1:N
    sum_1=0;
    for j=1:k
        A(k)=sum_1+s(j)*s(k-j+1);
        sum_1=A(k);
    end
    s(k+1)=-int(A(k),t,0,t);
end

sum=0;

for i=1:N+1
    approx_soln=sum+s(i);
    sum=approx_soln;
end

final_soln=sum;

final_subs=matlabFunction(final_soln);

t=0:0.001:10;

final_assess=final_subs(t);

plot(t,final_assess)

axis([0 10 0 1.2])

xlabel('$t$','FontSize',16,'interpreter','latex')
ylabel('$S(t)$','FontSize',14,'interpreter','latex')

legend({'$50^{th}$ order perturbation solution','Exact
solution'},'Interpreter','latex')
```

- Code for HAM-based solution using a polynomial base function:

```
function [] = ham_poly_com_exact(N,h)

x = sym('x', 'real');

y=sym(zeros(1,N+1));

t=zeros(1,N);

y(1)=x;

tic

for i=1:N
    term1=diff(y(i),x);
    sum1=0;
    for j=1:i
        sum1=sum1+y(j)*y(i-j+1);
    end
    term2=sum1;

if i==1
    t(i)=0;
else
    t(i)=1;
end

term=term1+term2-(1-t(i));

RHS=h*int(term,x,0,x);


y(i+1)=simplify(t(i)*y(i)+RHS);

i

end

   sum=0;
for k=1:N+1
    sum=sum+y(k);
end

final_soln=simplify(sum);

final_mat=matlabFunction(final_soln);

x=0:0.001:5;
```

```
HAM_soln=final_mat(x);

plot(x,HAM_soln,'-')

hold all

t=0:0.2:5;

plot(t,(exp(2*t)-1)./(exp(2*t)+1),'*')

toc

end
```

- Code for HAM-based solution using exponential base function:

```
function [] = ham_expo_com_exact(N,h)

x = sym('x', 'real');

y=sym(zeros(1,N+1));

t=zeros(1,N);

y(1)=1-exp(-x);

tic

for i=1:N

    term1=diff(y(i),x);

    sum1=0;

    for j=1:i
        sum1=sum1+y(j)*y(i-j+1);
    end
    term2=sum1;

if i==1
    t(i)=0;
else
    t(i)=1;
end

term=term1+term2-(1-t(i));

RHS=h*exp(-x)*int(term,x,0,x);
```

```
y(i+1)=simplify(t(i)*y(i)+RHS);

i

end

   sum=0;
for k=1:N+1
    sum=sum+y(k);
end

final_soln=simplify(sum);

final_mat=matlabFunction(final_soln);

x=0:0.001:5;

HAM_soln=final_mat(x);

plot(x,HAM_soln,'-')

toc

end
```

- Code for squared residual error calculation using a polynomial base function (the same can be done for the exponential):

```
function [] = ham_poly_res_err(N)

x = sym('x', 'real');

h = sym('h', 'real');

y=sym(zeros(1,N+1));

t=zeros(1,N);

y(1)=x;

tic

for i=1:N

    term1=diff(y(i),x);

    sum1=0;

    for j=1:i
        sum1=sum1+y(j)*y(i-j+1);
    end
    term2=sum1;
```

```
if i==1
    t(i)=0;
else
    t(i)=1;
end

term=term1+term2-(1-t(i));

RHS=h*int(term,x,0,x);


y(i+1)=simplify(t(i)*y(i)+RHS)

i

end

   sum=0;
for k=1:N+1
    sum=sum+y(k);
end


final_soln=simplify(sum);

res=diff(final_soln,x)+((final_soln)^2)-1;
sq_res=(res)^2;

%% Discrete error %%

sum_error=0;

for i=1:101

    sum_error=sum_error+subs(sq_res,x,(i-1)*((5-0)/(100)));

end

final_error=sum_error/(101);
error_h_now=matlabFunction(final_error);
error_min=fminbnd(error_h_now,-4,0)
h=-2:0.01:0;
em=error_h_now(h);
semilogy(h,em,'-')

error_value=subs(error_h_now,error_min)

toc
end
```

- Code for *h*-curve using a polynomial base function (the same can be done for the exponential):

```
function [] = ham_h_curve_poly(N)

x = sym('x', 'real');

h = sym('h', 'real');

y=sym(zeros(1,N+1));

t=zeros(1,N);

y(1)=x;

tic

for i=1:N

    term1=diff(y(i),x);

    sum1=0;

    for j=1:i
        sum1=sum1+y(j)*y(i-j+1);
    end

    term2=sum1;


if i==1
    t(i)=0;
else
    t(i)=1;
end

term=term1+term2-(1-t(i));

RHS=h*int(term,x,0,x);


y(i+1)=simplify(t(i)*y(i)+RHS)

i

end

   sum=0;
for k=1:N+1
    sum=sum+y(k);
end
toc
```

```
final_soln=simplify(sum);

Y1=diff(final_soln,x);

Y10 = subs(Y1, x, 1);
Y20=subs(final_soln,x,1);

Y11=matlabFunction(Y10);
Y12=matlabFunction(Y20);

h = -3:.001:2;
h1=Y11(h);
h2=Y12(h);

plot(h,h1,'-');
hold on
plot(h,h2,'--');

xlabel('$\hbar$','FontSize',16,'interpreter','latex')
ylabel('$S(1),~S^{\prime}
(1)$','FontSize',14,'interpreter','latex')

legend({'$S^{~\prime} (1)$','S(1)'},'Interpreter','latex')

axis([-1.43 0.05 -2 8])

end
```

REFERENCES

Abbasbandy, S., Shivanian, E., and Vajravelu, K. (2011). Mathematical properties of $\hbar$-curve in the frame work of the homotopy analysis method. *Communications in Nonlinear Science and Numerical Simulation, 16*(11), 4268–4275.

Adomian, G. (1994). *Solving frontier problems of physics: The decomposition method.* Boston, Kluwer Academic Publishers.

Armstrong, M. A. (1983). *Basic topology (Undergraduate texts in mathematics).* Springer, New York, NY.

Baker, G., and Graves-Morris, P. (1996). Padé approximants (2nd ed., *Encyclopedia of mathematics and its applications*). Cambridge University Press, Cambridge.

He, J. H. (1999). Homotopy perturbation technique. *Computer Methods in Applied Mechanics and Engineering, 178*(3–4), 257–262.

He, J. H. (2003). Homotopy perturbation method: A new nonlinear analytical technique. *Applied Mathematics and Computation, 135*(1), 73–79.

Liao, S. J. (1992). *The proposed homotopy analysis technique for the solution of nonlinear problems* (Doctoral dissertation, PhD. Thesis, Shanghai Jiao Tong University).

Liao, S. J. (2003). *Beyond perturbation: Introduction to the homotopy analysis method.* Boca Raton, Chapman & Hall/CRC Press.

Liao, S. (2010). An optimal homotopy-analysis approach for strongly nonlinear differential equations. *Communications in Nonlinear Science and Numerical Simulation, 15*(8), 2003–2016.

Liao, S. (2012). *Homotopy analysis method in nonlinear differential equations*. Higher Education Press, Beijing.

Lyapunov, A. M. (1992). The general problem of the stability of motion. *International Journal of Control*, *55*(3), 531–534.

Marinca, V., and Herişanu, N. (2008). Application of optimal homotopy asymptotic method for solving nonlinear equations arising in heat transfer. *International Communications in Heat and Mass Transfer*, *35*(6), 710–715.

Marinca, V., and Herisanu, N. (2015). *The optimal homotopy asymptotic method: Engineering applications*. Springer, Berlin/Heidelberg, Germany.

MATLAB. version 9.9 (R2020b). (2020). Natick, Massachusetts: The MathWorks Inc.

Nayfeh, A. H. (1981). *Introduction to perturbation techniques*. John Wiley & Sons, New York.

Olver, P. J. (1986). *Applications of lie groups to differential equations*. Springer, New York.

Singh, V. P. (1988). *Hydrologic systems: Vol. 1. Rainfall-runoff modeling*. Prentice Hall, Englewood Cliffs, NJ.

Wolfram Research, Inc., Mathematica, Version 12.0, Champaign, IL (2019).

FURTHER READING

Ali, L., Islam, S., Gul, T., Khan, I., and Dennis, L. C. C. (2016). New version of optimal homotopy asymptotic method for the solution of nonlinear boundary value problems in finite and infinite intervals. *Alexandria Engineering Journal*, *55*(3), 2811–2819.

Ayati, Z., and Biazar, J. (2015). On the convergence of homotopy perturbation method. *Journal of the Egyptian Mathematical Society*, *23*(2), 424–428.

Ghoreishi, M., Ismail, A. M., and Alomari, A. K. (2011). Comparison between homotopy analysis method and optimal homotopy asymptotic method for n-th order integro-differential equation. *Mathematical Methods in the Applied Sciences*, *34*(15), 1833–1842.

He, J. H. (2002). A note on delta-perturbation expansion method. *Applied Mathematics and Mechanics*, *23*(6), 634–638.

He, J. H. (2004). Comparison of homotopy perturbation method and homotopy analysis method. *Applied Mathematics and Computation*, *156*(2), 527–539.

He, J. H. (2008). Recent development of the homotopy perturbation method. *Topological Methods in Nonlinear Analysis*, *31*(2), 205–209.

He, J. H. (2010). A note on the homotopy perturbation method. *Thermal Science*, *14*(2), 565–568.

Hosseini, M. M., and Nasabzadeh, H. (2006). On the convergence of Adomian decomposition method. *Applied Mathematics and Computation*, *182*(1), 536–543.

Liao, S. (1997). Homotopy analysis method: A new analytical technique for nonlinear problems. *Communications in Nonlinear Science and Numerical Simulation*, *2*(2), 95–100.

Liao, S. (2004). On the homotopy analysis method for nonlinear problems. *Applied Mathematics and Computation*, *147*(2), 499–513.

Liao, S. (2005). Comparison between the homotopy analysis method and homotopy perturbation method. *Applied Mathematics and Computation*, *169*(2), 1186–1194.

Liao, S. (2009). Notes on the homotopy analysis method: Some definitions and theorems. *Communications in Nonlinear Science and Numerical Simulation*, *14*(4), 983–997.

Liao, S. (2010). On the relationship between the homotopy analysis method and Euler transform. *Communications in Nonlinear Science and Numerical Simulation*, *15*(6), 1421–1431.

Liao, S. (Ed.). (2013). *Advances in the homotopy analysis method*. World Scientific. https://www.worldscientific.com/worldscibooks/10.1142/8939#t=aboutBook

Liu, C. S. (2010). The essence of the homotopy analysis method. *Applied Mathematics and Computation*, *216*(4), 1299–1303.

Mabood, F. (2014). Comparison of optimal homotopy asymptotic method and homotopy per-
   turbation method for strongly non-linear equation. *Journal of the Association of Arab
   Universities for Basic and Applied Sciences, 16*, 21–26.
Nikishin, E. M. (1980). On simultaneous Padé approximants. *Matematicheskii Sbornik,
   155*(4), 499–519.
Odibat, Z. M. (2010). A study on the convergence of homotopy analysis method. *Applied
   Mathematics and Computation, 217*(2), 782–789.
Rach, R. (1987). On the Adomian (decomposition) method and comparisons with Picard's
   method. *Journal of Mathematical Analysis and Applications, 128*(2), 480–483.
Stahl, H. (1989). On the convergence of generalized Padé approximants. *Constructive
   Approximation, 5*(1), 221–240.
Turkyilmazoglu, M. (2011). Convergence of the homotopy perturbation method. *International
   Journal of Nonlinear Sciences and Numerical Simulation, 12*(1–8), 9–14.
Vajravelu, K., and Van Gorder, R. (2013). *Nonlinear flow phenomena and homotopy analysis.*
   Springer, Berlin.
Van Gorder, R. A., and Vajravelu, K. (2009). On the selection of auxiliary functions, opera-
   tors, and convergence control parameters in the application of the homotopy analysis
   method to nonlinear differential equations: A general approach. *Communications in
   Nonlinear Science and Numerical Simulation, 14*(12), 4078–4089.
Wazwaz, A. M. (1999). A reliable modification of Adomian decomposition method. *Applied
   Mathematics and Computation, 102*(1), 77–86.

Part II

Algebraic/Transcendental Equations

3 Numerical Solutions for the Colebrook Equation

3.1 INTRODUCTION

In open channels, such as rivers, canals, and ditches, flow occurs under the influence of gravity but is resisted by friction due to the bed surface and side surfaces. In pressurized pipes, flow occurs due to artificial pressure supplied to the pipe. In both channel flow and pipe flow, the flow energy comprises elevation (gravity) energy, potential energy (pressure), and kinetic energy (due to movement or velocity). During flow, the movement of water must overcome the resistance due to the surface by expending part of the flow energy. This expenditure of energy is termed energy loss. When energy is expressed in terms of head, this loss of energy becomes head loss. If the head loss is computed over a finite length of the flow conduit and when divided by the length, then this yields a slope, called a friction slope; sometimes it is also referred to as an energy slope. This slope is a fundamental parameter in flow hydraulics and has been expressed in different ways. Furthermore, this slope plays a fundamental role in the St. Venant or momentum equation, which comprises local acceleration, convective acceleration, acceleration due to the product of the pressure differential in space and gravitational acceleration, acceleration due to the product of the bed slope, acceleration due to gravity, acceleration due to the product of the frictional slope, and acceleration due to gravity. There are other acceleration terms, but they are too small and can be neglected. An order of magnitude analysis in channel flow shows that local and convective accelerations are of opposite signs and have the same order of magnitude, so they counterbalance each other. Of the remaining three terms, terms involving bed slope and frictional slope are much greater than the term involving the pressure differential and are, by far, the most important terms. Thus, the frictional slope plays a fundamental role in fluvial and pipe flow hydraulics.

Friction and friction slope have been quantified in a multitude of ways. In open channel hydraulics, friction is quantified through a factor called the friction factor or resistance coefficient. The most popular friction factors are Manning's and Chezy's friction factors. In pipe flow, the most popular are the Darcy-Weisbach friction factor, Hazen-Williams friction factor, and Colebrook friction factor, among others. The objective of this chapter is to solve the Colebrook equation for friction factor using numerical and homotopy-based Newton-like methods.

3.2 NEWTON-LIKE METHODS FOR NONLINEAR EQUATIONS USING HPM AND HAM

The Newton-Raphson (NR) method quickly produces an approximation to the root of an algebraic/transcendental equation. However, the method has drawbacks that limit its applicability. On the other hand, efficient numerical algorithms were

DOI: 10.1201/9781003368984-5

developed using the homotopy perturbation method (HPM) and homotopy analysis method (HAM) for solving nonlinear algebraic/transcendental equations numerically by Abbasbandy (2006) and Abbasbandy et al. (2007), respectively. We describe them in what follows.

3.2.1 Newton-Raphson Method

The NR method uses the idea that a continuous and differentiable function can be approximated by a straight line tangent to it. Suppose that we want to find the root of the following equation:

$$g(x) = 0 \tag{3.1}$$

near the point x_0. Figure 3.1 demonstrates the NR method geometrically.

In Figure 3.1, we consider the tangent line that passes through the point $P(x_n, g(x_n))$ and has slope $g'(x_n)$. The equation of this tangent is:

$$y - g(x_n) = g'(x_n)(x - x_n) \tag{3.2}$$

This line cuts the x-axis at a point where $y = 0$ and $x = x_{n+1}$. Therefore, we have:

$$0 - g(x_n) = g'(x_n)(x_{n+1} - x_n) \Rightarrow x_{n+1} = x_n - \frac{g(x_n)}{g'(x_n)} \tag{3.3}$$

Eq. (3.3) is the NR method, where a sequence of iterations $x_0, x_1, x_2, \ldots$ are produced that finally reach the root of the equation. It can be seen that if the initial guess is far from the root, then the method does not converge. Also, at any x_n where $g'(x_n)$ vanishes, the method does not work, as can be seen from Eq. (3.3).

3.2.2 HPM-Based Method

We suppose that α is a root of Eq. (3.1), i.e., $g(\alpha) = 0$. We further suppose that the function g is continuous, having continuous derivatives up to order 2, i.e., $g \in C^2$, on

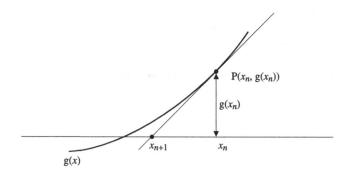

FIGURE 3.1 Pictorial demonstration of the Newton-Raphson method.

an interval containing the root α and $|g'(\alpha)| > 0$. Then, one can expand g in a Taylor series near x as follows:

$$g(x-\delta) = g(x) - \delta g'(x) + O(\delta^2) \tag{3.4}$$

We search for a small δ so that $g(x-\delta) \approx g(x) - \delta g'(x)$, and:

$$g(x-\delta) \approx 0 \Rightarrow g(x) - \delta g'(x) = 0 \Rightarrow \delta = \frac{g(x)}{g'(x)} \tag{3.5}$$

Therefore,

$$x - \delta = x - \frac{g(x)}{g'(x)} \tag{3.6}$$

Thus, the NR method can be developed as:

$$x_{n+1} = x_n - \frac{g(x_n)}{g'(x_n)} \tag{3.7}$$

Eq. (3.7) starts with x_0, which is the initial guess of the root. It can be shown that the iterative scheme in Eq. (3.7) converges if the initial guess of x_0 is sufficiently close to α (Atkinson and Han 2005). Also, it has local convergence properties. Going one step further, in a Taylor series expansion of g, we have:

$$g(x-\delta) = g(x) - \delta g'(x) + \frac{\delta^2}{2} g''(x) + O(\delta^3) \tag{3.8}$$

For small δ, we have:

$$g(x-\delta) \approx 0 \Rightarrow g(x) - \delta g'(x) + \frac{\delta^2}{2} g''(x) = 0 \tag{3.9}$$

Eq. (3.9) can be rearranged as follows:

$$\delta - \frac{\delta^2}{2} \frac{g''(x)}{g'(x)} = \frac{g(x)}{g'(x)} \tag{3.10}$$

or

$$A(\delta) = L(\delta) + N(\delta) = c \tag{3.11}$$

where $L(\delta) = \delta$ and $N(\delta) = \beta\delta^2$, in which $\beta = -\frac{1}{2} \frac{g''(x)}{g'(x)}$, are linear and nonlinear parts of the equation, respectively, and $c = \frac{g(x)}{g'(x)}$. When applying HPM, we construct a homotopy $\Phi(q)$ that satisfies:

$$(1-q)L[\Phi(q) - \delta_0] + q[A(\Phi(q)) - c] = 0 \tag{3.12}$$

where q is the embedding parameter that lies on $[0,1]$, $\phi(q)$ is the solution across q, $A(\Phi(q))$ is the operator of the equation, and δ_0 is the initial approximation to the final solution. Eq. (3.12) shows that:

$$\text{As } q = 0, \quad \mathcal{L}\big[\Phi(0) - \delta_0\big] = 0 \Rightarrow \Phi(0) - \delta_0 \Rightarrow \Phi(0) = \delta_0, \tag{3.13}$$

and

$$\text{As } q = 1, \quad A\big(\Phi(1)\big) - c = 0 \Rightarrow \Phi(1) = \delta \tag{3.14}$$

As q is small, i.e., $0 \leq q \leq 1$, one can express $\phi(q)$ as a series in terms of q, as follows:

$$\Phi(q) = \Phi_0 + q\Phi_1 + q^2\Phi_2 + q^3\Phi_3 + q^4\Phi_4 + \cdots \tag{3.15}$$

where Φ_m, $m \geq 1$ are the higher-order terms. For $q \to 1$, Eq. (3.16) produces the final solution as:

$$\delta = \lim_{q \to 1} \Phi(q) = \sum_{j=0}^{\infty} \Phi_i \tag{3.16}$$

Using Eq. (3.15), we have:

$$\begin{aligned}
(1-q)\mathcal{L}\big[\Phi(q) - \delta_0\big] &= (1-q)\big[\Phi(q) - \delta_0\big] \\
&= (1-q)\big[(\Phi_0 - \delta_0) + q\Phi_1 + q^2\Phi_2 + q^3\Phi_3 + \cdots\big] \\
&= (\Phi_0 - \delta_0) + q\big(\Phi_1 - (\Phi_0 - \delta_0)\big) + q^2(\Phi_2 - \Phi_1) \\
&\quad + q^3(\Phi_3 - \Phi_2) + \cdots
\end{aligned} \tag{3.17}$$

$$\begin{aligned}
A\big(\Phi(q)\big) - c &= \Phi + \beta\Phi^2 - c = \big[\Phi_0 + q\Phi_1 + q^2\Phi_2 + q^3\Phi_3 + \cdots\big] \\
&\quad + \beta\big[(\Phi_0)^2 + q(2\Phi_0\Phi_1) + q^2(\Phi_1^2 + 2\Phi_0\Phi_2) + q^3(2\Phi_0\Phi_3 + 2\Phi_1\Phi_2) + \cdots\big] \\
&- c = (\Phi_0 + \beta\Phi_0^2 - c) + q(\Phi_1 + 2\beta\Phi_0\Phi_1) + q^2\big[\Phi_2 + \beta(\Phi_1^2 + 2\Phi_0\Phi_2)\big] \\
&\quad + q^3\big[\Phi_3 + \beta(2\Phi_0\Phi_3 + 2\Phi_1\Phi_2)\big] + \cdots
\end{aligned} \tag{3.18}$$

Using Eqs. (3.17) and (3.18) and equating the like powers of q in Eq. (3.12), we have:

$$q^0: (\Phi_0 - \delta_0) = 0 \tag{3.19}$$

$$q^1: \Phi_1 - (\Phi_0 - \delta_0) + \Phi_0 + \beta\Phi_0^2 - c = 0 \tag{3.20}$$

$$q^2: \Phi_2 - \Phi_1 + \Phi_1 + 2\beta\Phi_0\Phi_1 = 0 \tag{3.21}$$

We always set:

$$\Phi_0 = \delta_0 = \frac{g(x)}{g'(x)} \ (=c) \qquad (3.22)$$

Using Eq. (3.22), we have from Eq. (3.20):

$$\Phi_1 = -\beta \Phi_0^2 \qquad (3.23)$$

Using Eq. (3.23), Eq. (3.21) gives:

$$\Phi_2 = -2\beta \Phi_0 \Phi_1 = 2\beta^2 \Phi_0^3 \qquad (3.24)$$

Proceeding in a like manner, one can obtain the $(m+1)$-th term approximation as follows:

$$\delta \approx \sum_{i=0}^{m} \Phi_i \qquad (3.25)$$

For $m = 0$,

$$\delta \approx \Phi_0 = c = \frac{g(x)}{g'(x)} \qquad (3.26)$$

Therefore,

$$\alpha = x - \delta \approx x - \frac{g(x)}{g'(x)} \qquad (3.27)$$

which produces:

$$x_{n+1} = x_n - \frac{g(x_n)}{g'(x_n)} \qquad (3.28)$$

which is the NR method.
    For $m = 1$,

$$\delta \approx \Phi_0 + \Phi_1 \qquad (3.29)$$

Using Eq. (3.22), we have from Eq. (3.23):

$$\Phi_1 = -\beta \Phi_0^2 = \frac{g^2(x) g''(x)}{2g'^3(x)} \qquad (3.30)$$

Therefore,

$$\alpha = x - \delta \approx x - \frac{g(x)}{g'(x)} - \frac{g^2(x) g''(x)}{2g'^3(x)} \qquad (3.31)$$

which produces:

$$x_{n+1} = x_n - \frac{g(x_n)}{g'(x_n)} - \frac{g^2(x_n)g''(x_n)}{2g'^3(x_n)} \tag{3.32}$$

which is known as the Householder's iteration scheme (Householder 1970).
    For $m = 2$,

$$\delta \approx \phi_0 + \phi_1 + \phi_2 \tag{3.33}$$

Using Eq. (3.22), we have from Eq. (3.24):

$$\Phi_2 = 2\beta^2\Phi_0{}^3 = \frac{g^3(x)g''^2(x)}{2g'^5(x)} \tag{3.34}$$

Therefore,

$$\alpha = x - \delta \approx x - \frac{g(x)}{g'(x)} - \frac{g^2(x)g''(x)}{2g'^3(x)} - \frac{g^3(x)g''^2(x)}{2g'^5(x)} \tag{3.35}$$

which produces:

$$x_{n+1} = x_n - \frac{g(x_n)}{g'(x_n)} - \frac{g^2(x_n)g''(x_n)}{2g'^3(x_n)} - \frac{g^3(x_n)g''^2(x_n)}{2g'^5(x_n)} \tag{3.36}$$

which is the HPM-based iteration scheme. A flowchart diagram containing the steps
of the method is given in Figure 3.2.

3.2.3 HAM-BASED METHOD

When applying HAM, we write Eq. (3.10) as follows:

$$\mathcal{N}[\Phi(q)] = 0 \tag{3.37}$$

The zeroth-order deformation equation can be constructed as follows:

$$(1-q)\mathcal{L}[\Phi(q) - \delta_0] - q\hbar H\mathcal{N}[\Phi(q)] = 0 \tag{3.38}$$

where q is the embedding parameter, $\Phi(q)$ is the representation of the solution
across q, δ_0 is the initial approximation, $\hbar$ is the auxiliary parameter, H is the auxiliary function, and $\mathcal{L}$ and $\mathcal{N}$ are the linear and nonlinear operators, respectively.
It can be seen from Eq. (3.38) that $\mathcal{L}[\Phi(q) - \delta_0]$ and $\hbar H\mathcal{N}[\Phi(q)]$ are homotopic.
More specifically, as $q = 0$, $\Phi(0) = \delta_0$, and $q = 1$, $\Phi(1) = \delta_0$, i.e., as q goes from 0 to
1, the unknown $\Phi(q)$ varies from the initial approximation δ_0 to the final solution δ.

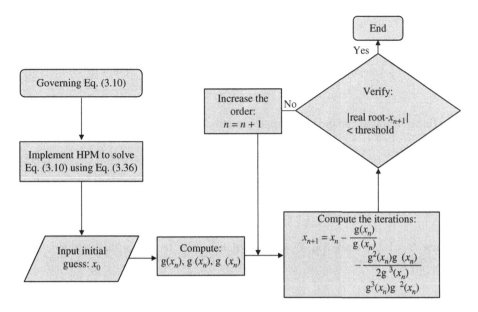

FIGURE 3.2   Flowchart for the HPM-based iteration scheme in Eq. (3.36).

The higher-order terms can be calculated from the higher-order deformation equations, given as follows:

$$\mathcal{L}\left[\delta_m - \chi_m \delta_{m-1}\right] = \hbar H R_m\left(\vec{\delta}_{m-1}\right), \ m = 1, 2, 3, \dots \tag{3.39}$$

where

$$\chi_m = \begin{cases} 0 & \text{when } m = 1, \\ 1 & \text{otherwise} \end{cases} \tag{3.40}$$

and

$$R_m\left(\vec{\delta}_{m-1}\right) = \frac{1}{(m-1)!} \left.\frac{\partial^{m-1} \mathcal{N}\left[\Phi(q)\right]}{\partial q^{m-1}}\right|_{q=0} \tag{3.41}$$

where δ_m, $m \geq 1$ are the higher-order terms. Now, the final solution can be obtained as follows:

$$\delta = \delta_0 + \sum_{m=1}^{\infty} \delta_m \tag{3.42}$$

The nonlinear function for the equation is selected as:

$$\mathcal{N}\left[\Phi(q)\right] = \Phi + \beta \Phi^2 - c \tag{3.43}$$

Therefore, using Eq. (3.43), the term R_m can be calculated from Eq. (3.41) as follows:

$$R_m\left(\vec{\delta}_{m-1}\right) = \delta_{m-1} + \beta \sum_{j=0}^{m-1} \delta_j \delta_{m-1-j} - \left(1-\chi_m\right)c \qquad (3.44)$$

The linear function and the initial approximation are chosen, respectively, as follows:

$$\mathcal{L}\left[\Phi(q)\right] = \Phi(q) \qquad (3.45)$$

$$\delta_0 = c = \frac{g(x)}{g'(x)} \qquad (3.46)$$

Setting $H = 1$ and using Eq. (3.45), the higher-order terms can be obtained from Eq. (3.39) as follows:

$$\delta_m = \left(\chi_m + \hbar\right)\delta_{m-1} + \hbar\beta \sum_{j=0}^{m-1} \delta_j \delta_{m-1-j} - \hbar\left(1-\chi_m\right)c, \ m = 1,2,3,\ldots \qquad (3.47)$$

Using Eq. (3.46), for $m = 1,2$, we have from Eq. (3.47):

$$\delta_1 = \hbar\delta_0 + \hbar\beta\delta_0^2 - \hbar c = \hbar\beta\delta_0^2 = -\frac{\hbar}{2}\frac{g^2(x)g''(x)}{g'^3(x)} \qquad (3.48)$$

$$\begin{aligned}\delta_2 &= \left(1+\hbar\right)\delta_1 + \hbar\beta\left(2\delta_0\delta_1\right) \\ &= -\frac{\hbar(1+\hbar)}{2}\frac{g^2(x)g''(x)}{g'^3(x)} + \frac{\hbar^2}{2}\frac{g^3(x)g''^2(x)}{g'^5(x)}\end{aligned} \qquad (3.49)$$

Finally, the $(M+1)$-th term approximation can be obtained as follows:

$$\delta \approx \sum_{m=0}^{M} \delta_m \qquad (3.50)$$

For $M = 0$,

$$\delta \approx \delta_0 = c = \frac{g(x)}{g'(x)} \qquad (3.51)$$

Therefore,

$$\alpha = x - \delta \approx x - \frac{g(x)}{g'(x)} \qquad (3.52)$$

which produces:

$$x_{n+1} = x_n - \frac{g(x_n)}{g'(x_n)} \tag{3.53}$$

which is the NR method and is the same as in Eq. (3.28).

For $M = 1$,

$$\delta \approx \delta_0 + \delta_1 \tag{3.54}$$

Therefore, using Eq. (3.48), we have:

$$\alpha = x - \delta \approx x - \frac{g(x)}{g'(x)} + \frac{\hbar}{2} \frac{g^2(x)g''(x)}{g'^3(x)} \tag{3.55}$$

which produces:

$$x_{n+1} = x_n - \frac{g(x_n)}{g'(x_n)} + \frac{\hbar}{2} \frac{g^2(x_n)g''(x_n)}{2g'^3(x_n)} \tag{3.56}$$

For $\hbar = -1$, Eq. (3.56) is known as the Householder's iteration scheme and is the same as Eq. (3.32).

For $M = 2$,

$$\delta \approx \delta_0 + \delta_1 + \delta_2 \tag{3.57}$$

Therefore, using Eq. (3.49), we have:

$$\alpha = x - \delta \approx x - \frac{g(x)}{g'(x)} + \frac{\hbar}{2} \frac{g^2(x)g''(x)}{g'^3(x)} + \frac{\hbar(1+\hbar)}{2} \frac{g^2(x)g''(x)}{g'^3(x)}$$
$$- \frac{\hbar^2}{2} \frac{g^3(x)g''^2(x)}{g'^5(x)} \tag{3.58}$$

which produces:

$$x_{n+1} = x_n - \frac{g(x_n)}{g'(x_n)} + \frac{\hbar(2+\hbar)}{2} \frac{g^2(x_n)g''(x_n)}{g'^3(x_n)} - \frac{\hbar^2}{2} \frac{g^3(x_n)g''^2(x_n)}{g'^5(x_n)} \tag{3.59}$$

which is the same as Eq. (3.36) when $\hbar = -1$.

The auxiliary parameter $\hbar$ in HAM plays a key role in controlling the convergence region and rate of the series (Liao 2012). In the NR and other methods, such as HPM, if the initial guess x_0 is not close to the root α, then the iteration does not converge; however, HAM can converge in such cases with the help of $\hbar$ (Abbasbandy et al. 2007). A suitable choice for $\hbar$ can be made by the so-called

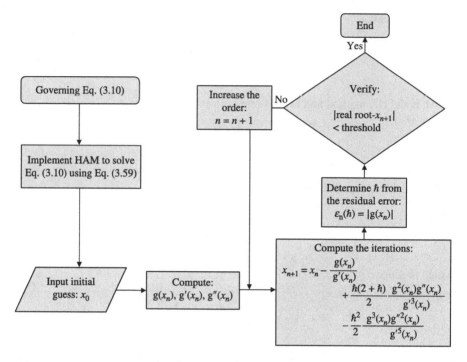

FIGURE 3.3   Flowchart for the HAM-based iteration scheme in Eq. (3.59).

$\hbar$-curve technique or by calculating the residual error (Liao 2012). The residual error can be calculated as:

$$\varepsilon_n(\hbar) = |g(x_n)| \tag{3.60}$$

where n is a particular order of approximation. Eq. (3.60) can be minimized to find a suitable $\hbar$ for which x_n converges to the root α. A flowchart diagram containing the steps of the method is provided in Figure 3.3. Some examples concerning the performances of the methods are shown next.

3.3   CONVERGENCE THEOREM OF THE HAM-BASED SOLUTION

The convergence theorem for the HAM-based solution given by Eq. (3.42) can be proved using the following theorems.

Theorem 3.1: *If the homotopy series $\Sigma_{m=0}^{\infty} \delta_m$ converges, then $R_m\left(\vec{\delta}_{m-1}\right)$ given by Eq. (3.44) satisfies the relation $\Sigma_{m=1}^{\infty} R_m\left(\vec{\delta}_{m-1}\right) = 0$.*

Proof: The linear operator for Eq. (3.10) was defined as follows:

$$\mathcal{L}[\delta] = \delta \tag{3.61}$$

According to Eq. (3.39), we obtain:

$$\mathcal{L}[\delta_1] = \hbar R_1\left(\vec{\delta}_0\right) \tag{3.62}$$

$$\mathcal{L}[\delta_2 - \delta_1] = \hbar R_2\left(\vec{\delta}_1\right) \tag{3.63}$$

$$\mathcal{L}[\delta_3 - \delta_2] = \hbar R_3\left(\vec{\delta}_2\right) \tag{3.64}$$

$$\mathcal{L}[\delta_m - \delta_{m-1}] = \hbar R_m\left(\vec{\delta}_{m-1}\right) \tag{3.65}$$

Adding all the terms, we get:

$$\mathcal{L}[\delta_m] = \hbar \sum_{k=1}^{m} R_k\left(\vec{\delta}_{k-1}\right) \tag{3.66}$$

As the series $\sum_{m=0}^{\infty} \delta_m$ is convergent, $\lim_{m\to\infty} \delta_m = 0$. Now, recalling the earlier summand and taking the limit, the required result follows as:

$$\hbar \sum_{k=1}^{\infty} R_k\left(\vec{\delta}_{k-1}\right) = \hbar \lim_{m\to\infty} \sum_{k=1}^{m} R_k\left(\vec{\delta}_{k-1}\right) = \lim_{m\to\infty} \mathcal{L}[\delta_m] = \lim_{m\to\infty} \delta_m = 0 \tag{3.67}$$

Theorem 3.2: *If $\hbar$ is so properly chosen that the series $\sum_{m=0}^{\infty} \delta_m$ converges absolutely to δ, then the homotopy series $\sum_{m=0}^{\infty} \delta_m$ satisfies the original governing Eq. (3.10).*

Proof: Using the *Cauchy product of two infinite series* defined in Chapter 2 in relation to Eq. (3.44), we get:

$$\sum_{m=1}^{\infty} \sum_{j=0}^{m-1} \delta_j \delta_{m-1-j} = \left(\sum_{m=0}^{\infty} \delta_m\right)^2 \tag{3.68}$$

Theorem 3.1 shows that if $\sum_{m=0}^{\infty} \delta_m$ converges, then $\sum_{m=1}^{\infty} R_m\left(\vec{\delta}_{m-1}\right) = 0$.

Therefore, substituting the earlier expressions in Eq. (3.44) and simplifying further lead to:

$$\sum_{m=0}^{\infty} \delta_m + \beta \left(\sum_{m=0}^{\infty} \delta_m\right)^2 - c = 0 \tag{3.69}$$

which is basically the original governing Eq. (3.10). Hence, the convergence result follows.

3.4 EXAMPLES

Here, we compare three methods, namely the NR, HPM (Eq. 3.36), and HAM (Eq. 3.59) methods, by applying these to four different examples.

Example 3.1

Let us consider the equation $x^3 - 2x^2 + 3x - 1 = 0$. The root is given by $\alpha = 0.4301597$ (up to seven decimal places). The initial guesses are chosen as $x_0 = 10$ and 100.

Example 3.2

Let us consider the equation $\exp(x) - 3x^2 + 5x - 8 = 0$. The root of the equation is given by $\alpha = 2.987780$ (up to six decimal places). The initial guesses are chosen as $x_0 = 10$ and 50.

Example 3.3

Let us consider the equation $x\exp(x^2) - \sin^2 x + 3\cos x + 5 = 0$ (Abbasbandy et al. 2007). The root of the equation is given by $\alpha = -1.20765$ (up to five decimal places). The initial guesses are chosen as $x_0 = -5$ and 10.

Example 3.4

Let us consider the equation $x^{1/5} - 1 = 0$. The root of the equation is given by $\alpha = 1$. The initial guesses are chosen as $x_0 = 2$ and 10.

The performances of all the methods are provided in Table 3.1 in terms of the iteration number taken to reach the desired root. For all the examples, it can be seen that the HPM- and HAM-based methods perform better than the usual NR

TABLE 3.1
Comparison of Iteration Numbers Taken by the Methods for the Selected Examples

Example	Initial Guess (x_0)	NR	HPM	HAM		
3.1	10	10	7	$6\,(\hbar = -2.2)$	$5\,(\hbar = -2.25)$	$7\,(\hbar = -2.3)$
	100	16	10	$8\,(\hbar = -2.2)$	$9\,(\hbar = -2.25)$	$9\,(\hbar = -2.3)$
3.2	20	23	13	$10\,(\hbar = -2)$	$8\,(\hbar = -2.2)$	$8\,(\hbar = -2.4)$
	50	53	28	$20\,(\hbar = -2)$	$18\,(\hbar = -2.2)$	$14\,(\hbar = -2.4)$
3.3	-5	30	16	$13\,(\hbar = -2)$	$10\,(\hbar = -2.2)$	$9\,(\hbar = -2.4)$
	10	Divergent	Divergent	$110\,(\hbar = -=0\ r)$	$103\,(\hbar = -1.10)$	$147\,(\hbar = -=7\ 0)$
3.4	2	7	5	$7\,(\hbar = -0.1)$	$6\,(\hbar = 0.1)$	$4\,(\hbar = 0.4)$
	10	Divergent	Divergent	$18\,(\hbar = 0.1)$	$12\,(\hbar = 0.2)$	$9\,(\hbar = 0.2)$

method. However, in some cases, it is observed that the initial guess significantly affects the performance and applicability of the methods, namely NR and HPM. On the other hand, a proper choice of $\hbar$ in HAM leads to convergent results even if one chooses a bad initial guess. For convenience, the MATLAB codes for each of the methods are provided at the end of this chapter.

3.5   APPLICATION TO THE COLEBROOK EQUATION

The Colebrook equation is used to estimate the Darcy-Weisbach friction factor (f) in fully developed pipe flows (Colebrook 1939). It expresses the friction factor as a function of the Reynolds number (Re) and relative roughness of pipe (ε/D). The equation is expressed as:

$$\frac{1}{\sqrt{f}} = -2\ln\left(\frac{2.51}{Re\sqrt{f}} + \frac{\varepsilon/D}{3.71}\right) \tag{3.70}$$

where D is the diameter of the pipe and ε is the average height of protrusion of the inner pipe surface. The Colebrook equation is assumed to describe the friction factor over the regime, $4000 \le Re \le 10^8$ and $0 \le \varepsilon/D \le 0.1$. Several attempts have been made to find an approximation to the Colebrook equation using different methodologies. For more details on the approximate analytical solutions to the Colebrook equation, one can refer to Zigrang and Sylvester (1982), Brkic (2011), Brkic and Cojbasic (2017), and Vatankhah (2018). Here, we derive the solution in terms of a series using the Lambert W function (Brkic 2012).
   We introduce:

$$x = \frac{1}{\sqrt{f}}, \ b = \frac{\varepsilon}{3.71D}, \ a = \frac{2.51}{Re} \tag{3.71}$$

Using this, Eq. (3.70) transforms to:

$$x = -2\ln(ax + b) \tag{3.72}$$

or

$$\exp\left(-\frac{x}{2}\right) = ax + b \tag{3.73}$$

Taking $p = \exp\left(-\frac{1}{2}\right)$, we get:

$$p^x = ax + b \tag{3.74}$$

Then, the solution can be given as (Brkic 2012):

$$x = -\frac{W\left(-\frac{\ln p}{a} p^{-\frac{b}{a}}\right)}{\ln p} - \frac{b}{a} \tag{3.75}$$

From Eq. (3.75), we have:

$$f = \cfrac{1}{\left(\cfrac{2W\left(\cfrac{\ln 10}{2a} 10^{\frac{b}{2a}} \right)}{\ln 10} - \cfrac{b}{a} \right)^2} \tag{3.76}$$

where W is the Lambert W function, which is defined as the inverse function of $f(W) = W exp(W)$ (Corless et al. 1996).

Here, we consider some test pairs $(Re, \varepsilon/D)$ to check the performances of the NR, HPM, and HAM methods in producing the friction factor, f, values. For comparison purposes, the actual value of f is estimated using Eq. (3.76). The HPM and HAM iterations are obtained using Eqs. (3.36) and (3.59), respectively. To check the effectiveness of the methods, two initial guesses are chosen, 0.1 and 0.001, for all the cases. The results are reported in Table 3.2. It can be observed from the table that both HPM and HAM can effectively produce convergent results. However, the HAM-based iteration scheme performs better through the auxiliary parameter $\hbar$. The auxiliary parameter is determined using the minimization of the residual error given by Eq. (3.60). For all the methods, the iteration can be stopped using a threshold value of $|x_{n+1} - x_n| < threshold$.

TABLE 3.2

Comparison of Iteration Numbers Taken by the Methods for the Selected Cases of the Colebrook Equation

$(Re, \varepsilon/D)$	Friction Factor, f	Initial Guess, x_0	Iteration Number		
			NR	HPM	HAM
$(4000, 0)$	0.00977	0.1	6	5	3 ($\hbar = -0.6$)
		0.001	7	5	3 ($\hbar = -2.6$)
$(10^4, 10^{-5})$	0.00735	0.1	6	4	3 ($\hbar = -0.55$)
		0.001	7	4	3 ($\hbar = -1.03$)
$(10^5, 10^{-4})$	0.00413	0.1	5	4	3 ($\hbar = -0.56$)
		0.001	5	4	3 ($\hbar = -0.54$)
$(10^6, 10^{-3})$	0.00383	0.1	3	3	3 ($\hbar = -0.5$)
		0.001	3	3	3 ($\hbar = -0.5$)
$(10^7, 10^{-2})$	0.00715	0.1	3	3	3 ($\hbar = -2$)
		0.001	3	3	3 ($\hbar = -2$)
$(10^8, 10^{-1})$	0.01917	0.1	3	3	3 ($\hbar = -1.5$)
		0.001	3	3	3 ($\hbar = -1.5$)

3.6   CONCLUDING REMARKS

When finding the roots of an algebraic/transcendental equation, one often adopts the classical Newton method or NR method. The NR method does not perform well in several situations, such as the choice of a bad initial guess, zeros of the function at the initial guess, etc. The HPM and HAM can provide Newton-like methods for solving nonlinear equations, which can overcome these drawbacks associated with the NR method. This chapter reviewed the iterative methods based on HPM and HAM by applying them to several examples. Also, the Colebrook equation is considered to check the performances of these methods over the NR method. It is observed that the homotopy-based methods can perform efficiently for all cases.

SUPPLEMENT TO CHAPTER 3

- Code for the Newton-Raphson method

```
% Newton Raphson method%
clc;
clear all;

syms y;
f=(y^(1/5))-1; % Example 3.4 %
f_1=diff(f,y); % Differentiation of f %
f_2=diff(f,y,2); % Two times differentiation of f %

N=6;
x=zeros(1,N+1);
x(1)=10;              % initial guess %

real_root=1; % Example 3.4 %

for i=1:N
c_1=subs(f,y,x(i));
c_2=subs(f_1,y,x(i));
c_3=subs(f_2,y,x(i));
x(i+1)=x(i)-(c_1/c_2); % NR method %

plot(i,abs(real_root-x(i)),'o')
hold all
i
end

final_soln=x(end);
```

- Code for the HPM-based iteration method

```
% HPM-based iteration scheme %
clc;
clear all;
```

```
syms y;
f=(y^(1/5))-1; % Example 3.4 %

f_1=diff(f,y); % Differentiation of f %
f_2=diff(f,y,2); % Two times differentiation of f %
N=3;
x=zeros(1,N+1);
x(1)=10;           % initial guess %

real_root=1; % Example 3.4 %

for i=1:N
c_1=subs(f,y,x(i));
c_2=subs(f_1,y,x(i));
c_3=subs(f_2,y,x(i));
x(i+1)=x(i)-(c_1/c_2)-((0.5*(c_3*c_1^2)/(c_2^3)))-
(0.5*((c_1^3*c_3^2)/(c_2^5))); % HPM-based iteration %

plot(i,abs(real_root-x(i)),'o')
hold all
i
end

final_soln=x(end);
```

- Code for the determination of $\hbar$ in HAM

```
% Determination of h %

clc;
clear all;
syms y h;
f=y^3-(2*y^2)+(3*y)-1; % Example 3.1 %

f_1=diff(f,y); % differentiation of f %
f_2=diff(f,y,2);  % two times differentiation of f %
N=4;
x=sym(zeros(1,N+1));
x(1)=10;              % initial guess %


for i=1:N
c_1=subs(f,y,x(:,i));
c_2=subs(f_1,y,x(:,i));
c_3=subs(f_2,y,x(:,i));
x(i+1)=x(i)-(c_1/c_2)-
((h*(2+h))*((0.5*(c_3*c_1^2)/(c_2^3))))-
((h^2)*(0.5*((c_1^3*c_3^2)/(c_2^5)))); % HAM-based
iteration %


i
end
```

```
soln_1=x(end);

final_soln=soln_1^3-(2*soln_1^2)+(3*soln_1)-1;

Y11=matlabFunction(abs(final_soln));

show_ans=fminbnd(Y11,-5,2) % minimization of residual
error over [-5,2] %
```

- Code for the HAM-based iteration method

```
% HAM-based iteration scheme %

syms y;
f=y^3-(2*y^2)+(3*y)-1; % Example 3.1 %
f_1=diff(f,y);
f_2=diff(f,y,2);
h=-1.740;
N=4;
x=zeros(1,N+1);
x(1)=10;

real_root=0.4301597; % Example 3.1 %


for i=1:N
c_1=subs(f,y,x(i));
c_2=subs(f_1,y,x(i));
c_3=subs(f_2,y,x(i));
x(i+1)=x(i)-(c_1/c_2)-
((h*(2+h))*((0.5*(c_3*c_1^2)/(c_2^3)))))-
((h^2)*(0.5*((c_1^3*c_3^2)/(c_2^5)))));

plot(i,abs(real_root-x(i)),'o')
hold all
i
end

final_soln=x(end)
```

REFERENCES

Abbasbandy, S. (2006). Modified homotopy perturbation method for nonlinear equations and comparison with Adomian decomposition method. *Applied Mathematics and Computation*, *172*(1), 431–438.

Abbasbandy, S., Tan, Y., and Liao, S. J. (2007). Newton-homotopy analysis method for nonlinear equations. *Applied Mathematics and Computation*, *188*(2), 1794–1800.

Atkinson, K., and Han, W. (2005). *Theoretical numerical analysis* (Vol. 39, pp. xviii-576). Springer, Berlin.

Brkić, D. (2011). Review of explicit approximations to the Colebrook relation for flow friction. *Journal of Petroleum Science and Engineering*, *77*(1), 34–48.

Brkić, D. (2012). Lambert W function in hydraulic problems. *Mathematica Balkanica*, *26*(3–4), 285–292.

Brkić, D., and Ćojbašić, Ž. (2017). Evolutionary optimization of Colebrook's turbulent flow friction approximations. *Fluids*, *2*(2), 15.

Colebrook, C. F. (1939). Turbulent flow in pipes with particular reference to the transition region between the smooth and rough pipe laws. *Journal of the Institution of Civil Engineers*, *11*(4), 133–156.

Corless, R. M., Gonnet, G. H., Hare, D. E., Jeffrey, D. J., and Knuth, D. E. (1996). On the Lambert W function. *Advances in Computational Mathematics*, *5*(1), 329–359.

Householder, A. S. (1970). *The numerical treatment of a single nonlinear equation.* McGraw-Hill, New York.

Liao, S. (2012). *Homotopy analysis method in nonlinear differential equations.* Higher Education Press, Beijing.

Vatankhah, A. R. (2018). Approximate analytical solutions for the Colebrook equation. *Journal of Hydraulic Engineering*, *144*(5), 06018007.

Zigrang, D. J., and Sylvester, N. D. (1982). Explicit approximations to the solution of Colebrook's friction factor equation. *AIChE Journal*, *28*(3), 514–515.

FURTHER READING

Awawdeh, F. (2010). On new iterative method for solving systems of nonlinear equations. *Numerical Algorithms*, *54*(3), 395–409.

Biberg, D. (2017). Fast and accurate approximations for the Colebrook equation. *Journal of Fluids Engineering*, *139*(3), 031401.

Brkić, D. (2011). W solutions of the CW equation for flow friction. *Applied Mathematics Letters*, *24*(8), 1379–1383.

Chun, C. (2006a). Construction of Newton-like iteration methods for solving nonlinear equations. *Numerische Mathematik*, *104*(3), 297–315.

Chun, C. (2006b). A new iterative method for solving nonlinear equations. *Applied Mathematics and Computation*, *178*(2), 415–422.

Chun, C. (2007). A method for obtaining iterative formulas of order three. *Applied Mathematics Letters*, *20*(11), 1103–1109.

Feng, X., and He, Y. (2007). High order iterative methods without derivatives for solving nonlinear equations. *Applied Mathematics and Computation*, *186*(2), 1617–1623.

Giustolisi, O., Berardi, L., and Walski, T. M. (2011). Some explicit formulations of Colebrook–White friction factor considering accuracy vs. computational speed. *Journal of Hydroinformatics*, *13*(3), 401–418.

Mikata, Y., and Walczak, W. S. (2016). Exact analytical solutions of the Colebrook-White equation. *Journal of Hydraulic Engineering*, *142*(2), 04015050.

Noor, M. A. (2010). Some iterative methods for solving nonlinear equations using homotopy perturbation method. *International Journal of Computer Mathematics*, *87*(1), 141–149.

Noor, M. A., and Khan, W. A. (2012). New iterative methods for solving nonlinear equation by using homotopy perturbation method. *Applied Mathematics and Computation*, *219*(8), 3565–3574.

Sonnad, J. R., and Goudar, C. T. (2004). Constraints for using Lambert W function-based explicit Colebrook–White equation. *Journal of Hydraulic Engineering*, *130*(9), 929–931.

Sonnad, J. R., and Goudar, C. T. (2006). Turbulent flow friction factor calculation using a mathematically exact alternative to the Colebrook–White equation. *Journal of Hydraulic Engineering*, *132*(8), 863–867.

Swamee, P. K., and Jain, A. K. (1976). Explicit equations for pipe-flow problems. *Journal of the Hydraulics Division*, *102*(5), 657–664.

Wu, Y., and Cheung, K. F. (2010). Two-parameter homotopy method for nonlinear equations. *Numerical Algorithms*, *53*(4), 555–572.

Part III

---

*Ordinary Differential Equations
(Single and System)*

4 Velocity Distribution in Smooth Uniform Open-Channel Flow

4.1 INTRODUCTION

In open-channel flow, velocity is a fundamental variable for characterizing flow, energy, momentum, sediment transport, pollutant transport, hydraulic geometry, and stability of hydraulic structures, as well as for a variety of design works. It varies in three dimensions (vertical, horizontal, and transverse) and is affected by channel morphology, characteristics of sediment in the channel, discharge coming from the watershed, vegetation, and channel works. At a point in a cross-section, the velocity varies along the vertical from the bed to the water surface. It increases from the bed to a certain point past the point corresponding to the average velocity and then decreases to the water surface due to air-water interaction or the resistance at the water surface. The equation describing the velocity of open-channel flow can be derived using Reynolds-averaged Navier-Stokes (RANS) equation. However, for practical applications, the RANS equation is simplified, or empirical equations are derived. This chapter solves a simplified version of the Navier-Stokes equation using homotopy-based methods.

4.2 VELOCITY MODEL

We consider a steady uniform two-dimensional turbulent flow in a polygonal channel, as shown in Figure 4.1. Using the continuity equation, the RANS equation in the streamwise direction x takes on the form (Yang et al. 2004; Absi 2011):

$$\frac{\partial(uv)}{\partial y} + \frac{\partial(uw)}{\partial z} = \nu\left(\frac{\partial^2 u}{\partial y^2} + \frac{\partial^2 u}{\partial z^2}\right) + \frac{\partial\left(-\overline{u'v'}\right)}{\partial y} + \frac{\partial\left(-\overline{u'w'}\right)}{\partial z} + g\sin\theta \qquad (4.1)$$

where y and z are the vertical and transverse directions, respectively; u, v, and w are the mean velocities along the x-, y-, and z-directions, respectively; u', v', and w' are the turbulent fluctuations along the x-, y-, and z-directions, respectively; ν is the kinematic viscosity of fluid; g is the acceleration due to gravity; and θ is the angle of the channel bed with the streamwise direction (see Figure 4.1(b)). Eq. (4.1) can also be rearranged as follows:

$$\frac{\partial}{\partial y}\left(uv - \frac{\tau_{xy}}{\rho}\right) + \frac{\partial}{\partial z}\left(uw - \frac{\tau_{xz}}{\rho}\right) = gS \qquad (4.2)$$

DOI: 10.1201/9781003368984-7

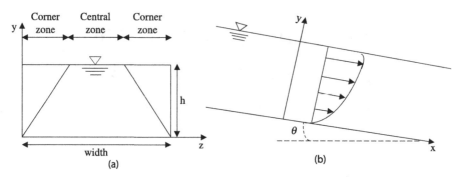

FIGURE 4.1 Definition sketch for a steady uniform open-channel flow: (a) cross-sectional view, and (b) flow configuration.

where $S = \sin\theta$ is the channel bed slope, $\tau_{xy} = \mu\dfrac{\partial u}{\partial y} - \rho\overline{u'v'}$, $\tau_{xz} = \mu\dfrac{\partial u}{\partial z} - \rho\overline{u'w'}$, and μ is the dynamic viscosity of fluid.

In an experimental study, Tracy (1963) reported that the terms containing the vertical gradient $\dfrac{\partial}{\partial y}$ contribute much compared to the horizontal gradient $\dfrac{\partial}{\partial z}$, i.e., $\dfrac{\partial}{\partial z}$ is negligible. Yang and McCorquodale (2004) showed that the importance of the terms containing these gradients was location-dependent, such as near the bed or the corner portion of the channel. Yang et al. (2004) concluded that the terms containing $\dfrac{\partial}{\partial y}$ dominated over the others. For large values of y, the viscous part of shear stress, $\mu\dfrac{\partial u}{\partial y}$, is small compared to $-\rho\overline{u'v'}$. Therefore, integration of Eq. (4.2) leads to (Yang et al. 2004; Absi 2008):

$$uv - \frac{\tau_{xy}}{\rho} = gyS + const. \tag{4.3}$$

The integral constant (*const.*) in Eq. (4.3) can be determined using the condition that at $y = 0$, $\tau_{xy} = \tau_b$, where τ_b is the local boundary shear stress defined as $\tau_b = \rho u_{*b}^2$. Thus, we have $const. = -\tau_b/\rho$, and:

$$uv - \frac{\tau_{xy}}{\rho} = gyS - \frac{\tau_b}{\rho} \tag{4.4}$$

Dividing Eq. (4.4) by the global shear velocity u_*, Eq. (4.4) can be arranged as follows:

$$-\frac{\overline{u'v'}}{u_*^2} = \left(1 - \frac{y}{h}\right) - \alpha_1\frac{y}{h}\frac{uv}{u_*^2} + c_1 \tag{4.5}$$

where α_1 and c_1 are constants independent of y, defined as $\alpha_1 = \dfrac{ghS}{u_*} - 1$. Therefore, Eq. (4.5) becomes:

$$-\frac{\overline{u'v'}}{u_*^2} = \left(1 - \frac{y}{h}\right) - \alpha_1\frac{y}{h} + \frac{uv}{u_*^2} \tag{4.6}$$

The third term on the right-hand side of Eq. (4.6) represents the effect of secondary currents. The component of secondary currents, v, in the outer region is generally in the downward direction. To that end, the term can be empirically modeled using a linear relation (Yang et al. 2004) as:

$$\frac{uv}{u_*^2} \approx -\alpha_2 \frac{y}{h} \tag{4.7}$$

where α_2 is a positive constant. Using Eq. (4.7), Eq. (4.6) becomes:

$$-\frac{\overline{u'v'}}{u_*^2} = \left(1 - \frac{y}{h}\right) - \alpha \frac{y}{h} \tag{4.8}$$

where $\alpha = \alpha_1 + \alpha_2$. As $u_*^2 < ghS$, α_1 takes on positive values, and hence, α is a positive constant. Using the *Boussinesq hypothesis*, the Reynolds shear stress can be given as:

$$-\overline{u'v'} = \varepsilon_t \frac{du}{dy} \tag{4.9}$$

where ε_t is the eddy viscosity whose profile can be modeled in different ways. However, Nezu and Nakagawa (1993) conducted experiments and reported that a parabolic profile can fit the measured values well. Therefore, eddy viscosity can be given as:

$$\varepsilon_t = \kappa u_* y \left(1 - \frac{y}{h}\right) \tag{4.10}$$

where $\kappa \approx 0.41$ is the von Karman constant. Using Eqs. (4.8)–(4.10), we can obtain:

$$\frac{du}{dy} = \frac{u_*}{\kappa y}\left(1 - \alpha \frac{\dfrac{y}{h}}{1 - \dfrac{y}{h}}\right) \tag{4.11}$$

Integrating Eq. (4.11) and considering the condition that the velocity is hypothetically zero at a distance, say $y = y_0$, we can get:

$$\frac{u}{u_*} = \frac{1}{\kappa}\left[\ln\left(\frac{y}{y_0}\right) + \alpha \ln\left(\frac{1 - \dfrac{y}{h}}{1 - \dfrac{y_0}{h}}\right)\right] \tag{4.12}$$

Since $\dfrac{y_0}{h} \ll 1$, we have $\dfrac{1 - \dfrac{y}{h}}{1 - \dfrac{y_0}{h}} \cong 1 - \dfrac{y}{h}$. Therefore, Eq. (4.12) takes on the form:

$$\frac{u}{u_*} = \frac{1}{\kappa}\left[\ln\left(\frac{y}{y_0}\right) + \alpha \ln\left(1 - \frac{y}{h}\right)\right] \tag{4.13}$$

Introducing the nondimensional variables, $U = \dfrac{u}{u_*}$, $\xi = \dfrac{y}{h}$, and $\xi_0 = \dfrac{y_0}{h}$, Eq. (4.13) becomes:

$$U = \frac{1}{\kappa}\left[\ln\left(\frac{\xi}{\xi_0}\right) + \alpha \ln\left(1-\xi\right)\right] \tag{4.14}$$

Eq. (4.14) is known as the dip-modified log law that predicts the velocity dip phenomenon with the second term on the right side of the equation. It can be seen that as $\alpha \to 0$, the equation becomes the classical *log-law*. Further, the second term plays an important role for the outer region of the channel, but it becomes negligible in the inner region, as $\ln\left(1-\dfrac{y}{h}\right) \approx 0$. Moreover, $\ln\left(1-\xi\right) \to \infty$ as $\xi \to 1$, which occurs because of the discrepancy arising from using Eq. (4.8) at the free surface. Thus, the velocity profile given by Eq. (4.14) is invalid in a very thin layer near the free surface, like the classical log-law, which becomes invalid at the boundary $y = 0$.

Yang et al. (2004) proposed an empirical formula, $\alpha(z) = 1.3\exp(-z/h)$, where z is the transverse distance from the side wall of the channel. On the channel axis at $z = b/2$, and therefore, $z/h = Ar/2$, b is the width of the channel and Ar is the aspect ratio of the channel defined as $Ar = b/h$. Therefore, we have $\alpha(Ar) = 1.3\exp(-Ar/2)$. A general formula can then be expressed as:

$$\alpha = a_1 \exp(-a_2 ArZ) \tag{4.15}$$

where $Z = z/(b/2)$ is the dimensionless transverse distance. It was found that the values of the coefficients are $a_1 = 1.3$ and $a_2 = 0.5$ (Yang et al. 2004). For wide-open channels $(Ar > 5)$, $\alpha \to 0$, and the dip-modified log-law Eq. (4.14) becomes the classical log-law profile. Our aim is to show the applicability of homotopy-based methods to solving Eq. (4.11), which is done in what follows.

4.3   HAM-BASED SOLUTION

When applying the homotopy analysis method (HAM), we rewrite Eq. (4.11) as follows:

$$\mathcal{N}[U] \equiv \kappa\xi(1-\xi)\frac{dU}{d\xi} + (1+\alpha)\xi - 1 = 0 \tag{4.16}$$

subject to the condition:

$$U(\xi = \xi_0) = 0 \tag{4.17}$$

When applying HAM, the zeroth-order deformation equation is constructed as follows:

$$(1-q)\mathcal{L}\big[\Phi(\xi;q) - U_0(\xi)\big] = q\hbar H(\xi)\mathcal{N}\big[\Phi(\xi;q)\big] \tag{4.18}$$

subject to the initial condition:

$$\Phi(\xi_0; q) = 0 \tag{4.19}$$

where q is the embedding parameter, $\Phi(\xi; q)$ is the representation of the solution across q, $U_0(\xi)$ is the initial approximation, $\hbar$ is the auxiliary parameter, $H(\xi)$ is the auxiliary function, and $\mathcal{L}$ and $\mathcal{N}$ are the linear and nonlinear operators, respectively. It may be noted that the governing Eq. (4.11) is linear in U. However, while describing the methodology, we still use the term 'nonlinear operator'. As stated in Chapter 2, this is done here and, in the book, to be consistent with the steps of the methodologies. In fact, such an introduction does not violate the applicability of the methods. The higher-order terms U_m for $m \geq 1$ can be calculated from the higher-order deformation equation, given as:

$$\mathcal{L}\left[U_m(\xi) - \chi_m U_{m-1}(\xi)\right] = \hbar H(\xi) R_m\left(\vec{U}_{m-1}\right),$$

$$m = 1, 2, 3, \dots \text{ subject to } U_m(\xi_0) = 0 \tag{4.20}$$

where

$$\chi_m = \begin{cases} 0 & \text{when } m = 1, \\ 1 & \text{otherwise} \end{cases} \tag{4.21}$$

and:

$$R_m\left(\vec{U}_{m-1}\right) = \frac{1}{(m-1)!} \frac{\partial^{m-1} \mathcal{N}\left[\Phi(\xi; q)\right]}{\partial q^{m-1}}\bigg|_{q=0} \tag{4.22}$$

Now, the final solution can be obtained as follows:

$$U(\xi) = U_0(\xi) + \sum_{m=1}^{\infty} U_m(\xi) \tag{4.23}$$

The nonlinear operator for the problem is selected as the equation itself, which can always be done for simplicity, given as:

$$\mathcal{N}\left[\Phi(\xi; q)\right] = \kappa \xi (1 - \xi) \frac{\partial \Phi(\xi; q)}{\partial \xi} + (1 + \alpha)\xi - 1 \tag{4.24}$$

Therefore, the term R_m can be calculated from Eq. (4.22) as follows:

$$R_m\left(\vec{U}_{m-1}\right) = \kappa \xi (1 - \xi) \frac{dU_{m-1}}{d\xi} + (1 + \alpha)\xi - 1 \tag{4.25}$$

We consider the following set of base functions to represent the solution of the present problem:

$$\left\{ \frac{1}{\kappa} \ln\left(\frac{\xi}{\xi_0} \right) + \xi^n \mid n = 0,12, 3,... \right\} \tag{4.26}$$

so that

$$U(\xi) = \frac{1}{\kappa} \ln\left(\frac{\xi}{\xi_0} \right) + \alpha_0 + \sum_{n=1}^{\infty} \alpha_n (\xi)^n \tag{4.27}$$

where α_0 and α_n are the coefficients of the series. Eq. (4.27) provides the so-called *rule of solution expression*. Following the rule of solution expression, the linear operator and the initial approximation are chosen, respectively, as follows:

$$\mathcal{L}\left[\Phi(\xi;q) \right] = \frac{\partial \Phi(\xi;q)}{\partial \xi} \text{ with the property } \mathcal{L}\left[C_2 \right] = 0 \tag{4.28}$$

$$U_0(\xi) = \frac{1}{\kappa} \ln\left(\frac{\xi}{\xi_0} \right) \tag{4.29}$$

where C_2 is an integral constant. It may be noted that Eq. (4.29) is a solution of the part of the original equation. Using Eq. (4.28), the higher-order terms can be obtained from Eq. (4.20) as follows:

$$U_m(\xi) = \chi_m U_{m-1}(\xi) + \hbar \int_{\xi_0}^{\xi} H(\xi) R_m \left(\vec{U}_{m-1} \right) d\xi + C_2, \ m = 1, 2, 3,... \tag{4.30}$$

where R_m is given by Eq. (4.25) and constant C_2 can be determined from the initial condition for higher-order deformation equations, i.e., $U_m\left(\xi = \xi_0 \right) = 0$, which simply yields $C_2 = 0$ for all $m \geq 1$.

Now, the auxiliary function $H(\xi)$ can be determined from the *rule of coefficient ergodicity*. Based on the rule of solution expression, the general form of $H(\xi)$ can be:

$$H(\xi) = \xi^\alpha \tag{4.31}$$

where α is an integer. Putting Eq. (4.31) into Eq. (4.30) and calculating some of U_m, we have:

$$R_1(U_0) = \alpha \xi \tag{4.32}$$

$$U_1(\xi) = \hbar \int_{\xi_0}^{\xi} \xi^\alpha (\alpha \xi) d\xi = \frac{\alpha \hbar}{\alpha + 2} \left(\xi^{\alpha+2} - \xi_0^{\alpha+2} \right) \tag{4.33}$$

$$R_2\left(\bar{U}_1\right) = \kappa\alpha\hbar\xi^{\alpha+2}\left(1-\xi\right)+\left(1+\alpha\right)\xi-1 \qquad (4.34)$$

$$U_2\left(\xi\right) = U_1\left(\xi\right) + \hbar\int_{\xi_0}^{\xi}\xi^{\alpha}\left(\kappa\alpha\hbar\xi^{\alpha+2}\left(1-\xi\right)+\left(1+\alpha\right)\xi-1\right)d\xi$$

$$= \frac{\alpha\hbar}{\alpha+2}\left(\xi^{\alpha+2}-\xi_0^{\alpha+2}\right)+\kappa\alpha\hbar^2\left[\left(\frac{\xi^{2\alpha+3}}{2\alpha+3}-\frac{\xi_0^{2\alpha+3}}{2\alpha+3}\right)-\left(\frac{\xi^{2\alpha+4}}{2\alpha+4}-\frac{\xi_0^{2\alpha+4}}{2\alpha+4}\right)\right]$$

$$+\left(1+\alpha\right)\hbar\left[\frac{\xi^{\alpha+2}}{\alpha+2}-\frac{\xi_0^{\alpha+2}}{\alpha+2}\right]-\hbar\left[\frac{\xi^{\alpha+1}}{\alpha+1}-\frac{\xi_0^{\alpha+1}}{\alpha+1}\right] \qquad (4.35)$$

Here, we can infer two observations. If we consider $\alpha \geq 1$, then some of the terms (such as ξ^{α}) do not appear in the solution expression in Eq. (4.26). This does not satisfy the *rule of coefficient ergodicity*, which tells us to include all the terms of the base function for the completeness of the solution. On the other hand, if we consider $\alpha \leq -2$, the terms (such as $\xi^{\alpha+1}$) appearing in the solution violate the *rule of solution expression* given by Eq. (4.26). Therefore, we are left with the only choices $\alpha = 0$ and $\alpha = -1$. We choose $\alpha = 0$, i.e., which results in $H\left(\xi\right) = 1$. Finally, the approximate solution can be obtained as follows:

$$U\left(\xi\right) \approx U_0\left(\xi\right) + \sum_{m=1}^{M}U_m\left(\xi\right) \qquad (4.36)$$

4.4   HPM-BASED SOLUTION

When applying the homotopy perturbation method (HPM), let us rewrite Eq. (4.16) in the following form:

$$\mathcal{N}\left(U\right) = f\left(\xi\right) \qquad (4.37)$$

We construct a homotopy $\Phi\left(\xi;q\right)$ that satisfies:

$$\left(1-q\right)\left[\mathcal{L}\left(\Phi\left(\xi;q\right)\right)-\mathcal{L}\left(U_0\left(\xi\right)\right)\right]+q\left[\mathcal{N}\left(\Phi\left(\xi;q\right)\right)-f\left(\xi\right)\right] = 0 \qquad (4.38)$$

where q is the embedding parameter that lies on $[0,1]$, $\mathcal{L}$ is the linear operator, and $U_0\left(\xi\right)$ is the initial approximation to the final solution. Eq. (4.38) shows that as $q = 0$, $\Phi\left(\xi;0\right) = U_0\left(\xi\right)$, and $q = 1$, $\Phi\left(\xi;1\right) = U\left(\xi\right)$. We now express $\Phi\left(\xi;q\right)$ as a series in terms of q, as follows:

$$\Phi\left(\xi;q\right) = \Phi_0 + q\Phi_1 + q^2\Phi_2 + q^3\Phi_3 + q^4\Phi_4 \ldots \qquad (4.39)$$

where Φ_m for $m \geq 1$ is the higher-order term. As $q \to 1$, Eq. (4.39) produces the final solution as:

$$U\left(\xi\right) = \lim_{q\to1}\Phi\left(\xi;q\right) = \sum_{i=0}^{\infty}\Phi_i \qquad (4.40)$$

For simplicity, we consider the linear and nonlinear operators as follows:

$$\mathcal{L}\big(\Phi(\xi;q)\big) = \frac{\partial \Phi(\xi;q)}{\partial \xi} \tag{4.41}$$

$$\mathcal{N}\big(\Phi(\xi;q)\big) = \kappa\xi(1-\xi)\frac{\partial \Phi(\xi;q)}{\partial \xi} + (1+\alpha)\xi - 1 \tag{4.42}$$

Using these expressions, Eq. (4.38) takes on the form:

$$(1-q)\left[\frac{\partial \Phi(\xi;q)}{\partial \xi} - \frac{dU_0}{d\xi}\right] + q\left[\kappa\xi(1-\xi)\frac{\partial \Phi(\xi;q)}{\partial \xi} + (1+\alpha)\xi - 1\right] = 0 \tag{4.43}$$

Using Eq. (4.39), we have:

$$(1-q)\left[\frac{\partial \Phi(\xi;q)}{\partial \xi} - \frac{dU_0}{d\xi}\right]$$

$$= (1-q)\left[\left(\frac{d\Phi_0}{d\xi} - \frac{dU_0}{d\xi}\right) + q\frac{d\Phi_1}{d\xi} + q^2\frac{d\Phi_2}{d\xi} + q^3\frac{d\Phi_3}{d\xi} + q^4\frac{d\Phi_4}{d\xi} + \cdots\right]$$

$$= \left(\frac{d\Phi_0}{d\xi} - \frac{dU_0}{d\xi}\right) + q\left(\frac{d\Phi_1}{d\xi} - \left(\frac{d\Phi_0}{d\xi} - \frac{dU_0}{d\xi}\right)\right) + q^2\left(\frac{d\Phi_2}{d\xi} - \frac{d\Phi_1}{d\xi}\right)$$

$$+ q^3\left(\frac{d\Phi_3}{d\xi} - \frac{d\Phi_2}{d\xi}\right) + q^4\left(\frac{d\Phi_4}{d\xi} - \frac{d\Phi_3}{d\xi}\right) + \cdots \tag{4.44}$$

$$\kappa\xi(1-\xi)\frac{\partial \Phi(\xi;q)}{\partial \xi} + (1+\alpha)\xi - 1$$

$$= \kappa\xi(1-\xi)\left(\frac{d\Phi_0}{d\xi} + q\frac{d\Phi_1}{d\xi} + q^2\frac{d\Phi_2}{d\xi} + q^3\frac{d\Phi_3}{d\xi} + q^4\frac{d\Phi_4}{d\xi} + \cdots\right)$$

$$+ (1+\alpha)\xi - 1 = \left[\kappa\xi(1-\xi)\frac{d\Phi_0}{d\xi} + (1+\alpha)\xi - 1\right] + q\left[\kappa\xi(1-\xi)\frac{d\Phi_1}{d\xi}\right]$$

$$+ q^2\left[\kappa\xi(1-\xi)\frac{d\Phi_2}{d\xi}\right] + q^3\left[\kappa\xi(1-\xi)\frac{d\Phi_3}{d\xi}\right] + \cdots \tag{4.45}$$

Using the boundary condition $U(\xi = \xi_0) = 0$, we have:

$$\Phi_0(\xi = \xi_0) = 0, \ \Phi_1(\xi = \xi_0) = 0, \ \Phi_2(\xi = \xi_0) = 0, \ldots \tag{4.46}$$

Using Eqs. (4.44) and (4.45) and equating the like powers of q in Eq. (4.43), the following system of differential equations is obtained:

$$\frac{d\Phi_0}{d\xi} - \frac{dU_0}{d\xi} = 0 \text{ subject to } \Phi_0(\xi = \xi_0) = 0 \tag{4.47}$$

$$\frac{d\Phi_1}{d\xi} - \left(\frac{d\Phi_0}{d\xi} - \frac{dU_0}{d\xi}\right) + \kappa\xi(1-\xi)\frac{d\Phi_0}{d\xi}$$

$$+(1+\alpha)\xi - 1 = 0 \text{ subject to } \Phi_1(\xi = \xi_0) = 0 \tag{4.48}$$

$$\frac{d\Phi_2}{d\xi} - \frac{d\Phi_1}{d\xi} + \kappa\xi(1-\xi)\frac{d\Phi_1}{d\xi} = 0 \text{ subject to } \Phi_2(\xi = \xi_0) = 0 \tag{4.49}$$

$$\frac{d\Phi_3}{d\xi} - \frac{d\Phi_2}{d\xi} + \kappa\xi(1-\xi)\frac{d\Phi_2}{d\xi} = 0 \text{ subject to } \Phi_3(\xi = \xi_0) = 0 \tag{4.50}$$

$$\frac{d\Phi_4}{d\xi} - \frac{d\Phi_3}{d\xi} + \kappa\xi(1-\xi)\frac{d\Phi_3}{d\xi} = 0 \text{ subject to } \Phi_4(\xi = \xi_0) = 0 \tag{4.51}$$

Proceeding in a like manner, one can arrive at the following recurrence relation:

$$\frac{d\Phi_m}{d\xi} = \left[1 - \kappa\xi(1-\xi)\right]\frac{d\Phi_{m-1}}{d\xi} \text{ subject to } \Phi_m(\xi = \xi_0) = 0 \text{ for } m \geq 2 \tag{4.52}$$

The initial approximation can be chosen as $\Phi_0 = \frac{1}{\kappa}\ln\left(\frac{\xi}{\xi_0}\right)$. Using this initial approximation, we can solve the equations iteratively using a symbolic computation software. We avoid those expressions here, as they are lengthy. Finally, the HPM-based solution can be approximated as follows:

$$U(\xi) \approx \sum_{i=0}^{M} \Phi_i \tag{4.53}$$

4.5　OHAM-BASED SOLUTION

When applying the optimal homotopy asymptotic method (OHAM), Eq. (4.16) can be written in the following form:

$$\mathcal{L}[U(\xi)] + \mathcal{N}[U(\xi)] + h(\xi) = 0 \tag{4.54}$$

We select $\mathcal{L}[U(\xi)] = \kappa\xi\frac{dU}{d\xi}$, $\mathcal{N}[U(\xi)] = -\kappa\xi^2\frac{dU}{d\xi} + (1+\alpha)\xi$, and $h(\xi) = -1$. With these considerations, the zeroth-order problem becomes:

$$\mathcal{L}(U_0(\xi)) + h(\xi) = 0 \text{ subject to } U_0(\xi = \xi_0) = 0 \tag{4.55}$$

Solving Eq. (4.55), one obtains:

$$U_0(\xi) = \frac{1}{\kappa}\ln\frac{\xi}{\xi_0} \tag{4.56}$$

To obtain $\mathcal{N}_0$, $\mathcal{N}_1$, $\mathcal{N}_2$, etc., the following can be used:

$$-\kappa\xi^2\frac{dU}{d\xi}+(1+\alpha)\xi = -\kappa\xi^2\left(\frac{dU_0}{d\xi}+q\frac{dU_1}{d\xi}+q^2\frac{dU_2}{d\xi}+q^3\frac{dU_3}{d\xi}+q^4\frac{dU_4}{d\xi}+\cdots\right)$$

$$+(1+\alpha)\xi = \left[-\kappa\xi^2\frac{dU_0}{d\xi}+(1+\alpha)\xi\right]+q\left[-\kappa\xi^2\frac{dU_1}{d\xi}\right]+q^2\left[-\kappa\xi^2\frac{dU_2}{d\xi}\right]$$

$$+q^3\left[-\kappa\xi^2\frac{dU_3}{d\xi}\right]+q^4\left[-\kappa\xi^2\frac{dU_4}{d\xi}\right]+\cdots \tag{4.57}$$

Using Eq. (4.57), the first-order problem reduces to:

$$\kappa\xi\frac{dU_1}{d\xi}=H_1\left(\xi,C_i\right)\left\{-\kappa\xi^2\frac{dU_0}{d\xi}+(1+\alpha)\xi\right\} \text{ subject to } U_1\left(\xi=\xi_0\right)=0 \tag{4.58}$$

We simply choose $H_1\left(\xi,C_1\right)=C_1$. Putting $k=2$, the second-order problem becomes:

$$\mathcal{L}\left[U_2\left(\xi,C_i\right)-U_1\left(\xi,C_i\right)\right]=H_2\left(\xi,C_i\right)\mathcal{N}_0\left(U_0\left(\xi\right)\right)$$

$$+H_1\left(\xi,C_i\right)\left[\mathcal{L}\left[U_1\left(\xi,C_i\right)\right]+\mathcal{N}_1\left[U_0\left(\xi\right),U_1\left(\xi,C_i\right)\right]\right] \text{ subject to } U_2\left(\xi=\xi_0\right)=0 \tag{4.59}$$

Further, we choose $H_2\left(\xi,C_i\right)=C_2$. Therefore, Eq. (4.59) becomes:

$$\kappa\xi\frac{dU_2}{d\xi}=\kappa\xi\frac{dU_1}{d\xi}+C_2\left\{-\kappa\xi^2\frac{dU_0}{d\xi}+(1+\alpha)\xi\right\}$$

$$+C_1\left[\kappa\xi\frac{dU_1}{d\xi}-\kappa\xi^2\frac{dU_1}{d\xi}\right] \text{ subject to } U_2\left(\xi=\xi_0\right)=0 \tag{4.60}$$

The terms of the series can be computed using the equations developed here. One can compute these terms without any difficulty using a symbolic computation software, such as MATLAB. Further, following Chapter 2, the higher-order terms can be computed in a similar manner. However, our aim is to produce an accurate solution with just two to three terms of an OHAM-based series. Therefore, we restrict our calculation up to $k=2$. Finally, the approximate solution can be found as:

$$U\left(\xi\right)\approx U_0\left(\xi\right)+U_1\left(\xi,C_1\right)+U_2\left(\xi,C_1,C_2\right) \tag{4.61}$$

where the terms are given by Eqs. (4.56), (4.58), and (4.60).

4.6 CONVERGENCE THEOREMS

The convergence of the HAM-based and OHAM-based solutions is proved theoretically using the following theorems.

4.6.1 CONVERGENCE THEOREM OF THE HAM-BASED SOLUTION

The convergence theorems for the HAM-based solution given by Eq. (4.16) can be proved using the following theorems.

Theorem 4.1: *If the homotopy series $\Sigma_{m=0}^{\infty} U_m(\xi)$ and $\Sigma_{m=0}^{\infty} U_m'(\xi)$ converge, then $R_m(\vec{U}_{m-1})$ given by Eq. (4.25) satisfies the relation $\Sigma_{m=1}^{\infty} R_m(\vec{U}_{m-1}) = 0$. Here, "'" denotes the derivative with respect to ξ.*

Proof: The proof of this theorem follows exactly from the previous chapters.

Theorem 4.2: *If $\hbar$ is so properly chosen that the series $\Sigma_{m=0}^{\infty} U_m(\xi)$ and $\Sigma_{m=0}^{\infty} U_m'(\xi)$ converge absolutely to $U(\xi)$ and $U'(\xi)$, respectively, then the homotopy series $\Sigma_{m=0}^{\infty} U_m(\xi)$ satisfies the original governing Eq. (4.16).*

Proof: Theorem 4.1 shows that if $\Sigma_{m=0}^{\infty} U_m(\xi)$ and $\Sigma_{m=0}^{\infty} U_m'(\xi)$ converge, then $\Sigma_{m=1}^{\infty} R_m(\vec{U}_{m-1}) = 0$.

Therefore, substituting the earlier expressions in Eq. (4.25) and simplifying further lead to:

$$\kappa\xi(1-\xi)\sum_{k=0}^{\infty} U_m' + (1+\alpha)\xi - 1 = 0 \qquad (4.62)$$

which is basically the original governing equation in Eq. (4.16). Furthermore, subject to the initial condition $U_0(\xi = \xi_0) = 0$ and the conditions for the higher-order deformation equation $U_m(\xi = \xi_0) = 0$, for $m \geq 1$, we easily obtain $\Sigma_{m=0}^{\infty} U_m(\xi = \xi_0) = 0$. Hence, the convergence result follows.

4.6.2 CONVERGENCE THEOREM OF THE OHAM-BASED SOLUTION

Theorem 4.3: *If the series $U_0(\xi) + \Sigma_{j=1}^{\infty} U_j(\xi, C_i)$, $i = 1, 2, \ldots, s$, converges, where $U_j(\xi, C_i)$ is governed by Eqs. (4.56), (4.58), and (4.60), then Eq. (4.61) is a solution of the original Eq. (4.16).*

Proof: The proof of this theorem follows exactly from Theorem 2.3 of Chapter 2.

4.7 RESULTS AND DISCUSSION

First, the numerical convergence of the HAM-based analytical solutions is established for a specific test case; then the solution is validated against the existing solution proposed by Yang et al. (2004). The derived analytical solution is tested under different physical conditions to evaluate the efficacy of the method. Finally, the HPM- and OHAM-based analytical solutions are validated by comparing them with the solution given by Yang et al. (2004).

4.7.1 Numerical Convergence and Validation of the HAM-Based Solution

The convergence of the HAM-based series solution depends on a suitable choice for the auxiliary parameter $\hbar$. For that purpose, the squared residual error regarding Eq. (4.16) can be calculated as:

$$\Delta_m = \int\limits_{\xi \in \Omega} \left(\mathcal{N}\left[U(\xi) \right] \right)^2 d\xi \qquad (4.63)$$

where $\Omega = \left[\xi_0, 1 \right]$ is the domain of the problem. The HAM-based series solution leads to the exact solution of the problem when Δ_m approaches zero. Therefore, for a particular order of approximation m, it is sufficient to minimize the corresponding Δ_m, which then yields an optimal value of parameter $\hbar$. Here, we consider a test case, where parameters are taken as $\alpha = 1.01$ and $\xi_0 = 0.0015$. Considering these values, we perform a test case, where the HAM-based solution is obtained. First, the squared residual error is calculated from Eq. (4.63), then the corresponding approximate solutions are obtained. Figure 4.2 shows the squared residual errors for a different order of approximation for the selected case. It can be seen from the figure that as the order of approximation increases, the corresponding residual error decreases. Thus, the adequacy of selecting the linear operator, initial approximation, and auxiliary function is established. Also, for a

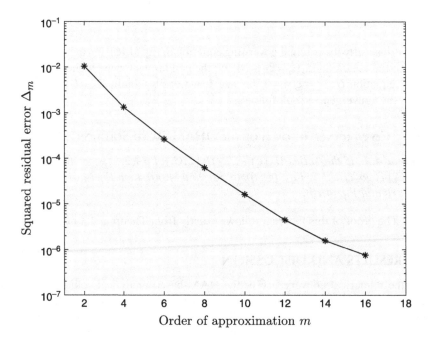

FIGURE 4.2 Squared residual error (Δ_m) versus different orders of approximation (m) of the HAM-based solution for the selected test case.

TABLE 4.1

Squared Residual Error (Δ_m) and Computational Time versus Different Orders of Approximation (m) for the Selected Test Case

Order of Approximation (m)	Squared Residual Error (Δ_m)	Computational Time (sec)
2	1.04744×10^{-2}	0.23672
4	1.33444×10^{-3}	0.48408
6	2.65078×10^{-4}	0.72147
8	6.26803×10^{-5}	0.96879
10	1.61925×10^{-5}	1.24704
12	4.49237×10^{-6}	1.50386
14	1.57463×10^{-6}	1.76218
16	7.54861×10^{-7}	2.07031

quantitative assessment, in Table 4.1, we provide the computational time taken by the computer to produce the corresponding order of approximation. The exact solution given by Yang et al. (2004) is compared with the 14th-order HAM-based solution in Figure 4.3. It can be seen that the HAM solution exactly matches the exact values, and a quantitative comparison is shown in Table 4.2, where a different order of approximation is considered and the corresponding solutions for

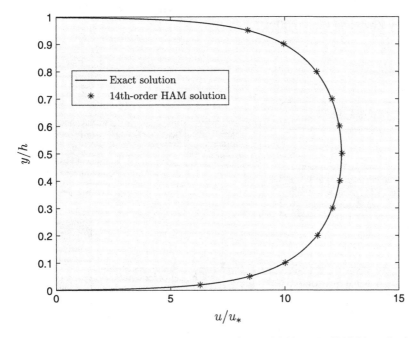

FIGURE 4.3 Comparison between the exact solution and 14th-order HAM-based solution for the selected test case [$\alpha = 1.01$ and $\xi_0 = 0.0015$].

TABLE 4.2

Comparison between HAM-Based Approximations and Exact Solution for the Selected Test Case

ξ	Exact Solution	HAM-Based Approximation		
		5th Order	10th Order	14th Order
0.02	6.2717	6.3054	6.2971	6.2923
0.05	8.4299	8.4901	8.4652	8.4552
0.10	9.9873	10.0588	10.0239	10.0129
0.20	11.3878	11.4605	11.4243	11.4134
0.30	12.0478	12.1204	12.0843	12.0734
0.40	12.3697	12.4406	12.4063	12.3953
0.50	12.4648	12.5279	12.5015	12.4904
0.60	12.3598	12.4135	12.3967	12.3854
0.70	12.0271	12.0778	12.0640	12.0527
0.80	11.3540	11.4045	11.3909	11.3794
0.90	9.9337	9.9934	9.9708	9.9585
0.95	8.3581	8.5814	8.4152	8.3890

some selected values of the normalized depth are reported. It is observed from the table that the higher the order of approximation, the better the accuracy. The flowchart diagram containing the steps of the method is given in Figure 4.4. All the computations reported here are performed using the BVPh 2.0 package developed by Zhao and Liao (2002).

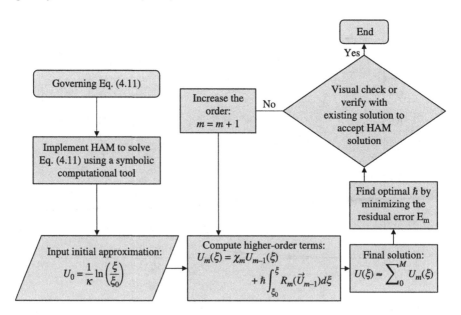

FIGURE 4.4   Flowchart for the HAM solution in Eq. (4.36).

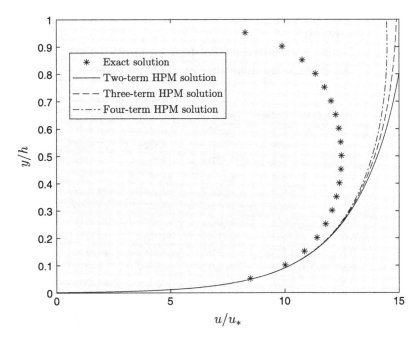

FIGURE 4.5 Comparison between the exact solution and two-, three-, and four-term HPM-based solutions for the selected test case [$\alpha = 1.01$ and $\xi_0 = 0.0015$].

4.7.2 VALIDATION OF THE HPM-BASED SOLUTION

The HPM-based analytical solution for the selected test case is validated against the analytical solution given by Yang et al. (2004). The two-, three-, and four-term HPM solutions are considered. It is seen that the HPM produces accurate results only for a restricted domain, specifically $\xi < 0.2$. Figure 4.5 compares the exact solution and the HPM approximations. It may be noted that one can achieve better results if the order of approximation is higher. However, for the purpose of this discussion, we have shown the results for lower-order approximations. The HPM-based values and the existing exact solution are compared numerically in Table 4.3. It may be noted that one can indeed obtain a more accurate solution by considering other types of linear operators and initial approximations of HPM. The flowchart diagram containing the steps of the method is given in Figure 4.6.

4.7.3 VALIDATION OF THE OHAM-BASED SOLUTION

For the assessment of the OHAM-based analytical solution, one needs to calculate the constants C_i. For that purpose, we calculate the residual as follows:

$$R(\xi, C_i) = \mathcal{L}\left[U_{OHAM}(\xi, C_i)\right] + \mathcal{N}\left[U_{OHAM}(\xi, C_i)\right] + h(\xi), \ i = 1, 2, \ldots, s \quad (4.64)$$

where $U_{OHAM}(\xi, C_i)$ is the approximate solution. When $R(\xi, C_i) = 0$, $U_{OHAM}(\xi, C_i)$ becomes the exact solution to the problem. One of the ways to obtain the optimal

TABLE 4.3

Comparison between HPM-Based Approximations and Exact Solution for the Selected Test Case

ξ	Exact Solution	HPM-Based Approximation		
		Two-Term	Three-Term	Four-Term
0.02	6.2717	6.3175	6.3173	6.3171
0.05	8.4299	8.5513	8.5501	8.5488
0.10	9.9873	10.2381	10.2332	10.2284
0.20	11.3878	11.9136	11.8943	11.8760
0.30	12.0478	12.8773	12.8347	12.7948
0.40	12.3697	13.5436	13.4690	13.4000
0.50	12.4648	14.0424	13.9269	13.8213
0.60	12.3598	14.4315	14.2661	14.1156
0.70	12.0271	14.7419	14.5169	14.3123
0.80	11.3540	14.9918	14.6969	14.4277
0.90	9.9337	15.1932	14.8169	14.4704
0.95	8.3581	15.2784	14.8566	14.4661

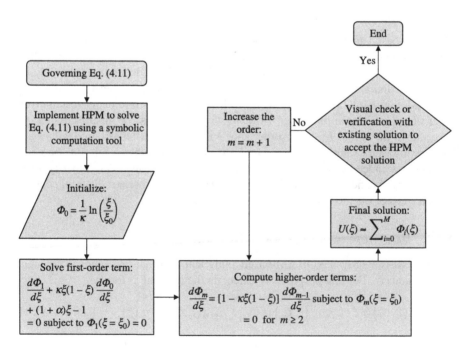

FIGURE 4.6　Flowchart for the HPM solution in Eq. (4.53).

C_i for which the solution converges is the minimization of squared residual error, i.e.,

$$J(C_i) = \int_{\xi \in D} R^2(\xi, C_i) d\xi, \ i = 1, 2, \dots, s \tag{4.65}$$

where $D = [\xi_0, 1]$ is the domain of the problem. The minimization of Eq. (4.65) leads to a system of algebraic equations as follows:

$$\frac{\partial J}{\partial C_1} = \frac{\partial J}{\partial C_2} = \dots = \frac{\partial J}{\partial C_s} = 0 \tag{4.66}$$

Solving this system, one can obtain the optimal values of the parameters. We obtain the optimal values using the MATLAB routine *fminsearch*, which minimizes an unconstrained multivariable function. Two- and three-term OHAM solutions are computed and compared in Figure 4.7 with the existing analytical solution of Yang et al. (2004). It can be seen that three terms of the OHAM-based series agree comparatively well with the corresponding analytical solution. One can indeed obtain more accurate solutions if the higher-order approximations are considered. For a quantitative assessment, we also compare the numerical values of solutions in Table 4.4. A flowchart diagram containing the steps of the method is given in Figure 4.8.

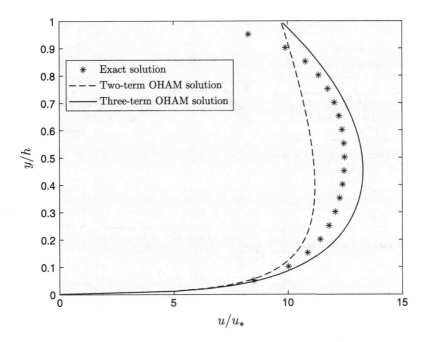

FIGURE 4.7 Comparison between the exact solution and two-, and three-term OHAM-based solutions for the selected test case [$\alpha = 1.01$ and $\xi_0 = 0.0015$].

TABLE 4.4

Comparison between OHAM-Based Approximations and Exact Solution for the Selected Test Case

ξ	Exact Solution	OHAM-Based Approximation	
		Two-Term	Three-Term
0.02	6.2717	6.2038	6.3599
0.05	8.4299	8.2539	8.6505
0.10	9.9873	9.6366	10.3996
0.20	11.3878	10.7113	12.0779
0.30	12.0478	11.0844	12.8821
0.40	12.3697	11.1702	13.2266
0.50	12.4648	11.0986	13.2412
0.60	12.3598	10.9274	12.9838
0.70	12.0271	10.6876	12.4853
0.80	11.3540	10.3974	11.7641
0.90	9.9337	10.0688	10.8319
0.95	8.3581	9.8928	10.2895

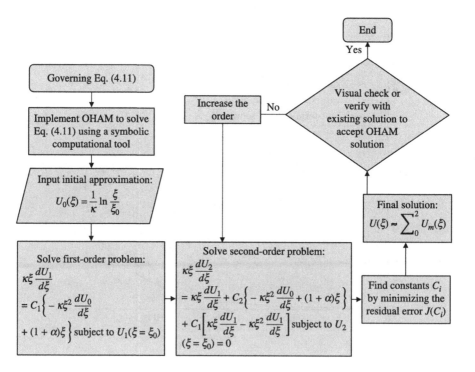

FIGURE 4.8 Flowchart for the OHAM solution in Eq. (4.61).

4.8   CONCLUDING REMARKS

The chapter describes the dip-modified log-law proposed by Yang et al. (2004) for measuring the streamwise velocity profile in open channels, including the dip phenomenon. The velocity model is derived by modeling the RANS equation. The model has several advantages over the classical log-law, which is invalid at the solid boundary and cannot predict the dip phenomenon. The governing RANS equation is solved using three different analytical methods, namely HAM, HPM, and OHAM. It is observed that both HAM and OHAM produce accurate solutions. Moreover, the OHAM can produce comparatively accurate results only with three terms of the series. On the other hand, the HPM provides a better solution only within a restricted domain. For all the methods, the classical logarithmic profile is used as an initial approximation for producing the higher-order terms of the series, which seems to be an appropriate choice for initial value problems. Different test cases are used to evaluate the stability of series solutions, and the quantitative assessment is carried out whenever needed. Further, theoretical as well as numerical convergence of the series solutions is provided.

REFERENCES

Absi, R. (2008). Comments on "turbulent velocity profile in fully-developed open channel flows". *Environmental Fluid Mechanics*, 8(4), 389–394.

Absi, R. (2011). An ordinary differential equation for velocity distribution and dip-phenomenon in open channel flows. *Journal of Hydraulic Research*, 49(1), 82–89.

Nezu, I., and Nakagawa, H. (1993). *Turbulence in open-channel flows*. Balkema, Rotterdam.

Tracy, H. J. (1963). *Turbulent flow in a three-dimensional channel* (Doctoral dissertation, Georgia Institute of Technology).

Yang, S. Q., and McCorquodale, J. A. (2004). Determination of boundary shear stress and Reynolds shear stress in smooth rectangular channel flows. *Journal of Hydraulic Engineering*, 130(5), 458–462.

Yang, S. Q., Tan, S. K., and Lim, S. Y. (2004). Velocity distribution and dip-phenomenon in smooth uniform open channel flows. *Journal of Hydraulic Engineering*, 130(12), 1179–1186.

Zhao, Y., and Liao, S. (2002). User Guide to BVPh 2.0. Available online: http://numericaltank. sjtu.edu.cn/BVPh.htm

FURTHER READING

Bonakdari, H., Larrarte, F., Lassabatere, L., and Joannis, C. (2008). Turbulent velocity profile in fully-developed open channel flows. *Environmental Fluid Mechanics*, 8(1), 1–17.

Guo, J. (2014). Modified log-wake-law for smooth rectangular open channel flow. *Journal of Hydraulic Research*, 52(1), 121–128.

Kundu, S., and Ghoshal, K. (2012). An analytical model for velocity distribution and dip-phenomenon in uniform open channel flows. *International Journal of Fluid Mechanics Research*, 39(5), 381–395.

Lassabatere, L., Pu, J. H., Bonakdari, H., Joannis, C., and Larrarte, F. (2013). Velocity distribution in open channel flows: Analytical approach for the outer region. *Journal of Hydraulic Engineering*, 139(1), 37–43.

Yang, S. Q. (2005). Interactions of boundary shear stress, secondary currents and velocity. *Fluid Dynamics Research*, 36(3), 121.

Yang, S. Q. (2010). Depth-averaged shear stress and velocity in open-channel flows. *Journal of Hydraulic Engineering*, 136(11), 952–958.

5 Sediment Concentration Distribution in Open-Channel Flow

5.1 INTRODUCTION

Sediment transport in open channels is of fundamental importance for river training and management, navigation, design of canals, habitat assessment, pollutant transport, stability of hydraulic geometry, health of ecosystems, and watershed management. Sediment transport involves bedload transport and suspended load transport. There is, however, a continual exchange of sediment between these two modes of transport. The transport itself involves three components: sediment concentration, sediment discharge, and sediment load. Basic to these components is sediment concentration, which varies in three dimensions (longitudinal, transverse, and vertical). The concentration is affected by sediment characteristics, discharge characteristics, river morphology, and watershed characteristics. For a given river cross-section, the concentration varies vertically and transversely. At a given point, the concentration profile varies from the bed to the water surface in the suspension zone. This vertical profile also varies with the density of sediment-laden flow and becomes complicated if the density is high because in high-sediment-laden flow, the settling velocity is hindered. The sediment concentration is often described by an advective-dispersion equation, which is a nonlinear partial differential equation of a parabolic type. This equation can be solved numerically, but analytical solutions have been obtained for simplified cases. This chapter presents linear governing equations of sediment concentration and solves them using homotopy-based methods and compares the derived solutions with exact solutions for simple cases.

5.2 SEDIMENT CONCENTRATION MODELS

The suspended sediment transport by fluid flow is governed by the advection-diffusion equation, which can be derived using the three-dimensional continuity equation along with *Fick's law*. For a 2D steady-uniform flow, the variation of sediment concentration occurs only in the vertical direction. The concentration decreases in the upward direction due to the gravitational effects on suspended sediment particles. Under this consideration, and substituting the vertical velocity component of fluid by the negative settling velocity, one can obtain the governing differential equation as follows (Julien 2010):

$$-\varepsilon_s \frac{dC}{dy} = \omega_s C \qquad (5.1)$$

where ε_s is the sediment diffusivity, y is the vertical distance from the channel bed, C is the volumetric sediment concentration at y, and ω_s is the settling

DOI: 10.1201/9781003368984-8

velocity of sediment. Eq. (5.1) represents the equilibrium of sediment suspension, which exists by balancing the upward diffusion due to turbulence (known as *entrainment flux*) and the downward settling due to gravity (called *deposition flux*). The sediment diffusion coefficient ε_s is generally approximated as $\varepsilon_s = \beta\varepsilon_t$, where ε_t is the turbulent diffusivity, and β is the coefficient of proportionality (van Rijn 1984). According to the *Boussinesq hypothesis*, the Reynolds shear stress τ can be given as:

$$\tau = \rho\varepsilon_t \frac{du}{dy} \tag{5.2}$$

where ρ is the mass density of fluid, and u is the streamwise velocity profile. Using Eq. (5.2), the sediment diffusivity can be obtained as:

$$\varepsilon_s = \left(\frac{\beta}{\rho}\right)\tau\left(\frac{du}{dy}\right)^{-1} \tag{5.3}$$

Using Eq. (5.3), the governing differential in Eq. (5.1) becomes:

$$\beta\tau\frac{dC}{dy} + \omega_s\rho\frac{du}{dy}C = 0 \tag{5.4}$$

One can see that to solve Eq. (5.4), one needs to know the profiles of shear stress τ and streamwise velocity u. Further, the initial condition can be given in two ways: (i) If we want to measure the concentration throughout the water column, $C = C_0$ at $y = 0$ can be prescribed, and (ii) for suspended sediment concentration, $C = C_a$ can be prescribed at a reference level, say $y = a$. We briefly describe the well-known Rouse equation (Rouse 1937) and models of Chiu et al. (2000) in what follows.

5.2.1 ROUSE EQUATION

Rouse (1937) derived the vertical distribution of suspended sediment concentration using the one-dimensional diffusion equation described earlier. The shear stress profile can be taken as:

$$\frac{\tau}{\tau_0} = 1 - \frac{y}{D} \tag{5.5}$$

where τ_0 is the bed shear stress, given as $\tau_0 = \rho u_*^2$, with u_* being the shear velocity. For the Rouse equation, the well-known Prandtl–von Karman velocity profile is considered, which is given as:

$$\frac{u}{u_*} = \frac{1}{\kappa}\ln\frac{y}{y_0} \tag{5.6}$$

where κ is the von Karman constant, typically assumed to be 0.4, and y_0 is the zero-velocity level. The velocity gradient can be then obtained as:

$$\frac{du}{dy} = \frac{u_*}{\kappa y}$$ (5.7)

Using Eqs. (5.5) and (5.7) and considering the condition, $C = C_a$ at $y = a$, the diffusion Eq. (5.4) can be solved, and the solution is given by:

$$\frac{C}{C_a} = \left(\frac{D-y}{y} \frac{a}{D-a} \right)^{R_0}$$ (5.8)

where $R_0 = \dfrac{\omega_s}{\beta \kappa u_*}$ is known as the *Rouse number*. Eq. (5.8) is famously known as the *Rouse equation* for measuring vertically distributed suspended sediment concentrations. However, this equation has several shortcomings in terms of accuracy because it is based on the log-law for velocity, which is inaccurate near the channel bed and near the water surface. Therefore, the Rouse equation is not useful for measuring depth-averaged sediment concentration because of its weakness in measuring near-bed sediment concentration.

5.2.2 CHIU ET AL. (2000) MODELS

Chiu et al. (2000) proposed a combined approach for developing sediment concentration distribution. The velocity profile was considered from an earlier study, which is based on the concept of information entropy together with the maximum entropy principle. Here, we briefly describe the entropic velocity distribution. Considering the time-averaged velocity u as a random variable and the corresponding Shannon entropy (Shannon 1948), the velocity profile was obtained as (Chiu 1987):

$$\frac{u}{u_{max}} = \frac{1}{M} \ln \left[1 + \left(\exp(M) - 1 \right) \frac{\xi - \xi_0}{\xi_{max} - \xi_0} \right]$$ (5.9)

where M is a parameter (commonly known as Chiu's parameter), u_{max} is the maximum velocity in a channel cross-section, ξ is the generalized coordinate combining the vertical (y) and transverse (z) directions, ξ_{max} is the maximum of ξ at which u_{max} occurs, and ξ_0 is the minimum value of ξ at which $u = 0$. The velocity Eq. (5.9) was derived, considering the following constraints:

$$\int_0^{u_{max}} f(u)\,du = 1$$ (5.10)

$$\int_0^{u_{max}} u f(u)\,du = \bar{u} = \frac{Q}{A}$$ (5.11)

where $f(u)$ is the probability density function (PDF) of the velocity u, $\bar{u}$ is the mean velocity, Q is the flow discharge, and A is the area of the cross-section. Eq. (5.10) is the normalization constraint, and Eq. (5.11) is the mean constraint that physically represents the hydrodynamic transport of mass through a channel cross-section of area A. Further, the cumulative distribution function (CDF) is given as:

$$F(u) = \text{Prob}(U \le u) = \int_0^u f(u)\,du = \frac{\xi - \xi_0}{\xi_{max} - \xi_0} \tag{5.12}$$

The expression $\dfrac{\xi - \xi_0}{\xi_{max} - \xi_0}$ in the velocity in Eq. (5.9) physically represents the fraction of the total cross-section in which the velocity is less than or equal to u. Based on the channel geometry and flow characteristics, this expression takes on different forms. For example, in wide rectangular channels, the expression becomes $\dfrac{By}{BD}$, where B is the channel width. In circular pipe flow, the expression takes on the form $\dfrac{\pi R^2 - \pi r^2}{\pi R^2} = 1 - \left(\dfrac{r}{R}\right)^2$, where R is the radius of the pipe, and r is the radial distance from the center of the pipe to a point where velocity is u. The velocity u shows a one-to-one relationship with the ξ–cordinate, and it increases with an increase in ξ. The generalized ξ–curves in a rectangular channel are shown in Figure 5.1 for both cases, when the maximum velocity occurs below the water surface $h > 0$ and at the water surface $(h \le 0)$.

For velocity distribution along a vertical, the following can be an expression of ξ:

$$\xi = \frac{y}{D-h} exp\left(1 - \frac{y}{D-h}\right) \tag{5.13}$$

where parameter h can be used to describe the location of maximum velocity. From Eq. (5.13), $\xi_0 = 0$. When the maximum velocity occurs below the water surface, i.e., $h > 0$, ξ_{max} can be obtained by calculating ξ at $y = D - h$, which implies $\xi_{max} = 1$. On the other hand, for maximum velocity occurring at the water surface, ξ_{max} can be obtained by substituting $y = D$. Introducing the concept of entropy and the generalized coordinate, it is possible to refine the modeling of open-channel flow velocity irrespective of the location of maximum velocity along a channel cross-section.

The entropy parameter M plays the central role in analyzing the velocity distribution in open channels. The stability of the PDF is tested through parameter M or the ratio of mean to maximum velocity that connects M. Further, it can be used as the basis for modeling velocity distribution under varying mean and maximum velocities and discharge. The relation between $\bar{u}/u_{max}$ and M can be given as:

$$\varphi = \frac{\bar{u}}{u_{max}} = \frac{exp(M)}{exp(M) - 1} - \frac{1}{M} \tag{5.14}$$

Investigations into M have shown that the parameter remains constant along a given channel cross-section regardless is the flow is steady or unsteady and varying

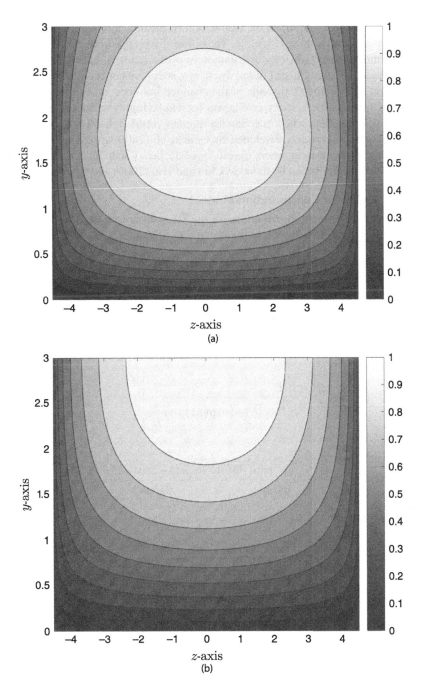

FIGURE 5.1 ξ−curves (a) $h > 0$, and (b) $h \leq 0$.

discharge. It further simplifies the calculation of mean velocity $\bar{u}$, as one needs to know only the maximum velocity u_{max}. The probabilistic approach introduced by Chiu et al. for modeling the velocity distribution in open channels takes into account all factors, such as velocity distribution; discharge; sediment concentration; and roughness, slope, and geometrical shape of the channel, but does not explicitly deal with individual factors. The velocity profile was seen to be highly accurate for measuring near-bed velocity data, where the classical log-law profile fails. Therefore, the sediment concentration models developed by Chiu et al. (2000) specifically focused on the near-bed concentration, where the effects of the heavy sediment concentration are observed. More details can be found in Chiu and Hsu (2006) and Chiu et al. (2000).

5.2.2.1 Sediment Concentration Model I

We consider the velocity distribution in Eq. (5.9) with $\xi = y/D$ and the shear stress profile in Eq. (5.5). Substitution of these in Eq. (5.4) gives:

$$\beta\tau_0\left(1-\frac{y}{D}\right)\frac{dC}{dy}+\omega_s\rho\frac{u_{max}\left(\exp(M)-1\right)}{MD\left[1+\left(\exp(M)-1\right)\frac{y}{D}\right]}$$

$$C=0 \text{ subject to } C(y=0)=C_0 \qquad (5.15)$$

Solving Eq. (5.15), one obtains:

$$\frac{C}{C_0}=\left[\frac{1-\frac{y}{D}}{1+\left(\exp(M)-1\right)\frac{y}{D}}\right]^{a_1} \qquad (5.16)$$

where

$$a_1=\frac{\omega_s u_{max}\left(1-\exp(-M)\right)}{\beta M u_*^2}=\frac{\omega_s\bar{u}\left(1-\exp(-M)\right)}{\beta M\varphi u_*^2}=a_2 a_3 \qquad (5.17)$$

in which

$$a_2=\frac{\left(1-\exp(-M)\right)}{M\varphi} \qquad (5.18)$$

and

$$a_3=\frac{\omega_s\bar{u}}{\beta u_*^2} \qquad (5.19)$$

The advantage of Eq. (5.16) over the Rouse equation is that it is capable of measuring the sediment concentration values at or near the channel bed.

5.2.2.2 Sediment Concentration Model II

Sediment concentration distribution modeling can be further refined. The refinement can be done by considering the effect of velocity distribution on sediment concentration through Eq. (5.3). Parameters M and h include the effects in greater detail, where h represents the effect of the location of maximum velocity. Using Eq. (5.9), the velocity gradient can be obtained as:

$$\frac{du}{dy} = \frac{\bar{u}\left(\exp(M)-1\right)}{\varphi M \xi_{max}} \left[\frac{1+\left(\exp(M)-1\right)\dfrac{\xi}{\xi_{max}}}{\dfrac{\xi}{y}\left(1-\dfrac{y}{D-h}\right)} \right]^{-1} \tag{5.20}$$

where ξ_{max} can be given as:

$$\xi_{max} = \begin{cases} 1 & \text{for } h \geq 0 \\ \xi(y=D) & \text{for } h < 0 \end{cases} \tag{5.21}$$

Chiu et al. (2000) proposed a refined shear stress profile that is compatible with velocity in Eq. (5.9), given as:

$$\frac{\tau}{\tau_0} = \frac{h}{D}\left(1-\frac{y}{D-h}\right) + \left(1-\frac{h}{D}\right)\left(1-\frac{y}{D-h}\right)^2 \tag{5.22}$$

Eq. (5.22) satisfies the boundary conditions at $y=0$, $\tau=\tau_0$ and at $y=D-h$, $\tau=0$, where the maximum velocity occurs. It may be noted that Eq. (5.22) becomes Eq. (5.5) when h/D approaches infinity, which is the case for one-dimensional flow. The previous equation includes the effect of a secondary current through the location of maximum velocity, which occurs in narrow open channels. Using Eqs. (5.20) and (5.22), the governing differential in Eq. (5.4) for the sediment concentration equation takes on the form:

$$\beta\tau_0 \left[\frac{h}{D}\left(1-\frac{y}{D-h}\right) + \left(1-\frac{h}{D}\right)\left(1-\frac{y}{D-h}\right)^2 \right]\frac{dC}{dy}$$

$$+ \omega_s \rho \, \frac{\bar{u}\left(\exp(M)-1\right)}{\varphi M \xi_{max}} \left[\frac{1+\left(\exp(M)-1\right)\dfrac{\xi}{\xi_{max}}}{\dfrac{\xi}{y}\left(1-\dfrac{y}{D-h}\right)} \right]^{-1}$$

$$C = 0 \text{ subject to } C(y=0) = C_0 \tag{5.23}$$

Arranging the terms and then solving Eq. (5.23), we obtain:

$$\frac{C}{C_0} = \exp\left[-a_1 I\left(\frac{y}{D},\frac{h}{D},M\right) \right] \tag{5.24}$$

where:

$$I\left(\frac{y}{D},\frac{h}{D},M\right) = \frac{\exp(M)}{\xi_{max}} \int_0^y \left[\frac{\dfrac{\tau}{\tau_0}\left(1+\left(\exp(M)-1\right)\dfrac{\xi}{\xi_{max}}\right)}{\dfrac{\xi}{y}\left(1-\dfrac{y}{D-h}\right)} \right]^{-1} y \tag{5.25}$$

The sediment concentration distribution in Eq. (5.24) was given by Chiu et al. (2000) in terms of an integral in Eq. (5.25). Our objective is to provide explicit analytical solutions to the models in Eqs. (5.15) and (5.23) using the homotopy-based methods, which is done in the next section. Before that, one can see that the assessment of the sediment concentration profiles requires the values of ω_s, $\bar{u}$, u_*, and β, other than the entropy parameter. The settling velocity ω_s can be calculated from a well-known formula given by Cheng (1997):

$$\omega_s = \frac{\nu}{d}\left(\sqrt{25+1.2d_*^2}-5\right)^{1.5} \tag{5.26}$$

where ν is the kinematic viscosity of fluid, d is the particle diameter, and d_* is the dimensionless particle diameter given by:

$$d_* = \left[\frac{(\rho_s-\rho)g}{\rho\nu^2}\right]^{1/3}d \tag{5.27}$$

in which ρ_s is the density of the sediment. The ratio of sediment to the turbulent diffusion coefficient, β, is often considered unity in the literature (van Rijn 1984). However, for accurate modeling of sediment concentration distribution, it needs to be estimated correctly, as it deviates from unity. Chiu et al. (2000) proposed an approach to obtain a better estimate of the parameter using a probabilistic approach, which provided a relationship between several physical factors and parameters. Details can be found in their work.

5.3 HAM-BASED ANALYTICAL SOLUTIONS

5.3.1 HAM SOLUTION FOR SEDIMENT CONCENTRATION MODEL I

For convenience, we first normalize Eq. (5.15) as follows:

$$\hat{C}=\frac{C}{C_0}, \; \hat{y}=\frac{y}{D} \tag{5.28}$$

Using Eq. (5.28), we can rearrange Eq. (5.15) as follows:

$$A_1\left(1-\hat{y}\right)\left[1+\left(\exp(M)-1\right)\hat{y}\right]\frac{d\hat{C}}{d\hat{y}}+\hat{C}=0 \text{ subject to } \hat{C}\left(\hat{y}=0\right)=1 \tag{5.29}$$

where $A_1 = \dfrac{\beta M u_*^2}{\omega_s u_{max}\left(\exp(M)-1\right)}$. Now, to apply the homotopy analysis method (HAM), the zeroth-order deformation equation is constructed as follows:

$$\left(1-q\right)\mathcal{L}\left[\Phi\left(\hat{y};q\right)-\hat{C}_0\left(\hat{y}\right)\right]=q\hbar H\left(\hat{y}\right)\mathcal{N}\left[\Phi\left(\hat{y};q\right)\right] \tag{5.30}$$

subject to the initial condition:

$$\Phi(0;q) = 1 \tag{5.31}$$

where q is the embedding parameter, $\Phi(\hat{y};q)$ is the representation of the solution across q, $\hat{C}_0(\hat{y})$ is the initial approximation, $\hbar$ is the auxiliary parameter, $H(\hat{y})$ is the auxiliary function, and $\mathcal{L}$ and $\mathcal{N}$ are the linear and nonlinear operators, respectively. The higher-order terms can be calculated from the higher-order deformation equation, given as follows:

$$\mathcal{L}\left[\hat{C}_m(\hat{y}) - \chi_m \hat{C}_{m-1}(\hat{y})\right] = \hbar H(\hat{y}) R_m\left(\vec{\hat{C}}_{m-1}\right),$$

$$m = 1,2,3,\ldots \text{ subject to } \hat{C}_m(0) = 0 \tag{5.32}$$

where:

$$\chi_m = \begin{cases} 0 & \text{when } m = 1, \\ 1 & \text{otherwise} \end{cases} \tag{5.33}$$

and:

$$R_m\left(\vec{\hat{C}}_{m-1}\right) = \frac{1}{(m-1)!} \frac{\partial^{m-1} \mathcal{N}\left[\Phi(\hat{y};q)\right]}{\partial q^{m-1}}\Bigg|_{q=0} \tag{5.34}$$

where $\hat{C}_m$ for $m \geq 1$ are the higher-order terms. Now, the final solution can be obtained as follows:

$$\hat{C}(\hat{y}) = \hat{C}_0(\hat{y}) + \sum_{m=1}^{\infty} \hat{C}_m(\hat{y}) \tag{5.35}$$

The nonlinear operator for the problem is selected as:

$$\mathcal{N}\left[\Phi(\hat{y};q)\right] = A_1(1-\hat{y})\left[1+(\exp(M)-1)\hat{y}\right]\frac{\partial \Phi(\hat{y};q)}{\partial \hat{y}} + \Phi(\hat{y};q) \tag{5.36}$$

Therefore, the term R_m can be calculated from Eq. (5.34) as:

$$R_m\left(\vec{\hat{C}}_{m-1}\right) = A_1(1-\hat{y})\left[1+(\exp(M)-1)\hat{y}\right]\frac{d\hat{C}_{m-1}}{d\hat{y}} + \hat{C}_{m-1} \tag{5.37}$$

We consider a polynomial base function to represent the solution of the present problem, which is given as:

$$\left\{\hat{y}^n \mid n = 0,12,\,3,\ldots\right\} \tag{5.38}$$

so that:

$$\hat{C}(\hat{y}) = \alpha_0 + \sum_{n=1}^{\infty} \alpha_n (\hat{y})^n \tag{5.39}$$

where α_0 and α_n are the coefficients of the series. Eq. (5.39) provides the so-called *rule of solution expression*. Following the rule of solution expression, the linear operator and the initial approximation are chosen, as follows:

$$\mathcal{L}\left[\Phi(\hat{y};q)\right] = \frac{\partial \Phi(\hat{y};q)}{\partial \hat{y}} \text{ with the property } \mathcal{L}[C_2] = 0 \tag{5.40}$$

$$\hat{C}_0(\hat{y}) = 1 \tag{5.41}$$

where C_2 is an integral constant. Using Eq. (5.40), the higher-order terms can be obtained from Eq. (5.32) as:

$$\hat{C}_m(\hat{y}) = \chi_m \hat{C}_{m-1}(\hat{y}) + \hbar \int_0^{\hat{y}} H(\hat{y}) R_m\left(\vec{C}_{m-1}\right) d\hat{y} + C_2, \ m = 1,2,3,\ldots \tag{5.42}$$

where R_m is given by Eq. (5.37), and constant C_2 can be determined from the initial condition for the higher-order deformation equations, i.e., $\hat{C}_m(\hat{y} = 0) = 0$, which simply yields $C_2 = 0$ for all $m \geq 1$.

Now, the auxiliary function $H(\hat{y})$ can be determined from the *rule of coefficient ergodicity*. Based on the rule of solution expression, the general form of $H(\hat{y})$ should be:

$$H(\hat{y}) = \hat{y}^{\alpha} \tag{5.43}$$

where α is an integer. Putting Eq. (5.43) into Eq. (5.42) and calculating some of $\hat{C}_m$, we have:

$$R_1\left(\hat{C}_0\right) = 1 \tag{5.44}$$

$$\hat{C}_1(\hat{y}) = \hbar \int_0^{\hat{y}} \hat{y}^{\alpha} \, d\hat{y} = \frac{\hbar}{\alpha+1} \hat{y}^{\alpha+1} \tag{5.45}$$

$$R_2\left(\vec{C}_1\right) = A_1 \hbar \hat{y}^{\alpha} (1-\hat{y})\left[1+(\exp(M)-1)\hat{y}\right] + \frac{\hbar}{\alpha+1} \hat{y}^{\alpha+1} \tag{5.46}$$

$$\hat{C}_2(\hat{y}) = \frac{\hbar}{\alpha+1} \hat{y}^{\alpha+1} + \hbar \int_0^{\hat{y}} \hat{y}^{\alpha} \left(A_1 \hbar \hat{y}^{\alpha} (1-\hat{y})\left[1+(\exp(M)-1)\hat{y}\right] + \frac{\hbar}{\alpha+1} \hat{y}^{\alpha+1} \right) d\hat{y} \tag{5.47}$$

Here, we can infer two observations. If we consider $\alpha \geq 1$, then the terms $\hat{y}^{\alpha}$ do not appear in the solution expression. This does not satisfy the *rule of coefficient ergodicity,* which tells us to include all the terms of the base function for the completeness of the solution. On the other hand, if we consider $\alpha \leq -2$, the terms appearing in the solution violate the *rule of solution expression.* Therefore, we are left with the only choices $\alpha = 0$ and $\alpha = -1$, out of which we choose $\alpha = 0$, i.e., $H(\hat{y}) = 1$. Finally, the approximate solution can be obtained as:

$$\hat{C}(\hat{y}) \approx \hat{C}_0(\hat{y}) + \sum_{m=1}^{M} \hat{C}_m(\hat{y}) \tag{5.48}$$

5.3.2 HAM SOLUTION FOR SEDIMENT CONCENTRATION MODEL II

Using the same normalization procedure introduced in Section 5.3.1, Eq. (5.23) can be rearranged as follows:

$$A_2 \left[1 + (\exp(M) - 1)\frac{\xi}{\xi_{max}} \right] \left[\hat{h} \left(1 - \frac{\hat{y}}{1 - \hat{h}} \right) + (1 - \hat{h}) \left(1 - \frac{\hat{y}}{1 - \hat{h}} \right)^2 \right] \frac{d\hat{C}}{d\hat{y}}$$

$$+ \frac{\xi}{D} \left(\frac{1}{\hat{y}} - \frac{1}{1 - \hat{h}} \right) \hat{C} = 0 \text{ subject to } \hat{C}(\hat{y} = 0) = 1 \tag{5.49}$$

where $A_2 = \dfrac{\beta \varphi M \xi_{max} u_*^2}{\omega_s \bar{u} D (\exp(M) - 1)}$. In a similar manner as with the case of sediment concentration model I, the zeroth-order and higher-order deformation equations can be constructed. Therefore, here we skip those steps. The nonlinear operator for the problem can be selected as:

$$N[\Phi(\hat{y}; q)] = A_2 \left[1 + (\exp(M) - 1)\frac{\xi}{\xi_{max}} \right] \left[\hat{h} \left(1 - \frac{\hat{y}}{1 - \hat{h}} \right) + (1 - \hat{h}) \left(1 - \frac{\hat{y}}{1 - \hat{h}} \right)^2 \right]$$

$$\times \frac{\partial \Phi(\hat{y}; q)}{\partial \hat{y}} + \frac{\xi}{D} \left(\frac{1}{\hat{y}} - \frac{1}{1 - \hat{h}} \right) \hat{C} \tag{5.50}$$

Therefore, term R_m can be calculated as:

$$R_m \left(\vec{\hat{C}}_{m-1} \right) = A_2 \left[1 + (\exp(M) - 1)\frac{\xi}{\xi_{max}} \right] \left[\hat{h} \left(1 - \frac{\hat{y}}{1 - \hat{h}} \right) + (1 - \hat{h}) \left(1 - \frac{\hat{y}}{1 - \hat{h}} \right)^2 \right]$$

$$\times \frac{d\hat{C}_{m-1}}{d\hat{y}} + \frac{\xi}{D} \left(\frac{1}{\hat{y}} - \frac{1}{1 - \hat{h}} \right) \hat{C} \tag{5.51}$$

The set of base functions is considered as follows:

$$\{ \hat{y}^n \mid n = 0, 1, 2 \dots \} \tag{5.52}$$

so that:

$$\hat{C}(\hat{y}) = a_0 + \sum_{n=1}^{\infty} a_n (\hat{y})^n \tag{5.53}$$

where a_0 and a_n are the coefficients of the series. Eq. (5.53) provides the so-called *rule of solution expression*. Following the description given in the previous section, the linear operator, initial approximation, and auxiliary function can be constructed the same in this case. Finally, the approximate solution can be obtained as:

$$\hat{C}(\hat{y}) \approx \hat{C}_0(\hat{y}) + \sum_{m=1}^{M} \hat{C}_m(\hat{y}) \tag{5.54}$$

5.4   HPM-BASED ANALYTICAL SOLUTIONS

5.4.1   HPM Solution for Sediment Concentration Model I

When applying the homotopy perturbation method (HPM), we rewrite Eq. (5.29) in the following form:

$$\mathcal{N}\left(\hat{C}\right) = f(\hat{y}) \tag{5.55}$$

We construct a homotopy $\Phi(\hat{y}; q)$ that satisfies:

$$(1-q)\left[\mathcal{L}\left(\Phi(\hat{y}; q)\right) - \mathcal{L}\left(\hat{C}_0(\hat{y})\right)\right] + q\left[\mathcal{N}\left(\Phi(\hat{y}; q)\right) - f(\hat{y})\right] = 0 \tag{5.56}$$

where q is the embedding parameter that lies on $[0,1]$, $\mathcal{L}$ is the linear operator, and $\hat{C}_0(\hat{y})$ is the initial approximation to the final solution. Eq. (5.56) shows that as $q = 0$, $\Phi(\hat{y}; 0) = \hat{C}_0(\hat{y})$ and as $q = 1$, $\Phi(\hat{y}; 1) = \hat{C}(\hat{y})$. Let us now express $\Phi(\hat{y}; q)$ as a series in terms of q, as follows:

$$\Phi(\hat{y}; q) = \Phi_0 + q\Phi_1 + q^2\Phi_2 + q^3\Phi_3 + q^4\Phi_4 \ldots \tag{5.57}$$

where Φ_m for $m \geq 1$ are the higher-order terms. As $q \to 1$, Eq. (5.57) produces the final solution as:

$$\hat{C}(\hat{y}) = \lim_{q \to 1} \Phi(\hat{y}; q) = \sum_{i=0}^{\infty} \Phi_i \tag{5.58}$$

For simplicity, we consider the linear and nonlinear operators as follows:

$$\mathcal{L}\left(\Phi(\hat{y}; q)\right) = \frac{\partial \Phi(\hat{y}; q)}{\partial \hat{y}} \tag{5.59}$$

$$\mathcal{N}\left(\Phi(\hat{y};q)\right) = A_1\left(1-\hat{y}\right)\left[1+\left(\exp(M)-1\right)\hat{y}\right]\frac{\partial\Phi(\hat{y};q)}{\partial\hat{y}} + \Phi(\hat{y};q) \qquad (5.60)$$

Using these expressions, Eq. (5.56) takes on the form:

$$(1-q)\left[\frac{\partial\Phi(\hat{y};q)}{\partial\hat{y}} - \frac{d\hat{C}_0}{d\hat{y}}\right]$$

$$+q\left[A_1\left(1-\hat{y}\right)\left[1+\left(\exp(M)-1\right)\hat{y}\right]\frac{\partial\Phi(\hat{y};q)}{\partial\hat{y}} + \Phi(\hat{y};q)\right] = 0 \qquad (5.61)$$

Using Eq. (5.57), we have:

$$(1-q)\left[\frac{\partial\Phi(\hat{y};q)}{\partial\hat{y}} - \frac{d\hat{C}_0}{d\hat{y}}\right]$$

$$= (1-q)\left[\left(\frac{d\Phi_0}{d\hat{y}} - \frac{d\hat{C}_0}{d\hat{y}}\right) + q\frac{d\Phi_1}{d\hat{y}} + q^2\frac{d\Phi_2}{d\hat{y}} + q^3\frac{d\Phi_3}{d\hat{y}} + q^4\frac{d\Phi_4}{d\hat{y}} + \cdots\right]$$

$$= \left(\frac{d\Phi_0}{d\hat{y}} - \frac{d\hat{C}_0}{d\hat{y}}\right) + q\left(\frac{d\Phi_1}{d\hat{y}} - \left(\frac{d\Phi_0}{d\hat{y}} - \frac{d\hat{C}_0}{d\hat{y}}\right)\right) + q^2\left(\frac{d\Phi_2}{d\hat{y}} - \frac{d\Phi_1}{d\hat{y}}\right)$$

$$+q^3\left(\frac{d\Phi_3}{d\hat{y}} - \frac{d\Phi_2}{d\hat{y}}\right) + q^4\left(\frac{d\Phi_4}{d\hat{y}} - \frac{d\Phi_3}{d\hat{y}}\right) + \cdots \qquad (5.62)$$

$$A_1\left(1-\hat{y}\right)\left[1+\left(\exp(M)-1\right)\hat{y}\right]\frac{\partial\Phi(\hat{y};q)}{\partial\hat{y}} + \Phi(\hat{y};q) = A_1\left(1-\hat{y}\right)$$

$$\times\left[1+\left(\exp(M)-1\right)\hat{y}\right]\left(\frac{d\Phi_0}{d\hat{y}} + q\frac{d\Phi_1}{d\hat{y}} + q^2\frac{d\Phi_2}{d\hat{y}} + q^3\frac{d\Phi_3}{d\hat{y}} + q^4\frac{d\Phi_4}{d\hat{y}} + \cdots\right)$$

$$+\left(\Phi_0 + q\Phi_1 + q^2\Phi_2 + q^3\Phi_3 + q^4\Phi_4 + \cdots\right)$$

$$= \left[A_1\left(1-\hat{y}\right)\left[1+\left(\exp(M)-1\right)\hat{y}\right]\frac{d\Phi_0}{d\hat{y}} + \Phi_0\right]$$

$$+q\left[A_1\left(1-\hat{y}\right)\left[1+\left(\exp(M)-1\right)\hat{y}\right]\frac{d\Phi_1}{d\hat{y}} + \Phi_1\right]$$

$$+q^2\left[A_1\left(1-\hat{y}\right)\left[1+\left(\exp(M)-1\right)\hat{y}\right]\frac{d\Phi_2}{d\hat{y}} + \Phi_2\right]$$

$$+q^3\left[A_1\left(1-\hat{y}\right)\left[1+\left(\exp(M)-1\right)\hat{y}\right]\frac{d\Phi_3}{d\hat{y}} + \Phi_3\right] + \cdots \qquad (5.63)$$

Using the boundary condition $\hat{C}(\hat{y}=0)=1$, we have:

$$\Phi_0(\hat{y}=0)=1,\ \Phi_1(\hat{y}=0)=0,\ \Phi_2(\hat{y}=0)=0,\ \dots. \tag{5.64}$$

Using Eqs. (5.62) and (5.63) and equating the like powers of q in Eq. (5.61), the following system of differential equations can be obtained:

$$\frac{d\Phi_0}{d\hat{y}}-\frac{d\hat{C}_0}{d\hat{y}}=0 \text{ subject to } \Phi_0(\hat{y}=0)=1 \tag{5.65}$$

$$\frac{d\Phi_1}{d\hat{y}}-\left(\frac{d\Phi_0}{d\hat{y}}-\frac{d\hat{C}_0}{d\hat{y}}\right)+A_1(1-\hat{y})\left[1+(\exp(M)-1)\hat{y}\right]\frac{d\Phi_0}{d\hat{y}}+\Phi_0=0$$

subject to $\Phi_1(\hat{y}=0)=0$ \hfill (5.66)

$$\frac{d\Phi_2}{d\hat{y}}-\frac{d\Phi_1}{d\hat{y}}+A_1(1-\hat{y})\left[1+(\exp(M)-1)\hat{y}\right]\frac{d\Phi_1}{d\hat{y}}+\Phi_1=0$$

subject to $\Phi_2(\hat{y}=0)=0$ \hfill (5.67)

$$\frac{d\Phi_3}{d\hat{y}}-\frac{d\Phi_2}{d\hat{y}}+A_1(1-\hat{y})\left[1+(\exp(M)-1)\hat{y}\right]\frac{d\Phi_2}{d\hat{y}}+\Phi_2=0$$

subject to $\Phi_3(\hat{y}=0)=0$ \hfill (5.68)

$$\frac{d\Phi_4}{d\hat{y}}-\frac{d\Phi_3}{d\hat{y}}+A_1(1-\hat{y})\left[1+(\exp(M)-1)\hat{y}\right]\frac{d\Phi_3}{d\hat{y}}+\Phi_3=0$$

subject to $\Phi_4(\hat{y}=0)=0$ \hfill (5.69)

Proceeding in a like manner, one can derive the following recurrence relation:

$$\frac{d\Phi_m}{d\hat{y}}=\left[1-A_1(1-\hat{y})\left[1+(\exp(M)-1)\hat{y}\right]\right]\frac{d\Phi_{m-1}}{d\hat{y}}-\Phi_{m-1}$$

subject to $\Phi_m(\hat{y}=0)=0$ for $m\geq2$ \hfill (5.70)

The initial approximation can be chosen as $\Phi_0=\dfrac{1-\dfrac{y}{D}}{1+(\exp(M)-1)\dfrac{y}{D}}$. Using this initial approximation, we can solve the equations iteratively using symbolic software. We do not include those expressions here, as they are lengthy. Finally, the HPM-based solution can be approximated as:

$$\hat{C}(\hat{y})\approx\sum_{i=0}^{M}\Phi_i \tag{5.71}$$

5.5　OHAM-BASED ANALYTICAL SOLUTIONS

5.5.1　OHAM Solution for Sediment Concentration Model I

When applying the optimal homotopy asymptotic method (OHAM), Eq. (5.29) can be written in the following form:

$$\mathcal{L}\big[\hat{C}(\hat{y})\big]+\mathcal{N}\big[\hat{C}(\hat{y})\big]+h(\hat{y})=0 \tag{5.72}$$

We select $\mathcal{L}\big[\hat{C}(\hat{y})\big]=\dfrac{d\hat{C}}{d\hat{y}}$, $\mathcal{N}\big[\hat{C}(\hat{y})\big]=\big\{A_1(1-\hat{y})\big[1+(\exp(M)-1)\hat{y}\big]-1\big\}\dfrac{d\hat{C}}{d\hat{y}}+\hat{C}$, and $h(\hat{y})=0$. With these considerations, the zeroth-order problem becomes:

$$\mathcal{L}\big(\hat{C}_0(\hat{y})\big)+h(\hat{y})=0 \text{ subject to } \hat{C}_0(\hat{y}=0)=1 \tag{5.73}$$

Solving Eq. (5.73), one obtains:

$$\hat{C}_0(\hat{y})=1 \tag{5.74}$$

To obtain $\mathcal{N}_0$, $\mathcal{N}_1$, $\mathcal{N}_2$, etc., the following can be used:

$$\big\{A_1(1-\hat{y})\big[1+(\exp(M)-1)\hat{y}\big]-1\big\}\frac{d\hat{C}}{d\hat{y}}+\hat{C}=\big\{A_1(1-\hat{y})\big[1+(\exp(M)-1)\hat{y}\big]-1\big\}$$

$$\times\left(\frac{d\hat{C}_0}{d\hat{y}}+q\frac{d\hat{C}_1}{d\hat{y}}+q^2\frac{d\hat{C}_2}{d\hat{y}}+q^3\frac{d\hat{C}_3}{d\hat{y}}+q^4\frac{d\hat{C}_4}{d\hat{y}}+\cdots\right)$$

$$+\big(\hat{C}_0+q\hat{C}_1+q^2\hat{C}_2+q^3\hat{C}_3+q^4\hat{C}_4+\cdots\big)=\left[\big\{A_1(1-\hat{y})\big[1+(\exp(M)-1)\hat{y}\big]-1\big\}\frac{d\hat{C}_0}{d\hat{y}}+\hat{C}_0\right]$$

$$+q\left[\big\{A_1(1-\hat{y})\big[1+(\exp(M)-1)\hat{y}\big]-1\big\}\frac{d\hat{C}_1}{d\hat{y}}+\hat{C}_1\right]$$

$$+q^2\left[\big\{A_1(1-\hat{y})\big[1+(\exp(M)-1)\hat{y}\big]-1\big\}\frac{d\hat{C}_2}{d\hat{y}}+\hat{C}_2\right]$$

$$+q^3\left[\big\{A_1(1-\hat{y})\big[1+(\exp(M)-1)\hat{y}\big]-1\big\}\frac{d\hat{C}_3}{d\hat{y}}+\hat{C}_3\right]$$

$$+q^4\left[\big\{A_1(1-\hat{y})\big[1+(\exp(M)-1)\hat{y}\big]-1\big\}\frac{d\hat{C}_4}{d\hat{y}}+\hat{C}_4\right]+\cdots \tag{5.75}$$

Using Eq. (5.75), the first-order problem reduces to:

$$\frac{d\hat{C}_1}{d\hat{y}}=H_1(\hat{y},C_i)\left\{\big\{A_1(1-\hat{y})\big[1+(\exp(M)-1)\hat{y}\big]-1\big\}\frac{d\hat{C}_0}{d\hat{y}}+\hat{C}_0\right\}$$

$$\text{subject to } \hat{C}_1(\hat{y}=0)=0 \tag{5.76}$$

The auxiliary functions can be chosen in many ways. Here, we select $H_1(\hat{y}, C_i) = C_1 + C_2\hat{y}$. Putting $k = 2$, the second-order problem becomes:

$$\mathcal{L}\left[\hat{C}_2(\hat{y}, C_i) - \hat{C}_1(\hat{y}, C_i)\right] = H_2(\hat{y}, C_i)\mathcal{N}_0\left(\hat{C}_0(\hat{y})\right) + H_1(\hat{y}, C_i)$$

$$\times\left[\mathcal{L}\left[\hat{C}_1(\hat{y}, C_i)\right] + \mathcal{N}_1\left[\hat{C}_0(\hat{y}), \hat{C}_1(\hat{y}, C_i)\right]\right] \text{ subject to } \hat{C}_2(\hat{y} = 0) = 0 \quad (5.77)$$

Further, we choose $H_2(\hat{y}, C_i) = H_1 + C_3\hat{y}^2$. Therefore, Eq. (5.77) becomes:

$$\frac{d\hat{C}_2}{d\hat{y}} = \frac{d\hat{C}_1}{d\hat{y}} + \left(C_1 + C_2\hat{y} + C_3\hat{y}^2\right)\left\{\left\{A_1\left(1 - \hat{y}\right)\left[1 + \left(\exp(M) - 1\right)\hat{y}\right] - 1\right\}\frac{d\hat{C}_0}{d\hat{y}} + \hat{C}_0\right\}$$

$$+ \left(C_1 + C_2\hat{y}\right)\left[\frac{d\hat{C}_1}{d\hat{y}} + \left\{A_1\left(1 - \hat{y}\right)\left[1 + \left(\exp(M) - 1\right)\hat{y}\right] - 1\right\}\frac{d\hat{C}_1}{d\hat{y}} + \hat{C}_1\right]$$

$$\text{subject to } \hat{C}_2(\hat{y} = 0) = 0 \quad\quad\quad (5.78)$$

The terms of the series can be computed using the equations developed here. One can compute these terms without any difficulty using symbolic computation software, such as MATLAB. Further, following Chapter 2, the higher-order terms can be computed in a similar manner. However, our aim is to produce an accurate solution with just two to three terms of the OHAM-based series. Therefore, we restrict our calculation up to $k = 2$. Finally, the approximate solution can be found as:

$$\hat{C}(\hat{y}) \approx \hat{C}_0(\hat{y}) + \hat{C}_1(\hat{y}, C_i) + \hat{C}_2(\hat{y}, C_i) \quad\quad (5.79)$$

where the terms are given by Eqs. (5.74), (5.76), and (5.78). It may be noted that we can derive the HPM- and OHAM-based solutions for the sediment concentration model II in a similar manner. However, here we avoid the case of model II, as the objective is to show how methodologies work for a problem.

5.6 CONVERGENCE THEOREMS

The convergence of the HAM-based and OHAM-based solutions is proved theoretically using the following theorems.

5.6.1 CONVERGENCE THEOREM OF HAM-BASED SOLUTION

The convergence theorems for the HAM-based solutions given by Eqs. (5.48) and (5.54) can be proved using the following theorems.

Theorem 5.1: *If the homotopy series $\sum_{m=0}^{\infty} \hat{C}_m(\hat{y})$ and $\sum_{m=0}^{\infty} \hat{C}_m{}'(\hat{y})$ converge, then $R_m\left(\vec{\hat{C}}_{m-1}\right)$ given by Eqs. (5.37) and (5.51) satisfies the relation $\sum_{m=1}^{\infty} R_m\left(\vec{\hat{C}}_{m-1}\right) = 0$. Here, ''' denotes the derivative with respect to $\hat{y}$.*

Proof: The proof of this theorem follows exactly from the previous chapters.

Theorem 5.2: *If ℏ is so properly chosen that the series* $\Sigma_{m=0}^{\infty}\hat{C}_m(\hat{y})$ *and* $\Sigma_{m=0}^{\infty}\hat{C}_m{}'(\hat{y})$
converge absolutely to $\hat{C}(\hat{y})$ *and* $\hat{C}'(\hat{y})$, *respectively, then the homotopy series*
$\Sigma_{m=0}^{\infty}\hat{C}_m(\hat{y})$ *satisfies the original governing Eq. (5.29).*

Proof: Theorem 5.1 shows that if $\Sigma_{m=0}^{\infty}\hat{C}_m(\hat{y})$ and $\Sigma_{m=0}^{\infty}\hat{C}_m{}'(\hat{y})$ converge, then
$\Sigma_{m=1}^{\infty}R_m\left(\vec{\hat{C}}_{m-1}\right)=0$.

Therefore, substituting the previous expressions in Eq. (5.37) and simplifying
further lead to:

$$A_1\left(1-\hat{y}\right)\left[1+\left(\exp(M)-1\right)\hat{y}\right]\sum_{k=0}^{\infty}\hat{C}_m{}'+\sum_{k=0}^{\infty}\hat{C}_m=0 \qquad (5.80)$$

which is basically the original governing equation in Eq. (5.29). Furthermore, subject
to the initial condition $\hat{C}_0\left(\hat{y}=0\right)=1$ and the conditions for the higher-order deforma-
tion equation $\hat{C}_m\left(\hat{y}=0\right)=0$ for $m\geq 1$, we easily obtain $\Sigma_{m=0}^{\infty}\hat{C}_m\left(\hat{y}=0\right)=1$. Hence,
the convergence result follows.

Theorem 5.3: *If ℏ is so properly chosen that the series* $\Sigma_{m=0}^{\infty}\hat{C}_m(\hat{y})$ *and* $\Sigma_{m=0}^{\infty}\hat{C}_m{}'(\hat{y})$
converge absolutely to $\hat{C}(\hat{y})$ *<u>and</u>* $\hat{C}'(\hat{y})$, *respectively, then the homotopy series*
$\Sigma_{m=0}^{\infty}\hat{C}_m(\hat{y})$ *satisfies the original governing Eq. (5.49).*

Proof: The proof of this theorem follows from Theorem 5.2.

5.6.2   CONVERGENCE THEOREM OF THE **OHAM**-BASED SOLUTION

Theorem 5.4: *If the series* $\hat{C}_0(\hat{y})+\Sigma_{j=1}^{\infty}\hat{C}_j(\hat{y},C_i)$, $i=1,2,\ldots,s$, *converges, where*
$\hat{C}_j(\hat{y},C_i)$ *is governed by Eqs. (5.74), (5.76), and (5.78), then Eq. (5.79) is a solution*
of the original Eq. (5.29).

Proof: The proof of this theorem can be followed exactly from Theorem 2.3 of
Chapter 2.

5.7   RESULTS AND DISCUSSION

First, the numerical convergence of the HAM-based analytical solutions is established for a specific test case. Then, the solution is validated over the existing solutions proposed by Chiu et al. (2000). The derived analytical solution is tested under
different physical conditions to check the efficiency of the method. Finally, the
HPM- and OHAM-based analytical solutions are validated by comparing them with
the solution given by Chiu et al. (2000).

5.7.1   NUMERICAL CONVERGENCE AND VALIDATION
OF THE HAM-BASED SOLUTION

The convergence of the HAM-based series solution depends on a suitable choice for the auxiliary parameter $\hbar$. For that purpose, the squared residual error with regard to Eq. (5.29) or Eq. (5.49) can be calculated as follows:

$$\Delta_m = \int_{\hat{y}\in\Omega} \left(\mathcal{N}\left[\hat{C}(\hat{y})\right]\right)^2 d\hat{y} \tag{5.81}$$

where $\Omega = [0,1]$ is the domain of the problem. The HAM-based series solution leads to the exact solution of the problem when Δ_m approaches zero. Therefore, for a particular order of approximation m, it is sufficient to minimize the corresponding Δ_m, which then yields an optimal value of parameter $\hbar$. Here, we consider two test cases: one is for sediment concentration model I, and the other is for model II. For model I, the parameter value A_1 is taken as 0.0011. It may be noted that the chosen value is feasible and is taken from a test case of Chiu et al. (2000). With this value, we perform a test case, where the HAM-based solution is found for model I. First, the squared residual errors are calculated from Eq. (5.77), and then the corresponding approximate solutions are obtained. Figure 5.2 shows the squared residual errors for a different order of approximation for the selected case. It can be seen from the figure that as the order of approximation increases, the corresponding residual error decreases. Thus, the adequacy of selecting the linear operator, initial approximation,

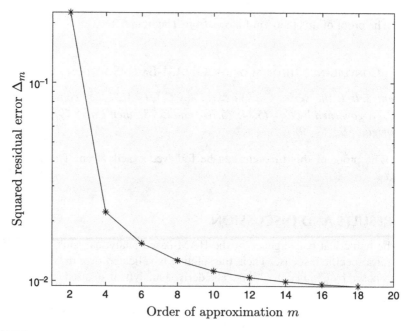

FIGURE 5.2   Squared residual error (Δ_m) versus a different order of approximations (m) of the HAM-based solution for the selected case (model I).

TABLE 5.1

Squared Residual Error (Δ_m) and Computational Time versus Different Orders of Approximations (m) for the Selected Case (Model I)

Order of Approximation (m)	Squared Residual Error (Δ_m)	Computational Time (sec)
2	0.22592	0.329
4	0.02222	0.658
6	0.01549	1.005
8	0.01268	1.390
10	0.01123	1.776
12	0.01042	2.175
14	0.00994	2.577
16	0.00963	3.008
18	0.00943	3.470

and the auxiliary function is established. Also, for a quantitative assessment, in Table 5.1, we provide the computational time taken by the computer to produce the corresponding order of approximations. Similarly, for model II, a test case is considered where the parameter value A_2 is taken as 0.0087. The squared residual errors for a different order of approximation for this case are depicted in Figure 5.3. The

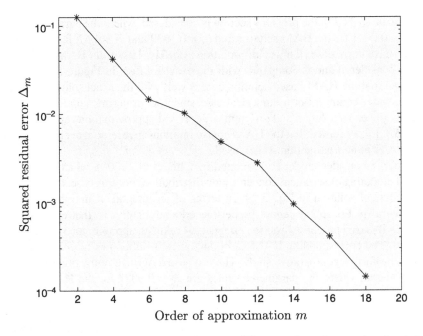

FIGURE 5.3 Squared residual error (Δ_m) versus different orders of approximation (m) of the HAM-based solution for the selected case (model II).

TABLE 5.2

Squared Residual Error (Δ_m) and Computational Time versus Different Orders of Approximation (m) for the Selected Case (Model II)

Order of Approximation (m)	Squared Residual Error (Δ_m)	Computational Time (sec)
2	0.1258	0.458
4	0.04222	0.746
6	0.01485	1.114
8	0.01022	1.253
10	0.00485	1.772
12	0.00278	2.014
14	0.00094	2.225
16	0.00041	2.681
18	0.00014	2.927

quantitative assessment is provided in Table 5.2. It is observed from the figure that as the order of approximation increases, the corresponding residual error decreases. All the computations reported here are performed using the BVPh 2.0 package developed by Zhao and Liao (2002).

The model I developed by Chiu et al. (2000) is considered here, who provided the exact solution in Eq. (5.16) for this case. The corresponding HAM-based analytical solution in Eq. (5.48) is considered to report a comparative study. For that purpose, a test case described in the previous section is considered, where the parameter values are $A_1 = 0.0011$ for the HAM solution and $a_2 = 0.1639$ and $a_3 = 4.57$. It may be noted that we have normalized Chiu et al.'s solution also. The 18th-order HAM-based solution is considered and is compared with the exact solution in Figure 5.4. It can be observed that the HAM-based solution agrees well with the exact solution throughout the water column. For a numerical assessment, considering some points in the domain, the exact solution and different-order HAM approximations are reported in Table 5.3. It is observed that the HAM approximations give a more accurate solution if the order of approximation is increased.

Now, we consider model II developed by Chiu et al. (2000), which is based on refined modeling of sediment concentration distribution. For this case, they provided the analytical solution in Eq. (5.24) in terms of an integral. It may be noted that the integral in Eq. (5.25) cannot be performed analytically. To that end, we use a MATLAB-based numerical routine *integral* to obtain an approximate solution to the integral. The corresponding HAM-based analytical solution in Eq. (5.54) is considered to perform a comparative study. The test case described in the previous section is considered, where the parameter values are $A_2 = 0.0087$ for the HAM solution, and $a_1 = 2.47$ and $\hat{h} = 0.2023$. We consider the 18th-order HAM-based solution and compare it with the analytical solution of Chiu et al. (2000) in Figure 5.5. It can be observed that the HAM-based solution agrees well with the exact solution throughout the water column. The numerical assessment is provided in Table 5.4. It is observed

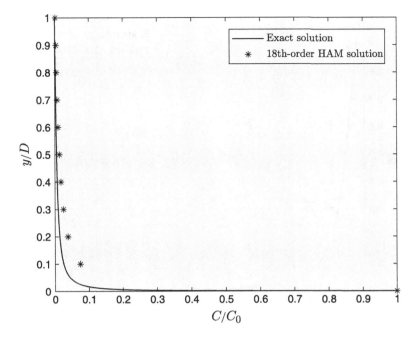

FIGURE 5.4   Comparison between the exact solution and 18th-order HAM-based solution for the selected case (model I) [$A_1 = 0.0011$, $a_2 = 0.1639$, and $a_3 = 4.57$].

TABLE 5.3

Comparison between HAM-Based Approximation and Exact Solution for the Selected Case (Model I)

$\hat{y}$	Exact Solution	HAM-Based Approximation		
		5th Order	10th Order	18th Order
0.0	1.0000	1.0000	1.0000	1.0000
0.1	0.0255	0.2677	0.1233	0.0742
0.2	0.0139	0.0832	0.0519	0.0378
0.3	0.0093	0.0426	0.0333	0.0248
0.4	0.0067	0.0303	0.0234	0.0176
0.5	0.0049	0.0224	0.0170	0.0128
0.6	0.0036	0.0159	0.0123	0.0094
0.7	0.0026	0.0106	0.0087	0.0066
0.8	0.0017	0.0062	0.0056	0.0044
0.9	0.0010	0.0022	0.0029	0.0037
1.0	0.0000	0.0031	0.0008	0.0001

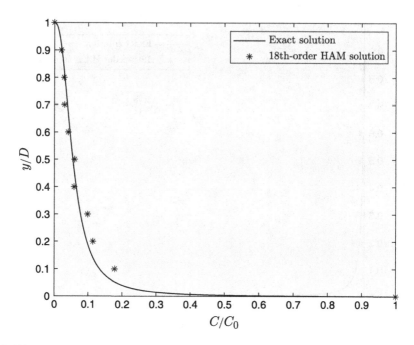

FIGURE 5.5 Comparison between the exact solution and 18th order HAM-based solution for the selected case (model II) [$A_2 = 0.0087$, $a_1 = 2.47$, and $\hat{h} = 0.2023$].

TABLE 5.4

Comparison between HAM-Based Approximation and Exact Solution for the Selected Case (Model II)

$\hat{y}$	Exact Solution	18th-Order HAM-Based Approximation
0.0	1.0000	1.0000
0.1	0.1343	0.1782
0.2	0.0965	0.1145
0.3	0.0774	0.0985
0.4	0.0646	0.0584
0.5	0.0547	0.0598
0.6	0.0463	0.0412
0.7	0.0387	0.0301
0.8	0.0310	0.0288
0.9	0.0222	0.0205
1.0	0.0000	0.0000

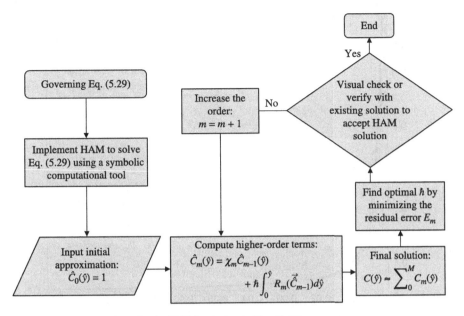

FIGURE 5.6   Flowchart for the HAM solution in Eq. (5.48).

that the 18th-order HAM approximation gives an accurate solution for the test case considered. A flowchart diagram containing the steps of the method for model I is given in Figure 5.6. In a similar way, the same can be given for model II also.

5.7.2   VALIDATION OF THE HPM-BASED SOLUTION

The HPM-based analytical solutions for the selected test case are validated over the analytical solution given by Chiu et al. (2000). First, we consider the sediment concentration model I. The two-, three-, and four-term HPM solutions are considered. It is seen that the HPM produces accurate results over the entire flow depth. Figure 5.7 compares the exact solution and the HPM approximations. To get a clear view of the different orders of approximations, the figure shows a restricted region of the domain, namely $C/C_0 \in [0,0.1]$. The HPM-based values and the existing exact solution are compared numerically in Table 5.5. A flowchart diagram containing the steps of the method is provided in Figure 5.8.

5.7.3   VALIDATION OF THE OHAM-BASED SOLUTION

For the assessment of the OHAM-based analytical solution, one needs to calculate the constants C_i. For that purpose, we calculate the residual as follows:

$$R\left(\hat{y},C_i\right)=\mathcal{L}\left[\hat{C}_{OHAM}\left(\hat{y},C_i\right)\right]+\mathcal{N}\left[\hat{C}_{OHAM}\left(\hat{y},C_i\right)\right]+h\left(\hat{y}\right), i=1,2,\dots,s \quad (5.82)$$

where $\hat{C}_{OHAM}\left(\hat{y},C_i\right)$ is the approximate solution. When $R\left(\hat{y},C_i\right)=0$, $\hat{C}_{OHAM}\left(\hat{y},C_i\right)$ becomes the exact solution to the problem. One of the ways to obtain the optimal

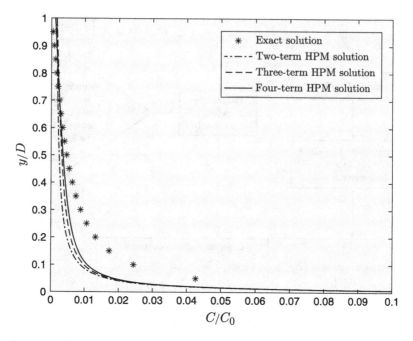

FIGURE 5.7 Comparison between the exact solution and two-, three-, and four-term HPM-based solutions for the selected case (model I) [$A_1 = 0.0011$, $a_2 = 0.1639$, and $a_3 = 4.57$].

TABLE 5.5

Comparison between HPM-Based Approximations and Exact Solution for the Selected Case (Model I)

$\hat{y}$	Exact Solution	HAM-Based Approximation		
		Two-Term Solution	Three-Term Solution	Four-Term Solution
0.0	1.0000	1.0000	1.0000	1.0000
0.1	0.0255	0.0087	0.0098	0.0108
0.2	0.0139	0.0047	0.0059	0.0067
0.3	0.0093	0.0034	0.0045	0.0052
0.4	0.0067	0.0028	0.0037	0.0042
0.5	0.0049	0.0024	0.0032	0.0036
0.6	0.0036	0.0022	0.0028	0.0030
0.7	0.0026	0.0020	0.0025	0.0025
0.8	0.0017	0.0018	0.0022	0.0020
0.9	0.0010	0.0017	0.0019	0.0016
1.0	0.0000	0.0016	0.0016	0.0012

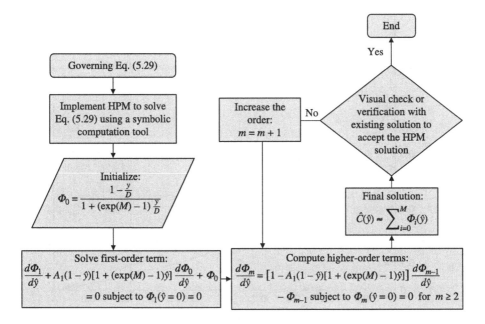

FIGURE 5.8   Flowchart for the HPM solution in Eq. (5.71).

C_i, for which the solution converges, is the minimization of squared residual error, i.e.,

$$J(C_i) = \int_{\hat{y} \in D} R^2(\hat{y}, C_i) d\hat{y}, \ i = 1, 2, \ldots, s \qquad (5.83)$$

where $D = [0,1]$ is the domain of the problem. The minimization of Eq. (5.83) leads to a system of algebraic equations as follows:

$$\frac{\partial J}{\partial C_1} = \frac{\partial J}{\partial C_2} = \cdots = \frac{\partial J}{\partial C_s} = 0 \qquad (5.84)$$

Solving this system, one can obtain the optimal values of the parameters. We obtain the optimal values using the MATLAB routine *fminsearch*, which minimizes an unconstrained multivariable function. Two- and three-term OHAM solutions are computed and compared in Figure 5.9 with the existing analytical solution of Chiu et al. (2000) to evaluate the effectiveness of the HAM approach. It can be seen that both of the OHAM-based approximations do not agree well with the exact solution. However, as one has great freedom in choosing the operators and the auxiliary functions, one may try other ways to obtain a more accurate solution. Also, one might expect a more accurate solution if the higher-order approximations are considered. For a quantitative assessment, we also compare the numerical values of the solutions in Table 5.6. A flowchart diagram containing the steps of the method is given in Figure 5.10.

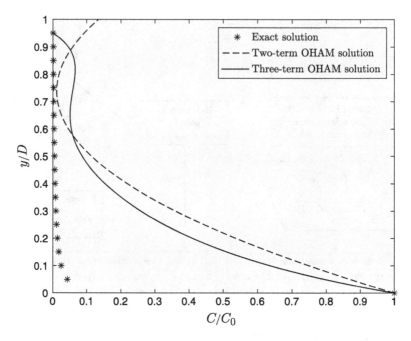

FIGURE 5.9 Comparison between the exact solution and two-, and three-term OHAM-based solutions for the selected case (model I) [$A_1 = 0.0011$, $a_2 = 0.1639$, and $a_3 = 4.57$].

TABLE 5.6

Comparison between OHAM-Based Approximations and the Exact Solution for the Selected Case (Model I)

$\hat{y}$	Exact Solution	OHAM-Based Approximation	
		Two-Term Solution	Three-Term Solution
0.0	1.0000	1.0000	1.0000
0.1	0.0255	0.7502	0.6369
0.2	0.0139	0.5368	0.4097
0.3	0.0093	0.3596	0.2584
0.4	0.0067	0.2187	0.1538
0.5	0.0049	0.1141	0.0862
0.6	0.0036	0.0457	0.0541
0.7	0.0026	0.0136	0.0533
0.8	0.0017	0.0179	0.0650
0.9	0.0010	0.0584	0.0451
1.0	0.0000	0.1351	-0.0870

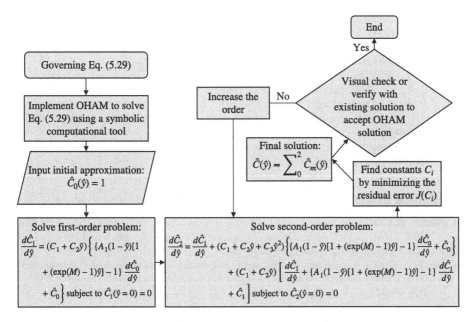

FIGURE 5.10 Flowchart for the OHAM solution in Eq. (5.79).

5.8 CONCLUDING REMARKS

We consider the sediment concentration models proposed by Chiu et al. (2000). The models have several advantages over the well-known Rouse equation, e.g., they can measure the concentration values starting from the channel bed to the water surface. Chiu et al. (2000) provided an exact analytical solution and an analytical solution in terms of an integral for models I and II, respectively. Here, we developed the HAM-based series solutions for both cases, where the solutions are provided explicitly in terms of a series. Different test cases are performed to check the stability of the series solutions, and a quantitative assessment is carried out whenever needed. It is seen that a polynomial set of base functions and a single-term linear operator can efficiently produce accurate solutions. On the other hand, the HPM- and OHAM-based analytical solutions are developed for the sediment concentration model I. It is seen that the HPM can produce a comparatively accurate solution, while the OHAM fails to perform efficiently. However, one has great freedom in choosing operators and the auxiliary functions; therefore, one may obtain more accurate solutions with other choices. Further, theoretical as well as numerical convergence of the series solutions is provided.

REFERENCES

Cheng, N. S. (1997). Simplified settling velocity formula for sediment particle. *Journal of Hydraulic Engineering*, *123*(2), 149–152.

Chiu, C. L. (1987). Entropy and probability concepts in hydraulics. *Journal of Hydraulic Engineering*, *113*(5), 583–599.

Chiu, C. L., and Hsu, S. M. (2006). Probabilistic approach to modeling of velocity distributions in fluid flows. *Journal of Hydrology, 316*(1–4), 28–42.

Chiu, C. L., Jin, W., and Chen, Y. C. (2000). Mathematical models of distribution of sediment concentration. *Journal of Hydraulic Engineering, 126*(1), 16–23.

Julien, P. Y. (2010). *Erosion and sedimentation.* Cambridge University Press, Cambridge, UK.

Rijn, L. C. V. (1984). Sediment transport, part II: Suspended load transport. *Journal of Hydraulic Engineering, 110*(11), 1613–1641.

Rouse, H. (1937). Modern conceptions of the mechanics of fluid turbulence. *Trans ASCE, 102*, 463–505.

Zhao, Y., and Liao, S. (2002). User Guide to BVPh 2.0. Available online: http://numericaltank. sjtu.edu.cn/BVPh.htm

FURTHER READING

Camenen, B. (2007). *A unified sediment transport formulation for coastal inlet application* (Vol. 7, No. 1). [US Army Corps of Engineers, Engineer Research and Development Center], Coastal and Hydraulics Laboratory.

Choo, T. H. (2000). An efficient method of the suspended sediment-discharge measurement using entropy concept. *Water Engineering Research: International Journal of KWRA, 1*(2), 95–105.

Christiansen, J. E. (1935). Distribution of silt in open channels. *Eos, Transactions American Geophysical Union, 16*(2), 478–485.

Coleman, N. L. (1986). Effects of suspended sediment on the open-channel velocity distribution. *Water Resources Research, 22*(10), 1377–1384.

Graf, W. H., and Cellino, M. (2002). Suspension flows in open channels; Experimental study. *Journal of Hydraulic Research, 40*(4), 435–447.

Jaynes, E. T. (1957a). Information theory and statistical mechanics. *Physical Review, 106*(4), 620.

Jaynes, E. T. (1957b). Information theory and statistical mechanics. II. *Physical Review, 108*(2), 171.

Kumbhakar, M., Ghoshal, K., and Singh, V. P. (2020). Two-dimensional distribution of streamwise velocity in open channel flow using maximum entropy principle: Incorporation of additional constraints based on conservation laws. *Computer Methods in Applied Mechanics and Engineering, 361*, 112738.

O'Brien, M. P. (1933). Review of the theory of turbulent flow and its relation to sediment-transportation. *Eos, Transactions American Geophysical Union, 14*(1), 487–491.

Shannon, C. E. (1948). A mathematical theory of communication. *The Bell System Technical Journal, 27*(3), 379–423.

Simons, D. B., and Şentürk, F. (1992). *Sediment transport technology: Water and sediment dynamics.* Water Resources Publication.

Vanoni, V. A. (Ed.). (2006, March). *Sedimentation engineering.* American Society of Civil Engineers.

Xingkui, W., and Ning, Q. (1989). Turbulence characteristics of sediment-laden flow. *Journal of Hydraulic Engineering, 115*(6), 781–800.

Yang, C. T. (1973). Incipient motion and sediment transport. *Journal of the Hydraulics Division, 99*(10), 1679–1704.

Zagustin, K. (1968). Sediment distribution in turbulent flow. *Journal of Hydraulic Research, 6*(2), 163–172.

6 Richards Equation under Gravity-Driven Infiltration and Constant Rainfall Intensity

6.1 INTRODUCTION

Infiltration is defined by the rate of water entry into the soil at the surface, and water moving down is termed percolation. The rate of infiltration is primarily influenced by antecedent soil moisture, soil texture and structure, and land use and land cover. As long as there is enough water, say rainfall intensity, to satisfy the soil infiltration capacity, more water or greater intensity does not increase the rate of infiltration. If the intensity of rainfall is less than the soil infiltration capacity rate, then the rate of infiltration is the same as rainfall intensity.

Infiltration is most accurately described by the Richards equation, which is obtained by coupling the Darcy-Buckingham flux law with the continuity equation. The flux law describes the flux as a function of the hydraulic head, which consists of the capillary head and gravity head. The proportionality is characterized by hydraulic conductivity, which is a function of soil moisture content, which in turn is a function of the capillary head.

The Richards equation is a nonlinear equation, and its analytical solution can be obtained only for very simplified cases. Baiamonte (2020) formulated the governing equations for computing infiltration, and three methods are described in this chapter to derive analytical solutions of these equations.

6.2 GOVERNING EQUATION AND HAM-BASED SOLUTION

Referring to Figure 6.1 of Baiamonte (2020), the continuity equation can be expressed as follows:

$$AZd\theta = (iA - Q)dt \tag{6.1}$$

where A is the cross-sectional area, Z is the unit height (1 m), θ is the actual volumetric soil water content, t is the time, i is the rainfall intensity, and Q is the outflow discharge at the bottom of the soil layer. Eq. (6.1) is based on mass conservation, i.e., the water volume exiting the jet in time dt is equal to the water volume removed from the tank at the same time dt during a rainfall event with constant intensity i.

DOI: 10.1201/9781003368984-9

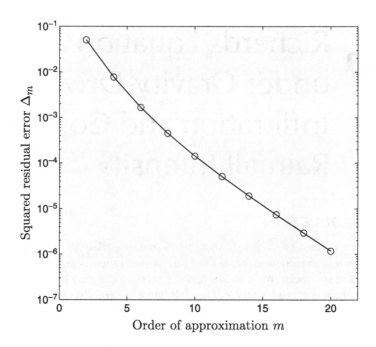

FIGURE 6.1 Squared residual error (Δ_m) versus different orders of approximation (m) of the HAM-based solution for the selected case.

6.2.1 TORRICELLI'S LAW

Following Toricelli's law, the soil infiltration rate can be obtained as follows:

$$f = \frac{Q}{A} = c_e \frac{A_0}{A} (2g\Theta)^{\frac{1}{2}} = c_e r_*^2 (2g\Theta)^{\frac{1}{2}} \cong K_s \Theta^{\frac{1}{2}} \tag{6.2}$$

where $\Theta = \theta/\theta_s$, θ_s is the saturated volumetric water content, g is the acceleration of gravity, A_0 is the cross-section of the hole, r_* is the ratio of the radius of the circular hole to that of the water tank, c_e is the contraction coefficient that is close to unity, and $K_s = r_*^2 (2g)^{\frac{1}{2}}$ is the saturated soil hydraulic conductivity.

The soil infiltration rate and rainfall intensity can be normalized as follows:

$$f_* = \frac{f}{K_s} = \cong \Theta^{\frac{1}{2}}, \rho = \frac{i}{K_s} \tag{6.3}$$

The time can be nondimensionalized using $\tau = \dfrac{t}{t_c}$, where $t_c = \dfrac{Z\theta_s}{K_s} = \dfrac{Z\theta_s}{r_*^2 (2g)^{\frac{1}{2}}}$. Using these nondimensional quantities, the governing equation can be obtained from Eq. (6.1) in dimensionless form as follows:

$$\frac{d\Theta}{d\tau} = \rho - \Theta^{\frac{1}{2}} \tag{6.4}$$

The initial condition can be given as $\Theta(\tau = \tau_0) = \Theta_0$. There can be two cases: $\rho = 0$ and $\rho \neq 0$.

Case I: $\rho = 0$ corresponds to the recession of the infiltration process, which starts at the end of the rainfall at time τ_r, i.e., $\tau_0 = \tau_r$ and $\Theta_0 = \Theta_r$, with Θ_r being the normalized water content at the end of the rainfall period. Using these quantities, Eq. (6.4) can be solved as follows:

$$\int_{\tau_r}^{\tau} d\tau = \int_{\Theta_r}^{\Theta} \frac{d\Theta}{-\Theta^{\frac{1}{2}}} \Rightarrow \tau = \tau_r + 2\left(\Theta_r^{\frac{1}{2}} - \Theta^{\frac{1}{2}}\right) \tag{6.5}$$

The time required for the zero-flux condition can be obtained by setting $\Theta = 0$ in Eq. (6.5), which is given as follows:

$$\tau_{zf} = \tau_r + 2\Theta_r^{\frac{1}{2}} \tag{6.6}$$

Case II: The case corresponding to $\rho \neq 0$ is solved using the homotopy analysis method (HAM). While using HAM, there is no need to distinguish the cases, as the problem can be solved within a generalized framework. To apply HAM, we rearrange Eq. (6.4) together with the initial condition as follows:

$$\left(\rho - \Theta^{\frac{1}{2}}\right)\frac{d\tau}{d\Theta} - 1 = 0 \text{ subject to } \tau(\Theta = \Theta_0) = \tau_0 \tag{6.7}$$

The zeroth-order deformation equation is constructed as follows:

$$(1-q)\mathcal{L}\left[\Phi(\Theta; q) - \tau_0(\Theta)\right] = q\hbar H(\Theta)\mathcal{N}\left[\Phi(\Theta; q)\right] \tag{6.8}$$

subject to the initial condition:

$$\Phi(\Theta_0; q) = \tau_0 \tag{6.9}$$

where q is the embedding parameter, $\Phi(\Theta; q)$ is the representation of the solution across q, $\tau_0(\Theta)$ is the initial approximation, $\hbar$ is the auxiliary parameter, $H(\Theta)$ is the auxiliary function, and $\mathcal{L}$ and $\mathcal{N}$ are the linear and nonlinear operators, respectively. The higher-order terms can be calculated from the higher-order deformation equation given as follows:

$$\mathcal{L}\left[\tau_m(\Theta) - \chi_m \tau_{m-1}(\Theta)\right] = \hbar H(\Theta) R_m(\vec{\tau}_{m-1}),$$
$$m = 1, 2, 3, \ldots \text{ subject to } \tau_m(\Theta_0) = 0 \tag{6.10}$$

where:

$$\chi_m = \begin{cases} 0 & \text{when } m = 1, \\ 1 & \text{otherwise} \end{cases} \tag{6.11}$$

and:

$$R_m\left(\vec{\tau}_{m-1}\right) = \frac{1}{(m-1)!}\frac{\partial^{m-1}\mathcal{N}\left[\Phi(\Theta;q)\right]}{\partial q^{m-1}}\Bigg|_{q=0} \tag{6.12}$$

where τ_m for $m \geq 1$ are the higher-order terms. Now, the final solution can be obtained as follows:

$$\tau(\Theta) = \tau_0(\Theta) + \sum_{m=1}^{\infty}\tau_m(\Theta) \tag{6.13}$$

The nonlinear operator for the problem is selected as:

$$\mathcal{N}\left[\Phi(\Theta;q)\right] = \left(\rho - \Theta^{\frac{1}{2}}\right)\frac{\partial\Phi(\Theta;q)}{\partial\Theta} - 1 \tag{6.14}$$

Therefore, term R_m can be calculated from Eq. (6.12) as follows:

$$R_m\left(\vec{\tau}_{m-1}\right) = \left(\rho - \Theta^{\frac{1}{2}}\right)\frac{d\tau_{m-1}}{d\Theta} - 1 \tag{6.15}$$

Based on the nature of the equation and the analytical solution given by Baiamonte (2020), the following set of base functions is considered to represent the solution:

$$\left\{\Theta^{\frac{n}{2}} \mid n = 2, 3, 4, 5, \ldots\right\} \tag{6.16}$$

so that:

$$\tau(\Theta) = a_0 + \sum_{n=2}^{\infty}a_n(\Theta)^{\frac{n}{2}} \tag{6.17}$$

where a_0 and a_n are the coefficients of the series. Eq. (6.17) provides the so-called *rule of solution expression*. Following the rule of solution expression, the linear operator and the initial approximation are chosen, respectively, as follows:

$$\mathcal{L}\left[\Phi(\Theta;q)\right] = \frac{\partial\Phi(\Theta;q)}{\partial\Theta} \text{ with the property } \mathcal{L}[C_2] = 0 \tag{6.18}$$

$$\tau_0(\Theta) = \tau_0 \tag{6.19}$$

where C_2 is an integral constant. Using Eq. (6.18), the higher-order terms can be obtained from Eq. (6.10) as follows:

$$\tau_m(\Theta) = \chi_m \tau_{m-1}(\Theta) + \hbar \int_{\Theta_0}^{\Theta} H(\Theta) R_m(\vec{\tau}_{m-1}) d\Theta + C_2, \quad m = 1, 2, 3, \ldots \quad (6.20)$$

where R_m is given by Eq. (6.15), and constant C_2 can be determined from the initial condition for the higher-order deformation equations, i.e., $\tau_m(\Theta_0) = 0$, which simply yields $C_2 = 0$ for all $m \geq 1$.

Now, the auxiliary function $H(\Theta)$ can be determined from the *rule of coefficient ergodicity*. Based on the rule of solution expression and Eq. (6.17), the general form of $H(\Theta)$ should be:

$$H(\Theta) = \Theta^{\frac{\alpha}{2}} \quad (6.21)$$

where α is an integer. Putting Eq. (6.21) into Eq. (6.20) and calculating some of τ_m, we have:

$$R_1(\tau_0) = -1 \quad (6.22)$$

$$\tau_1(\Theta) = -\hbar \int_{\Theta_0}^{\Theta} \Theta^{\frac{\alpha}{2}} d\Theta = -\frac{\hbar}{\frac{\alpha}{2} + 1}\left(\Theta^{\frac{\alpha}{2}} - \Theta_0^{\frac{\alpha}{2}}\right) \quad (6.23)$$

$$R_2(\vec{\tau}_1) = -\hbar\left(\rho - \Theta^{\frac{1}{2}}\right)\Theta^{\frac{\alpha}{2}} - 1 \quad (6.24)$$

$$\tau_2(\Theta) = \tau_1(\Theta) + \hbar \int_{\Theta_0}^{\Theta} \Theta^{\frac{\alpha}{2}}\left[-\hbar\left(\rho - \Theta^{\frac{1}{2}}\right)\Theta^{\frac{\alpha}{2}} - 1\right] d\Theta = -\frac{2\hbar}{\frac{\alpha}{2} + 1}\left(\Theta^{\frac{\alpha}{2}} - \Theta_0^{\frac{\alpha}{2}}\right)$$

$$-\frac{\rho\hbar^2}{\alpha + 1}\left(\Theta^{\alpha+1} - \Theta_0^{\alpha+1}\right) + \frac{\hbar^2}{\alpha + \frac{3}{2}}\left(\Theta^{\alpha+\frac{3}{2}} - \Theta_0^{\alpha+\frac{3}{2}}\right) \quad (6.25)$$

Here, we infer two observations. If we consider $\alpha \geq 1$, then some of the terms (such as Θ^{α}) do not appear in the solution expression. This does not satisfy the *rule of coefficient ergodicity*, which allows us to include all the terms of the base function for the completeness of the solution. On the other hand, if we consider $\alpha \leq -1$, the terms appearing in the solution violate the *rule of solution expression*. Therefore, we are left with the only choice $\alpha = 0$, which results in $H(\Theta) = 1$. Finally, the approximate solution can be obtained as follows:

$$\tau(\Theta) \approx \tau_0(\Theta) + \sum_{m=1}^{M} \tau_m(\Theta) \quad (6.26)$$

6.2.2 Brooks and Corey's Hydraulic Conductivity Function

In the previous section, we used Toricelli's law for the hydraulic conductivity function, which cannot describe conductivity in realistic situations. To that end, here we consider the soil hydraulic conductivity function proposed by Brooks and Corey (1964) given as follows:

$$f \cong K_s \Theta^{\frac{1}{c}} \tag{6.27}$$

where c is the pore connectivity index. It can be seen that for $c = 2$, Eq. (6.27) specializes into Toricelli's law. Following the same procedure described in the previous section, the time variable is nondimensionalized as $\tau = \dfrac{t}{t_c}$, where $t_c = \dfrac{Z\theta_s}{K_s} = \dfrac{Z\theta_s}{r_*^2 (2g)^{\frac{1}{c}}}$.

Using these quantities, the governing equation can be obtained from Eq. (6.1) in dimensionless form as follows:

$$\frac{d\Theta}{d\tau} = \rho - \Theta^{\frac{1}{c}} \tag{6.28}$$

The initial condition can be given as $\Theta(\tau = \tau_0) = \Theta_0$. There can be two cases: $\rho = 0$ and $\rho \neq 0$.

Case I: For $\rho = 0$, the same approach described in the previous section can be adopted. Eq. (6.28) can be solved as follows:

$$\int_{\tau_r}^{\tau} d\tau = \int_{\Theta_r}^{\Theta} \frac{d\Theta}{-\Theta^{\frac{1}{c}}} \Rightarrow \tau = \tau_r + \frac{c}{c-1}\left(\Theta_r^{\frac{1}{c}} - \Theta^{\frac{1}{c}}\right) \tag{6.29}$$

Case II: In a manner similar to the case of Toricelli's law, the case corresponding to $\rho \neq 0$ is solved using HAM. For that, first, we rearrange Eq. (6.28) together with the initial condition as follows:

$$\left(\rho - \Theta^{\frac{1}{c}}\right)\frac{d\tau}{d\Theta} - 1 = 0 \text{ subject to } \tau(\Theta = \Theta_0) = \tau_0 \tag{6.30}$$

Now, the zeroth-order and higher-order deformation equations can be constructed in a similar manner as done in the previous section. Therefore, here we skip those steps. The nonlinear operator for the equation is selected as:

$$\mathcal{N}\left[\Phi(\Theta; q)\right] = \left(\rho - \Theta^{\frac{1}{c}}\right)\frac{\partial\Phi(\Theta; q)}{\partial\Theta} - 1 \tag{6.31}$$

Therefore, term R_m can be calculated as follows:

$$R_m\left(\vec{\tau}_{m-1}\right) = \left(\rho - \Theta^{\frac{1}{c}}\right)\frac{d\tau_{m-1}}{d\Theta} - 1 \tag{6.32}$$

Following the case of $c = 2$, the set of base functions is considered as follows:

$$\left\{ \Theta^{\frac{n}{c}} \mid n = 2, 3, 4, 5, \ldots \right\} \tag{6.33}$$

so that:

$$\tau(\Theta) = a_0 + \sum_{n=2}^{\infty} a_n (\Theta)^{\frac{n}{c}} \tag{6.34}$$

where a_0 and a_n are the coefficients of the series. Eq. (6.34) provides the so-called *rule of solution expression*. Following the description given in the previous section, the linear operator, initial approximation, and auxiliary function can be constructed the same in this case. Finally, the approximate solution can be obtained as:

$$\tau(\Theta) \approx \tau_0(\Theta) + \sum_{m=1}^{M} \tau_m (\Theta) \tag{6.35}$$

6.3 HPM-BASED SOLUTION

It can be seen that Eq. (6.7) is a special case of Eq. (6.30) for $c = 2$. We rewrite Eq. (6.30) in the following form:

$$\mathcal{N}(\tau) = f(\Theta) \tag{6.36}$$

We construct a homotopy $\Phi(\Theta; q)$ that satisfies:

$$(1-q)\left[\mathcal{L}(\Phi(\Theta; q)) - \mathcal{L}(\tau_0(\Theta)) \right] + q\left[\mathcal{N}(\Phi(\Theta; q)) - f(\Theta) \right] = 0 \tag{6.37}$$

where q is the embedding parameter that lies in $[0,1]$, $\mathcal{L}$ is the linear operator, and $\tau_0(\Theta)$ is the initial approximation to the final solution. Eq. (6.37) shows that as $q = 0$, $\Phi(\Theta; 0) = \tau_0(\Theta)$ and as $q = 1$, $\Phi(\Theta; 1) = \tau(\Theta)$. We now express $\Phi(\Theta; q)$ as a series in terms of q as follows:

$$\Phi(\Theta; q) = \Phi_0 + q\Phi_1 + q^2\Phi_2 + q^3\Phi_3 + q^4\Phi_4 \ldots \tag{6.38}$$

where Φ_0 for $m \geq 1$ are the higher-order terms. As $q \to 1$, Eq. (6.38) produces the final solution as:

$$\tau(\Theta) = \lim_{q \to 1} \Phi(\Theta; q) = \sum_{i=0}^{\infty} \Phi_i \tag{6.39}$$

For simplicity, we consider the linear and operators as follows:

$$\mathcal{L}(\Phi(\Theta; q)) = \frac{\partial \Phi(\Theta; q)}{\partial \Theta} \tag{6.40}$$

$$\mathcal{N}\left(\Phi\left(\Theta;q\right)\right)=\left(\rho-\Theta^{\frac{1}{c}}\right)\frac{\partial\Phi\left(\Theta;q\right)}{\partial\Theta}-1 \qquad (6.41)$$

Using these expressions, Eq. (6.37) takes on the form:

$$\left(1-q\right)\left[\frac{\partial\Phi\left(\Theta;q\right)}{\partial\Theta}-\frac{d\tau_0}{d\Theta}\right]+q\left[\left(\rho-\Theta^{\frac{1}{c}}\right)\frac{\partial\Phi\left(\Theta;q\right)}{\partial\Theta}-1\right]=0 \qquad (6.42)$$

Using Eq. (6.38), we have:

$$\left(1-q\right)\left[\frac{\partial\Phi\left(\Theta;q\right)}{\partial\Theta}-\frac{d\tau_0}{d\Theta}\right]$$

$$=\left(1-q\right)\left[\left(\frac{d\Phi_0}{d\Theta}-\frac{d\tau_0}{d\tau}\right)+q\frac{d\Phi_1}{d\Theta}+q^2\frac{d\Phi_2}{d\Theta}+q^3\frac{d\Phi_3}{d\Theta}+q^4\frac{d\Phi_4}{d\Theta}+\cdots\right]$$

$$=\left(\frac{d\Phi_0}{d\Theta}-\frac{d\tau_0}{d\Theta}\right)+q\left(\frac{d\Phi_1}{d\Theta}-\left(\frac{d\Phi_0}{d\Theta}-\frac{d\tau_0}{d\Theta}\right)\right)+q^2\left(\frac{d\Phi_2}{d\Theta}-\frac{d\Phi_1}{d\Theta}\right)$$

$$+q^3\left(\frac{d\Phi_3}{d\Theta}-\frac{d\Phi_2}{d\Theta}\right)+q^4\left(\frac{d\Phi_4}{d\Theta}-\frac{d\Phi_3}{d\Theta}\right)+\cdots \qquad (6.43)$$

$$\left(\rho-\Theta^{\frac{1}{c}}\right)\frac{\partial\Phi\left(\Theta;q\right)}{\partial\Theta}-1$$

$$=\left(\rho-\Theta^{\frac{1}{c}}\right)\left(\frac{d\Phi_0}{d\Theta}+q\frac{d\Phi_1}{d\Theta}+q^2\frac{d\Phi_2}{d\Theta}+q^3\frac{d\Phi_3}{d\Theta}+q^4\frac{d\Phi_4}{d\Theta}+\cdots\right)-1$$

$$=\left[\left(\rho-\Theta^{\frac{1}{c}}\right)\frac{d\Phi_0}{d\Theta}-1\right]+q\left(\rho-\Theta^{\frac{1}{c}}\right)\frac{d\Phi_1}{d\Theta}+q^2\left(\rho-\Theta^{\frac{1}{c}}\right)\frac{d\Phi_2}{d\Theta}+q^3\left(\rho-\Theta^{\frac{1}{c}}\right)\frac{d\Phi_3}{d\Theta}+\cdots$$

$$\qquad (6.44)$$

Using the initial condition $\tau\left(\Theta=\Theta_0\right)=\tau_0$, we have:

$$\Phi_0\left(\Theta_0\right)=\tau_0,\ \Phi_1\left(\Theta_0\right)=0,\ \Phi_2\left(\Theta_0\right)=0,\ldots \qquad (6.45)$$

Using Eqs. (6.43) and (6.44) and equating the like powers of q in Eq. (6.42), the following system of differential equations is obtained:

$$\frac{d\Phi_0}{d\Theta}-\frac{d\tau_0}{d\Theta}=0 \text{ subject to } \Phi_0\left(\Theta_0\right)=\tau_0 \qquad (6.46)$$

$$\frac{d\Phi_1}{d\Theta}-\left(\frac{d\Phi_0}{d\Theta}-\frac{d\tau_0}{d\Theta}\right)+\left[\left(\rho-\Theta^{\frac{1}{c}}\right)\frac{d\Phi_0}{d\Theta}-1\right]=0 \text{ subject to } \Phi_1\left(\Theta_0\right)=0 \quad (6.47)$$

$$\frac{d\Phi_2}{d\Theta} - \frac{d\Phi_1}{d\Theta} + \left(\rho - \Theta^{\frac{1}{c}}\right)\frac{d\Phi_1}{d\Theta} = 0 \text{ subject to } \Phi_2(\Theta_0) = 0 \qquad (6.48)$$

$$\frac{d\Phi_3}{d\Theta} - \frac{d\Phi_2}{d\Theta} + \left(\rho - \Theta^{\frac{1}{c}}\right)\frac{d\Phi_2}{d\Theta} = 0 \text{ subject to } \Phi_3(\Theta_0) = 0 \qquad (6.49)$$

$$\frac{d\Phi_4}{d\Theta} - \frac{d\Phi_3}{d\Theta} + \left(\rho - \Theta^{\frac{1}{c}}\right)\frac{d\Phi_3}{d\Theta} = 0 \text{ subject to } \Phi_4(\Theta_0) = 0 \qquad (6.50)$$

Proceeding in a like manner, one can arrive at the following recurrence relation:

$$\frac{d\Phi_m}{d\Theta} = \left[1 - \left(\rho - \Theta^{\frac{1}{c}}\right)\right]\frac{d\Phi_{m-1}}{d\Theta} \text{ subject to } \Phi_m(\Theta_0) = 0 \text{ for } m \geq 2 \qquad (6.51)$$

The initial approximation can be chosen as $\Phi_0 = \tau_0$. Using this initial approximation, we can solve the equations iteratively using symbolic software. We do not include those expressions here, as they are lengthy. Finally, the HPM-based solution can be approximated as follows:

$$\tau(\Theta) \approx \sum_{i=0}^{M} \Phi_i \qquad (6.52)$$

6.4   OHAM-BASED SOLUTION

When applying the optimal homotopy asymptotic method (OHAM), Eq. (6.30) can be written in the following form:

$$\mathcal{L}[\tau(\Theta)] + \mathcal{N}[\tau(\Theta)] + h(\Theta) = 0 \qquad (6.53)$$

where $\mathcal{L}[\tau(\Theta)] = \rho\frac{d\tau}{d\Theta}$, $N[\tau(\Theta)] = -\Theta^{\frac{1}{c}}\frac{d\tau}{d\Theta} - 1$, and $h(\Theta) = 0$. With these considerations, the zeroth-order problem becomes:

$$\mathcal{L}(\tau_0(\Theta)) + h(\Theta) = 0 \text{ subject to } \tau_0(\Theta_0) = \tau_0 \qquad (6.54)$$

Solving Eq. (6.54), one obtains:

$$\tau_0(\Theta) = \tau_0 \qquad (6.55)$$

To obtain $\mathcal{N}_0$, $\mathcal{N}_1$, $\mathcal{N}_2$, etc., the following can be used:

$$\left(-\Theta^{\frac{1}{c}}\right)\frac{\partial\tau}{\partial\Theta} - 1 = \left(-\Theta^{\frac{1}{c}}\right)\left(\frac{d\tau_0}{d\Theta} + q\frac{d\tau_1}{d\Theta} + q^2\frac{d\tau_2}{d\Theta} + q^3\frac{d\tau_3}{d\Theta} + q^4\frac{d\tau_4}{d\Theta} + \cdots\right)$$

$$-1 = \left[\left(-\Theta^{\frac{1}{c}}\right)\frac{d\tau_0}{d\Theta} - 1\right] + q\left(-\Theta^{\frac{1}{c}}\right)\frac{d\tau_1}{d\Theta} + q^2\left(-\Theta^{\frac{1}{c}}\right)\frac{d\tau_2}{d\Theta} + q^3\left(-\Theta^{\frac{1}{c}}\right)\frac{d\tau_3}{d\Theta} + \cdots \quad (6.56)$$

Using Eq. (6.56), the first-order problem reduces to:

$$\rho\frac{d\tau_1}{d\Theta} = H_1(\Theta,C_i)\left\{\left(-\Theta^{\frac{1}{c}}\right)\frac{d\tau_0}{d\Theta} - 1\right\} \text{ subject to } \tau_1(\Theta_0) = 0 \quad (6.57)$$

We simply choose $H_1(\Theta,C_1) = C_1$. Putting $k = 2$, the second-order problem becomes:

$$\mathcal{L}\left[\tau_2(\Theta,C_i) - \tau_1(\Theta,C_i)\right] = H_2(\Theta,C_i)\mathcal{N}_0\left(\tau_0(\Theta)\right) + H_1(\Theta,C_i)$$
$$\times\left[\mathcal{L}\left[\tau_1(\Theta,C_i)\right] + \mathcal{N}_1\left[\tau_0(\Theta),\tau_1(\Theta,C_i)\right]\right] \text{ subject to } \tau_2(\Theta_0) = 0 \quad (6.58)$$

Further, we choose $H_2(\Theta,C_i) = C_2$. Therefore, Eq. (6.58) becomes:

$$\rho\frac{d\tau_2}{d\Theta} = \rho\frac{d\tau_1}{d\Theta} + C_2\left\{\left(-\Theta^{\frac{1}{c}}\right)\frac{d\tau_0}{d\Theta} - 1\right\}$$
$$+ C_1\left[\rho\frac{d\tau_1}{d\Theta} + \left(-\Theta^{\frac{1}{c}}\right)\frac{d\tau_1}{d\Theta}\right] \text{ subject to } \tau_2(\Theta) = 0 \quad (6.59)$$

The terms of the series can be computed using the equations developed here. One can compute these terms without any difficulty using symbolic computation software, such as MATLAB. Further, following Chapter 2, the higher-order terms can be computed in a similar manner. However, our aim is to produce an accurate solution with just two to three terms of the OHAM-based series. Therefore, we restrict our calculation up to $k = 2$. Finally, the approximate solution can be found as:

$$\tau(\Theta) \approx \tau_0(\Theta) + \tau_1(\Theta,C_1) + \tau_2(\Theta,C_1,C_2) \quad (6.60)$$

where the terms are given by Eqs. (6.55), (6.57), and (6.59).

6.5 CONVERGENCE THEOREMS

The convergence of the HAM-based and OHAM-based solutions is proved theoretically using the following theorems.

6.5.1 CONVERGENCE THEOREM OF HAM-BASED SOLUTIONS

The convergence theorems for the HAM-based solutions given by Eqs. (6.26) and (6.35) can be proved using the following theorems.

Theorem 6.1: *If the homotopy series $\sum_{m=0}^{\infty} \tau_m(\Theta)$ and $\sum_{m=0}^{\infty} \tau_m'(\Theta)$ converge, then $R_m(\vec{\tau}_{m-1})$ given by Eqs. (6.15) and (6.32) satisfy the relation $\sum_{m=1}^{\infty} R_m(\vec{\tau}_{m-1}) = 0$. Here, "'" denotes the derivative with respect to Θ.*

Proof: The proof of this theorem follows exactly from the previous chapters.

Theorem 6.2: *If $\hbar$ is so properly chosen that the series $\sum_{m=0}^{\infty} \tau_m(\Theta)$ and $\sum_{m=0}^{\infty} \tau_m'(\Theta)$ converge absolutely to $\tau(\Theta)$ and $\tau'(\Theta)$, respectively, then the homotopy series $\sum_{m=0}^{\infty} \tau_m(\Theta)$ satisfies the original governing Eq. (6.7).*

Proof: Theorem 6.1 shows that if $\sum_{m=0}^{\infty} \tau_m(\Theta)$ and $\sum_{m=0}^{\infty} \tau_m'(\Theta)$ converge, then $\sum_{m=1}^{\infty} R_m(\vec{\tau}_{m-1}) = 0$.

Therefore, substituting the previous expressions in Eq. (6.15) and simplifying further lead to:

$$\left(\rho - \Theta^{\frac{1}{2}}\right) \sum_{k=0}^{\infty} \tau_m' - 1 = 0 \tag{6.61}$$

which is basically the original governing equation in Eq. (6.7). Furthermore, subject to the initial condition $\tau_0(\Theta_0) = \tau_0$ and the conditions for the higher-order deformation equation $\tau_m(\Theta_0) = 0$ for $m \geq 1$, we easily obtain $\sum_{m=0}^{\infty} \tau_0(\Theta_0) = \tau_0$. Hence, the convergence result follows.

Theorem 6.3: *If $\hbar$ is so properly chosen that the series $\sum_{m=0}^{\infty} \tau_m(\Theta)$ and $\sum_{m=0}^{\infty} \tau_m'(\Theta)$ converge absolutely to $\tau(\Theta)$ and $\tau'(\Theta)$, respectively, then the homotopy series $\sum_{m=0}^{\infty} \tau_m(\Theta)$ satisfies the original governing Eq. (6.30).*

Proof: The proof of this theorem follows from Theorem 6.3.

6.5.2 CONVERGENCE THEOREM OF OHAM-BASED SOLUTION

Theorem 6.4: *If the series $\tau_0(\Theta) + \sum_{j=1}^{\infty} \tau_j(\Theta, C_i)$, $i = 1, 2, \ldots, s$ converges, where $\tau_j(\Theta, C_i)$ are governed by Eqs. (6.55), (6.57), and (6.59), then Eq. (6.60) is a solution of the original Eq. (6.30).*

Proof: The proof of this theorem follows exactly from Theorem 2.3 of Chapter 2.

6.6   RESULTS AND DISCUSSION

First, the numerical convergence of the HAM-based analytical solution is established for a specific test case, and then the solution is validated over the existing analytical solution derived by Baiamonte (2020). Then, HPM- and OHAM-based analytical solutions are validated over the existing analytical solution. Finally, the derived analytical solution based on HAM is tested under different physical conditions to check the efficiency of the method.

6.6.1   Numerical Convergence and Validation
of the HAM-Based Solution

The convergence of the HAM-based series solution depends on a suitable choice for the auxiliary parameter $\hbar$. For that purpose, the squared residual error with regard to Eq. (6.7) or Eq. (6.30) can be calculated as follows:

$$\Delta_m = \int_{\Theta \in \Omega} \left(\mathcal{N}\left[\tau(\Theta)\right] \right)^2 d\Theta \tag{6.62}$$

where Ω is the domain of the problem. The HAM-based series solution leads to the exact solution of the equation when Δ_m tends to zero. Therefore, for a particular order of approximation m, it is sufficient to minimize the corresponding Δ_m, which then yields an optimal value of parameter $\hbar$. Here, we consider a test case, which is the case under Toricelli's law with $\rho = 1.2$, $\Theta_0 = 0 = \tau_0$. For this case, the solution given by Baiamonte (2020) reads as follows:

$$\tau = 2\left[\rho \log \left| \frac{\rho - \sqrt{\Theta_0}}{\rho - \sqrt{\Theta}} \right| - \left(\sqrt{\Theta} - \sqrt{\Theta_0} \right) \right] \tag{6.63}$$

Figure 6.1 shows the squared residual errors for a different order of approximation for the selected case. It can be seen from the figure that as the order of approximation increases, the corresponding residual error decreases. Thus, the adequacy of selecting the linear operator, initial approximation, and auxiliary function is established. Also, for a quantitative assessment, numerical results are reported in Table 6.1 along with the computational time taken by the computer to produce the corresponding order of approximation. For the selected case, comparison between the analytical solution in Eq. (6.63) and the 20th-order HAM-based approximate solution is shown in Figure 6.2. An excellent agreement between the computed and observed values can be found from both figures. In addition, we choose the 5th, 10th, and 20th orders of approximation and compare them with the corresponding analytical solution in Table 6.2. It can be observed from the table that the higher the order of approximation, the better the accuracy. All the computations are performed using the BVPh 2.0 package developed by Zhao and Liao (2002). A flowchart diagram containing the steps of the method is provided in Figure 6.3.

TABLE 6.1

Squared Residual Error (Δ_m) and Computational Time versus Different Orders of Approximation (m) for the Selected Case

Order of Approximation (m)	Squared Residual Error (Δ_m)	Computational Time (sec)
2	5.07×10^{-2}	0.843
4	7.78×10^{-3}	1.795
6	1.67×10^{-3}	2.768
8	4.53×10^{-4}	3.919
10	1.44×10^{-4}	5.063
12	5.09×10^{-5}	6.145
14	1.91×10^{-5}	7.407
16	7.43×10^{-6}	8.802
18	2.94×10^{-6}	9.916
20	1.18×10^{-6}	11.050

6.6.2 VALIDATION OF HPM-BASED SOLUTION

The HPM-based analytical solution for the selected test case is validated over the analytical solution given by Baiamonte (2020). Figure 6.4 shows a comparison between the two-term, three-term, and four-term HPM-based solutions and the corresponding numerical solution. It is observed from the figure that the HPM-based

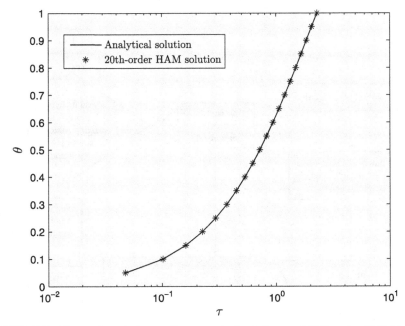

FIGURE 6.2 Comparison between the numerical solution and 20th-order HAM-based solution for the selected case.

TABLE 6.2

Comparison between HAM-Based Approximation and Numerical Solution for the Selected Case

Θ	Numerical Solution	HAM-Based Approximation		
		5th Order	10th Order	20th Order
0.0	0.0000	0.0000	0.0000	0.0000
0.1	0.1017	0.1056	0.1012	0.1016
0.2	0.2247	0.2288	0.2243	0.2247
0.3	0.3676	0.3717	0.3672	0.3676
0.4	0.5321	0.5362	0.5317	0.5321
0.5	0.7213	0.7253	0.7209	0.7213
0.6	0.9397	0.9430	0.9393	0.9397
0.7	1.1941	1.1942	1.1936	1.1940
0.8	1.4941	1.4845	1.4933	1.4941
0.9	1.8547	1.8201	1.8521	1.8547
1.0	2.3002	2.2076	2.2899	2.3001

solution is accurate only for some specific values in the domain. As the HPM does not provide an accurate solution for the equation over the entire domain, we have considered solution sup to $\Theta = 0.5$ for a quantitative assessment. The HPM-based solution and the existing analytical solution are compared in Table 6.3. A flowchart diagram containing the steps of the method is given in Figure 6.5.

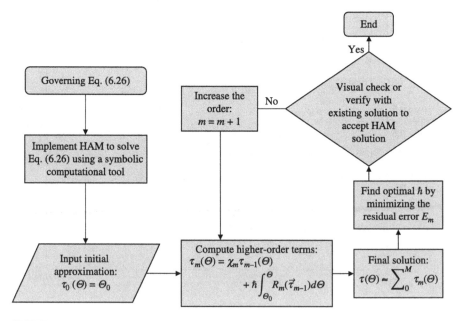

FIGURE 6.3   Flowchart for the HAM solution in Eq. (6.35).

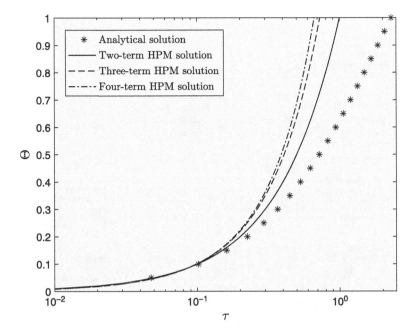

FIGURE 6.4 Comparison between the analytical solution and two-, three-, and four-term HPM-based solutions for the selected case.

TABLE 6.3

Comparison between HPM-Based Approximation and Analytical Solution for the Selected Case

Θ	Analytical Solution	HPM-Based Approximation		
		Two-Term	Three-Term	Four-Term
0.00	0.0000	0.0000	0.0000	0.0000
0.05	0.0477	0.0500	0.0475	0.0477
0.10	0.1017	0.1000	0.1011	0.1016
0.15	0.1607	0.1500	0.1587	0.1605
0.20	0.2247	0.2000	0.2196	0.2238
0.25	0.2936	0.2500	0.2833	0.2913
0.30	0.3676	0.3000	0.3495	0.3627
0.35	0.4470	0.3500	0.4180	0.4381
0.40	0.5321	0.4000	0.4887	0.5172
0.45	0.6234	0.4500	0.5612	0.6000
0.50	0.7213	0.5000	0.6357	0.6864

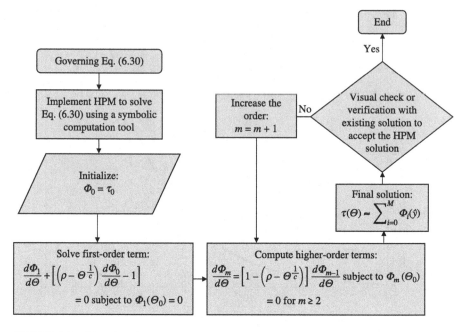

FIGURE 6.5   Flowchart for the HPM solution in Eq. (6.52).

6.6.3 VALIDATION OF OHAM-BASED SOLUTION

For the assessment of the OHAM-based analytical solution, one needs to calculate constants for C_i. For that purpose, we calculate the residual as follows:

$$R(\Theta, C_i) = \mathcal{L}\left[\tau_{OHAM}(\Theta, C_i)\right] + \mathcal{N}\left[\tau_{OHAM}(\Theta, C_i)\right] + h(\Theta), i = 1, 2, \ldots, s \quad (6.64)$$

where $\tau_{OHAM}(\Theta, C_i)$ is the approximate solution. When $R(\Theta, C_i) = 0$, $\tau_{OHAM}(\Theta, C_i)$ becomes the exact solution to the equation. One of the ways to obtain the optimal C_i for which the solution converges is the minimization of squared residual error, i.e.,

$$J(C_i) = \int_{\Theta \in D} R^2(\Theta, C_i) d\Theta, \, i = 1, 2, \ldots, s \quad (6.65)$$

where $D = [0,1]$ is the domain of the equation. The minimization of Eq. (6.65) leads to a system of algebraic equations as follows:

$$\frac{\partial J}{\partial C_1} = \frac{\partial J}{\partial C_2} = \cdots = \frac{\partial J}{\partial C_s} = 0 \quad (6.66)$$

Solving this system, one can obtain the optimal values of parameters. We obtain the optimal values using the MATLAB routine *fminsearch*, which minimizes an unconstrained multivariable function. Using these values, two- and three-term OHAM solutions are computed and compared in Figure 6.6 with the existing analytical

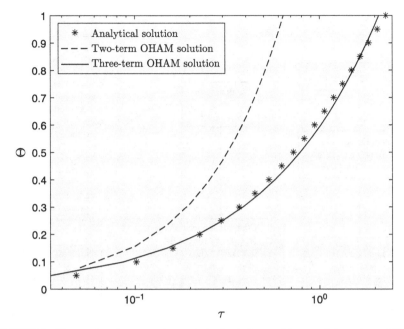

FIGURE 6.6 Comparison between the analytical solution and two- and three-term OHAM-based solutions for the selected case.

solution of Baiamonte (2020) to see the effectiveness of the OHAM. It can be seen that just three terms of the OHAM-based series agree well with the corresponding analytical solution to the problem. For a quantitative assessment, we also compare the numerical values of the solutions in Table 6.4. A flowchart diagram containing the steps of the method is given in Figure 6.7.

TABLE 6.4

Comparison between OHAM-Based Approximation and Analytical Solution for the Selected Case

Θ	Numerical Solution	OHAM-Based Approximation	
		Two-Term	Three-Term
0.0	0.0000	0.0000	0.0000
0.1	0.1017	0.1569	0.0863
0.2	0.2247	0.3137	0.2200
0.3	0.3676	0.4706	0.3846
0.4	0.5321	0.6274	0.5742
0.5	0.7213	0.7843	0.7854
0.6	0.9397	0.9411	1.0159
0.7	1.1941	1.0980	1.2639
0.8	1.4941	1.2549	1.5282
0.9	1.8547	1.4117	1.8076
1.0	2.3002	1.5686	2.1015

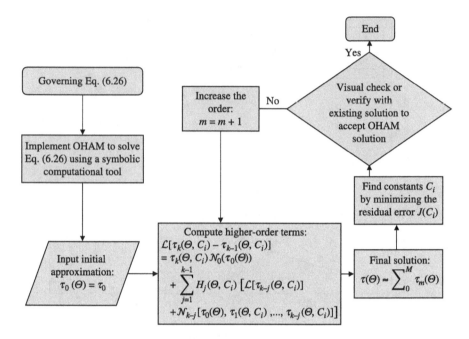

FIGURE 6.7 Flowchart for the OHAM solution in Eq. (6.60).

6.6.4 BEHAVIOR OF THE SOLUTION

The analytical solutions based on three different methods were discussed in the previous sections. Henceforth, the discussion here is based on the HAM-based analytical solution only. First, we discuss the volumetric water content under Toricelli's law. The time to reach equilibrium, i.e., the time needed to achieve the steady-state condition of the infiltration process, τ_{eq}, was discussed by Baiamonte (2020) based on the analytical solution given. For $\rho \leq 1$, τ_{eq} cannot be achieved in a finite time such that the water content asymptotically attains an equilibrium value Θ_∞. On the other hand, for $\rho > 1$, τ_{eq} is achieved in a finite time at the soil saturation condition, which can be derived by imposing $\Theta = \Theta_{eq} = 1$. For the cases of ρ and under zero antecedent soil moisture conditions $\Theta_0 = 0$, Baiamonte (2020) provided the following equations:

$$\tau_{eq} = 2\left[\rho \log\left|\frac{1}{1-\beta}\right| - \beta\rho\right], \rho \leq 1, \Theta_{eq} = (\beta\rho)^2 \tag{6.67}$$

$$\tau_{eq} = 2\left[\rho \log\left|\frac{\rho}{\rho-1}\right| - 1\right], \rho > 1, \Theta_{eq} = 1 \tag{6.68}$$

where β can be chosen as 0.95.

For different values of ρ under $\Theta_0 = 0$, the HAM-based analytical solution in Eq. (6.22) is assessed for the determination of normalized water content. For all the

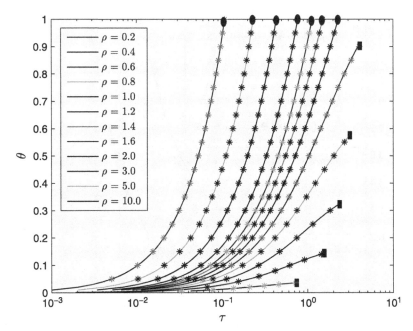

FIGURE 6.8 Temporal variation of normalized water content for different values of ρ. [The continuous line represents the 15th-order HAM-based solution, the asterisk represents the analytical solution given by Baiamonte (2020), ■ represents τ_{eq} for ρ > 1, and ● represents τ_{eq} for ρ ≤ 1].

cases considered, the 15th-order HAM-based approximation is selected, as it could sufficiently minimize the squared residual error. In Figure 6.8, the analytical solution given by Baiamonte (2020) and the HAM-based approximation are compared, where it can be seen that the HAM-based solution validates the solutions exactly. It is observed that for ρ > 1, the equilibrium is achieved in a finite time. On the other hand, for ρ ≤ 1, Θ the equilibrium condition is asymptotically attained.

Now, we discuss the volumetric water content under Brooks and Corey's hydraulic conductivity function. For that, the HAM-based analytical solution obtained here is compared with the analytical solution derived by Baiamonte (2020). The analytical solution proposed by him is as follows:

$$\tau = \frac{1}{\rho}\left[\Theta F_1\left(1, c, 1+c, \frac{\Theta^{1/c}}{\rho}\right) - \Theta_0 F_1\left(1, c, 1+c, \frac{\Theta_0^{1/c}}{\rho}\right)\right] \qquad (6.69)$$

where F_1 is the hypergeometric function defined as $F_1(a, b, c, z) = \sum_{n=0}^{\infty} \frac{(a)_n (b)_n}{(c)_n} \frac{z^n}{n!}$

where $(a)_n$ is the Pochhammer symbol (Abramowitz and Stegun 1972). In Figure 6.9, the HAM-based analytical solution and the analytical solution of Baiamonte (2020) are compared for a different set of values of the parameters. It can be observed from the figure that for all the cases, the proposed solutions given by HAM match exactly the analytical solution.

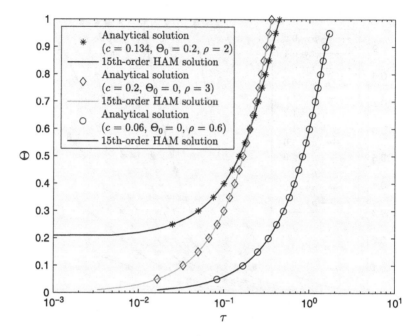

FIGURE 6.9 Temporal variation of normalized water content for different sets of values of the parameters: (i) $(\rho, c, \Theta_0) = (2, 0.134, 0.2)$, (ii) $(\rho, c, \Theta_0) = (3, 0.2, 0)$, and (iii) $(\rho, c, \Theta_0) = (0.6, 0.06, 0)$ [The continuous line represents the 15th-order HAM-based solution, and discrete values represent the analytical solutions given by Baiamonte (2020)].

6.7 CONCLUDING REMARKS

Here, we derive analytical solutions of the Richards equation using HAM, the homotopy perturbation method (HPM), and OHAM. The governing equations are considered from the model of Baiamonte (2020), who provided exact analytical solutions for those equations. Under different physical conditions, the solutions obtained using HAM are assessed and compared with Baiamonte's solution to check the efficiency. It is found that only a few terms of the HAM-based solutions exactly match the existing analytical solutions. On the other hand, the solutions based on HPM do not provide accurate values for the entire domain. Further, it is observed that only three terms of the OHAM-based series solution agree well with the corresponding analytical solution. Also, for the considered equation, it is found that a simple set of based functions is sufficient to provide accurate solutions. The theoretical as well as numerical convergence of the series solutions is provided.

REFERENCES

Abramowitz, M., and Stegun, I. A. (1972). *Handbook of mathematical functions.* Dover Publications, New York, NY.

Baiamonte, G. (2020). Analytical solution of the Richards equation under gravity-driven infiltration and constant rainfall intensity. *Journal of Hydrologic Engineering,* 25(7), 04020031.

Brooks, R., and Corey, A. (1964). *Hydraulic properties of porous media: Hydrology paper no. 3*. Colorado State University, Fort Collins, CO.

Zhao, Y., and Liao, S. (2002). User guide to BVPh 2.0. *School of naval architecture, ocean and civil engineering* (Vol. 40). Shanghai.

FURTHER READING

Assouline, S. (2005). On the relationships between the pore size distribution index and characteristics of the soil hydraulic functions. *Water Resources Research, 41*(7), W07019.

Assouline, S. (2013). Infiltration into soils: Conceptual approaches and solutions. *Water Resources Research, 49*(4), 1755–1772.

Baiamonte, G., D'Asaro, F., and Calvo, R. (2019). Gravity-driven infiltration and subsidence phenomena in *Posidonia oceanica* residues. *Journal of Hydrologic Engineering, 24*(6), 04019016.

Baiamonte, G., and Singh, V. P. (2016). Analytical solution of kinematic wave time of concentration for overland flow under green-Ampt infiltration. *Journal of Hydrologic Engineering, 21*(3), 04015072.

Batu, V. (1978). Steady infiltration from single and periodic strip sources. *Soil Science Society of America Journal, 42*(4), 544–549.

Broadbridge, P., Daly, E., and Goard, J. (2017). Exact solutions of the Richards equation with nonlinear plant-root extraction. *Water Resources Research, 53*(11), 9679–9691.

Broadbridge, P., and White, I. (1987). Time to ponding: Comparison of analytic, quasi-analytic, and approximate predictions. *Water Resources Research, 23*(12), 2302–2310.

Broadbridge, P., and White, I. (1988). Constant rate rainfall infiltration: A versatile nonlinear model: 1. Analytic solution. *Water Resources Research, 24*(1), 145–154.

de Willigen, P., van Dam, J. C., Javaux, M., and Heinen, M. (2012). Root water uptake as simulated by three soil water flow models. *Vadose Zone Journal, 11*(3).

James, W. P., Warinner, J., and Reedy, M. (1992). Application of the green-Ampt infiltration equation to watershed modeling 1. *JAWRA Journal of the American Water Resources Association, 28*(3), 623–635.

Milly, P. C. D. (1985). Stability of the green-Ampt profile in a delta function soil. *Water Resources Research, 21*(3), 399–402.

Parlange, J. Y. (1971). Theory of water-movement in soils: 2. One-dimensional infiltration. *Soil Science, 111*(3), 170–174.

Raats, P. A. C. (1976). Analytical solutions of a simplified flow equation. *Transactions of the ASAE, 19*(4), 683–0689.

Sander, G. C., Parlange, J. Y., Kühnel, V., Hogarth, W. L., Lockington, D., and O'kane, J. P. J. (1988). Exact nonlinear solution for constant flux infiltration. *Journal of Hydrology, 97*(3–4), 341–346.

Vand, A. S., Sihag, P., Singh, B., and Zand, M. (2018). Comparative evaluation of infiltration models. *KSCE Journal of Civil Engineering, 22*(10), 4173–4184.

Xing, X., Wang, H., and Ma, X. (2018). Brooks–Corey modeling by one-dimensional vertical infiltration method. *Water, 10*(5), 593.

7 Error Equation for Unsteady Uniform Flow

7.1 INTRODUCTION

Channel flow can be unsteady nonuniform, unsteady uniform, steady nonuniform, or steady uniform flow. The most accurate description of channel flow is given by the St. Venant equations (Singh 1995). For many cases, reasonably accurate solutions can be given by two approximations of the St. Venant equations, which are the kinematic wave and diffusion wave. The question then arises about the error made by these two approximations. By comparing with the St. Venant equations, errors of kinematic wave, diffusion wave approximations of steady nonuniform flow equations, and unsteady uniform flow equations have been derived in terms of the Riccati equation, which is a nonlinear equation (Singh 1995). The error equation for uniform unsteady flow is derived analytically using the homotopy analysis method (HAM), homotopy perturbation method (HPM), and optimal homotopy asymptotic method (OHAM).

7.2 GOVERNING EQUATION

The error equation for unsteady uniform flow can be given as follows:

$$\frac{dE}{d\tau} = C_0(\tau) + C_1(\gamma, \tau)E + C_2(\gamma, \tau)E^2 \tag{7.1}$$

where

$$C_0(\tau) = \frac{3}{2\tau} \tag{7.2}$$

$$C_1(\gamma, \tau) = \frac{3}{2\tau}\left(1 - \frac{2}{3}\gamma^{0.5}\tau^{0.5}\right) \tag{7.3}$$

$$C_2(\gamma, \tau) = -\frac{\gamma^{0.5}}{2\tau^{0.5}} \tag{7.4}$$

where E is the relative error, τ is the dimensionless time, and γ is a dimensionless parameter. Eq. (7.1) is a Riccati equation (Singh 1995). The initial condition for the equation is given as:

$$E(1) = 0 \tag{7.5}$$

DOI: 10.1201/9781003368984-10

The objective is to find an analytical solution to Eq. (7.1) subject to the initial condition in Eq. (7.5). To that end, Eqs. (7.2)–(7.4) are substituted into Eq. (7.1) and rearranged, leading to the governing equation as follows:

$$2\tau\frac{dE}{d\tau} - 3 - 3\left(1 - \frac{2}{3}\gamma^{0.5}\tau^{0.5}\right)E + \gamma^{0.5}\tau^{0.5}E^2 = 0 \text{ subject to } E(1) = 0 \quad (7.6)$$

Eq. (7.6) is solved analytically using the three methods as described in what follows.

7.3   STANDARD HAM-BASED SOLUTION

When applying HAM, the zeroth-order deformation equation for Eq. (7.6) is constructed as follows:

$$(1-q)\mathcal{L}\left[\Phi(\tau;q) - E_0(\tau)\right] = q\hbar H(\tau)\mathcal{N}\left[\Phi(\tau;q)\right] \quad (7.7)$$

subject to the initial condition:

$$\Phi(1;q) = 0 \quad (7.8)$$

where q is the embedding parameter, $\Phi(\tau;q)$ is the representation of the solution across q, $E_0(\tau)$ is the initial approximation, $\hbar$ is the auxiliary parameter, $H(\tau)$ is the auxiliary function, and $\mathcal{L}$ and $\mathcal{N}$ are the linear and nonlinear operators, respectively. The higher-order terms can be calculated from the following higher-order deformation equation:

$$\mathcal{L}\left[E_m(\tau) - \chi_m E_{m-1}(\tau)\right] = \hbar H(\tau) R_m\left(\vec{E}_{m-1}\right), \; m = 1,2,3,\ldots \text{ subject to } E_m(1) = 0 \quad (7.9)$$

where

$$\chi_m = \begin{cases} 0 & \text{when } m = 1, \\ 1 & \text{otherwise} \end{cases} \quad (7.10)$$

and

$$R_m\left(\vec{E}_{m-1}\right) = \frac{1}{(m-1)!}\frac{\partial^{m-1}\mathcal{N}\left[\Phi(\tau;q)\right]}{\partial q^{m-1}}\bigg|_{q=0} \quad (7.11)$$

where E_m for $m \geq 1$ are the higher-order terms. Now, the final solution can be obtained as follows:

$$E(\tau) = E_0(\tau) + \sum_{m=1}^{\infty} E_m(\tau) \quad (7.12)$$

The nonlinear operator for the problem is considered as:

$$N\left[\Phi(\tau;q)\right]=2\tau\frac{\partial\Phi(\tau;q)}{\partial\tau}-3-3\left(1-\frac{2}{3}\gamma^{0.5}\tau^{0.5}\right)\Phi(\tau;q)+\gamma^{0.5}\tau^{0.5}\Phi(\tau;q)^{2} \quad (7.13)$$

Therefore, term R_m can be calculated as follows:

$$R_m\left(\vec{E}_{m-1}\right)=2\tau\frac{dE_{m-1}}{d\tau}-3\left(1-\chi_m\right)-3\left(1-\frac{2}{3}\gamma^{0.5}\tau^{0.5}\right)E_{m-1}+\gamma^{0.5}\tau^{0.5}\sum_{j=0}^{m-1}E_jE_{m-1-j} \quad (7.14)$$

To apply HAM, the following set of base functions is considered to represent the solution:

$$\left\{\tau^{0.5(1+n)}\mid n=-1,0,1,\ 2,\ 3...\right\} \quad (7.15)$$

so that

$$E(\tau)=\sum_{n=-1}^{\infty}a_n\tau^{0.5(1+n)} \quad (7.16)$$

where a_n is the coefficient of the series. Eq. (7.15) provides the so-called *rule of solution expression*. Following the rule of solution expression, the linear operator and the initial approximation are chosen, respectively, as follows:

$$\mathcal{L}\left[\Phi(\tau;q)\right]=\frac{\partial\Phi(\tau;q)}{\partial\tau}\ \text{with the property}\ \mathcal{L}[C_2]=0 \quad (7.17)$$

$$E_0(\tau)=0 \quad (7.18)$$

where C_2 is an integral constant. Using Eq. (7.17), the higher-order terms can be obtained from Eq. (7.9) as follows:

$$E_m(\tau)=\chi_m E_{m-1}(\tau)+\hbar\int_1^{\tau}H(x)R_m\left(\vec{E}_{m-1}\right)dx+C_2,\ m=1,2,3,... \quad (7.19)$$

where R_m is given by Eq. (7.14), and the constant C_2 can be determined from the initial condition for the higher-order deformation equations, i.e., $E_m(1)=0$, which simply yields $C_2=0$ for all $m\geq 1$.

Now, the auxiliary function $H(\tau)$ can be determined from the *rule of coefficient ergodicity*. Based on the rule of solution expression and Eq. (7.16), the general form of $H(\tau)$ should be:

$$H(\tau)=\tau^{0.5(1+\alpha)} \quad (7.20)$$

where α is an integer. Putting Eq. (7.20) into Eq. (7.19) and calculating some of E_m, we have:

$$R_1(E_0) = -3 \tag{7.21}$$

$$E_1(\tau) = -3\hbar \int_1^\tau \tau^{0.5(1+\alpha)} d\tau = -\frac{3\hbar}{0.5(1+\alpha)+1}\left(\tau^{0.5(1+\alpha)+1} - 1\right) \tag{7.22}$$

$$R_2\left(\vec{E}_1\right) = 2\tau \frac{dE_1}{d\tau} - 3\left(1 - \frac{2}{3}\gamma^{0.5}\tau^{0.5}\right)E_1 + 2\gamma^{0.5}\tau^{0.5}E_0E_1$$

$$= -6\hbar\tau^{0.5(1+\alpha)+1} + \frac{9\hbar}{0.5(1+\alpha)+1}\left(1 - \frac{2}{3}\gamma^{0.5}\tau^{0.5}\right)\left(\tau^{0.5(1+\alpha)+1} - 1\right) \tag{7.23}$$

$$E_2(\tau) = -\frac{3\hbar}{0.5(1+\alpha)+1}\left(\tau^{0.5(1+\alpha)+1} - 1\right)$$

$$+ \hbar \int_1^\tau \tau^{0.5(1+\alpha)}\left[-6\hbar\tau^{0.5(1+\alpha)+1} + \frac{9\hbar}{0.5(1+\alpha)+1}\left(1 - \frac{2}{3}\gamma^{0.5}\tau^{0.5}\right)\left(\tau^{0.5(1+\alpha)+1} - 1\right)\right]d\tau \tag{7.24}$$

Here, we can infer two observations. If we consider $\alpha \geq -1$, then some of the terms (such as $\tau^{0.5}$) do not appear in the solution expression. This does not satisfy the *rule of coefficient ergodicity*, which allows us to include all the terms of the base function for the completeness of the solution. On the other hand, if we consider $\alpha \leq -3$, the terms appearing in the solution violate the *rule of solution expression*. Therefore, we are left with the only choice $\alpha = -2$, which results in $H(\tau) = \tau^{-0.5}$. Finally, the approximate solution can be obtained as follows:

$$E(\tau) \approx E_0(\tau) + \sum_{m=1}^M E_m(\tau) \tag{7.25}$$

The convergence of the HAM-based series solution depends on a suitable choice for the auxiliary parameter $\hbar$. For that purpose, the squared residual error of Eq. (7.6) can be calculated as follows:

$$\Delta_m = \int_{\tau \in \Omega} \left(N[E(\tau)]\right)^2 d\tau \tag{7.26}$$

where Ω is the domain of the independent variable. For a particular order of approximation m, the corresponding Δ_m is minimized, which yields an optimal value of parameter $\hbar$, and the series solution leads to its convergence. Here, we consider a test case where the parameter γ is selected as 1 and domain $\Omega = [1,50]$. Figure 7.1 shows the squared residual error for a different order of approximation of the HAM-based solution. It can be seen from the figure that as the order of approximation increases,

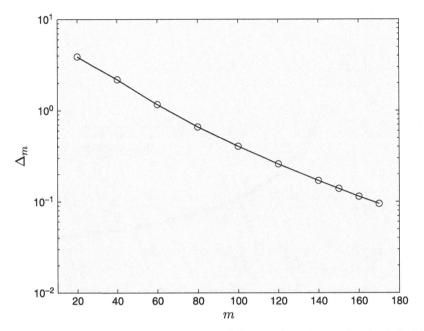

FIGURE 7.1 Squared residual error (Δ_m) versus different orders of approximation (m) of the HAM-based solution for $\gamma = 1$.

the corresponding residual error decreases. However, it is observed that the convergence is very slow, and the method does not produce a reasonable error even for the 170th order of approximation. For a quantitative assessment, numerical results are reported in Table 7.1, where computational time is also included. Both the figure and the table suggest that the method performs slowly despite an appropriate choice of

TABLE 7.1

Squared Residual Error (Δ_m) and Computational Time versus Different Orders of Approximation (m) for $\gamma = 1$

Order of Approximation (m)	Squared Residual Error (Δ_m)	Computational Time (sec)
20	3.889	5.05
40	2.177	14.21
60	1.162	38.02
80	0.660	74.73
100	0.403	122.53
120	0.258	191.13
140	0.170	291.89
150	0.139	352.89
160	0.114	424.38
170	0.095	506.91

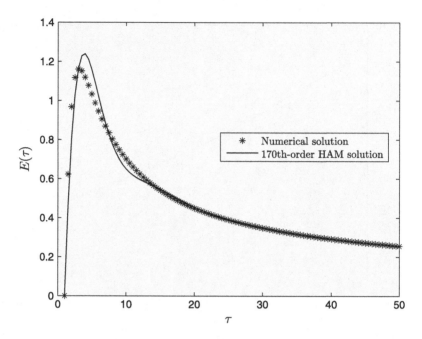

FIGURE 7.2 Comparison between the numerical solution and 170th-order HAM-based solution for $\gamma = 1$.

the base functions for the equation under consideration. Further, the computational time increases exponentially with the order of approximation. All the computations reported here are performed using the MATHEMATICA package BVPh 2.0 developed by Zhao and Liao (2002). In Figure 7.2, comparison between a numerical solution and the 170th-order HAM-based approximate solution is shown, and numerical results are reported in Table 7.2 for some selected values of the independent variable. The comparison shows that the method performs poorly and cannot produce satisfactory results even for the 170th order of approximation. Finally, these shortcomings suggest a need to refine the method to develop an efficient solution for the equation term. To that end, the method is modified and discussed in the next section.

7.4 MODIFIED HAM-BASED SOLUTION

The difficulty in solving the governing equation efficiently by the standard HAM was discussed in the previous section. To overcome the difficulty, here, we modify the method. First, the time domain is divided into several subdomains, i.e., $\tau = \bigcup_{i=1}^{n} [\tau_{i-1}, \tau_i]$, where τ_0 is the starting point of the domain, i.e., $\tau_0 = 1$, and τ_n is the end point to be considered. Then, the HAM-based solution is developed for each of the subdomains, which may be called a subdomain solution. Now, the zeroth-order deformation equation can be constructed as follows:

$$(1-q)\mathcal{L}_i\left[\Phi_i(\tau;q) - E_{i,0}(\tau)\right] = q\hbar_i H_i(\tau)\mathcal{N}\left[\Phi_i(\tau;q)\right], \, i = 1,2,3,\ldots,n \quad (7.27)$$

TABLE 7.2

Comparison between HAM-Based Approximation and Numerical Solution for $\gamma = 1$

τ	Numerical Solution	170th-Order HAM-Based Approximation
1.00	0.0000	0.0000
2.75	1.1485	1.1070
4.50	1.0771	1.2110
6.25	0.9252	0.9871
8.00	0.8037	0.7773
9.75	0.7118	0.6597
11.50	0.6411	0.6057
13.25	0.5855	0.5750
15.00	0.5405	0.5458
20.00	0.4494	0.4551
30.00	0.3481	0.3483
40.00	0.2918	0.2909
50.00	0.2552	0.2529

subject to the initial condition:

$$\Phi_i(\delta_i; q) = \beta_i, \ i = 1, 2, 3, \ldots, n \tag{7.28}$$

where subscript i represents the quantity for the i-th subproblem, q is the embedding parameter, $\Phi_i(\tau; q)$ is the representation of the solution across q, $E_{i,0}(\tau)$ is the initial approximation, $\hbar_i$ is the auxiliary parameter, $H_i(\tau)$ is the auxiliary function, and $\mathcal{L}_i$ and $\mathcal{N}$ are the linear and nonlinear operators, respectively. It can be seen that the nonlinear operator remains the same for each of the subdomains, as the original governing equation does not change. The parameter δ_i is the starting point of the domain for the i-th subdomain, and β_i is the value of the error at that point. For example, for the first subdomain, δ_1 and β_1 are 1 and 0, respectively, while for the next subdomain δ_2 is the end point of the domain considered for the first subdomain and β_2 is the value of the error at that point, which can be determined from the solution of the first subdomain. The higher-order terms can be calculated from the following higher-order deformation equation:

$$\mathcal{L}_i\left[E_{i,m}(\tau) - \chi_m E_{i,m-1}(\tau)\right] = \hbar_i H_i(\tau) R_{i,m}\left(\vec{E}_{i,m-1}\right),$$
$$m = 1, 2, 3, \ldots \text{ subject to } E_{i,m}(\delta_i) = 0; \ i = 1, 2, 3, \ldots, n \tag{7.29}$$

where

$$\chi_m = \begin{cases} 0 & \text{when } m = 1, \\ 1 & \text{otherwise} \end{cases} \tag{7.30}$$

and

$$R_{i,m}\left(\vec{E}_{i,m-1}\right) = \frac{1}{(m-1)!} \frac{\partial^{m-1} \mathcal{N}\left[\Phi_i\left(\tau; q\right)\right]}{\partial q^{m-1}}\bigg|_{q=0} \tag{7.31}$$

where $E_{i,m}$ are the higher-order terms. Now, the final solution for a particular subdomain can be obtained as follows:

$$E_i\left(\tau\right) = E_{i,0}\left(\tau\right) + \sum_{m=1}^{\infty} E_{i,m}\left(\tau\right) \tag{7.32}$$

Using the nonlinear operator as the original governing equation, term $R_{i,m}$ can be calculated as follows:

$$R_{i,m}\left(\vec{E}_{i,m-1}\right) = 2\tau \frac{dE_{i,m-1}}{d\tau} - 3\left(1 - \chi_m\right)$$
$$-3\left(1 - \frac{2}{3}\gamma^{0.5}\tau^{0.5}\right)E_{i,m-1} + \gamma^{0.5}\tau^{0.5} \sum_{j=0}^{m-1} E_{i,j} E_{i,m-1-j} \tag{7.33}$$

While solving the equation using standard HAM in the previous section, it was observed that the selected set of base functions is a suitable choice. Taking this into account, as the equation does not change for any of the subdomains, the same set of base functions is considered here. Hence, the linear operator and the initial approximation are chosen, respectively, as follows:

$$\mathcal{L}_i\left[\Phi_i\left(\tau; q\right)\right] = \frac{\partial \Phi_i\left(\tau; q\right)}{\partial \tau} \text{ with the property } \mathcal{L}_i\left[C_2\right] = 0 \tag{7.34}$$

$$E_{i,0}\left(\tau\right) = \beta_i \tag{7.35}$$

where C_2 is an integral constant. Using Eq. (7.34), the higher-order terms can be obtained from Eq. (7.29) as follows:

$$E_{i,m}\left(\tau\right) = \chi_m E_{i,m-1}\left(\tau\right) + \hbar_i \int_{\delta_i}^{\tau} H_i\left(x\right) R_{i,m}\left(\vec{E}_{i,m-1}\right) dx + C_2, \ m = 1,2,3,\dots \tag{7.36}$$

where $R_{i,m}$ is given by Eq. (7.33), and the constant C_2 can be determined from the initial condition for the higher-order deformation equations, i.e., $E_{i,m}\left(\delta_i\right) = 0$, which simply yields $C_2 = 0$ for all $m \geq 1$.

Following the discussion in the previous section, the auxiliary function $H_i\left(\tau\right)$ can be determined as $H_i\left(\tau\right) = \tau^{-0.5}$. Therefore, the approximate solution for a particular subdomain can be obtained as follows:

$$E_i\left(\tau\right) \approx E_{i,0}\left(\tau\right) + \sum_{m=1}^{M} E_{i,m}\left(\tau\right) \tag{7.37}$$

Finally, the approximate solution for the original equation can be obtained as:

$$E(\tau) = \begin{cases} E_1(\tau) \text{ for } \tau \in [1,\tau_1] \\ \quad . \\ \quad . \\ \quad . \\ E_n(\tau) \text{ for } \tau \in [\tau_{n-1},\tau_n] \end{cases} \tag{7.38}$$

For a suitable choice of the auxiliary parameter $\hbar_i$, the following squared residual error for a particular subdomain can be minimized:

$$\Delta_{i,m} = \int_{\tau \in \Omega_i} \left(\mathcal{N}\left[E_i(\tau) \right] \right)^2 d\tau \tag{7.39}$$

where Ω_i is the i-th subdomain, and $\Omega = \bigcup_{i=1}^{n} \Omega_i$. Here, we consider a test case where the parameter γ is selected as 1 and domain $\Omega = [1,50]$. The domain is now divided into three subdomains by decomposing the domain as $\Omega = [1,5] \cup [5,30] \cup [30,50]$. For a specific value of γ, say $\gamma = 1$, the squared residual error $\Delta_{i,m}$ is calculated for each of the subdomains with a different order of approximation, and is presented in Figure 7.3. It can be seen from the figure that for each of the subdomains, the residual error decreases with the increasing order of approximation. For all the cases, the

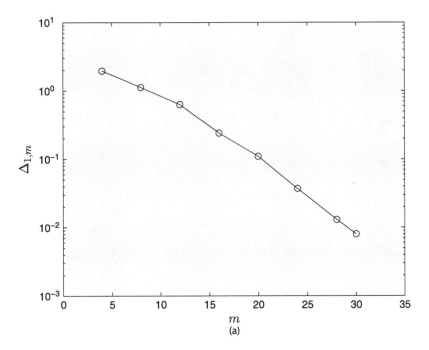

FIGURE 7.3 Squared residual error (Δ_m) versus different orders of approximation (m) of the HAM-based solution with $\gamma = 1$ for (a) $\tau \in [1,5]$, (b) $\tau \in [5,30]$, and (c) $\tau \in [30,50]$. (*Continued*)

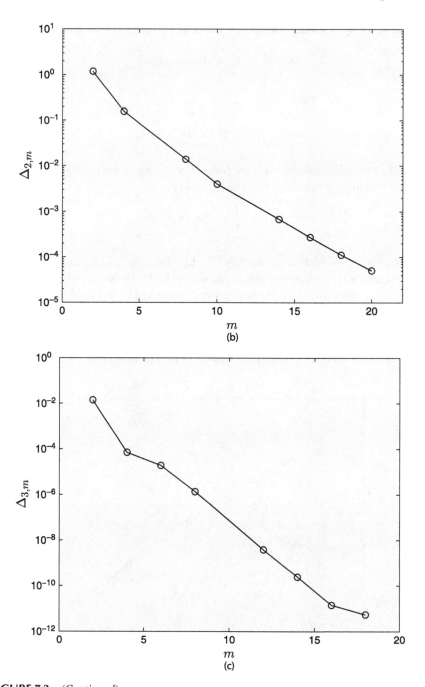

FIGURE 7.3   (*Continued*)

TABLE 7.3

Squared Residual Error (Δ_m) and Computational Time versus Different Orders of Approximation (m) for the First Subdomain $\tau \in [1, 5]$ with $\gamma = 1$

Order of Approximation (m)	Squared Residual Error (Δ_m)	Computational Time (sec)
4	1.954	1.08
8	1.123	2.32
12	0.631	3.73
16	0.240	5.36
20	0.110	7.09
24	0.037	9.04
28	0.013	11.24
30	0.008	12.47

residual error as well as the computational time are reported in Tables 7.3–7.5. It can be noted from both figures and tables that the modified method not only reduces the computational time but also enhances the accuracy of the solution as compared to the standard HAM method in the previous section.

Now, the modified method-based approximate solution is compared with a numerical solution for different values of parameter γ, specifically $\gamma = 1, 2, 3.5$, and 5. For that purpose, the original domain is converted into three subdomains, and HAM is applied to each of these subdomains. The solution at the end point of the domain in the first subdomain determines the initial condition for the next subdomain. Using this, the approximate solutions are obtained and compared in

TABLE 7.4

Squared Residual Error (Δ_m) and Computational Time versus Different Orders of Approximation (m) for the Second Subdomain $\tau \in [5, 30]$ with $\gamma = 1$

Order of Approximation (m)	Squared Residual Error (Δ_m)	Computational Time (sec)
2	1.184	1.29
4	0.156	2.63
8	0.014	5.43
10	0.004	6.94
14	6.81×10^{-4}	10.10
16	2.70×10^{-4}	11.74
18	1.12×10^{-4}	13.41
20	5.05×10^{-5}	15.11

TABLE 7.5

Squared Residual Error (Δ_m) and Computational Time versus Different Orders of Approximation (m) for the Third Subdomain $\tau \in [30, 50]$ with $\gamma = 1$

Order of Approximation (m)	Squared Residual Error (Δ_m)	Computational Time (sec)
2	0.014	1.88
4	7.03×10^{-5}	3.77
6	1.90×10^{-5}	5.79
8	1.37×10^{-6}	7.77
12	3.82×10^{-9}	11.90
14	2.45×10^{-10}	14.03
16	1.42×10^{-11}	16.22
18	5.41×10^{-12}	18.51

Figures 7.4(a)–(c). It can be noted that for the last two subdomains, only the 18th- or 20th- order approximations provide excellent accuracy, and thus reduce the efficiency as well as time complexity of the main equation. A quantitative assessment is carried out and present in Tables 7.6–7.8. For the present equation, we divide the domain into three subdomains, and it results in a satisfactory result in terms of accuracy and efficiency. Indeed, depending on the equation in hand, one can decompose the domain as per requirement. A flowchart diagram containing the steps of the method is provided in Figure 7.5.

7.5 HPM-BASED SOLUTION

We write Eq. (7.6) in the following form:

$$\mathcal{N}(E) = f(\tau) \tag{7.40}$$

We construct a homotopy $\Phi(\tau; q)$ that satisfies:

$$(1-q)\left[\mathcal{L}(\Phi(\tau;q)) - \mathcal{L}(E_0(\tau))\right] + q\left[\mathcal{N}(\Phi(\tau;q)) - f(\tau)\right] = 0 \tag{7.41}$$

where q is the embedding parameter that lies in $[0,1]$, $\mathcal{L}$ is the linear operator, and $E_0(\tau)$ is the initial approximation to the final solution. Eq. (7.41) shows that as $q = 0$, $\Phi(\tau;0) = E_0(\tau)$ and as $q = 1$, $\Phi(\tau;1) = E(\tau)$. We now express $\Phi(\tau;q)$ as a series in terms of q, as follows:

$$\Phi(\tau;q) = \Phi_0 + q\Phi_1 + q^2\Phi_2 + q^3\Phi_3 + q^4\Phi_4 \ldots \tag{7.42}$$

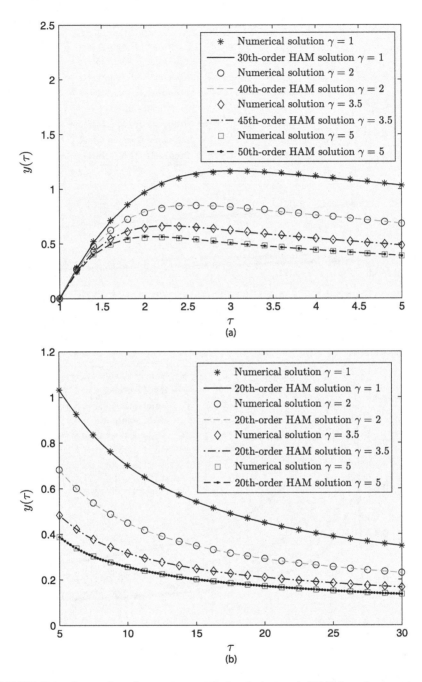

FIGURE 7.4 Comparison between numerical solution and HAM-based approximate solution with $\gamma = 1.0$, 2.0, 3.5, and 5.0 for (a) $\tau \in [1,5]$, (b) $\tau \in [5,30]$, (c) $\tau \in [30,50]$, and (d) $\tau \in [1,50]$. (*Continued*)

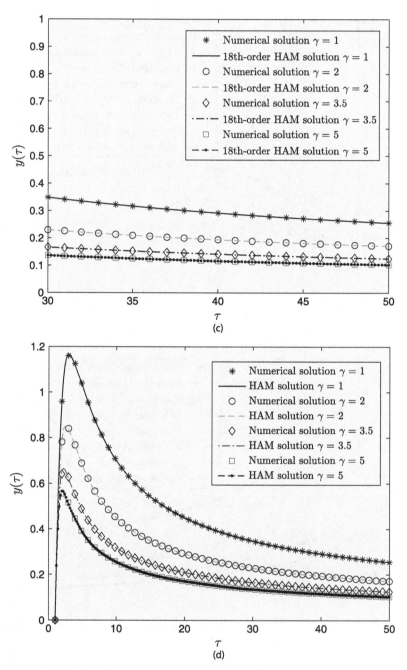

FIGURE 7.4   *(Continued)*

TABLE 7.6
Comparison between HAM-Based Approximation and Numerical Solution for the First Subdomain $\tau \in [1,5]$

	$\gamma = 1.0$		$\gamma = 2.0$		$\gamma = 3.5$		$\gamma = 5.0$	
τ	Numerical Solution	30th-Order HAM Approximation	Numerical Solution	40th-Order HAM Approximation	Numerical Solution	45th-Order HAM Approximation	Numerical Solution	50th-Order HAM Approximation
1.0	0.000	0.000	0.000	0.000	0.000	0.000	0.000	0.000
1.5	0.622	0.613	0.557	0.553	0.496	0.494	0.455	0.455
2.0	0.968	0.973	0.787	0.795	0.642	0.653	0.555	0.565
2.5	1.118	1.130	0.847	0.852	0.654	0.656	0.546	0.545
3.0	1.161	1.166	0.838	0.834	0.625	0.619	0.512	0.505
3.5	1.152	1.147	0.804	0.799	0.586	0.581	0.475	0.472
4.0	1.119	1.110	0.763	0.761	0.548	0.548	0.442	0.443
4.5	1.077	1.072	0.721	0.722	0.514	0.516	0.413	0.415
5.0	1.032	1.032	0.683	0.684	0.483	0.485	0.388	0.389

TABLE 7.7

Comparison between HAM-Based Approximation and Numerical Solution for the Second Subdomain $\tau \in [5,30]$

	$\gamma = 1.0$		$\gamma = 2.0$		$\gamma = 3.5$		$\gamma = 5.0$	
τ	Numerical Solution	20th-Order HAM Approximation	Numerical Solution	20th-Order HAM Approximation	Numerical Solution	20th-Order HAM Approximation	Numerical Solution	20th-Order HAM Approximation
5	1.032	1.032	0.683	0.683	0.483	0.483	0.388	0.388
7	0.869	0.869	0.561	0.560	0.394	0.393	0.316	0.316
9	0.748	0.747	0.479	0.479	0.337	0.337	0.272	0.272
11	0.660	0.659	0.422	0.423	0.299	0.299	0.241	0.242
13	0.593	0.593	0.380	0.381	0.270	0.270	0.219	0.219
15	0.541	0.541	0.348	0.348	0.248	0.248	0.201	0.201
17	0.499	0.499	0.322	0.322	0.230	0.230	0.187	0.187
19	0.464	0.464	0.301	0.300	0.216	0.215	0.176	0.176
21	0.436	0.435	0.283	0.283	0.204	0.204	0.166	0.167
23	0.411	0.411	0.268	0.268	0.193	0.194	0.158	0.158
25	0.390	0.390	0.255	0.255	0.184	0.185	0.151	0.151
27	0.372	0.371	0.244	0.244	0.176	0.176	0.144	0.144
30	0.348	0.348	0.229	0.229	0.166	0.166	0.136	0.136

TABLE 7.8
Comparison between HAM-Based Approximation and Numerical Solution for the Third Subdomain $\tau \in [30,50]$

τ	$\gamma = 1.0$		$\gamma = 2.0$		$\gamma = 3.5$		$\gamma = 5.0$	
	Numerical Solution	18th-Order HAM Approximation	Numerical Solution	18th-Order HAM Approximation	Numerical Solution	18th-Order HAM Approximation	Numerical Solution	18th-Order HAM Approximation
30	0.348	0.348	0.229	0.229	0.166	0.166	0.136	0.136
32	0.334	0.334	0.221	0.221	0.160	0.160	0.131	0.131
34	0.322	0.322	0.213	0.213	0.155	0.155	0.127	0.127
36	0.311	0.311	0.206	0.206	0.150	0.150	0.123	0.123
38	0.301	0.301	0.200	0.200	0.145	0.145	0.119	0.119
40	0.292	0.292	0.194	0.194	0.141	0.141	0.116	0.116
42	0.283	0.283	0.189	0.188	0.138	0.138	0.113	0.113
44	0.275	0.275	0.184	0.184	0.134	0.134	0.110	0.110
46	0.268	0.268	0.179	0.179	0.131	0.131	0.108	0.108
48	0.261	0.261	0.175	0.175	0.128	0.128	0.105	0.105
50	0.255	0.255	0.171	0.171	0.125	0.125	0.103	0.103

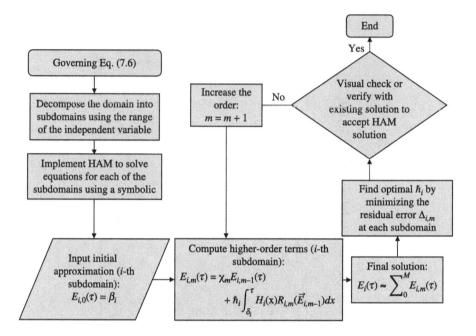

FIGURE 7.5   Flowchart for the modified HAM solution in Eq. (7.38).

where Φ_m for $m \geq 1$ are the higher-order terms. As $q \to 1$, Eq. (7.42) produces the final solution as:

$$E(\tau) = \lim_{q \to 1} \Phi(\tau; q) = \sum_{i=0}^{\infty} \Phi_i \qquad (7.43)$$

For simplicity, we consider the linear and nonlinear operators as follows:

$$\mathcal{L}\big(\Phi(\tau; q)\big) = 2\tau \frac{\partial \Phi(\tau; q)}{\partial \tau} - 3 \qquad (7.44)$$

$$\mathcal{N}\big(\Phi(\tau; q)\big) = 2\tau \frac{\partial \Phi(\tau; q)}{\partial \tau} - 3 - 3\left(1 - \frac{2}{3}\gamma^{0.5}\tau^{0.5}\right)\Phi(\tau; q) + \gamma^{0.5}\tau^{0.5}\Phi(\tau; q)^2 \quad (7.45)$$

Using these expressions, Eq. (7.41) takes on the form:

$$2\tau(1-q)\left[\frac{\partial \Phi(\tau; q)}{\partial \tau} - \frac{dE_0}{d\tau}\right]$$

$$+ q\left[2\tau \frac{\partial \Phi(\tau; q)}{\partial \tau} - 3 - 3\left(1 - \frac{2}{3}\gamma^{0.5}\tau^{0.5}\right)\Phi(\tau; q) + \gamma^{0.5}\tau^{0.5}\Phi(\tau; q)^2\right] = 0 \qquad (7.46)$$

The following expression can be used for finding the terms of the series:

$$\left(\Phi_0 + q\Phi_1 + q^2\Phi_2 + q^3\Phi_3 + q^4\Phi_4 \cdots\right)^2 = \left(\Phi_0\right)^2 + q\left(2\Phi_0\Phi_1\right)$$
$$+ q^2\left(\Phi_1^2 + 2\Phi_0\Phi_2\right) + q^3\left(2\Phi_0\Phi_3 + 2\Phi_1\Phi_2\right) + \cdots \tag{7.47}$$

Using Eq. (7.47), we have:

$$(1-q)\left[\frac{\partial\Phi(\tau;q)}{\partial\tau} - \frac{dE_0}{d\tau}\right]$$

$$= (1-q)\left[\left(\frac{d\Phi_0}{d\tau} - \frac{dE_0}{d\tau}\right) + q\frac{d\Phi_1}{d\tau} + q^2\frac{d\Phi_2}{d\tau} + q^3\frac{d\Phi_3}{d\tau} + q^4\frac{d\Phi_4}{d\tau} + \cdots\right]$$

$$= \left(\frac{d\Phi_0}{d\tau} - \frac{dE_0}{d\tau}\right) + q\left(\frac{d\Phi_1}{d\tau} - \left(\frac{d\Phi_0}{d\tau} - \frac{dE_0}{d\tau}\right)\right) + q^2\left(\frac{d\Phi_2}{d\tau} - \frac{d\Phi_1}{d\tau}\right)$$

$$+ q^3\left(\frac{d\Phi_3}{d\tau} - \frac{d\Phi_2}{d\tau}\right) + q^4\left(\frac{d\Phi_4}{d\tau} - \frac{d\Phi_3}{d\tau}\right) + \cdots \tag{7.48}$$

$$2\tau\frac{\partial\Phi(\tau;q)}{\partial\tau} - 3 - 3\left(1 - \frac{2}{3}\gamma^{0.5}\tau^{0.5}\right)\Phi(\tau;q) + \gamma^{0.5}\tau^{0.5}\Phi(\tau;q)^2$$

$$= 2\tau\left(\frac{d\Phi_0}{d\tau} + q\frac{d\Phi_1}{d\tau} + q^2\frac{d\Phi_2}{d\tau} + q^3\frac{d\Phi_3}{d\tau} + \cdots\right)$$

$$- 3 - 3\left(1 - \frac{2}{3}\gamma^{0.5}\tau^{0.5}\right)\left\{\Phi_0 + q\Phi_1 + q^2\Phi_2 + q^3\Phi_3 + \cdots\right\}$$

$$+ \gamma^{0.5}\tau^{0.5}\left\{\left(\Phi_0\right)^2 + q\left(2\Phi_0\Phi_1\right) + q^2\left(\Phi_1^2 + 2\Phi_0\Phi_2\right) + q^3\left(2\Phi_0\Phi_3 + 2\Phi_1\Phi_2\right) + \cdots\right\}$$

$$= \left\{2\tau\frac{d\Phi_0}{d\tau} - 3 - 3\left(1 - \frac{2}{3}\gamma^{0.5}\tau^{0.5}\right)\Phi_0 + \gamma^{0.5}\tau^{0.5}\Phi_0^2\right\}$$

$$+ q\left[2\tau\frac{d\Phi_1}{d\tau} - 3\left(1 - \frac{2}{3}\gamma^{0.5}\tau^{0.5}\right)\Phi_1 + 2\gamma^{0.5}\tau^{0.5}\Phi_0\Phi_1\right]$$

$$+ q^2\left[2\tau\frac{d\Phi_2}{d\tau} - 3\left(1 - \frac{2}{3}\gamma^{0.5}\tau^{0.5}\right)\Phi_2 + \gamma^{0.5}\tau^{0.5}\left(\Phi_1^2 + 2\Phi_0\Phi_2\right)\right]$$

$$+ q^3\left[2\tau\frac{d\Phi_3}{d\tau} - 3\left(1 - \frac{2}{3}\gamma^{0.5}\tau^{0.5}\right)\Phi_3 + \gamma^{0.5}\tau^{0.5}\left(2\Phi_0\Phi_3 + 2\Phi_1\Phi_2\right)\right] + \cdots \tag{7.49}$$

Using the initial condition $\Phi(\tau = 1) = 0$, we have:

$$\Phi_0(1) = 0, \ \Phi_1(1) = 0, \ \Phi_2(1) = 0, \ \ldots . \tag{7.50}$$

Using Eqs. (7.48) and (7.49) and equating the like powers of q in Eq. (7.46), the following system of differential equations is obtained:

$$2\tau\left(\frac{d\Phi_0}{d\tau} - \frac{dE_0}{d\tau}\right) = 0 \text{ subject to } \Phi_0(1) = 0 \qquad (7.51)$$

$$2\tau\left(\frac{d\Phi_1}{d\tau} - \left(\frac{d\Phi_0}{d\tau} - \frac{dE_0}{d\tau}\right)\right) + 2\tau\frac{d\Phi_0}{d\tau} - 3 - 3\left(1 - \frac{2}{3}\gamma^{0.5}\tau^{0.5}\right)\Phi_0$$
$$+ \gamma^{0.5}\tau^{0.5}\Phi_0{}^2 = 0 \text{ subject to } \Phi_1(1) = 0 \qquad (7.52)$$

$$2\tau\left(\frac{d\Phi_2}{d\tau} - \frac{d\Phi_1}{d\tau}\right) + 2\tau\frac{d\Phi_1}{d\tau} - 3\left(1 - \frac{2}{3}\gamma^{0.5}\tau^{0.5}\right)\Phi_1$$
$$+ 2\gamma^{0.5}\tau^{0.5}\Phi_0\Phi_1 = 0 \text{ subject to } \Phi_2(1) = 0 \qquad (7.53)$$

$$2\tau\left(\frac{d\Phi_3}{d\tau} - \frac{d\Phi_2}{d\tau}\right) + 2\tau\frac{d\Phi_2}{d\tau} - 3\left(1 - \frac{2}{3}\gamma^{0.5}\tau^{0.5}\right)\Phi_2$$
$$+ \gamma^{0.5}\tau^{0.5}\left(\Phi_1{}^2 + 2\Phi_0\Phi_2\right) = 0 \text{ subject to } \Phi_3(1) = 0 \qquad (7.54)$$

$$2\tau\left(\frac{d\Phi_4}{d\tau} - \frac{d\Phi_3}{d\tau}\right) + 2\tau\frac{d\Phi_3}{d\tau} - 3\left(1 - \frac{2}{3}\gamma^{0.5}\tau^{0.5}\right)\Phi_3$$
$$+ \gamma^{0.5}\tau^{0.5}\left(2\Phi_0\Phi_3 + 2\Phi_1\Phi_2\right) = 0 \text{ subject to } \Phi_4(1) = 0 \qquad (7.55)$$

The initial approximation can be chosen as $E_0 = \frac{3}{2}\ln\tau$, which is indeed a solution of part of the original equation. Using this initial approximation, we can solve the equations iteratively using symbolic software. We do not include those expressions here, as they are lengthy. Finally, the HPM-based solution can be approximated as follows:

$$E(\tau) \approx \sum_{i=0}^{M}\Phi_i \qquad (7.56)$$

The previous sections, which develop the HAM-based solution for the equation, show that one can obtain a good approximate solution if the domain is subdivided into several parts before applying the methodology. Therefore, this is the same for the HPM/OHAM methods. However, we will restrict our analysis to a small region of the domain, as the objective is to show the application of these methods. The HPM-based analytical solution in Eq. (7.56) is validated over a numerical solution obtained using the function *ode45* from MATLAB. Figure 7.6 shows the comparison between the two-term, three-term, and four-term HPM-based solutions and the corresponding numerical solution over $\tau \in [1,3]$. It is observed from the figure that only the two-term HPM solution produces good approximations to the actual solution. The solution starts deviating when the parameter γ takes on higher values. Furthermore, following the previous section, to achieve an accurate solution over the entire domain of τ, one can indeed subdivide the domain into several subdomains and then apply HPM to each of the subdomains. For a quantitative assessment, we consider the two-term order HPM approximation and report it in Table 7.9

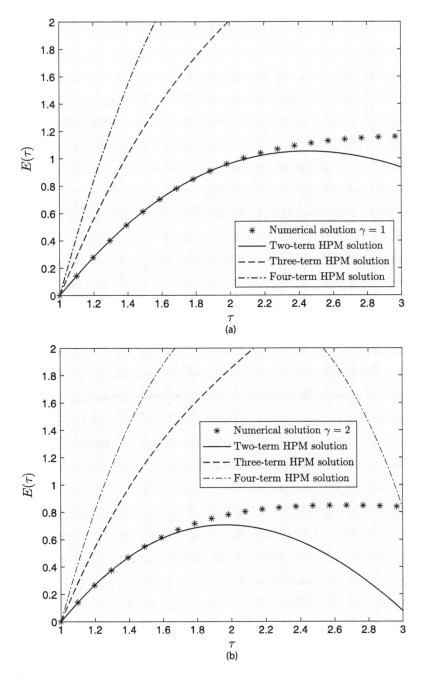

FIGURE 7.6 Comparison between the numerical solution and two-, three-, and four-term HPM-based solutions with (a) $\gamma = 1.0$, (b) $\gamma = 2.0$, (c) $\gamma = 2.5$, and (d) $\gamma = 5.0$. (*Continued*)

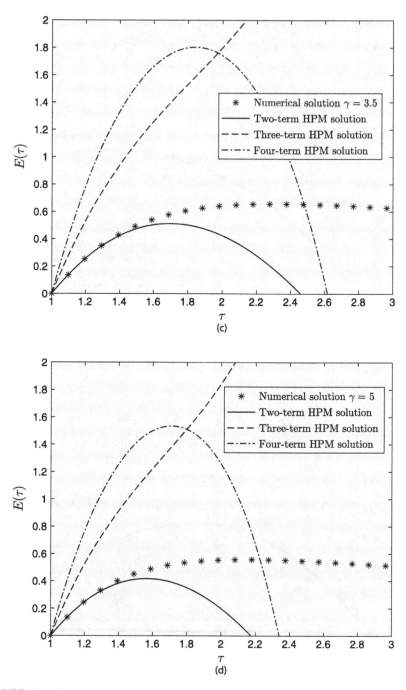

FIGURE 7.6 *(Continued)*

TABLE 7.9

Comparison between Two-Term HPM-Based Approximation and Numerical Solution for $\tau \in [1,3]$

τ	γ = 1.0		γ = 2.0		γ = 3.5		γ = 5.0	
	Numerical Solution	Two-Term HPM Approximation	Numerical Solution	Two-Term HPM Approximation	Numerical Solution	Two-Term HPM Approximation	Numerical Solution	Two-Term HPM Approximation
1.0	0.0000	0.0000	0.0000	0.0000	0.0000	0.000	0.0000	0.0000
1.2	0.2821	0.2819	0.2700	0.2700	0.2575	0.2567	0.2482	0.2462
1.4	0.5207	0.5208	0.4764	0.4747	0.4338	0.4239	0.4037	0.3833
1.6	0.7121	0.7130	0.6239	0.6133	0.5447	0.5035	0.4919	0.4156
1.8	0.8591	0.8593	0.7234	0.6891	0.6090	0.5014	0.5366	0.3513
2.0	0.9679	0.9624	0.7867	0.7065	0.6423	0.4244	0.5549	0.1988
2.2	1.0455	1.0253	0.8239	0.6705	0.6557	0.2793	0.5576	-0.0336
2.4	1.0985	1.0512	0.8427	0.5855	0.6566	0.0721	0.5514	-0.3385
2.6	1.1324	1.0429	0.8488	0.4557	0.6498	-0.1915	0.5403	-0.7093
2.8	1.1520	1.0032	0.8463	0.2850	0.6385	-0.5067	0.5268	-1.1399
3.0	1.1606	0.9346	0.8381	0.0767	0.6247	-0.8690	0.5121	-1.6254

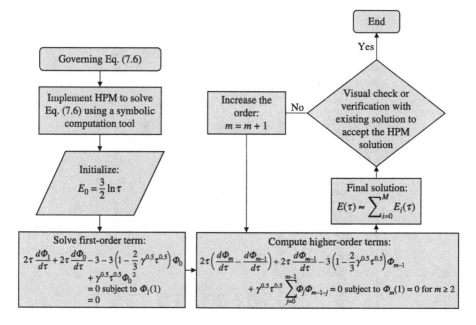

FIGURE 7.7 Flowchart for the HPM solution in Eq. (7.56).

by comparing it with a numerical solution. A flowchart containing the steps of the method is given in Figure 7.7.

7.6 OHAM-BASED SOLUTION

When applying OHAM, the approximate Eq. (7.6) can be written in the following form:

$$\mathcal{L}\big[E(\tau)\big] + \mathcal{N}\big[E(\tau)\big] + h(\tau) = 0 \tag{7.57}$$

where $\mathcal{L}\big[E(\tau)\big] = 2\tau\dfrac{\partial E}{\partial \tau} - 3E$, $\mathcal{N}\big[E(\tau)\big] = -3\left(-\dfrac{2}{3}\gamma^{0.5}\tau^{0.5}\right)E + \gamma^{0.5}\tau^{0.5}E^2$, and $h(\tau) = -3$. With these considerations, the zeroth-order problem becomes:

$$\mathcal{L}\big(E_0(\tau)\big) + h(\tau) = 0 \text{ subject to } E_0(1) = 0 \tag{7.58}$$

Solving Eq. (7.54), one obtains:

$$E_0(\tau) = \tau^{3/2} - 1 \tag{7.59}$$

To obtain $\mathcal{N}_0$, $\mathcal{N}_1$, $\mathcal{N}_2$, etc., the following can be used:

$$-3\left(-\frac{2}{3}\gamma^{0.5}\tau^{0.5}\right)E + \gamma^{0.5}\tau^{0.5}E^2 = -3\left(-\frac{2}{3}\gamma^{0.5}\tau^{0.5}\right)\left\{E_0 + qE_1 + q^2E_2 + q^3E_3 + \cdots\right\}$$

$$+ \gamma^{0.5}\tau^{0.5}\left\{(E_0)^2 + q(2E_0E_1) + q^2\left(E_1^2 + 2E_0E_2\right) + q^3\left(2E_0E_3 + 2E_1E_2\right) + \cdots\right\}$$

$$= \left\{-3\left(-\frac{2}{3}\gamma^{0.5}\tau^{0.5}\right)E_0 + \gamma^{0.5}\tau^{0.5}E_0^2\right\} + q\left[-3\left(-\frac{2}{3}\gamma^{0.5}\tau^{0.5}\right)E_1 + 2\gamma^{0.5}\tau^{0.5}E_0E_1\right]$$

$$+ q^2\left[-3\left(-\frac{2}{3}\gamma^{0.5}\tau^{0.5}\right)E_2 + \gamma^{0.5}\tau^{0.5}\left(E_1^2 + 2E_0E_2\right)\right]$$

$$+ q^3\left[-3\left(-\frac{2}{3}\gamma^{0.5}\tau^{0.5}\right)E_3 + \gamma^{0.5}\tau^{0.5}\left(2E_0E_3 + 2E_1E_2\right)\right] + \cdots \quad (7.60)$$

Using Eq. (7.60), the first-order problem reduces to:

$$2\tau\frac{dE_1}{d\tau} - 3E_1 = H_1(\tau,C_i)\left\{-3\left(-\frac{2}{3}\gamma^{0.5}\tau^{0.5}\right)E_0 + \gamma^{0.5}\tau^{0.5}E_0^2\right\} \text{ subject to } E_1(1) = 0 \quad (7.61)$$

We choose $H_1(\tau,C_i) = C_1 + C_2\tau^{0.5}$. Putting $k = 2$, the second-order problem becomes:

$$\mathcal{L}\left[E_2(\tau,C_i) - E_1(\tau,C_i)\right] = H_2(\tau,C_i)\mathcal{N}_0\left(E_0(\tau)\right)$$
$$+ H_1(\tau,C_i)\left[\mathcal{L}\left[E_1(\tau,C_i)\right] + \mathcal{N}_1\left[E_0(\tau),E_1(\tau,C_i)\right]\right] \text{ subject to } E_2(1) = 0 \quad (7.62)$$

Further, we choose $H_2(\tau,C_i) = C_3$. Therefore, Eq. (7.62) becomes:

$$2\tau\frac{dE_2}{d\tau} = 2\tau\frac{dE_1}{d\tau} + C_3\left\{-3\left(1 - \frac{2}{3}\gamma^{0.5}\tau^{0.5}\right)E_0 + \gamma^{0.5}\tau^{0.5}E_0^2\right\}$$

$$+ \left(C_1 + C_2\tau^{0.5}\right)\left[2\tau\frac{dE_1}{d\tau} - 3E_1 - 3\left(-\frac{2}{3}\gamma^{0.5}\tau^{0.5}\right)E_1 + 2\gamma^{0.5}\tau^{0.5}E_0E_1\right]$$

$$\text{subject to } E_2(1) = 0 \quad (7.63)$$

It can be seen that we do not write the explicit forms of the terms here, as those are lengthy and may not be appropriate to keep up the flow. Indeed, one can compute these terms without any difficulty using symbolic computation software, such as MATLAB. Further, following Chapter 2, the higher-order terms can be computed in a similar manner. However, our aim is to produce an accurate solution with just two to three terms of the OHAM-based series. Therefore, we restrict our calculation up to $k = 2$. Finally, the approximate solution can be found as:

$$E(\tau) \approx E_0(\tau) + E_1(\tau,C_1,C_2) + E_2(\tau,C_1,C_2,C_3) \quad (7.64)$$

where terms are given by Eqs. (7.59), (7.61), and (7.63).

For the assessment of the OHAM-based analytical solution, one needs to calculate the constant C_i. For that purpose, we calculate the residual as follows:

$$R(\tau, C_i) = \mathcal{L}[E_{OHAM}(\tau, C_i)] + \mathcal{N}[E_{OHAM}(\tau, C_i)] + h(\tau),\ i = 1, 2, \ldots, s \quad (7.65)$$

where $E_{OHAM}(\tau, C_i)$ is the approximate solution. When $R(\tau, C_i) = 0$, $E_{OHAM}(\tau, C_i)$ becomes the exact solution to the equation. One of the ways to obtain the optimal C_i for which the solution converges is the minimization of squared residual error, i.e.,

$$J(C_i) = \int_{\tau \in D} R^2(\tau, C_i)\, d\tau,\ i = 1, 2, \ldots, s \quad (7.66)$$

where D is the domain of the equation. The minimization of Eq. (7.66) leads to a system of algebraic equations as follows:

$$\frac{\partial J}{\partial C_1} = \frac{\partial J}{\partial C_2} = \cdots = \frac{\partial J}{\partial C_s} = 0 \quad (7.67)$$

Solving this system, one can obtain the optimal values of the parameters. We obtain the optimal values using the MATLAB routine *fminsearch*, which minimizes an unconstrained multivariable function. Using these values, two- and three-term OHAM solutions are computed and compared in Figure 7.8 for $\gamma = 1$ with a

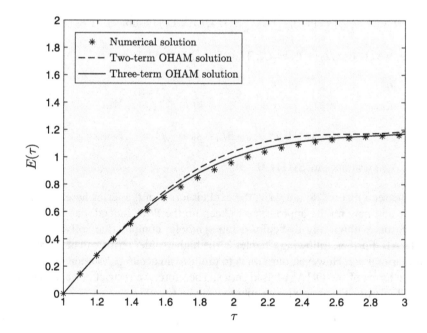

FIGURE 7.8　Comparison between numerical solution and two-, three-term OHAM-based solution with $\gamma = 1.0$.

TABLE 7.10

Comparison between OHAM-Based Approximation and Numerical Solution for $\tau \in [1,3]$

		$\gamma = 1.0$	
	Numerical	Two-Term OHAM	Three-Term OHAM
τ	Solution	Approximation	Approximation
1.0	0.0000	0.0000	0.0000
1.2	0.2821	0.2880	0.2854
1.4	0.5207	0.5437	0.5330
1.6	0.7121	0.7576	0.7349
1.8	0.8591	0.9247	0.8895
2.0	0.9679	1.0439	1.0001
2.2	1.0455	1.1186	1.0731
2.4	1.0985	1.1561	1.1173
2.6	1.1324	1.1686	1.1420
2.8	1.1520	1.1722	1.1567
3.0	1.1606	1.1879	1.1702

numerical solution to see the effectiveness of the method. For a quantitative assessment, we consider OHAM approximation and report it in Table 7.10 by comparing it with a numerical solution. It can be seen that the OHAM solution provides excellent approximation to the solution. A flowchart diagram containing the steps of the methods is provided in Figure 7.9.

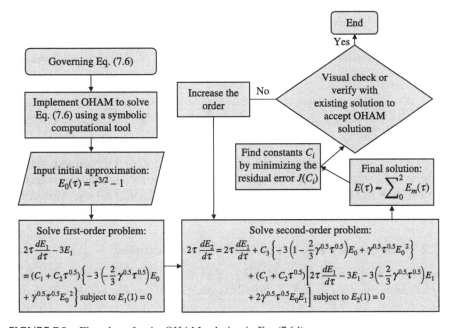

FIGURE 7.9 Flowchart for the OHAM solution in Eq. (7.64).

7.7 CONVERGENCE THEOREMS

The convergence of the HAM- and OHAM-based solutions is proved theoretically using the following theorems.

7.7.1 CONVERGENCE THEOREM OF HAM-BASED SOLUTION

Theorem 7.1: If the homotopy series $\Sigma_{m=0}^{\infty} E_{i,m}(\tau)$ and $\Sigma_{m=0}^{\infty} E_{i,m}{}'(\tau)$ converge, then $R_{i,m}\left(\vec{E}_{m-1}\right)$ given by Eq. (7.33) satisfies the relation $\Sigma_{m=1}^{\infty} R_{i,m}\left(\vec{E}_{m-1}\right) = 0$, $i=1,2,3,...,n$. Here, ''' denotes the derivative with respect to τ.

Proof: The proof of this theorem follows exactly from similar theorems of the previous chapters.

Theorem 7.2: If $\hbar_i$ is so properly chosen that the series $\Sigma_{m=0}^{\infty} E_{i,m}(\tau)$ and $\Sigma_{m=0}^{\infty} E_{i,m}{}'(\tau)$ converge absolutely to $E_i(\tau)$ and $E_i{}'(\tau)$, respectively, then the homotopy series $\Sigma_{m=0}^{\infty} E_{i,m}(\tau)$ satisfies the original governing Eq. (7.6).

Proof: Using the Cauchy product defined in Theorem 2.2 of Chapter 2 in relation to Eq. (7.33), we get:

$$\sum_{m=1}^{\infty}\sum_{j=0}^{m-1} E_{i,j} E_{i,m-j-1} = \left(\sum_{m=0}^{\infty} E_{i,m}\right)^2 \tag{7.68}$$

Theorem 7.1 shows that if $\Sigma_{m=0}^{\infty} E_{i,m}(\tau)$ and $\Sigma_{m=0}^{\infty} E_{i,m}{}'(\tau)$ converge, then $\Sigma_{m=1}^{\infty} R_m\left(\vec{E}_{m-1}\right) = 0$.

Therefore, substituting the earlier expressions in Eq. (7.33) and simplifying further lead to:

$$2\tau\sum_{m=0}^{\infty} E_{i,m}' - 3\sum_{m=0}^{\infty}(1-\chi_{m+1}) - 3\left(1-\frac{2}{3}\gamma^{0.5}\tau^{0.5}\right)\sum_{m=0}^{\infty} E_{i,m} + \gamma^{0.5}\tau^{0.5}\left(\sum_{m=0}^{\infty} E_{i,m}\right)^2 = 0 \tag{7.69}$$

which is basically the original governing equation in Eq. (7.6). Furthermore, subject to the initial condition $E_0\left(\delta_i\right) = \beta_i$ and the conditions for the higher-order deformation equation $E_m\left(\delta_i\right) = 0$, for $m \geq 1$, we easily obtain $\Sigma_{m=0}^{\infty} E_m\left(\delta_i\right) = \beta_i$. Hence, the convergence result follows.

7.7.2 CONVERGENCE THEOREM OF OHAM-BASED SOLUTION

Theorem 7.3: If the series $E_0(\tau) + \Sigma_{j=1}^{\infty} E_j(\tau, C_i)$, $i = 1,2,...,s$ converges, where $E_j(\tau, C_i)$ are governed by Eqs. (7.59), (7.61), and (7.63), then Eq. (7.64) is a solution of the original Eq. (7.6).

Proof: The proof of this theorem follows exactly from Theorem 2.3 of Chapter 2.

7.8   CONCLUDING REMARKS

A modified method based on HAM is applied here to derive an efficient analytical approximate solution for the error equation for unsteady uniform flow. It is observed that the standard HAM cannot produce a satisfactory result even for a large number of iterations and also requires a lot of computational time. To overcome these, a modification is made where the original equation domain can be converted by dividing the domain into several smaller subdomains. HAM is then applied for each of these subdomains, which produces an efficient analytical approximation to the original equation. In addition, the new method reduces the time complexity of the equation. Considering different values of the parameter present in the equation, several test cases are performed to establish the adequacy of the method. Moreover, the HPM- and OHAM-based solutions are also developed for the considered equation. The solutions obtained using these methods are analyzed for a small region of the domain. Indeed, one can follow the same method that is proposed for HAM to obtain more accurate solutions. It is seen that two-term HPM can provide a relatively accurate solution to the original equation. However, both two- and three-term OHAM-based solutions produce excellent approximations. It is suggested to use this method for different kinds of equations in hydrology and water resources.

REFERENCES

Singh, V. P. (1995). *Kinematic wave modeling in water resources: Surface water hydrology.* John Wiley, New York, NY.

Zhao, Y., and Liao, S. (2002). User Guide to BVPh 2.0. School of naval architecture, ocean and civil engineering, Shanghai, *40*.

FURTHER READING

Abbasbandy, S. (2006). Iterated He's homotopy perturbation method for quadratic Riccati differential equation. *Applied Mathematics and Computation*, *175*(1), 581–589.

Abbasbandy, S. (2006). Homotopy perturbation method for quadratic Riccati differential equation and comparison with Adomian's decomposition method. *Applied Mathematics and Computation*, *172*(1), 485–490.

Aminikhah, H., and Hemmatnezhad, M. (2010). An efficient method for quadratic Riccati differential equation. *Communications in Nonlinear Science and Numerical Simulation*, *15*(4), 835–839.

Jafari, H., and Firoozjaee, M. A. (2010). Multistage homotopy analysis method for solving non-linear Riccati differential equations. *IJRRAS*, *4*(2), 128–132.

Riccati, H. P., Mabood, F., Izanl, A., Ismai, M., and Hashim, I. (2013). Application of optimal homotopy asymptotic method for the approximate solution of Riccati equation. *Sains Malaysiana*, *42*(6), 863–867.

Shah, N. A., Ahmad, I., Omar, B., Abouelregal, A. E., and Ahmad, H. (2020). Multistage optimal homotopy asymptotic method for the nonlinear Riccati ordinary differential equation in nonlinear physics. *Applied Mathematics and Information Sciences*, *14*(6), 1–7.

Tan, Y., and Abbasbandy, S. (2008). Homotopy analysis method for quadratic Riccati differential equation. *Communications in Nonlinear Science and Numerical Simulation*, *13*(3), 539–546.

Vahidi, A. R., Azimzadeh, Z., and Didgar, M. (2013). An efficient method for solving Riccati equation using homotopy perturbation method. *Indian Journal of Physics*, *87*(5), 447–454.

8 Spatially Varied Flow Equations

8.1 INTRODUCTION

A steady, gradually varied channel flow equation is needed to construct flow profiles. Depending on the slope, normal depth, initial flow depth, and downstream boundary condition, 12 different kinds of flow profiles develop (Cruise et al. 2007). These profiles are classified based on the hydraulic slope and the relation between normal depth and critical depth for a specific flow discharge. The hydraulic slope is classified as a mild slope designated by symbol M; a steep slope designated by symbol S; a horizontal slope designated by symbol H; a critical slope designated by symbol C; and an adverse slope designated by symbol A. For the mild slope (the flow depth is greater than the critical depth), three types of flow profiles can develop, designated as M-1, where flow is subcritical and the depth is greater than the normal depth; M-2, where flow is subcritical and the depth is less than the normal depth, and the depth is greater than the critical depth but tends to approach the critical flow and depth; and M-3, where flow is supercritical and the depth is less than the critical depth but tends to approach the critical flow and depth. For the steep slope (the normal flow depth is less than the critical depth), three types of flow profiles can develop, designated as S-1, where the flow is supercritical and the depth is less than the critical depth but tends to become subcritical; S-2, where the flow is supercritical and the flow depth is less than the critical depth but tends to approach the normal depth; and S-3, where the flow is supercritical and the depth is less than the critical depth but tends to approach the normal depth. For the horizontal slope (the normal depth is not defined but the critical depth is) the profiles include H-2, where the flow is subcritical and tends to approach the critical depth, and H-3, where the flow is supercritical but approaches the critical depth. For the critical slope (the critical depth is equal to the normal depth), the profiles include C-1, where the flow depth is critical but becomes subcritical, and C-3, where the flow is supercritical but tends to be critical. For the adverse slope (the critical depth is defined but not the normal depth), the profile types are C-2, where the flow is subcritical but tends to be critical, and C-3, where the flow is supercritical but tends to be critical. These flow profiles may be modified if the channel is subjected to a lateral inflow, which can be uniform, decreasing, or increasing. A flow profile with uniform lateral flow constitutes the subject matter of this chapter under a mild slope.

DOI: 10.1201/9781003368984-11

8.2 GOVERNING EQUATION

The governing equation for a steady, spatially varied flow with uniformly increasing discharge can be given as (Chow 1959, 1969; Gill 1977):

$$\frac{dy}{dx} = \frac{S_0 - S_f - \dfrac{2\beta Q q_*}{gA^2}}{1 - \dfrac{\beta Q^2}{gA^2 y}} \tag{8.1}$$

where LHS stands for the slope of the water surface with respect to the channel bed; S_0 and S_f are the channel bed slope and energy slope, respectively; Q is the flow discharge at x; $q_* = \dfrac{dQ}{dx} = \dfrac{Q_0}{L}$; Q_0 is the discharge at the downstream location; L is the length of the channel; g is the acceleration due to gravity; A is the cross-sectional area of flow; y is the flow depth at location x; and β is the momentum coefficient that accounts for the nonuniformity in the distribution of velocity. Figure 8.1 shows a schematic diagram of spatially varied flow. As the objective of this work is to find a better analytical approximate solution to the governing equation, for simplicity, the momentum coefficient β is assumed to be unity.

It may be convenient to nondimensionalize Eq. (8.1) by defining dimensionless variables as follows:

$$Q = \frac{Q_0 x}{L}, \ x_* = \frac{x}{L}, \ y_* = \frac{y}{y_d}, \ A = by = by_d y_*, \ F_0 = \frac{Q_0}{b\sqrt{gy_d^3}}, \ G = \frac{S_0 L}{y_d} \tag{8.2}$$

Using Eq. (8.2), Eq. (8.1) can be written as:

$$\frac{dy_*}{dx_*} = \frac{Gy_*^3 - \dfrac{S_f L y_*^3}{y_d} - 2x_* y_* F_0^2}{y_*^3 - F_0^2 x_*^2} \tag{8.3}$$

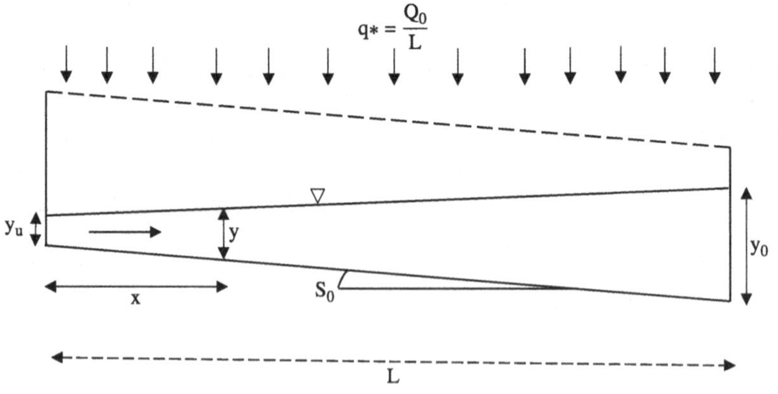

$$q* = \frac{Q_0}{L}$$

FIGURE 8.1 Schematic diagram of spatially varied flow.

Using the Darcy-Weisbach equation for hydraulic friction, we get $S_f = \dfrac{f}{8} \dfrac{F_0^2 x_*^2}{y_*^3}$. Therefore, Eq. (8.3) takes on the form:

$$\frac{dy_*}{dx_*} = \frac{Gy_*^3 - \dfrac{G_f F_0^2 x_*^2}{8} - 2x_* y_* F_0^2}{y_*^3 - F_0^2 x_*^2} \tag{8.4}$$

where $G_f = \dfrac{fL}{y_d}$, F_0 is the Froude number at the downstream location, b is the constant width of the channel, y_d is the depth at the downstream section, and f is the Darcy-Weisbach coefficient of friction. The initial condition for Eq. (8.4) can be given as $y = y_d$ at $x = L$, i.e., $y_* = 1$ at $x_* = 1$. For brevity, henceforth, the asterisk will be dropped in the subsequent parts of the chapter.

8.3   HAM-BASED SOLUTION

Considering different flow cases, analytical solutions of the governing equation based on the homotopy analysis method (HAM) are derived in the following. First, we describe HAM in general form and then apply it to specific forms of the equation.

8.3.1   GENERAL METHODOLOGY

To solve Eq. (8.4) using HAM, we construct the zeroth-order deformation equation as follows:

$$(1-q)\mathcal{L}\left[\Phi(x;q) - y_0(x)\right] = q\hbar H(x)\mathcal{N}\left[\Phi(x;q)\right] \tag{8.5}$$

subject to the initial condition:

$$\Phi(1;q) = 1 \tag{8.6}$$

where q is the embedding parameter, $\Phi(x;q)$ is the representation of solution across q, $y_0(x)$ is the initial approximation, $\hbar$ is the auxiliary parameter, $H(x)$ is the auxiliary function, and $\mathcal{L}$ and $\mathcal{N}$ are the linear and nonlinear operators, respectively. The higher-order terms can be calculated from the following higher-order deformation equation:

$$\mathcal{L}\left[y_m(x) - \chi_m y_{m-1}(x)\right] = \hbar H(x) R_m(\vec{y}_{m-1}), \ m = 1, 2, 3, \dots \text{ subject to } y_m(0) = 0 \tag{8.7}$$

where

$$\chi_m = \begin{cases} 0 & \text{when } m = 1, \\ 1 & \text{otherwise} \end{cases} \tag{8.8}$$

and

$$R_m\left(\vec{y}_{m-1}\right) = \frac{1}{(m-1)!} \frac{\partial^{m-1} \mathcal{N}\left[\Phi(x;q)\right]}{\partial q^{m-1}} \Bigg|_{q=0} \tag{8.9}$$

where y_m for $m \geq 1$ is the higher-order term. Now, the final solution can be obtained as follows:

$$y(x) = y_0(x) + \sum_{m=1}^{\infty} y_m(x) \tag{8.10}$$

8.3.2 HAM-BASED SOLUTION FOR SUBCRITICAL FLOW

Here, we consider a subcritical flow for both frictionless and frictional cases. A subcritical flow occurs when the actual water depth is greater than the critical depth and the flow is dominated by the gravitational and friction forces. Mathematically, for a subcritical flow, the Froude number is less than 1.

8.3.2.1 Frictionless Case

Setting $G_f = 0$, the governing equation for this case becomes:

$$\frac{dy}{dx} = \frac{Gy^3 - 2xyF_0^2}{y^3 - F_0^2 x^2} \tag{8.11}$$

To solve Eq. (8.11) using HAM, the nonlinear operator can be chosen as:

$$\mathcal{N}\left[\Phi(x;q)\right] = \left[\Phi(x;q)^3 - F_0^2 x^2\right] \frac{\partial \Phi(x;q)}{\partial x} - G\Phi(x;q)^3 + 2xF_0^2 \Phi(x;q) \tag{8.12}$$

Using Eq. (8.12), the term $R_m\left(\vec{S}_{m-1}\right)$ can be obtained from Eq. (8.9) in closed form as follows:

$$R_m\left(\vec{y}_{m-1}\right) = \sum_{j=0}^{m-1} y_{m-1-j} \sum_{k=0}^{j} y_{j-k} \sum_{l=0}^{k} y_l \frac{dy_{k-l}}{dx} - F_0^2 x^2 \frac{dy_{m-1}}{dx}$$

$$- G\sum_{j=0}^{m-1} y_{m-1-j} \sum_{k=0}^{j} y_k y_{j-k} + 2xF_0^2 y_{m-1} \tag{8.13}$$

For Eq. (8.11), we choose a set of polynomial base functions:

$$\left\{x^n \mid n = 0,1, 2, 3...\right\} \tag{8.14}$$

so that

$$y(x) = \sum_{n=0}^{\infty} a_n x^n \tag{8.15}$$

where a_n is the coefficient of the series. Equation (8.15) provides the so-called *rule of solution expression*. Following the rule of solution expression, the linear operator and the initial approximation are chosen, respectively, as follows:

$$\mathcal{L}\left[\Phi(x;q)\right] = \frac{\partial \Phi(x;q)}{\partial x} \text{ with property } \mathcal{L}[C_2] = 0 \qquad (8.16)$$

$$y_0(x) = x \qquad (8.17)$$

where C_2 is an integral constant. Using Eq. (8.16), the higher-order terms can be obtained from Eq. (8.7) as follows:

$$y_m(x) = \chi_m y_{m-1}(x) + \hbar \int_1^x H(\tau) R_m(\bar{y}_{m-1}) d\tau + C_2, \ m = 1,2,3,\dots \qquad (8.18)$$

where R_m is given by Eq. (8.13), and constant C_2 can be determined from the initial condition for the higher-order deformation equations, i.e., $y_m(1) = 0$, which simply yields $C_2 = 0$ for all $m \geq 1$.

Now, the auxiliary function $H(x)$ can be determined from the *rule of coefficient ergodicity*. Based on the rule of solution expression and Eq. (8.15), the general form of $H(x)$ should be

$$H(x) = x^\alpha \qquad (8.19)$$

where α is an integer. Substituting Eq. (8.19) into Eq. (8.18) and calculating some of y_m, we can infer two observations. It can be seen that if we consider $\alpha \geq 2$, then term x^α will be missing in the solution expression. This does not satisfy the *rule of coefficient ergodicity*, which allows us to include all the terms of the base function for the completeness of the solution. On the other hand, if we consider $\alpha \leq -1$, then the terms appearing in the solution violate the *rule of solution expression*. Therefore, we are left with two choices, $\alpha = 0$ or 1, and can choose either of them. Here, we take $\alpha = 0$, i.e., $H(x) = 1$. Determination of the auxiliary function $H(x)$ can be achieved following the *rule of solution expression* and the *rule of coefficient ergodicity*, as was already done.

Finally, the approximate solution can be obtained as follows:

$$y(x) = y_0(x) + \sum_{m=1}^{M} y_m(x) \qquad (8.20)$$

8.3.2.2 Frictional Case

The governing equation for this case becomes:

$$\frac{dy}{dx} = \frac{Gy^3 - \dfrac{G_f F_0^2 x^2}{8} - 2xy F_0^2}{y^3 - F_0^2 x^2} \qquad (8.21)$$

To solve Eq. (8.21) using HAM, the nonlinear operator can be chosen as:

$$N\left[\Phi(x;q)\right]=\left[\Phi(x;q)^3-F_0^2x^2\right]\frac{\partial\Phi(x;q)}{\partial x}-G\Phi(x;q)^3$$
$$+\frac{G_fF_0^2x^2}{8}+2xF_0^2\Phi(x;q) \tag{8.22}$$

Using Eq. (5.22), the term $R_m\left(\vec{y}_{m-1}\right)$ can be obtained from Eq. (8.9) in closed form as follows:

$$R_m\left(\vec{y}_{m-1}\right)=\sum_{j=0}^{m-1}y_{m-1-j}\sum_{k=0}^{j}y_{j-k}\sum_{l=0}^{k}y_l\frac{dy_{k-l}}{dx}-F_0^2x^2\frac{dy_{m-1}}{dx}$$
$$-G\sum_{j=0}^{m-1}y_{m-1-j}\sum_{k=0}^{j}y_ky_{j-k}+\frac{G_fF_0^2x^2}{8}\left(1-\chi_{m-1}\right)+2xF_0^2y_{m-1} \tag{8.23}$$

For the governing Eq. (8.21), the set of base functions, initial approximation, linear operator, and auxiliary function are chosen the same as with the frictional case.

8.4　HPM-BASED ANALYTICAL SOLUTION

Here, we derive the homotopy perturbation method (HPM) based analytical solutions for both flow cases.

8.4.1　FRICTIONLESS CASE

We rewrite Eq. (8.11) as follows:

$$\left(y^3-F_0^2x^2\right)\frac{dy}{dx}-Gy^3+2xyF_0^2=0 \tag{8.24}$$

Equation (8.24) can be thought of as the following form:

$$N(y)=f(x) \tag{8.25}$$

We construct a homotopy $\Phi(x;q)$ that satisfies:

$$(1-q)\left[\mathcal{L}\left(\Phi(x;q)\right)-\mathcal{L}\left(y_0(x)\right)\right]+q\left[N\left(\Phi(x;q)\right)-f(x)\right]=0 \tag{8.26}$$

where q is the embedding parameter that lies on $[0,1]$, $\mathcal{L}$ is the linear operator, and $y_0(x)$ is the initial approximation to the final solution. Equation (8.26) shows that as $q=0$, $\Phi(x;0)=y_0(x)$ and as $q=1$, $\Phi(x;1)=y(x)$. We now express $\Phi(x;q)$ as a series in terms of q as follows:

$$\Phi(x;q)=\Phi_0+q\Phi_1+q^2\Phi_2+q^3\Phi_3+q^4\Phi_4\ldots \tag{8.27}$$

where Φ_m for $m \geq 1$ is the higher-order term. As $q \to 1$, Eq. (8.27) produces the final solution as:

$$y(x) = \lim_{q \to 1} \Phi(x;q) = \sum_{i=0}^{\infty} \Phi_i \tag{8.28}$$

For simplicity, we consider the single-term linear and nonlinear operators as follows:

$$\mathcal{L}\big(\Phi(x;q)\big) = \frac{\partial \Phi(x;q)}{\partial x} \tag{8.29}$$

$$\mathcal{N}\big(\Phi(x;q)\big) = \big(\Phi(x;q)^3 - F_0^2 x^2\big)\frac{\partial \Phi(x;q)}{\partial x} - G\Phi(x;q)^3 + 2x\Phi(x;q)F_0^2 \tag{8.30}$$

Using these expressions, Eq. (8.26) takes on the form:

$$(1-q)\left[\frac{\partial \Phi(x;q)}{\partial x} - \frac{dy_0}{dx}\right]$$
$$+ q\left[\big(\Phi(x;q)^3 - F_0^2 x^2\big)\frac{\partial \Phi(x;q)}{\partial x} - G\Phi(x;q)^3 + 2x\Phi(x;q)F_0^2\right] = 0 \tag{8.31}$$

The following expression can be used for finding the terms of the series:

$$\big(\Phi_0 + q\Phi_1 + q^2\Phi_2 + q^3\Phi_3 + q^4\Phi_4 \cdots\big)^3 = \big(\Phi_0\big)^3 + q\big(3\Phi_0^2\Phi_1\big)$$
$$+ q^2\big(3\Phi_0^2\Phi_2 + 3\Phi_1^2\Phi_0\big) + q^3\big(3\Phi_0^2\Phi_3 + 6\Phi_0\Phi_1\Phi_2 + \Phi_1^3\big) + \cdots \tag{8.32}$$

Using Eq. (8.32), we have

$$(1-q)\left[\frac{\partial \Phi(x;q)}{\partial x} - \frac{dy_0}{dx}\right]$$
$$= (1-q)\left[\left(\frac{d\Phi_0}{dx} - \frac{dy_0}{dx}\right) + q\frac{d\Phi_1}{dx} + q^2\frac{d\Phi_2}{dx} + q^3\frac{d\Phi_3}{dx} + q^4\frac{d\Phi_4}{dx} + \cdots\right]$$
$$= \left(\frac{d\Phi_0}{dx} - \frac{dy_0}{dx}\right) + q\left(\frac{d\Phi_1}{dx} - \left(\frac{d\Phi_0}{dx} - \frac{dy_0}{dx}\right)\right) + q^2\left(\frac{d\Phi_2}{dx} - \frac{d\Phi_1}{dx}\right)$$
$$+ q^3\left(\frac{d\Phi_3}{dx} - \frac{d\Phi_2}{dx}\right) + q^4\left(\frac{d\Phi_4}{dx} - \frac{d\Phi_3}{dx}\right) + \cdots \tag{8.33}$$

$$\big(\Phi(x;q)^3 - F_0^2 x^2\big)\frac{\partial \Phi(x;q)}{\partial x} = \left[\big(\Phi_0^3 - F_0^2 x^2\big) + q\big(3\Phi_0^2\Phi_1\big) + q^2\big(3\Phi_0^2\Phi_2 + 3\Phi_1^2\Phi_0\big)\right.$$
$$+ q^3\big(3\Phi_0^2\Phi_3 + 6\Phi_0\Phi_1\Phi_2 + \Phi_1^3\big) + \cdots\left]\left(\frac{d\Phi_0}{dx} + q\frac{d\Phi_1}{dx} + q^2\frac{d\Phi_2}{dx} + q^3\frac{d\Phi_3}{dx} + \cdots\right)\right.$$
$$= \frac{d\Phi_0}{dx}\big(\Phi_0^3 - F_0^2 x^2\big) + q\left[\frac{d\Phi_1}{dx}\big(\Phi_0^3 - F_0^2 x^2\big) + 3\Phi_0^2\Phi_1\frac{d\Phi_0}{dx}\right] + q^2\left[\frac{d\Phi_2}{dx}\big(\Phi_0^3 - F_0^2 x^2\big)\right.$$
$$+ \frac{d\Phi_0}{dx}\big(3\Phi_0^2\Phi_2 + 3\Phi_1^2\Phi_0\big) + 3\Phi_0^2\Phi_1\frac{d\Phi_1}{dx}\right] + q^3\left[\frac{d\Phi_3}{dx}\big(\Phi_0^3 - F_0^2 x^2\big)\right.$$
$$+ \frac{d\Phi_0}{dx}\big(3\Phi_0^2\Phi_3 + 6\Phi_0\Phi_1\Phi_2 + \Phi_1^3\big) + \frac{d\Phi_1}{dx}\big(3\Phi_0^2\Phi_2 + 3\Phi_1^2\Phi_0\big) + 3\Phi_0^2\Phi_1\frac{d\Phi_2}{dx}\right] + \cdots \tag{8.34}$$

$$2x\Phi(x;q)F_0^2 - G\Phi(x;q)^3 = 2xF_0^2\left(\Phi_0 + q\Phi_1 + q^2\Phi_2 + q^3\Phi_3 + \cdots\right)$$
$$-G\left[\left(\Phi_0\right)^3 + q\left(3\Phi_0^2\Phi_1\right) + q^2\left(3\Phi_0^2\Phi_2 + 3\Phi_1^2\Phi_0\right) + q^3\left(3\Phi_0^2\Phi_3 + 6\Phi_0\Phi_1\Phi_2 + \Phi_1^3\right) + \cdots\right]$$
$$= \left(2x\Phi_0F_0^2 - G\Phi_0^3\right) + q\left[2x\Phi_1F_0^2 - 3\Phi_0^2\Phi_1G\right] + q^2\left[2x\Phi_2F_0^2 - G\left(3\Phi_0^2\Phi_2 + 3\Phi_1^2\Phi_0\right)\right]$$
$$+ q^3\left[2x\Phi_3F_0^2 - G\left(3\Phi_0^2\Phi_3 + 6\Phi_0\Phi_1\Phi_2 + \Phi_1^3\right)\right] + \cdots \tag{8.35}$$

Using the initial condition $y(x=1)=1$, we have

$$\Phi_0(1)=1,\ \Phi_1(1)=0,\ \Phi_2(1)=0,\ \ldots. \tag{8.36}$$

Using Eqs. (8.33)–(8.36) and equating the like powers of q in Eq. (8.31), the following system of differential equations is obtained:

$$\frac{d\Phi_0}{dx} - \frac{dy_0}{dx} = 0 \text{ subject to } \Phi_0(1)=1 \tag{8.37}$$

$$\frac{d\Phi_1}{dx} - \left(\frac{d\Phi_0}{dx} - \frac{dy_0}{dx}\right) + \frac{d\Phi_0}{dx}\left(\Phi_0^3 - F_0^2x^2\right) + \left(2x\Phi_0F_0^2 - G\Phi_0^3\right)$$
$$= 0 \text{ subject to } \Phi_1(1)=0 \tag{8.38}$$

$$\frac{d\Phi_2}{dx} - \frac{d\Phi_1}{dx} + \frac{d\Phi_1}{dx}\left(\Phi_0^3 - F_0^2x^2\right) + 3\Phi_0^2\Phi_1\frac{d\Phi_0}{dx} + 2x\Phi_1F_0^2$$
$$- 3\Phi_0^2\Phi_1G = 0 \text{ subject to } \Phi_2(1)=0 \tag{8.39}$$

$$\frac{d\Phi_3}{dx} - \frac{d\Phi_2}{dx} + \frac{d\Phi_2}{dx}\left(\Phi_0^3 - F_0^2x^2\right) + \frac{d\Phi_0}{dx}\left(3\Phi_0^2\Phi_2 + 3\Phi_1^2\Phi_0\right) + 3\Phi_0^2\Phi_1\frac{d\Phi_1}{dx}$$
$$+ 2x\Phi_2F_0^2 - G\left(3\Phi_0^2\Phi_2 + 3\Phi_1^2\Phi_0\right) = 0 \text{ subject to } \Phi_3(1)=0 \tag{8.40}$$

$$\frac{d\Phi_4}{dx} - \frac{d\Phi_3}{dx} + \frac{d\Phi_3}{dx}\left(\Phi_0^3 - F_0^2x^2\right) + \frac{d\Phi_0}{dx}\left(3\Phi_0^2\Phi_3 + 6\Phi_0\Phi_1\Phi_2 + \Phi_1^3\right)$$
$$+ \frac{d\Phi_1}{dx}\left(3\Phi_0^2\Phi_2 + 3\Phi_1^2\Phi_0\right) + 3\Phi_0^2\Phi_1\frac{d\Phi_2}{dx} + 2x\Phi_3F_0^2$$
$$- G\left(3\Phi_0^2\Phi_3 + 6\Phi_0\Phi_1\Phi_2 + \Phi_1^3\right) = 0 \text{ subject to } \Phi_4(1)=0 \tag{8.41}$$

The initial approximation can be chosen as $\Phi_0 = 1 + G(x-1)$, which is indeed a solution of the part of the original equation. Using this initial approximation, we can solve the equations iteratively using symbolic software. We avoid those expressions here, as they are lengthy. Finally, the HPM-based solution can be approximated as follows:

$$y(x) \approx \sum_{i=0}^{M}\Phi_i \tag{8.42}$$

8.4.2 FRICTIONAL CASE

For this case, Eq. (8.21) is rewritten as:

$$\left(y^3 - F_0^2 x^2\right)\frac{dy}{dx} - Gy^3 + 2xyF_0^2 + \frac{G_f F_0^2 x^2}{8} = 0 \tag{8.43}$$

The subsequent derivation follows exactly as in Section 5.4.1. Therefore, here we avoid those steps. The linear operator and initial approximation are chosen the same as with the frictionless case. The nonlinear operator is chosen as follows:

$$\mathcal{N}\left(\Phi(x;q)\right) = \left(\Phi(x;q)^3 - F_0^2 x^2\right)\frac{\partial\Phi(x;q)}{\partial x}$$

$$- G\Phi(x;q)^3 + 2x\Phi(x;q)F_0^2 + \frac{G_f F_0^2 x^2}{8} \tag{8.44}$$

Using these expressions, Eq. (8.26) takes on the form:

$$(1-q)\left[\frac{\partial\Phi(x;q)}{\partial x} - \frac{dy_0}{dx}\right]$$

$$+ q\left[\left(\Phi(x;q)^3 - F_0^2 x^2\right)\frac{\partial\Phi(x;q)}{\partial x} - G\Phi(x;q)^3 + 2x\Phi(x;q)F_0^2 + \frac{G_f F_0^2 x^2}{8}\right] = 0 \quad (8.45)$$

One can see that the equation for the frictional case adds only one extra term. Therefore, using Eqs. (8.33)–(8.36) and equating the like powers of q in Eq. (8.45), all other equations except the one are the same as in the previous section. The different equation is given as follows:

$$\frac{d\Phi_1}{dx} - \left(\frac{d\Phi_0}{dx} - \frac{dy_0}{dx}\right) + \frac{d\Phi_0}{dx}\left(\Phi_0^3 - F_0^2 x^2\right) + \left(2x\Phi_0 F_0^2 - G\Phi_0^3\right)$$

$$+ \frac{G_f F_0^2 x^2}{8} = 0 \text{ subject to } \Phi_1(1) = 0 \tag{8.46}$$

However, because Eq. (5.46) is different, the higher-order terms will be different from the expressions obtained in the previous section. The terms can be easily obtained using symbolic software. We avoid those expressions here, as they are lengthy. Finally, the HPM-based solution can be approximated as follows:

$$y(x) \approx \sum_{i=0}^{M}\Phi_i \tag{8.47}$$

8.5 OHAM-BASED ANALYTICAL SOLUTION

Here, we derive the optimal homotopy asymptotic method (OHAM) based analytical solutions for both flow cases.

8.5.1 FRICTIONLESS CASE

For applying OHAM, the approximate Eq. (8.24) can be written in the following form:

$$\mathcal{L}[y(x)] + \mathcal{N}[y(x)] + h(x) = 0 \tag{8.48}$$

where $\mathcal{L}[y(x)] = \dfrac{dy}{dx}$, $\mathcal{N}[y(x)] = (y^3 - F_0^2 x^2 - 1)\dfrac{dy}{dx} - Gy^3 + 2xyF_0^2$, and $h(x) = 0$.

With these considerations, the zeroth-order problem becomes:

$$\mathcal{L}(y_0(x)) + h(x) = 0 \text{ subject to } y_0(1) = 1 \tag{8.49}$$

Solving Eq. (8.49), one obtains:

$$y_0(x) = 1 \tag{8.50}$$

It may be noted that one can use Eqs. (88.32), (8.34), and (8.35) from the previous section to obtain the values $\mathcal{N}_0$, $\mathcal{N}_1$, $\mathcal{N}_2$, etc. The first-order problem reduces to:

$$\frac{dy_1}{dx} = H_1(x, C_i)\left\{\frac{dy_0}{dx}(y_0^3 - F_0^2 x^2 - 1) + (2xy_0 F_0^2 - Gy_0^3)\right\} \text{ subject to } y_1(1) = 0 \tag{8.51}$$

Let us simply choose $H_1(x, C_1) = C_1$. Using $k = 2$, the second-order problem becomes:

$$\mathcal{L}[y_2(x, C_i) - y_1(x, C_i)] = H_2(x, C_i)\mathcal{N}_0(y_0(x))$$
$$+ H_1(x, C_i)\left[\mathcal{L}[y_1(x, C_i)] + \mathcal{N}_1[y_0(x), y_1(x, C_i)]\right] \text{ subject to } y_2(1) = 0 \tag{8.52}$$

Further, we choose $H_2(t, C_i) = C_2$. Therefore, Eq. (8.52) becomes:

$$\frac{dy_2}{dx} = \frac{dy_1}{dx} + C_2\left\{\frac{dy_0}{dx}(y_0^3 - F_0^2 x^2 - 1) + (2xy_0 F_0^2 - Gy_0^3)\right\}$$
$$+ C_1\left[\frac{dy_1}{dx} + \frac{dy_1}{dx}(y_0^3 - F_0^2 x^2 - 1) + 3y_0^2 y_1 \frac{dy_0}{dx} + 2xy_1 F_0^2 - 3y_0^2 y_1 G\right]$$
$$\text{subject to } y_2(1) = 0 \tag{8.53}$$

We do not write the explicit forms of the terms here, as those are lengthy and may not be appropriate to keep up the flow. Indeed, one can compute these terms without any difficulty using symbolic computation software, such as MATLAB. Further, following Chapter 2, the higher-order terms can be computed in a similar manner. However, our aim is to produce an accurate solution with just two to three terms of OHAM-based series. Therefore, we restrict our calculation up to $k = 2$. Finally, the approximate solution can be found as:

$$y(x) \approx y_0(x) + y_1(x, C_1) + y_2(x, C_1, C_2) \tag{8.54}$$

where the terms are given by Eqs. (8.50), (8.51), and (8.53).

8.5.2   FRICTIONAL CASE

The OHAM-based solution for this case can be developed using the same procedure as in the previous section. However, the operators take on the forms $\mathcal{L}[y(x)] = \dfrac{dy}{dx}$, $\mathcal{N}[y(x)] = (y^3 - F_0^2 x^2 - 1)\dfrac{dy}{dx} - Gy^3 + 2xyF_0^2 + \dfrac{G_f F_0^2 x^2}{8}$, and $h(x) = 0$. The zeroth-order representation produces the same solution:

$$y_0(x) = 1 \tag{8.55}$$

However, due to the new expression of the nonlinear operator, $\mathcal{N}_0$, $\mathcal{N}_1$, $\mathcal{N}_2$, etc., change. As a result of this, the first-order problem reduces to:

$$\frac{dy_1}{dx} = H_1(x, C_i) \left\{ \frac{dy_0}{dx} \left(y_0^3 - F_0^2 x^2 - 1 \right) + \left(2xy_0 F_0^2 - Gy_0^3 + \frac{G_f F_0^2 x^2}{8} \right) \right\}$$

$$\text{subject to } y_1(1) = 0 \tag{8.56}$$

Let us simply choose $H_1(x, C_i) = C_1$ and $H_2(x, C_i) = C_2$. Then, the second-order representation becomes:

$$\frac{dy_2}{dx} = \frac{dy_1}{dx} + C_2 \left\{ \frac{dy_0}{dx} \left(y_0^3 - F_0^2 x^2 - 1 \right) + \left(2xy_0 F_0^2 - Gy_0^3 + \frac{G_f F_0^2 x^2}{8} \right) \right\}$$

$$+ C_1 \left[\frac{dy_1}{dx} + \frac{dy_1}{dx} \left(y_0^3 - F_0^2 x^2 - 1 \right) + 3y_0^2 y_1 \frac{dy_0}{dx} + 2xy_1 F_0^2 - 3y_0^2 y_1 G \right]$$

$$\text{subject to } y_2(1) = 0 \tag{8.57}$$

Using these equations, the approximations $y_1(x, C_1)$ and $y_2(x, C_1, C_2)$ can be obtained.

8.6   CONVERGENCE THEOREMS

The convergence of the HAM-based and OHAM-based solutions is proved theoretically using the following theorems.

8.6.1   CONVERGENCE THEOREM OF THE HAM-BASED SOLUTION

The convergence theorems for the HAM-based solution for Eqs. (8.11) and (8.21) can be proved using the following theorems.

Theorem 8.1: If the homotopy series $\Sigma_{m=0}^{\infty} y_m(x)$ and $\Sigma_{m=0}^{\infty} y_m'(x)$ converge, then $R_m(\vec{y}_{m-1})$ given by Eqs. (8.13) and (8.23) satisfies the relation $\Sigma_{m=1}^{\infty} R_m(\vec{y}_{m-1}) = 0$. Here "'" denotes the derivative with respect to x.

Proof: The proof of this theorem follows exactly from the previous chapters.

Theorem 8.2: If $\hbar$ is so properly chosen that the series $\Sigma_{m=0}^{\infty} y_m(x)$ and $\Sigma_{m=0}^{\infty} y_m{}'(x)$ converge absolutely to $y(x)$ and $y'(x)$, respectively, then the homotopy series $\Sigma_{m=0}^{\infty} y_m(x)$ satisfies the original governing Eq. (8.11).

Proof: Using the Cauchy product defined in Theorem 2.2 of Chapter 2 in relation to Eq. (8.13), we get:

$$\sum_{m=1}^{\infty}\sum_{j=0}^{m-1} y_{m-1-j} \sum_{k=0}^{j} y_k y_{j-k} = \left(\sum_{m=0}^{\infty} y_m\right)^3 \tag{8.58}$$

$$\sum_{m=1}^{\infty}\sum_{j=0}^{m-1} y_{m-1-j} \sum_{k=0}^{j} y_{j-k} \sum_{l=0}^{k} y_l \frac{dy_{k-l}}{dx} = \left(\sum_{m=0}^{\infty} y_m\right)^3 \left(\sum_{k=0}^{\infty} y_k'\right) \tag{8.59}$$

Theorem 8.1 shows that if $\Sigma_{m=0}^{\infty} y_m(x)$ and $\Sigma_{m=0}^{\infty} y_m'(x)$ converge, then $\Sigma_{m=1}^{\infty} R_m(\vec{y}_{m-1}) = 0$.

Therefore, substituting the previous expressions in Eq. (8.13) and simplifying further lead to:

$$\left(\sum_{m=0}^{\infty} y_m\right)^3 \left(\sum_{k=0}^{\infty} y_k'\right) - F_0^2 x^2 \sum_{m=0}^{\infty} y_m' - G\left(\sum_{k=0}^{\infty} y_k\right)^3 + 2xF_0^2 \sum_{m=0}^{\infty} y_m = 0 \tag{8.60}$$

which is basically the original governing equation in Eq. (8.11). Furthermore, subject to the initial condition $y_0(1) = 1$ and the conditions for the higher-order deformation equation $y_m(1) = 0$, for $m \geq 1$, we easily obtain $\Sigma_{m=0}^{\infty} y_m(1) = 1$. Hence, the convergence result follows.

Theorem 8.3: If $\hbar$ is properly chosen so that the series $\Sigma_{m=0}^{\infty} y_m(x)$ and $\Sigma_{m=0}^{\infty} y_m{}'(x)$ converge absolutely to $y(x)$ and $y'(x)$, respectively, then the homotopy series $\Sigma_{m=0}^{\infty} y_m(x)$ satisfies the original governing Eq. (8.21).

Proof: Substituting the identities in Eqs. (8.58) and (8.59) and simplifying further lead to:

$$\left(\sum_{m=0}^{\infty} y_m\right)^3 \left(\sum_{k=0}^{\infty} y_k'\right) - F_0^2 x^2 \sum_{m=0}^{\infty} y_m' - G\left(\sum_{k=0}^{\infty} y_k\right)^3$$
$$+ \frac{G_f F_0^2 x^2}{8} \sum_{m=0}^{\infty} (1-\chi_{m+1}) + 2xF_0^2 \sum_{m=0}^{\infty} y_m = 0 \tag{8.61}$$

which is basically the original governing equation Eq. (8.21). Furthermore, subject to the initial condition $y_0(1) = 1$ and the conditions for the higher-order deformation equation $y_m(1) = 0$, for $m \geq 1$, we easily obtain $\Sigma_{m=0}^{\infty} y_m(1) = 1$. Hence, the convergence result follows.

8.6.2 CONVERGENCE THEOREM OF THE OHAM-BASED SOLUTION

Theorem 8.4: If the series $y_0(x) + \Sigma_{j=1}^{\infty} y_j(x, C_i)$, $i = 1, 2, \ldots, s$ converges, where $y_j(x, C_i)$ are governed by Eqs. (8.50), (8.51), and (8.53), then Eq. (8.54) is a solution of the original Eq. (8.11).

Proof: The proof of this theorem follows exactly from Theorem 2.3 of Chapter 2.

Theorem 8.5: If the series $y_0(x) + \Sigma_{j=1}^{\infty} y_j(x, C_i)$, $i = 1, 2, \ldots, s$ converges, where $y_j(x, C_i)$ are governed by Eqs. (8.55)–(8.57), then Eq. (8.54) is a solution of the original Eq (8.21).

Proof: The proof of this theorem follows exactly from Theorem 2.3 of Chapter 2.

8.7 RESULTS AND DISCUSSION

First, we validate the HAM-, HPM-, and OHAM-based analytical solutions for test cases. Then, the improved analytical solutions obtained here are compared with the perturbation-based solutions given by Gill (1977).

8.7.1 VALIDATION OF THE HAM-BASED SOLUTION

The convergence of the HAM-based series solution depends on a suitable choice for the auxiliary parameter $\hbar$. For that purpose, the squared residual error with regard to Eq. (8.11) or (8.21) can be calculated as follows:

$$\Delta_m = \int\limits_{x \in \Omega} \left(\mathcal{N}[y(x)] \right)^2 dt \qquad (8.62)$$

where Ω is the domain of the equation, which is $[0,1]$. The HAM-based series solution leads to the exact solution of the equation once Δ_m tends to its minimum value. Therefore, for a particular order of approximation m, it is sufficient to minimize the corresponding Δ_m, which then yields an optimal value of parameter $\hbar$. Here, we consider two test cases for both frictionless and frictional governing equations. The physical parameters are chosen as $G = 0$, $F_0 = 0.2$ and $G = 0$, $F_0 = 0.2$, $G_f = 0.8$ for frictionless and frictional cases, respectively. Figure 8.2 (a) and (b) shows the squared residual errors for a different order of approximation for frictionless and frictional cases, respectively. It can be seen from the figures that as the order of approximation increases, the corresponding residual errors decrease, which shows the stability of the method. Also, for a quantitative assessment, numerical results are reported in Table 8.1. For both cases considered here, comparisons between numerical solutions and the 20th-order HAM-based approximate solutions are shown in Figure 8.3. An excellent agreement between the computed and observed values can be seen from both figures. Also, a comparison is shown numerically in Table 8.2 considering some selected values in the domain. All the computations reported in this chapter are performed using the MATHEMATICA package BVPh 2.0 developed by Zhao and Liao (2002). A flowchart diagram containing

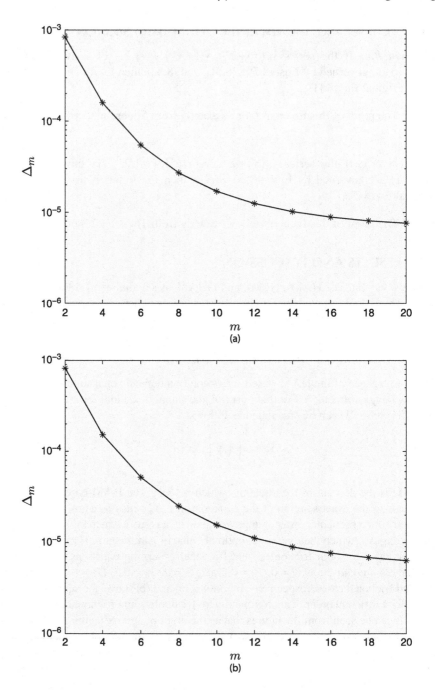

FIGURE 8.2 Squared residual error (Δ_m) versus different orders of approximation (m) of the HAM-based solution for (a) frictionless and (b) frictional cases.

TABLE 8.1

Squared Residual Errors (Δ_m) versus Different Orders of Approximation (m) for Frictionless and Frictional Cases

Order of Approximation (m)	Squared Residual Error (Δ_m)	
	Frictionless Case (Eq. 8.11)	Frictional Case (Eq. 8.21)
2	8.35×10^{-4}	8.18×10^{-4}
4	1.59×10^{-4}	1.53×10^{-4}
6	5.48×10^{-5}	5.21×10^{-5}
8	2.71×10^{-5}	2.52×10^{-5}
10	1.70×10^{-5}	1.56×10^{-5}
12	1.25×10^{-5}	1.12×10^{-5}
14	1.02×10^{-5}	8.97×10^{-6}
16	8.88×10^{-6}	7.66×10^{-6}
18	8.09×10^{-6}	6.85×10^{-6}
20	7.62×10^{-6}	6.33×10^{-6}

the steps of the method for solving Eq. (8.11) is provided in Figure 8.4. The same can be done for Eq. (8.21) also.

8.7.2 VALIDATION OF THE HPM-BASED SOLUTION

The HPM-based analytical solutions for both cases are validated against a numerical solution obtained using *ode45* routine of MATLAB. Figure 8.5 shows a comparison between two-term, three-term, and four-term HPM-based solutions and the corresponding numerical solution. It is observed from the figure that even three-term HPM solution produces accurate values while the two-term solution slightly deviates from the numerical solution. For a quantitative assessment, we consider three-term order approximation and report it in Table 8.3 by comparing it with a numerical solution. A flowchart containing the steps of the method for solving Eq. (8.11) is given in Figure 8.6. The same can be done for Eq. (8.21) also.

8.7.3 VALIDATION OF THE OHAM-BASED SOLUTION

For the assessment of the OHAM-based analytical solution, one needs to calculate the constants C_i. For that purpose, we calculate the residual as follows:

$$R(x,C_i) = \mathcal{L}\left[y_{OHAM}(x,C_i)\right] + \mathcal{N}\left[y_{OHAM}(x,C_i)\right] + h(x), \ i = 1,2,\ldots,s \quad (8.63)$$

where $y_{OHAM}(x,C_i)$ is the approximate solution. When $R(x,C_i) = 0$, $y_{OHAM}(x,C_i)$ becomes the exact solution to the problem. One of the ways to obtain the optimal

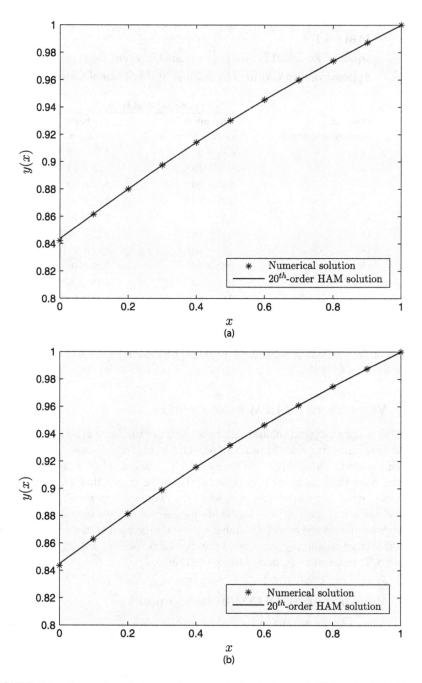

FIGURE 8.3 Comparison between the numerical solution and 20th-order HAM-based solution for (a) frictionless, and (b) frictional cases.

TABLE 8.2

Comparison between HAM-Based Approximations and Numerical Solutions for Frictionless and Frictional Cases

| | Frictionless Case (Eq. 8.11) | | Frictional Case (Eq. 8.21) | |
| | Numerical | 20th-Order HAM-Based | Numerical | 20th-Order HAM-Based |
x	Solution	Approximation	Solution	Approximation
1.0	1.0000	1.0000	1.0000	1.0000
0.9	0.9872	0.9872	0.9876	0.9876
0.8	0.9739	0.9739	0.9746	0.9746
0.7	0.9600	0.9600	0.9609	0.9609
0.6	0.9454	0.9454	0.9465	0.9465
0.5	0.9302	0.9302	0.9314	0.9314
0.4	0.9142	0.9142	0.9156	0.9156
0.3	0.8975	0.8975	0.8989	0.8989
0.2	0.8800	0.8800	0.8814	0.8815
0.1	0.8616	0.8618	0.8630	0.8632
0.0	0.8421	0.8436	0.8435	0.8449

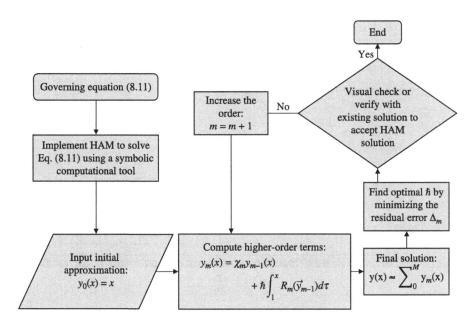

FIGURE 8.4 Flowchart for the HAM solution in Eq. (8.20).

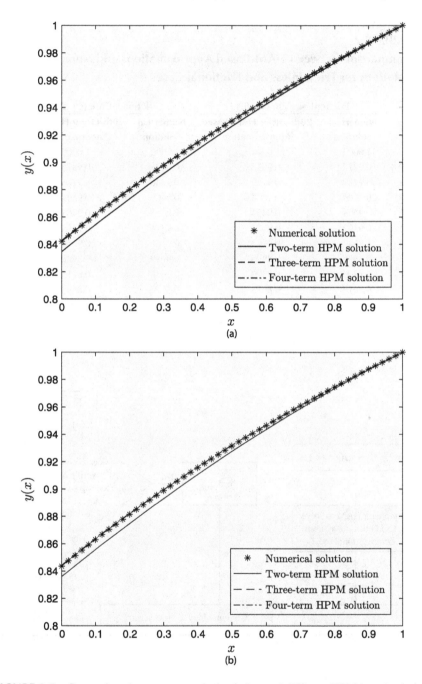

FIGURE 8.5 Comparison between numerical solution and different HPM-based solutions for (a) frictionless and (b) frictional cases.

TABLE 8.3

Comparison between HPM-Based Approximations and Numerical Solutions for Frictionless and Frictional Cases

	Frictionless Case (Eq. 8.11)		Frictional Case (Eq. 8.21)	
x	Numerical Solution	Three-Term HPM-Based Approximation	Numerical Solution	Three-Term HPM-Based Approximation
1.0	1.0000	1.0000	1.0000	1.0000
0.9	0.9872	0.9873	0.9876	0.9877
0.8	0.9739	0.9741	0.9746	0.9748
0.7	0.9600	0.9603	0.9609	0.9612
0.6	0.9454	0.9458	0.9465	0.9470
0.5	0.9302	0.9307	0.9314	0.9320
0.4	0.9142	0.9147	0.9156	0.9162
0.3	0.8975	0.8980	0.8989	0.8995
0.2	0.8800	0.8803	0.8814	0.8819
0.1	0.8616	0.8617	0.8630	0.8633
0.0	0.8421	0.8422	0.8435	0.8438

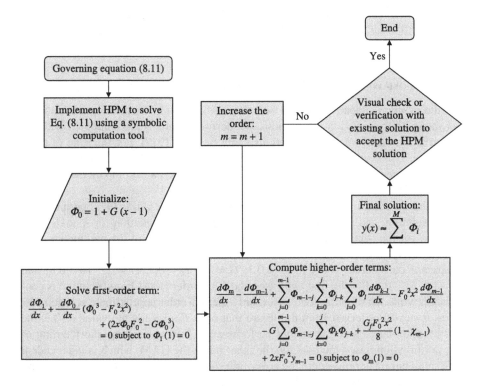

FIGURE 8.6   Flowchart for the HPM solution in Eq. (8.42).

C_i for which the solution converges is the minimization of squared residual error, i.e.,

$$J(C_i) = \int_{x \in D} R^2(x, C_i) dx, \ i = 1, 2, \ldots, s \tag{8.64}$$

where $D = [0,1]$ is the domain of the problem. The minimization of Eq. (8.64) leads to a system of algebraic equations as follows:

$$\frac{\partial J}{\partial C_1} = \frac{\partial J}{\partial C_2} = \cdots = \frac{\partial J}{\partial C_s} = 0 \tag{8.65}$$

Solving this system, one can obtain the optimal values of parameters. We obtain the optimal values using the MATLAB routine *fminsearch*, which minimizes an unconstrained multivariable function. Using these values, two- and three-term OHAM solutions are computed and compared in Figure 8.7 with a numerical solution to see the effectiveness of the OHAM method. It can be seen that even with a few terms of the OHAM-based series, the solution agrees well with the corresponding numerical solution to the equation. For a quantitative assessment, we also compare the numerical values of the solutions in Table 8.4. A flowchart containing the steps of the method for solving Eq. (8.11) is provided in Figure 8.8. The same can be done for Eq. (8.21) also.

8.7.4 COMPARISON WITH GILL'S (1977) SOLUTION

As the primary objective of this work is to find an improved analytical solution for the spatially varied flow equations, the derived solutions are compared with the perturbation-based analytical solutions given by Gill (1977). For the sake of completeness, Gill's solutions are provided briefly in Appendices A and B. Here, we consider some selected values of parameters G, F_0, and G_f and compare the solutions for both frictionless and frictional cases.

Figure 8.9 compares the HAM-based analytical solution, Gill's (1977) perturbation-based solution, and numerical solution of the governing equation for spatially varied flow for the frictionless case, with some selected values of the parameters. The numerical solution is obtained using a fourth-order Runge-Kutta method. The four cases are considered: $G = 0.8$, $F_0 = 0.3$; $G = 1.2$, $F_0 = 0.5$; $G = 1.5$, $F_0 = 0.8$; and $G = 1.8$, $F_0 = 0.9$. For the first two cases, the 20th-order HAM-based solution is considered, while for the other two cases, only the 6th-order approximate solution suffices. It can be observed from the figure that once F_0 exceeds the value of 0.5, the perturbation-based solution cannot perform well. This is a serious shortcoming of the perturbation-based method. On the other hand, the HAM, which is independent of a small parameter present in the governing equation, can provide an accurate solution over the entire domain of the equation.

For the frictional case, the value of parameter G_f is taken as 0.8. Figure 8.10 compares the HAM-based analytical solution, Gill's (1977) perturbation-based solution,

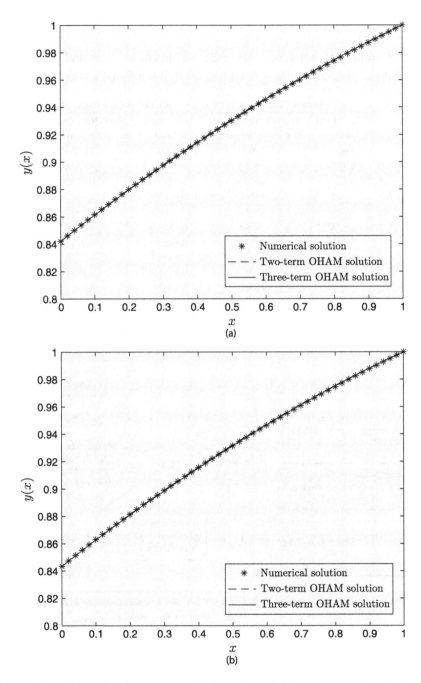

FIGURE 8.7 Comparison between numerical solution and different OHAM-based solutions for (a) frictionless and (b) frictional cases.

TABLE 8.4

Comparison between OHAM-Based Approximations and Numerical Solutions for Frictionless and Frictional Cases

	Frictionless Case (Eq. 8.11)		Frictional Case (Eq. 8.21)	
x	Numerical Solution	Three-Term OHAM-Based Approximation	Numerical Solution	Three-Term OHAM-Based Approximation
1.0	1.0000	1.0000	1.0000	1.0000
0.9	0.9872	0.9875	0.9876	0.9879
0.8	0.9739	0.9743	0.9746	0.9750
0.7	0.9600	0.9603	0.9609	0.9612
0.6	0.9454	0.9456	0.9465	0.9467
0.5	0.9302	0.9301	0.9314	0.9314
0.4	0.9142	0.9140	0.9156	0.9153
0.3	0.8975	0.8971	0.8989	0.8985
0.2	0.8800	0.8795	0.8814	0.8809
0.1	0.8616	0.8612	0.8630	0.8627
0.0	0.8421	0.8422	0.8435	0.8436

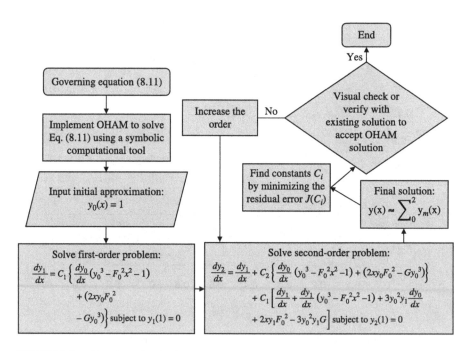

FIGURE 8.8   Flowchart for the OHAM solution in Eq. (8.54).

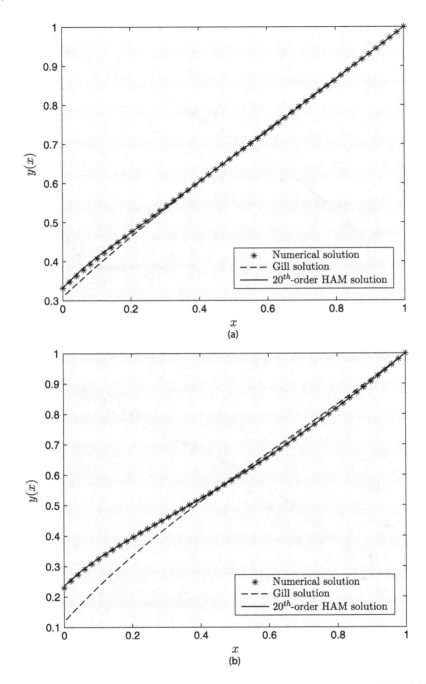

FIGURE 8.9 Comparison between HAM-based analytical solution, Gill's (1977) perturbation-based solution, and numerical solution for frictionless case with the parameter values: (a) $G = 0.8$, $F_0 = 0.3$, (b) $G = 1.2$, $F_0 = 0.5$, (c) $G = 1.5$, $F_0 = 0.8$, and (d) $G = 1.8$, $F_0 = 0.9$. (*Continued*)

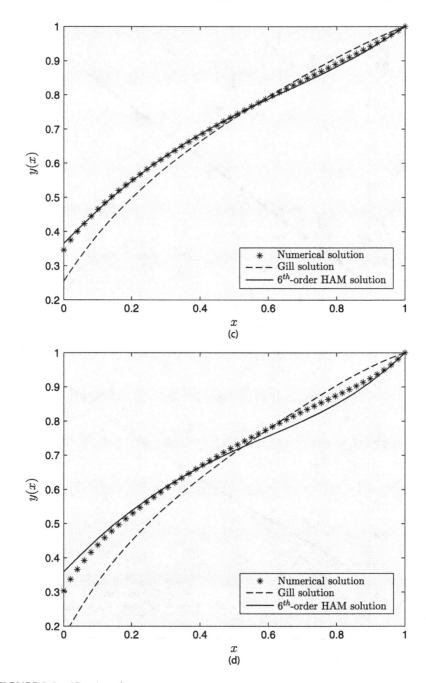

FIGURE 8.9 (*Continued*)

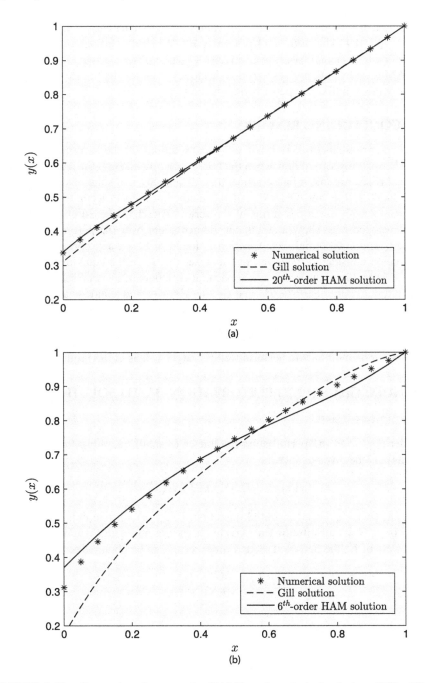

FIGURE 8.10 Comparison between the HAM-based analytical solution, Gill's (1977) perturbation-based solution, and numerical solution for the frictional case with the parameter values: (a) $G = 0.8$, $F_0 = 0.3$, and (b) $G = 1.8$, $F_0 = 0.9$.

and numerical solution with two sets of values for parameters $G = 0.8$, $F_0 = 0.3$ and $G = 1.8$, $F_0 = 0.9$. The 20th- and 6th-order HAM-based solutions are considered for the first and second cases, respectively. It can be observed from the figure that the HAM-based solution gives excellent accuracy even for the higher values of the Froude number, unlike the perturbation method.

8.8   CONCLUDING REMARKS

This chapter presents three types of analytical solutions for the governing equations of steady, spatially varied flow when the flow is subcritical. Both the frictionless and frictional cases are taken into consideration. The analytical solutions are obtained using HAM, HPM, and OHAM. All the methods produce accurate solutions of the equation with just a few terms of the series. The HAM-based analytical solutions developed here show a significant improvement over the perturbation-based solutions. The polynomial base function is found to be an appropriate choice for the equation considered; however, one may seek to solve the equation using other kinds of base functions, such as exponential, fractional, etc. For selected values of the physical parameters G, F_0, and G_f, the HAM-based solutions are compared with both Gill's perturbation-based solutions and numerical solutions based on the Runge-Kutta fourth-order method. The approximate solutions obtained using HAM show an excellent agreement with numerical solutions for all the cases, while the perturbation method fails to perform well for larger values of the Froude number.

APPENDIX: GILL'S (1977) PERTURBATION-BASED SOLUTION

A: FRICTIONLESS CASE

Considering F_0^2 as the perturbation parameter (a small quantity), the solution to Eq. (8.11) can be given as follows:

$$y(x) = 1 + F_0^2 y_1(x) + F_0^4 y_2(x) + \cdots \qquad (A.1)$$

Using Eq. (A.1) in the given initial condition, i.e., $y(x) = 1$, one obtains $y_1(1) = 0$, $y_2(1) = 0$, ... Now, substituting Eq. (A.1) in the governing Eq. (8.11) and equating the like powers of F_0, the first- and second-order terms can be obtained as (Gill 1977):

$$y_1(x) = 1 - \frac{G}{F_0^2} + \frac{G}{F_0^2} x - x^2 \qquad (A.2)$$

$$y_2(x) = \frac{G}{3F_0^2} - \frac{1}{2} + 2\left(1 - \frac{G}{F_0^2}\right) x^2 + \frac{5G}{3F_0^2} x^3 - \frac{3}{2} x^4 \qquad (A.3)$$

The final approximate solution is given as:

$$y(x) \approx \left(1 + F_0^2 - G + \frac{GF_0^2}{3} - \frac{F_0^4}{2}\right) + Gx + F_0^2\left(2F_0^2 - 1 - 2G\right) x^2$$

$$+ \frac{5GF_0^2}{3} x^3 - \frac{3F_0^4}{2} x^4 \qquad (A.4)$$

B: Frictional Case

The procedure for obtaining the approximate solution for this case is exactly the same as in Appendix A, only instead of Eq. (8.11), we need to consider Eq. (8.21). Using a similar methodology, the first- and second-order terms can be obtained as:

$$y_1(x) = 1 - \frac{G}{F_0^2} + \frac{G_f}{24} + \frac{G}{F_0^2}x - x^2 - \frac{G_f}{24}x^3 \tag{A.5}$$

$$y_2(x) = \frac{G}{3F_0^2} - \frac{1}{2} + \frac{G_f}{40} - \frac{GG_f}{32F_0^2} + \frac{G_f^2}{384} + \left(2 - \frac{2G}{F_0^2} + \frac{G_f}{12}\right)x^2$$

$$+ \left(\frac{5G}{3F_0^2} - \frac{G_f}{8} + \frac{GG_f}{8F_0^2} - \frac{G_f^2}{192}\right)x^3 - \left(\frac{3}{2} + \frac{3GG_f}{32F_0^2}\right)x^4 + \frac{G_f}{60}x^5 + \frac{G_f^2}{384}x^6 \tag{A.6}$$

The final approximate solution is given as:

$$y(x) \approx 1 + F_0^2 y_1(x) + F_0^4 y_2(x) + \cdots \tag{A.7}$$

where $y_1(x)$ and $y_2(x)$ are given by Eqs. (A.5) and (A.6), respectively.

REFERENCES

Chow, V. T. (1959). *Open channel hydraulics*. McGraw-Hill Book Co. Inc., New York, NY.

Chow, V. T. (1969). Spatially varied flow equations. *Water Resources Research*, *5*(5), 1124–1128.

Cruise, J. F., Sherif, M. M., and Singh, V. P. (2007). *Elementary hydraulics*. Thomson and Nelson, Toronto.

Gill, A. (1977). Perturbation solution of spatially varied flow in open channels. *Journal of Hydraulic Research*, *15*(4), 337–350.

Zhao, Y., and Liao, S. (2002). User Guide to BVPh 2.0. School of naval architecture, ocean and civil engineering, Shanghai, *40*.

FURTHER READING

Chipongo, K., and Khiadani, M. (2017). Spatially varied open-channel flow with increasing discharge equation. *Journal of Hydraulic Engineering*, *143*(3), 04016089.

Guarga, R., and Lara, M. (1984). Trapezoidal channels with increasing discharge. *Journal of Engineering Mechanics*, *110*(7), 1050–1059.

Hager, W. H. (1983). Open channel hydraulics of flows with increasing discharge. *Journal of Hydraulic Research*, *21*(3), 177–193.

Henderson, F. M. (1966). *Open channel flow*. The Macmillan Company, New York, NY.

Kouchakzadeh, S., Kholghi, M. K., and Vatankhah Mohammad-Abadi, A. R. (2002). Spatially varied flow in non-prismatic channels. II: Numerical solution and experimental verification. *Irrigation and Drainage: The Journal of the International Commission on Irrigation and Drainage*, *51*(1), 51–60.

Kouchakzadeh, S., and Mohammad-Abadi, A. V. (2002). Spatially varied flow in non-prismatic channels. I: Dynamic equation. *Irrigation and Drainage: The Journal of the International Commission on Irrigation and Drainage*, *51*(1), 41–50.

Kouchakzadeh, S., Vatankhah, A. R., and Townsend, R. D. (2001). A modified perturbation solution procedure for spatially-varied flows. *Canadian Water Resources Journal*, *26*(3), 399–416.

Li, W. H. (1955). Open channels with nonuniform discharge. *Transactions of the American Society of Civil Engineers*, *120*(1), 255–274.

Maranzoni, A. (2021). Analysis of the water surface profiles of spatially varied flow with increasing discharge using the method of singular points. *Journal of Hydraulic Research*, *59*(5), 791–809.

9 Modeling of a Nonlinear Reservoir

9.1 INTRODUCTION

In systems approach to rainfall runoff modeling, the usual practice is to represent a watershed by reservoirs, which can be linear or nonlinear (Singh 1988). Depending on the watershed topography, the reservoirs are arranged as a network consisting of reservoirs in series or in series and parallel. For simplicity, reservoirs are assumed to be linear, forming the basis of the unit hydrograph theory. In reality, however, the reservoirs are nonlinear. Although nonlinear reservoirs can also be arranged as a network, the analytical solution becomes intractable, and numerical solutions are the only resort. However, it is possible to derive approximate solutions using homotopy analysis methods. The objective of this chapter is to present these approximate solutions for a single nonlinear reservoir.

9.2 GOVERNING EQUATION AND ANALYTICAL SOLUTION

In Chapter 2, after nondimensionalizing, we obtain the governing equation for the modeling of a nonlinear reservoir as follows:

$$\frac{dS_*}{dt_*} + S_*{}^x = I_* \text{ subject to } S_*(t_* = 0) = 0 \tag{9.1}$$

where $I_* = \begin{cases} 1 \text{ for } 0 \leq t_* < T_* \\ 0 \text{ for } t_* > T_* \end{cases}$ and x is an exponent greater than 1 and is not equal

to 1. Here inflow, which is rainfall excess, is considered as a single pulse occurring for the duration defined by $0 \leq t_* < T_*$. Later we will consider the case where rainfall excess is constituted by a series of pulses. For $t_* > T_*$, Eq. (9.1) takes on the form:

$$\frac{dS_*}{dt_*} + S_*{}^x = 0 \tag{9.2}$$

which can be solved analytically using the separation of variable technique. Solving Eq. (9.2), one obtains:

$$S_* = \left[(1-x)(C_1 - t_*)\right]^{\frac{1}{1-x}} \tag{9.3}$$

where C_1 is an integral constant that can be determined using the solution of Eq. (9.1) for $0 \leq t_* < T_*$.

For $0 \leq t_* < T_*$, we solved Eq. (9.1) using the homotopy analysis method (HAM) in the previous chapter for a specific value of the exponent, $x = 2$. Here, we aim to solve

DOI: 10.1201/9781003368984-12

the equation for an arbitrary value of the exponent. It can be seen that the exponent is a noninteger value, whose presence makes Eq. (9.1) a highly nonlinear differential equation. The application of HAM directly to Eq. (9.1) creates a difficulty in obtaining the higher-order terms, as the homotopy derivative for the nonlinear term will be rather complex (see Case 7, Section 2.6, Chapter 2). For that reason, we approximate the nonlinear term using the Padé approximation to a relatively weaker form and then solve the approximate governing equation using HAM. For $0 \leq t_* < T_*$, dropping the asterisk for simplicity, Eq. (9.1) can be written as:

$$\frac{dS}{dt} + S^x = 1 \text{ subject to } S(t=0) = 0 \tag{9.4}$$

The initial condition is given as $S(t=0)=0$. Therefore, we have to obtain the Padé approximation around a point other than $t=0$, as the higher-order terms of the Taylor series at that point become undefined for the nonlinear term S^x. To that end, the $[m,n]$-order Padé approximation of S^x around $t=1$ can be written as:

$$S^x = \frac{\sum_{i=0}^{m} A_i (S-1)^i}{1 + \sum_{j=1}^{n} B_j (S-1)^j} \tag{9.5}$$

where A_i and B_j are the coefficients. Before proceeding further, we need to select an appropriate order for the approximation made in Eq. (9.5) to assess the approximation. Here, we consider [2] and [3] order approximations and determine which of them works better. As mentioned earlier, the determination of coefficients is cumbersome for hand calculations. Therefore, we use the MATHEMATICA command *PadeApproximant* to obtain the coefficients. Following this, the [2] and [3] order approximations are obtained as:

$$\left[S^x\right]_{2,2} = \frac{1 + \frac{1}{2}(2+x)(S-1) + \frac{1}{12}(2+3x+x^2)(S-1)^2}{1 + \frac{1}{2}(2-x)(S-1) + \frac{1}{12}(2-3x+x^2)(S-1)^2} \tag{9.6}$$

$$\left[S^x\right]_{3,3} = \frac{1 + \frac{1}{2}(3+x)(S-1) + \frac{1}{10}(6+5x+x^2)(S-1)^2 + \frac{1}{120}(6+11x+6x^2+x^3)(S-1)^3}{1 + \frac{1}{2}(3-x)(S-1) + \frac{1}{10}(6-5x+x^2)(S-1)^2 + \frac{1}{120}(-6+11x-6x^2+x^3)(S-1)^3} \tag{9.7}$$

For particular values of exponent x, Figure 9.1(a) and (b) shows a comparison between the original function and its corresponding Padé approximants of order $[2]$ and $[3]$, respectively. It can be seen from the figure that the higher the approximation, the better the accuracy. Also, it can be concluded that the [3] order is a suitable choice for

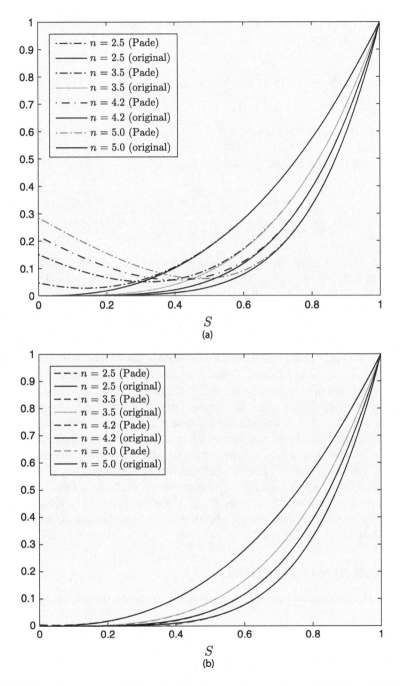

FIGURE 9.1 Comparison of S^x and its Padé approximants of orders (a) [2] and (b) [3] for $x = 2.5, 3.5, 4.2,$ and 5.0.

the approximation. Therefore, choosing $m = 3 = n$ in Eq. (9.5), the governing equation can be rearranged as:

$$\left[1 + \sum_{j=1}^{3} B_j (S-1)^j\right]\frac{dS}{dt} + \sum_{i=0}^{3} A_i (S-1)^i - \left[1 + \sum_{j=1}^{3} B_j (S-1)^j\right] = 0$$

$$\text{subject to } S(t=0) = 0 \qquad\qquad\qquad (9.8)$$

Equation (9.8) can further be simplified as:

$$\left[B_0^* + B_1^* S + B_2^* S^2 + B_3^* S^3\right]\frac{dS}{dt} + A_0^* + A_1^* S + A_2^* S^2 + A_3^* S^3$$

$$= 0 \text{ subject to } S(t=0) = 0 \qquad\qquad\qquad (9.9)$$

where B_i^* and A_i^* are the coefficients involving B_i and A_i. Mathematically, $B_0^* = 1 - B_1 + B_2 - B_3$, $B_1^* = B_1 - 2B_2 + 3B_3$, $B_2^* = B_2 - 3B_3$, $B_3^* = B_3$, $A_0^* = A_0 - A_1 + A_2 - A_3 - B_0^*$, $A_1^* = A_1 - 2A_2 + 3A_3 - B_1^*$, $A_2^* = A_2 - 3A_3 - B_2^*$, and $A_3^* = A_3 - B_3^*$.

We have already seen the accuracy of [3] order Padé approximation for the non-linear term. Let us now see how the approximate differential in Eq. (9.9) works in comparison to the original Eq. (9.4). Thus, we solve both differential equations using a numerical method, namely, the Runge-Kutta fourth-order method, for which the built-in MATLAB script *ode45* is used. Figure 9.2 (a) and (b) shows a comparison between the numerical solution of the original Eq. (9.4) and that of the approximate Eq. (9.9) with [3] order Padé approximant for the nonlinear term for some selected values of exponent $x = 2.5, 3.5, 4.2$, and 5.0. For the sake of completeness, we have shown the full solution that includes the part for $t > T$. It can be seen from the figure that the chosen approximation produces an accurate solution for the governing equation. Therefore, it can be concluded that differential equations having noninteger power nonlinearity (or a similar kind, such as logarithmic, transcendental, etc.) can be first converted to a relatively weaker nonlinear form using the Padé approximation technique.

We now proceed to our primary objective of solving the approximate equation using HAM.

9.3   HAM-BASED SOLUTION

To apply HAM, we construct the zeroth-order deformation equation as follows:

$$(1-q)\mathcal{L}\left[\Phi(t;q) - S_0(t)\right] = q\hbar H(t)\mathcal{N}\left[\Phi(t;q)\right] \qquad\qquad (9.10)$$

subject to the initial condition:

$$\Phi(0;q) = 0 \qquad\qquad\qquad (9.11)$$

where q is the embedding parameter, $\Phi(t;q)$ is the representation of the solution across q, $S_0(t)$ is the initial approximation, $\hbar$ is the auxiliary parameter, $H(t)$ is the

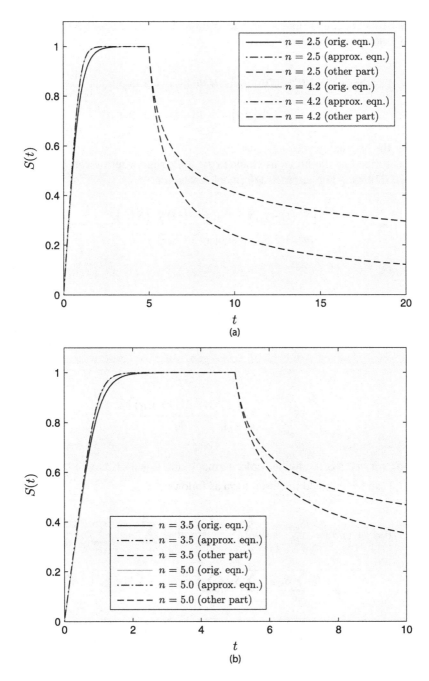

FIGURE 9.2 Comparison of numerical solutions of the original Eq. (9.4) and the approximate Eq. (9.9) for (a) $x = 2.5$, 4.2 and (b) $x = 3.5$, 5.0.

auxiliary function, and $\mathcal{L}$ and $\mathcal{N}$ are the linear and nonlinear operators, respectively. The nonlinear operator can be chosen as:

$$\mathcal{N}\left[\Phi(t;q)\right]=\left[B_0^* + B_1^*\Phi(t;q)+B_2^*\Phi(t;q)^2+B_3^*\Phi(t;q)^3\right]\frac{\partial\Phi(t;q)}{\partial t}$$

$$+A_0^* + A_1^*\Phi(t;q)+A_2^*\Phi(t;q)^2+A_3^*\Phi(t;q)^3 \tag{9.12}$$

which is the original governing equation.

As described in the previous chapter, the higher-order terms can be calculated from the following higher order deformation equation:

$$\mathcal{L}\left[S_m(t)-\chi_m S_{m-1}(t)\right]=\hbar H(t) R_m\left(\vec{S}_{m-1}\right),$$

$$m=1,2,3,\ldots, \text{ subject to } S_m(0)=0 \tag{9.13}$$

where

$$\chi_m = \begin{cases} 0 & \text{when } m=1, \\ 1 & \text{otherwise} \end{cases} \tag{9.14}$$

and

$$R_m\left(\vec{S}_{m-1}\right)=\frac{1}{(m-1)!}\frac{\partial^{m-1}\mathcal{N}\left[\Phi(t;q)\right]}{\partial q^{m-1}}\bigg|_{q=0} \tag{9.15}$$

where S_m for $m\geq 1$ is the higher-order terms. Using Eqs. (9.12) and (9.15), the term $R_m\left(\vec{S}_{m-1}\right)$ can be obtained in closed form as follows:

$$R_m\left(\vec{S}_{m-1}\right)=B_0^*\frac{dS_{m-1}}{dt}+B_1^*\sum_{j=0}^{m-1}S_j\frac{dS_{m-1-j}}{dt}+B_2^*\sum_{j=0}^{m-1}S_{m-1-j}\sum_{k=0}^{j}S_k\frac{dS_{j-k}}{dt}$$

$$+B_3^*\sum_{j=0}^{m-1}S_{m-1-j}\sum_{k=0}^{j}S_{j-k}\sum_{l=0}^{k}S_l\frac{dS_{k-l}}{dt}+A_0^*\left(1-\chi_m\right)+A_1^*S_{m-1}$$

$$+A_2^*\sum_{j=0}^{m-1}S_j S_{m-1-j}+A_3^*\sum_{j=0}^{m-1}S_{m-1-j}\sum_{k=0}^{j}S_k S_{j-k} \tag{9.16}$$

Equation (9.16) can be obtained using the guidelines discussed in Section 2.6 of Chapter 2. The structure of this equation may look complicated but is easy to obtain using the properties of the homotopy derivative. Further, in the context of HAM, these closed-form expressions may look difficult to compute, but it is emphasized here again that with the development of efficient symbolic computation software, such as MATHEMATICA, it is easy to do computations. This issue will be discussed in more detail later. Now, the final solution is obtained as follows:

$$S(t) = S_0(t) + \sum_{m=1}^{\infty} S_m(t) \tag{9.17}$$

As mentioned in the previous chapter, the convergence of the series solution using HAM depends on the choice of linear operator, auxiliary function, and auxiliary parameter. Liao (2012) proposed fundamental rules for making that choice. First, we choose a set of base functions to represent the solution. For a given differential equation, the first choice to represent its solution that comes to mind is a polynomial base function, as it is the most obvious one and also easy to compute. Indeed, HAM provides us with the freedom of choosing several kinds of base functions. However, for simplicity, here we choose a polynomial function and see how it works. Therefore, we consider the set of base functions:

$$\{t^n \mid n = 1, 2, 3...\} \tag{9.18}$$

so that:

$$S(t) = \sum_{n=1}^{\infty} a_n t^n \tag{9.19}$$

Equation (9.18) provides the so-called *rule of solution expression*. It is interesting to note that the chosen set of base functions in Eq. (9.17) is always consistent with the given initial condition, i.e., it always satisfies the initial condition $S(t = 0) = 0$.

In accordance with the rule of solution expression, we choose the following linear operator and the initial approximation, respectively:

$$\mathcal{L}[\Phi(t;q)] = \frac{\partial \Phi(t;q)}{\partial t} \text{ with the property } \mathcal{L}[C_2] = 0 \tag{9.20}$$

$$S_0(t) = t \tag{9.21}$$

where C_2 is an integral constant. Using Eq. (9.20), the higher-order terms can be obtained from Eq. (9.13) as follows:

$$S_m(t) = \chi_m S_{m-1}(t) + \hbar \mathcal{L}^{-1}\left[H(t) R_m\left(\vec{S}_{m-1}\right)\right] + C_2$$

$$= \chi_m S_{m-1}(t) + \hbar \int_0^t H(\tau) R_m\left(\vec{S}_{m-1}\right) d\tau + C_2, \; m = 1,2,3,... \tag{9.22}$$

where R_m is given by Eq. (9.16), and the constant C_2 can be determined from the initial condition for the higher order deformation equations, i.e., $S_m(0) = 0$, which simply yields $C_2 = 0$ for all $m \geq 1$.

Now, the auxiliary function $H(t)$ can be determined from the *rule of coefficient ergodicity*. Based on the rule of solution expression and Eq. (9.21), the general form of $H(t)$ should be:

$$H(t) = t^{\alpha} \tag{9.23}$$

where α is an integer. Substituting Eq. (9.23) into Eq. (9.22) and computing some of S_m, we can infer two observations. If we consider $\alpha \geq 2$, then the term t^{α} will be missing in the solution expression. This does not satisfy the *rule of coefficient ergodicity*, which allows us to include all the terms of the base function for the completeness of the solution. On the other hand, if we consider $\alpha \leq -1$, the terms appearing in the solution violate the *rule of solution expression*. Therefore, we are left with two choices, $\alpha = 0$ or 1, and can choose either of them. Here, we take $\alpha = 0$, i.e., $H(t) = 1$. The auxiliary function $H(t)$ can be determined, following the *rule of solution expression* and the *rule of coefficient ergodicity*, as already done. For this, one needs to calculate the terms of the series solution using HAM. To avoid calculating the terms by hand, which is a difficult task for most of the equations, it is suggested to use MATHEMATICA.

Equation (9.22) now takes on the form:

$$S_m(t) = \chi_m S_{m-1}(t) + \hbar \int_0^t R_m\left(\vec{S}_{m-1}\right) d\tau, \ m = 1, 2, 3, \ldots \tag{9.24}$$

Using the initial approximation in Eq. (9.21) and expanding the summation in Eq. (9.16), the first-order term of the series can be obtained from Eq. (9.24) as:

$$R_1\left(\vec{S}_0\right) = B_0^* \frac{dS_0}{dt} + B_1^* S_0 \frac{dS_0}{dt} + B_2^* S_0^2 \frac{dS_0}{dt} + B_3^* S_0^3 \frac{dS_0}{dt} + A_0^* + A_1^* S_0$$
$$+ A_2^* S_0^2 + A_3^* S_0^3 = B_0^* + B_1^* t + B_2^* t^2 + B_3^* t^3 + A_0^* + A_1^* t + A_2^* t^2 + A_3^* t^3 \tag{9.25}$$

$$S_1(t) = \hbar \int_0^t R_1\left(\vec{S}_0\right) d\tau$$
$$= \hbar \left[B_0^* t + B_1^* \frac{t^2}{2} + B_2^* \frac{t^3}{3} + B_3^* \frac{t^4}{4} + A_0^* t + A_1^* \frac{t^2}{2} + A_2^* \frac{t^3}{3} + A_3^* \frac{t^4}{4} \right] \tag{9.26}$$

The other terms can be calculated in a like manner. Finally, the approximate solution can be obtained as:

$$S(t) = S_0(t) + \sum_{m=1}^{M} S_m(t) \tag{9.27}$$

Based on the fundamental rules developed by Liao (2012), Eq. (9.27) was obtained. As mentioned in the previous chapter, the auxiliary parameter $\hbar$ plays the role of monitoring the rate and region of convergence of the series solution in Eq. (9.27).

9.4   HOMOTOPY PERTURBATION METHOD (HPM)-BASED SOLUTION

Now, we aim to solve Eq. (9.4) using the homotopy perturbation method (HPM). It may be noted that the purpose is to compare the solutions using different methods. Therefore, we apply HPM for a specific value of the exponent, say $x = 5$. To that end, the governing equation is written in the following form:

$$\mathcal{N}(S) = f(t) \tag{9.28}$$

We construct a homotopy $\Phi(t;q)$, which satisfies:

$$(1-q)\left[\mathcal{L}\big(\Phi(t;q)\big) - \mathcal{L}\big(S_0(t)\big)\right] + q\left[\mathcal{N}\big(\Phi(t;q)\big) - f(t)\right] = 0 \tag{9.29}$$

where q is the embedding parameter that lies on $[0,1]$, $\mathcal{L}$ is the linear operator, and $S_0(t)$ is the initial approximation to the final solution. Equation (9.29) shows that as $q = 0$, $\Phi(t;0) = S_0(t)$ and as $q = 1$, $\Phi(t;1) = S(t)$. We now express $\Phi(t;q)$ as a series in terms of q, as follows:

$$\Phi(t;q) = \Phi_0 + q\Phi_1 + q^2\Phi_2 + q^3\Phi_3 + q^4\Phi_4\ldots \tag{9.30}$$

where Φ_m for $m \geq 1$ are the higher-order terms. As $q \to 1$, Eq. (9.30) produces the final solution as:

$$S(t) = \lim_{q \to 1} \Phi(t;q) = \sum_{i=0}^{\infty} \Phi_i \tag{9.31}$$

It was seen in the previous chapter that exponential base functions produce a faster convergent series. Therefore, we consider the following forms for the operators:

$$\mathcal{L}\big(\Phi(t;q)\big) = \frac{\partial \Phi(t;q)}{\partial t} + \Phi(t;q) \tag{9.32}$$

$$\mathcal{N}\big(\Phi(t;q)\big) = \frac{\partial \Phi(t;q)}{\partial t} + \Phi^5(t;q) - 1 \tag{9.33}$$

Using these expressions, Eq. (9.29) takes on the form:

$$(1-q)\left[\frac{\partial \Phi(t;q)}{\partial t} + \Phi(t;q) - \frac{dS_0}{dt} - S_0\right]$$
$$+ q\left[\frac{\partial \Phi(t;q)}{\partial t} + \Phi^5(t;q) - 1\right] = 0 \tag{9.34}$$

Substituting Eq. (9.30) into Eq. (9.34) and the initial condition $S(0) = 0$, one can obtain:

$$(1-q)\left[\left(\frac{d\Phi_0}{dt} + q\frac{d\Phi_1}{dt} + q^2\frac{d\Phi_2}{dt} + q^3\frac{d\Phi_3}{dt} + q^4\frac{d\Phi_4}{dt} + \cdots\right)\right.$$

$$+\left(\Phi_0 + q\Phi_1 + q^2\Phi_2 + q^3\Phi_3 + q^4\Phi_4\cdots\right) - \frac{dS_0}{dt} - S_0\Bigg]$$

$$+q\left[\left(\frac{d\Phi_0}{dt} + q\frac{d\Phi_1}{dt} + q^2\frac{d\Phi_2}{dt} + q^3\frac{d\Phi_3}{dt} + q^4\frac{d\Phi_4}{dt} + \cdots\right)\right.$$

$$+\left(\Phi_0 + q\Phi_1 + q^2\Phi_2 + q^3\Phi_3 + q^4\Phi_4\cdots\right)^5 - 1\Bigg] = 0 \qquad (9.35)$$

and:

$$\Phi_0(0) = 0, \ \Phi_1(0) = 0, \ \Phi_2(0) = 0, \ \dots. \qquad (9.36)$$

The nonlinear term can be expanded as follows:

$$\left(\Phi_0 + q\Phi_1 + q^2\Phi_2 + q^3\Phi_3 + q^4\Phi_4\cdots\right)^5 = (\Phi_0)^5 + q\left(5\Phi_0^4\Phi_1\right)$$

$$+q^2\left(5\Phi_0^4\Phi_2 + 10\Phi_0^3\Phi_1^2\right) + q^3\left(5\Phi_0^4\Phi_3 + 20\Phi_0^3\Phi_1\Phi_2 + 10\Phi_0^2\Phi_1^3\right)$$

$$+q^4\left(5\Phi_0^4\Phi_4 + 20\Phi_0^3\Phi_1\Phi_3 + 10\Phi_0^3\Phi_2^2 + 30\Phi_0^2\Phi_1^2\Phi_2 + 5\Phi_0\Phi_1^4\right) + \cdots \qquad (9.37)$$

Using Eq. (9.37) and equating the like powers of q in Eq. (9.35), the following system of differential equations is obtained:

$$\frac{d\Phi_0}{dt} + \Phi_0 - \frac{dS_0}{dt} - S_0 = 0 \text{ subject to } \Phi_0(0) = 0 \qquad (9.38)$$

$$\frac{d\Phi_1}{dt} + \Phi_1 - \left(\frac{d\Phi_0}{dt} + \Phi_0 - \frac{dS_0}{dt} - S_0\right) + \frac{d\Phi_0}{dt} + \Phi_0^5 - 1$$

$$= 0 \text{ subject to } \Phi_1(0) = 0 \qquad (9.39)$$

$$\frac{d\Phi_2}{dt} + \Phi_2 - \left(\frac{d\Phi_1}{dt} + \Phi_1\right) + \frac{d\Phi_1}{dt} + 5\Phi_0^4\Phi_1 = 0 \text{ subject to } \Phi_2(0) = 0 \quad (9.40)$$

$$\frac{d\Phi_3}{dt} + \Phi_3 - \left(\frac{d\Phi_2}{dt} + \Phi_2\right) + \frac{d\Phi_2}{dt} + 5\Phi_0^4\Phi_2 + 10\Phi_0^3\Phi_1^2$$

$$= 0 \text{ subject to } \Phi_3(0) = 0 \qquad (9.41)$$

$$\frac{d\Phi_4}{dt} + \Phi_4 - \left(\frac{d\Phi_3}{dt} + \Phi_3\right) + \frac{d\Phi_3}{dt} + 5\Phi_0^4\Phi_3 + 20\Phi_0^3\Phi_1\Phi_2 + 10\Phi_0^2\Phi_1^3$$

$$= 0 \text{ subject to } \Phi_4(0) = 0 \qquad (9.42)$$

Taking the initial approximation $S_0 = 1 - \exp(-t)$, we can solve the equations iteratively using symbolic software. We avoid those expressions here, as they are lengthy. Finally, the HPM-based solution can be approximated as follows:

$$S(t) \approx \sum_{i=0}^{M} \Phi_i \tag{9.43}$$

9.5 OPTIMAL HOMOTOPY ASYMPTOTIC METHOD (OHAM)-BASED SOLUTION

Here, we apply the optimal homotopy asymptotic method (OHAM) to the problem under consideration. Similar to HPM, we consider the exponent $x = 5$. When applying OHAM, the original Eq. (9.4) can be written in the following form:

$$\mathcal{L}[S(t)] + \mathcal{N}[S(t)] + h(t) = 0 \tag{9.44}$$

where $\mathcal{L}[S(t)] = \dfrac{dS}{dt} + S$, $N[S(t)] = S^5(t) - S$, and $h(t) = -1$. Here we choose the linear operator to be of a specific form because of the expectation for a faster series approximation, as it produces the solution in terms of exponential functions. With these considerations, the zeroth-order problem becomes:

$$\mathcal{L}(S_0(t)) + h(t) = 0 \text{ subject to } S_0(0) = 0 \tag{9.45}$$

Solving Eq. (9.45), one obtains:

$$S_0(t) = 1 - \exp(-t) \tag{9.46}$$

The first-order problem reduces to:

$$\frac{dS_1}{dt} + S_1 = H_1(t, C_i)\left(S_0^5 - S_0\right) \text{ subject to } S_1(0) = 0 \tag{9.47}$$

Let us simply choose $H_1(t, C_1) = C_1$. Putting $k = 2$, the second-order problem becomes:

$$\mathcal{L}[S_2(t, C_i) - S_1(t, C_i)] = H_2(t, C_i)\mathcal{N}_0(S_0(t))$$
$$+ H_1(t, C_i)[\mathcal{L}[S_1(t, C_i)] + \mathcal{N}_1[S_0(t), S_1(t, C_i)]] \text{ subject to } S_2(0) = 0 \tag{9.48}$$

Expanding $\mathcal{N}(\Phi(t, C_i; q))$, we obtain the coefficient of q as $\mathcal{N}_1[S_0(t), S_1(t, C_i)] = 5S_0(t)^4 S_1(t, C_i) - S_1(t, C_i)$. Further, we choose $H_2(t, C_i) = C_2$. Therefore, Eq. (9.48) becomes:

$$\frac{dS_2}{dt} + S_2 = \frac{dS_1}{dt} + S_1 + C_2\left(S_0^5 - S_0\right)$$
$$+ C_1\left[\frac{dS_1}{dt} + S_1 + 5S_0(t)^4 S_1(t, C_i) - S_1(t, C_i)\right] \text{ subject to } S_2(0) = 0 \tag{9.49}$$

It can be seen that we do not write the explicit forms of the terms here, as those are lengthy and may not be appropriate to keep up the flow. Indeed, one can compute these terms without any difficulty using symbolic computation software, such as MATLAB. Further, following Chapter 2, the higher-order terms can be computed in a similar manner. However, our aim is to produce an accurate solution with just two to three terms of the OHAM-based series. Therefore, we restrict our calculation up to $k = 2$. Finally, the approximate solution can be found as:

$$S(t) \approx S_0(t) + S_1(t, C_1) + S_2(t, C_1, C_2) \tag{9.50}$$

where the terms are given by Eqs. (9.46), (9.47), and (9.49).

9.6   CONVERGENCE THEOREMS

Here, the convergence of the HAM-based and OHAM-based solutions are proved theoretically using the following theorems.

9.6.1   CONVERGENCE THEOREM OF THE HAM-BASED SOLUTION IN EQ. (9.27)

Theorem 9.1: If the homotopy series $\Sigma_{m=0}^{\infty} S_m(t)$ and $\Sigma_{m=0}^{\infty} S_m'(t)$ converge, then $R_m(\vec{S}_{m-1})$ given by Eq. (9.16) satisfies the relation $\Sigma_{m=1}^{\infty} R_m(\vec{S}_{m-1}) = 0$. [Here '$'$' denotes the derivative with respect to t.]

Proof: The proof of this theorem follows exactly from Theorem 2.1 of Chapter 2.

Theorem 9.2: If $\hbar$ is so properly chosen that the series $\Sigma_{m=0}^{\infty} S_m(t)$ and $\Sigma_{m=0}^{\infty} S_m'(t)$ converge absolutely to $S(t)$ and $S'(t)$, respectively, then the homotopy series $\Sigma_{m=0}^{\infty} S_m(t)$ satisfies the original governing Eq. (9.9).

Proof: Using the Cauchy product defined in Theorem 2.2 of Chapter 2 in relation to Eq. (9.16), we get:

$$\sum_{m=1}^{\infty} \sum_{j=0}^{m-1} S_j S_{m-1-j} = \left(\sum_{m=0}^{\infty} S_m \right)^2 \tag{9.51}$$

$$\sum_{m=1}^{\infty} \sum_{j=0}^{m-1} S_j S_{m-1-j}' = \left(\sum_{m=0}^{\infty} S_m \right) \left(\sum_{k=0}^{\infty} S_k' \right) \tag{9.52}$$

$$\sum_{m=1}^{\infty} \sum_{j=0}^{m-1} S_{m-1-j} \sum_{k=0}^{j} S_k S_{j-k} = \left(\sum_{m=0}^{\infty} S_m \right)^3 \tag{9.53}$$

$$\sum_{m=1}^{\infty}\sum_{j=0}^{m-1} S_{m-1-j} \sum_{k=0}^{j} S_k S'_{j-k} = \left(\sum_{m=0}^{\infty} S_m\right)^2 \left(\sum_{k=0}^{\infty} S'_k\right) \qquad (9.54)$$

$$\sum_{m=1}^{\infty}\sum_{j=0}^{m-1} S_{m-1-j} \sum_{k=0}^{j} S_{j-k} \sum_{l=0}^{k} S_l S'_{k-l} = \left(\sum_{m=0}^{\infty} S_m\right)^3 \left(\sum_{k=0}^{\infty} S'_k\right) \qquad (9.55)$$

Theorem 9.1 shows that if $\sum_{m=0}^{\infty} S_m(t)$ and $\sum_{m=0}^{\infty} S_m{}'(t)$ converge then $\sum_{m=1}^{\infty} R_m\left(\vec{S}_{m-1}\right) = 0$.

Therefore, substituting the earlier expressions in Eq. (9.16) and simplifying further lead to:

$$B_0{}^* \sum_{m=0}^{\infty} S'_m + B_1{}^* \left(\sum_{m=0}^{\infty} S'_m\right)\left(\sum_{k=0}^{\infty} S_k\right) + B_2{}^* \left(\sum_{m=0}^{\infty} S'_m\right)\left(\sum_{k=0}^{\infty} S_k\right)^2$$

$$+ B_3{}^* \left(\sum_{m=0}^{\infty} S'_m\right)\left(\sum_{k=0}^{\infty} S_k\right)^3 + A_0{}^* \sum_{m=0}^{\infty} (1-\chi_{m+1}) + A_1{}^* \sum_{m=0}^{\infty} S_m$$

$$+ A_2{}^* \left(\sum_{k=0}^{\infty} S_k\right)^2 + A_3{}^* \left(\sum_{k=0}^{\infty} S_k\right)^3 = 0 \qquad (9.56)$$

which is basically the original governing equation in Eq. (9.9). Furthermore, subject to the initial condition $S_0(0) = 0$ and the conditions for the higher order deformation equation $S_m(0) = 0$, for $m \geq 1$, we easily obtain $\sum_{m=0}^{\infty} S_m(0) = 0$. Hence, the convergence result follows.

9.6.2 Convergence Theorem of the OHAM-Based Solution in Eq. (9.50)

Theorem 9.3: If the series $S_0(t) + \sum_{j=1}^{\infty} S_j(t, C_i)$, $i = 1, 2, \ldots, s$ converges, where $S_j(t, C_i)$ are governed by Eqs. (9.46), (9.47), and (9.49), then Eq. (9.50) is a solution of the original Eq. (9.4).

Proof: The proof of this theorem follows exactly from Theorem 3 of Chapter 2.

9.7 RESULTS AND DISCUSSION

Here, first we validate the solutions obtained using HAM, HPM, and OHAM. Then, we compare them to get a comparative idea about the performances of the methods.

9.7.1 Validation of the HAM-Based Solution

In this work, the original nonlinear equation was converted to a relatively weaker form using the Padé approximation technique. As mentioned earlier, the coefficients of the approximation can be obtained using the MATHEMATICA command

TABLE 9.1

Padé Approximant Coefficients for Selected Values of the Exponent

Exponent	Padé Approximation Coefficients							
(x)	B_0^*	B_1^*	B_2^*	B_3^*	A_0^*	A_1^*	A_2^*	A_3^*
2.5	231	99	−11	1	−230	−110	110	230
3.5	−85.8	28.6	−7.8	1	86.8	−36.4	36.4	−86.8
4.2	−27.48	19.02	−6.75	1	28.48	−25.78	25.78	−28.48
5.0	−14	14	−6	1	15	−20	20	−15

PadeApproximant. Here, we consider some selected values for exponent x, say $x = 2.5, 3.5, 4.2$, and 5, and find the corresponding Padé approximation for the non-linear term. The coefficients obtained for the approximation are given in Table 9.1. With the selected values of the exponent, we check how the HAM-based solution works as compared to the numerical solution. It can be seen that the auxiliary parameter $\hbar$ needs to be determined for a particular order of approximation in order to assess the HAM-based solution. For that, as mentioned in the previous chapter, the squared residual error is calculated as follows:

$$\Delta_m = \int_{t \in \Omega} \left(\mathcal{N}\left[S(t) \right] \right)^2 dt \qquad (9.57)$$

The series solution obtained using HAM leads to the exact solution of the problem once Δ_m tends to its minimum value. Therefore, for a particular order of approximation m, the corresponding Δ_m is minimized, which yields an optimal value of parameter $\hbar$. We consider a test case $x = 2.5$ and see the behavior of error Δ_m for a different order of approximation. Figure 9.3 shows the squared residual error for a different order of approximation. It can be seen from the figure that as the order of approximation increases, the corresponding residual error decreases, which shows the stability of the method. Also, the numerical results are reported in Table 9.2. For the test case considered, we compute HAM-based analytical solutions for a different order of approximation and compare them with the corresponding numerical solution. The MATLAB script *ode45* is used to obtain the numerical solution of the problem, and the HAM-based approximation is achieved using the MATHEMATICA package BVPh 2.0 developed by Zhao and Liao (2002).

Unlike the previous chapter, here, the computation of HAM is performed in MATHEMATICA instead of MATLAB. It may be noted that as the nonlinear equation becomes more complex in nature, MATLAB starts performing slowly. For that reason, we suggest using MATHEMATICA (specifically, BVPh 2.0) to avoid any unnecessary difficulty with the computation. Figure 9.4 shows a comparison between the numerical solution and 10th-, 20th-, 30th-, and 45th-order approximations of the HAM-based solution for the exponent $x = 2.5$ over the interval $t \in [0,3]$. The other part of the solution for $t \geq T$ is not included here, as the primary objective is to see the efficiency of the HAM-based solution. One can indeed compute that

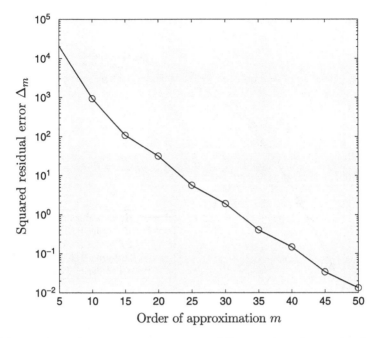

FIGURE 9.3 Squared residual error (Δ_m) versus different orders of approximation (m) of the HAM-based approximation for the exponent $x = 2.5$.

using the initial condition calculated from the HAM-based solution at the end point $t = 3$. For a quantitative assessment, the solutions are compared in Table 9.3. It can be seen from both the figure and the table that as the order of approximation increases, the HAM-based solution becomes more accurate. Specifically, the 45th-order HAM-based approximation shows excellent agreement with the numerical solution. For the

TABLE 9.2

Squared Residual Error (Δ_m) versus Different Orders of Approximation (m) for Exponent $x = 2.5$

Order of Approximation (m)	Squared Residual Error (Δ_m)
5	20768.1000
10	936.8700
15	109.1954
20	31.4290
25	5.6280
30	1.9093
35	0.4038
40	0.1484
45	0.0344
50	0.0133

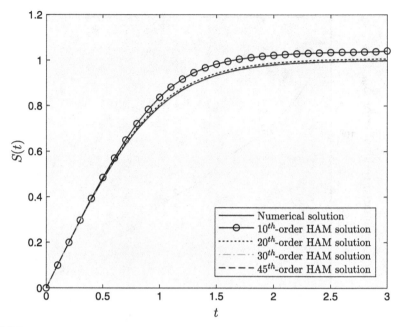

FIGURE 9.4 Comparison between the numerical solution and HAM-based solution for a different order of approximation for the exponent $x = 2.5$.

other values of the exponent, $x = 3.5$, 4.2, and 5, the approximate solutions are compared with the corresponding numerical solutions, as shown in Figures 9.5, 9.6, and 9.7, respectively. It can be seen from the figures that the HAM-based solution provides excellent accuracy for all the cases considered. Overall, the homotopy analysis method, together with the Padé approximation, seems to be an efficient analytical

TABLE 9.3

Comparison between HAM-Based Approximation and Numerical Solution

t	Numerical Solution	HAM-Based Approximation			
		25th Order	30th Order	40th Order	45th Order
0.0	0.0000	0.0000	0.0000	0.0000	0.0000
0.3	0.2955	0.2959	0.2957	0.2956	0.2955
0.6	0.5568	0.5593	0.5582	0.5573	0.5571
0.9	0.7494	0.7538	0.7517	0.7501	0.7498
1.2	0.8691	0.8735	0.8713	0.8697	0.8694
1.5	0.9349	0.9389	0.9369	0.9354	0.9351
1.8	0.9684	0.9723	0.9704	0.9690	0.9687
2.1	0.9849	0.9888	0.9869	0.9854	0.9852
2.4	0.9928	0.9967	0.9948	0.9934	0.9931
2.7	0.9966	1.0005	0.9986	0.9972	0.9969
3.0	0.9984	1.0023	1.0004	0.9993	0.9987

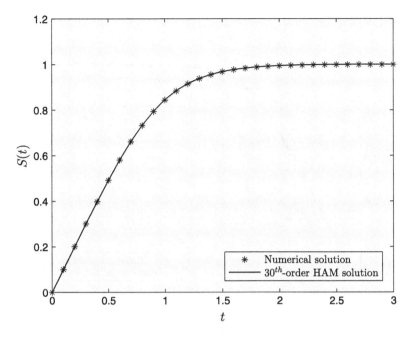

FIGURE 9.5 Comparison between the numerical solution and 30th-order HAM-based solution for the exponent $x = 3.5$.

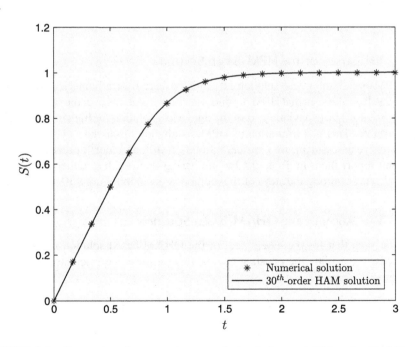

FIGURE 9.6 Comparison between the numerical solution and 30th-order HAM-based solution for the exponent $x = 4.2$.

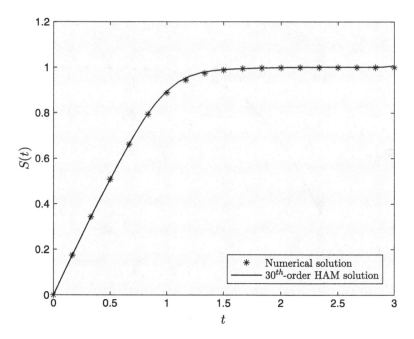

FIGURE 9.7 Comparison between the numerical solution and 30th-order HAM-based solution for the exponent $x = 5.0$.

method for solving nonlinear differential equations with noninteger power nonlinearity. A flowchart containing the steps of the method is provided in Figure 9.8.

9.7.2 VALIDATION OF THE HPM-BASED SOLUTION

The HPM-based analytical solution is validated here over a numerical solution in Figure 9.9. It can be seen that HPM produces accurate solutions over the range $t \in [0,1]$, but the solution starts deviating from the numerical solution for higher values of the time variable. This is a limitation of HPM, as already discussed in Chapter 2. For a quantitative assessment, we consider second-, third-, and fourth-order approximations and report them in Table 9.4 by comparing them with a numerical solution. A flowchart containing the steps of the method is given in Figure 9.10.

9.7.3 VALIDATION OF THE OHAM-BASED SOLUTION

It can be seen that for the assessment of the OHAM-based solution in Eq. (9.50), one needs to calculate the constant C_i. For that purpose, we calculate the residual as follows:

$$R(t,C_i) = \mathcal{L}\big[S_{\text{OHAM}}(t,C_i)\big] + \mathcal{N}\big[S_{\text{OHAM}}(t,C_i)\big] + h(t), i = 1,2,\ldots,s \quad (9.58)$$

where $S_{\text{OHAM}}(t,C_i)$ is the approximate solution. When $R(t,C_i) = 0$, $S_{\text{OHAM}}(t,C_i)$ becomes the exact solution to the problem. But in nonlinear problems, it is rare for

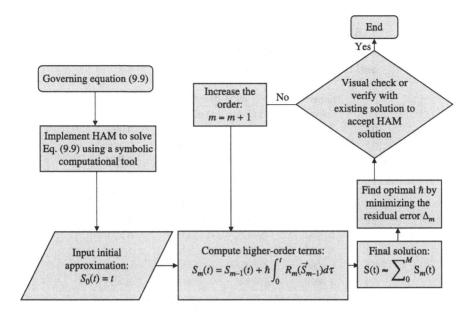

FIGURE 9.8   Flowchart for the HAM solution in Eq. (9.27).

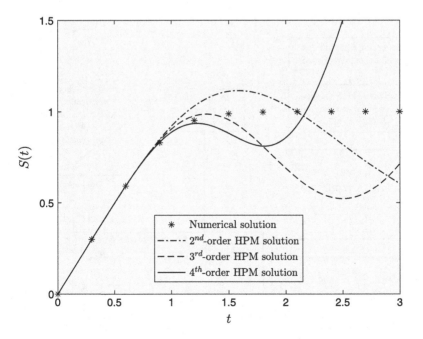

FIGURE 9.9   Comparison between the numerical solution and HPM-based solution for the exponent $x = 5.0$.

TABLE 9.4

Comparison between HPM-Based Approximation and Numerical Solution $(x = 5)$

t	Numerical Solution	HPM-Based Approximation		
		2nd Order	3rd Order	4th Order
0.0	0.0000	0.0000	0.0000	0.0000
0.3	0.2999	0.2996	0.2999	0.2999
0.6	0.5925	0.5915	0.5932	0.5928
0.9	0.8296	0.8484	0.8420	0.8329
1.2	0.9506	1.0304	0.9765	0.9342
1.5	0.9882	1.1108	0.9542	0.8806
1.8	0.9973	1.0916	0.8091	0.8106
2.1	0.9994	0.9982	0.6350	0.9249
2.4	0.9999	0.8661	0.5303	1.3182
2.7	1.0000	0.7276	0.5540	1.9089
3.0	1.0000	0.6061	0.7145	2.4985

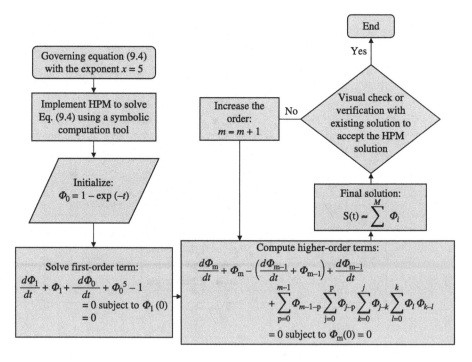

FIGURE 9.10 Flowchart for the HPM solution in Eq. (9.43).

$R(t, C_i)$ to be zero. However, we can minimize it to obtain an accurate solution. One of the ways to obtain the optimal C_i for which the solution converges is the minimization of the squared residual error, i.e.,

$$J(C_i) = \int_{t \in D} R^2(t, C_i) dt, \ i = 1, 2, \ldots, s \tag{9.59}$$

where D is the domain of the problem. The minimization of Eq. (9.59) leads to a system of algebraic equations as follows:

$$\frac{\partial J}{\partial C_1} = \frac{\partial J}{\partial C_2} = \cdots = \frac{\partial J}{\partial C_s} = 0 \tag{9.60}$$

Solving this system, one can obtain the optimal values of the parameters. We obtain the optimal values using the MATLAB routine *fminsearch*, which minimizes an unconstrained multivariable function. Using these values, two- and three-term OHAM solutions are computed and compared in Figure 9.11 with a numerical solution to check the effectiveness of the proposed approach. It can be seen that even with three terms of the OHAM-based series, the solution agrees well with the corresponding numerical solution to the problem. For a quantitative assessment, we consider two- and three-term approximations and report them in Table 9.5 by comparing them with a numerical solution. A flowchart containing the steps of the method is provided in Figure 9.12.

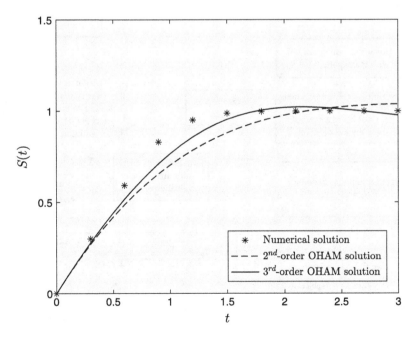

FIGURE 9.11 Comparison between the numerical solution and OHAM-based solution for the exponent $x = 5.0$.

TABLE 9.5

Comparison between OHAM-Based Approximation and Numerical Solution $(x = 5)$

| | | OHAM-Based Approximation | |
| | Numerical | Two-Term | Three-Term |
t	Solution	Solution	Solution
0.0	0.0000	0.0000	−0.0000
0.3	0.2999	0.2699	0.2787
0.6	0.5925	0.4861	0.5165
0.9	0.8296	0.6560	0.7110
1.2	0.9506	0.7853	0.8573
1.5	0.9882	0.8799	0.9536
1.8	0.9973	0.9459	1.0049
2.1	0.9994	0.9895	1.0217
2.4	0.9999	1.0164	1.0159
2.7	1.0000	1.0314	0.9982
3.0	1.0000	1.0383	0.9768

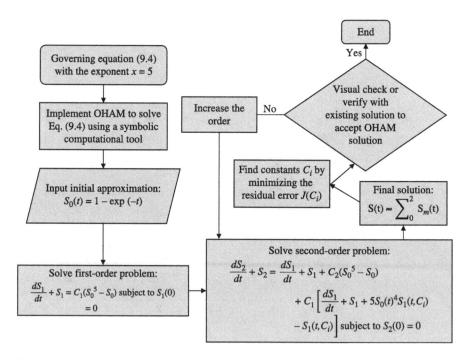

FIGURE 9.12 Flowchart for the OHAM solution in Eq. (9.50).

9.8 CONCLUDING REMARKS

This chapter proposes three different kinds of analytical solutions of the governing equation for modeling a nonlinear reservoir using the homotopy-based methods. The solutions are developed using HAM, HPM, and OHAM. The governing differential equation is a highly nonlinear equation having noninteger power nonlinearity; therefore, it is difficult to handle using any of the existing analytical methods. To overcome this difficulty, first, the nonlinear term is converted to a weakly nonlinear term as a ratio of two polynomials, and then the resulting approximate equation is solved using HAM. The polynomial base function is found to be an appropriate choice for the considered problem; however, one may seek to solve the problem using other kinds of base functions, such as exponential, fractional, etc. For a selected set of values for the exponent x, the HAM-based solutions are compared with the corresponding numerical solutions based on the Runge-Kutta fourth-order method. The approximate solutions show an excellent agreement with the numerical solutions. For HPM and OHAM, we calculate the solution for a specific value of the exponent, namely $x = 5$. It is seen that OHAM can produce an efficient solution even for three terms of the series. All in all, the present work shows the potential of the nonperturbation approaches for solving nonlinear differential equations with noninteger power nonlinearity.

REFERENCES

Singh, V. P. (1988). *Hydrologic systems: Vol. 1 rainfall-runoff modeling.* Prentice Hall, Englewood Cliffs, NJ.
Liao, S. (2012). *Homotopy analysis method in nonlinear differential equations.* Higher Education Press, Beijing.
Zhao, Y., and Liao, S. (2002). User Guide to BVPh 2.0. School of naval architecture, ocean and civil engineering, Shanghai, *40.*

FURTHER READING

Akbarzade, M., and Ganji, D. D. (2010). Coupled method of homotopy perturbation method and variational approach for solution to nonlinear cubic-quintic duffing oscillator. *Advances in Theoretical and Applied Mechanics, 3*(7), 329–337.
Sedighi, H. M., and Daneshmand, F. (2014). Nonlinear transversely vibrating beams by the homotopy perturbation method with an auxiliary term. *Journal of Applied and Computational Mechanics, 1*(1), 1–9.
Sedighi, H. M., Shirazi, K. H., and Attarzadeh, M. A. (2013). A study on the quintic nonlinear beam vibrations using asymptotic approximate approaches. *Acta Astronautica, 91,* 245–250.
Sedighi, H. M., Shirazi, K. H., and Zare, J. (2012). An analytic solution of transversal oscillation of quintic non-linear beam with homotopy analysis method. *International Journal of Non-Linear Mechanics, 47*(7), 777–784.
Suleman, M., and Wu, Q. (2015). Comparative solution of nonlinear quintic cubic oscillator using modified homotopy perturbation method. Advances in Mathematical Physics, *2015.*

10 Nonlinear Muskingum Method for Flood Routing

10.1 INTRODUCTION

The flood routing technique is crucial for many engineering practices, such as the design of several environmental and water resources projects. This technique can be achieved accurately by solving the system of the St. Venant equations, which is represented by a nonlinear partial differential equation (PDE) system and in general cannot be solved analytically. However, the advanced numerical techniques can be used to find an approximate solution (Taylor and Davis 1973; Scarlatos 1982). There are simpler methods based on approximations of the St. Venant equations, such as diffusion wave approximation and kinematic wave approximation, which are also quite popular. However, these approximations also require numerical solutions for realistic inflow and boundary conditions. The simplest flood routing method is the Muskingum method, which has been popular because its simplicity since its inception in the mid-1930s. This method has both a linear form and nonlinear form. The nonlinear form also requires numerical solutions and will be discussed in this chapter.

10.2 GOVERNING EQUATION

The Muskingum method is based on the mass balance equation. Its core idea is that the rate of change of water storage must equal the difference between inflow and outflow discharges, which can be represented by the following differential equation:

$$\frac{dS}{dt} = I(t) - Q(t) \text{ subject to } Q(0) = Q_0 \tag{10.1}$$

where S is the storage; I and Q are the rate of inflow and outflow, respectively; and t is the time. The subscript (0) denotes the value of the variable at the initial time. The rate of inflow is always given. Therefore, based on its form, Eq. (10.1) contains two unknown variables, namely S and Q. Further, an additional equation relating inflow and outflow rates to storage can be proposed as follows (Singh and Scarlatos 1987):

$$S = k\left[\alpha I + (1-\alpha)Q\right] \tag{10.2}$$

or:

$$S = k_1\left[\alpha_1 I^m + (1-\alpha_1)Q^m\right] \tag{10.3}$$

or:

$$S = k_2 \left[\alpha_2 I + (1 - \alpha_2) Q \right]^p \tag{10.4}$$

where k, k_1, k_2, α, α_1, α_2, m, and p are the parameters. Parameter k for the linear case given by Eq. (10.2) can be assumed as a constant and can be represented as the average reach travel time. For the nonlinear case given by Eqs. (10.3) and (10.4), parameters k_1 and k_2 are dimensional parameters. Parameters α, α_1, and α_2 act as weighted coefficients. The theoretical values for parameters m and p can be 0.60 or 0.67, depending on the formulae of Manning or Chezy. A detailed discussion of these parameters is given by Singh and Scarlatos (1987).

With the help of empirical relations given by Eqs. (10.2)–(10.4), one can transform the governing Eq. (10.1) to a first-order ordinary differential equation (ODE) with a single unknown Q. Substituting Eqs. (10.2)–(10.4) into the governing equation and then rearranging the terms, we obtain the following differential equations:

$$\frac{dQ}{dt} + \frac{Q}{k(1-\alpha)} = \frac{I}{k(1-\alpha)} - \frac{\alpha}{1-\alpha}\frac{dI}{dt} \tag{10.5}$$

$$\frac{dQ}{dt} + \frac{Q^{2-m}}{k_1(1-\alpha_1)m} - \frac{Q^{1-m}\left(I - k_1\alpha_1 m I^{m-1}\dfrac{dI}{dt}\right)}{k_1(1-\alpha_1)m} = 0 \tag{10.6}$$

$$\frac{dD}{dt} + \frac{D^{2-p}}{k_2(1-\alpha_2)p} - \frac{D^{1-p}}{k_2(1-\alpha_2)p} = 0 \text{ where } D = \alpha_2 I + (1-\alpha_2)Q \tag{10.7}$$

It can be seen that Eq. (10.5) is a linear differential equation, while the other two equations are nonlinear. The objective is to find the analytical solutions of these equations, which is carried out in the next section.

10.3 ANALYTICAL SOLUTIONS

10.3.1 ANALYTICAL SOLUTION OF EQ. (10.5)

Let us rewrite the equation here together with the initial condition:

$$\frac{dQ}{dt} + \frac{Q}{k(1-\alpha)} = \frac{I}{k(1-\alpha)} - \frac{\alpha}{1-\alpha}\frac{dI}{dt} \text{ subject to } Q(0) = Q_0 \tag{10.8}$$

Eq. (10.8) is a first-order linear equation in Q. It has been extensively studied in the literature, which includes the analytical solution using the Laplace transform method (Diskin 1967; Singh and McCann 1980). However, due to the simplicity of Eq. (10.8) and the development of powerful mathematical software, such as MATLAB, here we provide the direct solution to the equation. The integrating factor (IF) corresponding to Eq. (10.8) can be obtained as:

$$\text{IF} = \exp\left(\int \frac{dt}{k(1-\alpha)}\right) = \exp\left(\frac{t}{k(1-\alpha)}\right) \qquad (10.9)$$

Now, multiplying Eq. (10.8) by this IF and then integrating, we have:

$$Q = \exp\left(-\frac{t}{k(1-\alpha)}\right)\int \exp\left(\frac{t}{k(1-\alpha)}\right)\left[\frac{I}{k(1-\alpha)} - \frac{\alpha}{1-\alpha}\frac{dI}{dt}\right]dt$$

$$+ C_1\exp\left(-\frac{t}{k(1-\alpha)}\right) \qquad (10.10)$$

The integral constant C_1 can be determined from the given condition, i.e., $Q(0) = Q_0$, which is as follows:

$$C_1 = Q_0 - \int \exp\left(\frac{t}{k(1-\alpha)}\right)\left[\frac{I}{k(1-\alpha)} - \frac{\alpha}{1-\alpha}\frac{dI}{dt}\right]dt\bigg|_{t=0} \qquad (10.11)$$

Substituting Eq. (10.11) into Eq. (10.10), the final solution can be obtained as follows:

$$Q = \exp\left(-\frac{t}{k(1-\alpha)}\right)\int \exp\left(\frac{t}{k(1-\alpha)}\right)\left[\frac{I}{k(1-\alpha)} - \frac{\alpha}{1-\alpha}\frac{dI}{dt}\right]dt$$

$$+ \exp\left(-\frac{t}{k(1-\alpha)}\right)\left[Q_0 - \int \exp\left(\frac{t}{k(1-\alpha)}\right)\left[\frac{I}{k(1-\alpha)} - \frac{\alpha}{1-\alpha}\frac{dI}{dt}\right]dt\bigg|_{t=0}\right] \qquad (10.12)$$

10.3.2 HAM-BASED ANALYTICAL SOLUTION FOR EQ. (10.6)

The equation for this case takes the following form after rearrangement:

$$\frac{dQ}{dt} + \frac{Q^{1-m}}{k_1(1-\alpha_1)m}\left(Q - I + k_1\alpha_1 mI^{m-1}\frac{dI}{dt}\right) = 0 \text{ subject to } Q(0) = Q_0 \quad (10.13)$$

It can be seen that the equation contains a noninteger exponent of the dependent variable, which makes Eq. (10.13) a highly nonlinear differential equation. Following the discussion in previous chapters, the direct application of the homotopy analysis method (HAM) to Eq. (10.13) creates some difficulty in obtaining the higher-order terms. Therefore, we first approximate the nonlinear term using the Padé approximation and then solve the approximate governing equation using HAM. The equation also contains the term I^{m-1}, which though unknown, can lead to serious difficulty in computation due to the noninteger exponent. Thus, we need to construct the Padé approximant for terms Q^{1-m}, I^{m-1}. The $[m,n]$ order Padé approximation of a term x^{1-a} around $x = n_0$ can be written as:

$$\left[x^{1-a}\right]_{m,n} = \frac{\sum_{i=0}^{m} A_i \left(x - n_0\right)^i}{1 + \sum_{j=1}^{n} B_j \left(x - n_0\right)^j} \tag{10.14}$$

The Padé approximation for x^{a-1} will be simply the reciprocal of Eq. (10.14). As examples, here we calculate [2,2] and [3,3] order approximants and provide them as follows:

$$\left[x^{1-a}\right]_{2,2} = \frac{n_0^{1-a} - \frac{1}{2}(-3+a)n_0^{-a}\left(-n_0+x\right) + \frac{1}{12}\left(6-5a+a^2\right)n_0^{-1-a}\left(-n_0+x\right)^2}{1 + \frac{(1+a)(-n_0+x)}{2n_0} + \frac{a(1+a)(-n_0+x)^2}{12n_0^2}} \tag{10.15}$$

$$\left[x^{1-a}\right]_{3,3} = \frac{\begin{array}{c}n_0^{1-a} - \frac{1}{2}(-4+a)n_0^{-a}\left(-n_0+x\right) + \frac{1}{10}\left(12-7a+a^2\right)n_0^{-1-a}\left(-n_0+x\right)^2 \\ -\frac{1}{120}\left(-24+26a-9a^2+a^3\right)n_0^{-2-a}\left(-n_0+x\right)^3\end{array}}{1 + \frac{(2+a)(-n_0+x)}{2n_0} + \frac{(1+a)(2+a)(-n_0+x)^2}{10n_0^2} + \frac{a(1+a)(2+a)(-n_0+x)^3}{120n_0^3}} \tag{10.16}$$

Considering different values of exponent a and keeping in mind the relevance with the Muskingum method, we test the accuracy of the approximation in Figure 10.1. It can be noted from Eqs. (10.8) and (10.9) that point n_0 cannot be taken as zero because of the structure of the approximations. Therefore, we consider the point as $n_0 = 0.5$. Figures 15.1a and 15.1b plot the original function and its [2,2] and [3,3] order approximants over the domain [0,10] and [0,1], respectively. It can be seen that both approximants are accurate for a certain region of the domain, specifically the accuracy breaks down for larger values of the variable. Also, as the domain becomes larger, the higher-order approximant is relatively more accurate than the lower one – this is indeed a characteristic of Padé approximation. Further, it can be seen that an accurate approximant can be obtained for a smaller domain. The reason behind this is that the Padé approximation is the ratio of a two-series expansion, and the accuracy of the approximation depends on the order of magnitude of the variable. These suggest that the rate of inflow I and outflow Q need to be normalized in order to have an accurate solution while using Padé approximation for the nonlinear terms.

Eq. (10.13) is normalized as follows:

$$\hat{Q} = \frac{Q}{I_{max}}, \quad \hat{I} = \frac{I}{I_{max}}, \quad \text{and } \hat{t} = \frac{t}{t_{right}},$$

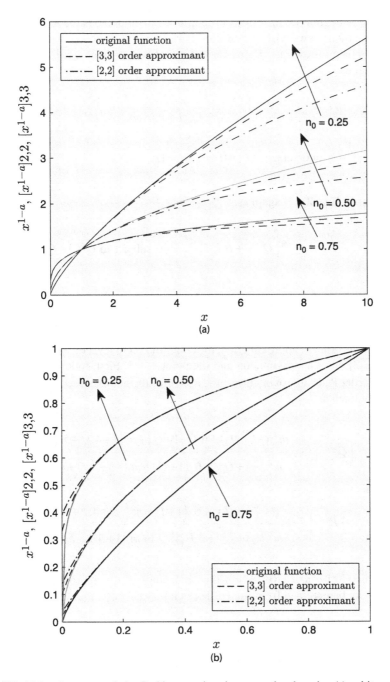

FIGURE 10.1 Accuracy of the Padé approximation over the domain: (a) arbitrary, e.g. [0, 10], and (b) [0, 1].

where I_{max} is the maximum value of the inflow rate and t_{right} is the maximum value of the time interval we wish to consider. Parameters I_{max} and t_{right} can be estimated from the given inflow rate. Using these normalized variables, Eq. (10.13) can be rearranged as:

$$\frac{I_{max}}{t_{right}}\frac{d\hat{Q}}{d\hat{t}}+\frac{I_{max}^{1-m}\hat{Q}^{1-m}}{k_1(1-\alpha_1)m}\left[I_{max}\hat{Q}-I_{max}\hat{I}+k_1\alpha_1mI_{max}^{m-1}\hat{I}^{m-1}\left(\frac{I_{max}}{t_{right}}\right)\frac{d\hat{I}}{d\hat{t}}\right]$$

$$=0 \text{ subject to } \hat{Q}(0)=\frac{Q_0}{I_{max}}=\widehat{Q_0} \qquad (10.17)$$

We rearrange Eq. (10.17) again as:

$$\frac{d\hat{Q}}{d\hat{t}}+A_1\hat{Q}^{1-m}\left[\hat{Q}-\hat{I}+B_1\hat{I}^{m-1}\frac{d\hat{I}}{d\hat{t}}\right]=0 \text{ subject to } \hat{Q}(0)=\widehat{Q_0} \qquad (10.18)$$

where $A_1 = \dfrac{I_{max}^{1-m}t_{right}}{k_1(1-\alpha_1)m}$ and $B_1 = \dfrac{k_1\alpha_1mI_{max}^{m-1}}{t_{right}}$. It was seen from Figure 10.1 that both the [2,2] and [3,3] order approximants are accurate for the nonlinear terms, and the higher the order of approximation, the better the accuracy. For simplicity, here we use the [2,2] order approximation and the point $n_0 = 1$. First, following Eq. (10.15), the [2,2] order Padé approximant around $n_0 = 1$ for the nonlinear terms $\hat{Q}^{1-m}$ and $\hat{I}^{1-m}$ can be obtained as:

$$\left[\hat{Q}^{1-m}\right]_{2,2} = \frac{m^2\left(-1+\hat{Q}\right)^2+6\,\hat{Q}\left(1+\hat{Q}\right)+m\left(1+4\,\hat{Q}-5\,\hat{Q}^2\right)}{m^2\left(-1+\hat{Q}\right)^2+6\left(1+\hat{Q}\right)+m\left(-5+4\,\hat{Q}+\hat{Q}^2\right)} \qquad (10.19)$$

$$\left[\hat{I}^{m-1}\right]_{2,2} = \frac{m^2\left(-1+\hat{I}\right)^2+6\left(1+\hat{I}\right)+m\left(-5+4\,\hat{I}+\hat{I}^2\right)}{m^2\left(-1+\hat{I}\right)^2+6\,\hat{I}\left(1+\hat{I}\right)+m\left(1+4\,\hat{I}-5\,\hat{I}^2\right)} \qquad (10.20)$$

Using Eqs. (10.19) and (10.20) and then rearranging, the approximate governing equation can be obtained from Eq. (10.18) as follows:

$$f_1\left(\hat{t}\right)\left[\alpha_0+\alpha_1\hat{Q}+\alpha_2\hat{Q}^2\right]\frac{d\hat{Q}}{d\hat{t}}+A_1\left(\beta_0+\beta_1\hat{Q}+\beta_2\hat{Q}^2\right)\left[f_1\left(\hat{t}\right)\hat{Q}+f_2\left(\hat{t}\right)\right]=0$$

$$\underset{\Longrightarrow}{\overset{\text{simplifying}}{\Longrightarrow}} f_1\left(\hat{t}\right)\left[\alpha_0+\alpha_1\hat{Q}+\alpha_2\hat{Q}^2\right]\frac{d\hat{Q}}{d\hat{t}}$$

$$+A_1\left[\beta_0f_2\left(\hat{t}\right)+\left(\beta_0f_1\left(\hat{t}\right)+\beta_1f_2\left(\hat{t}\right)\right)\hat{Q}+\left(\beta_1f_1\left(\hat{t}\right)+\beta_2f_2\left(\hat{t}\right)\right)\hat{Q}^2+\beta_2f_1\left(\hat{t}\right)\hat{Q}^3\right]=0 \quad (10.21)$$

where $\alpha_0 = \beta_2 = m^2 - 5m + 6$, $\alpha_1 = \beta_1 = -2m^2 + 4m + 6$, $\alpha_2 = \beta_0 = m^2 + m$, $f_1(\hat{t})$ and $f_2(\hat{t})$ are the known functions given as $f_1(\hat{t}) = \beta_0 + \beta_1\hat{I} + \beta_2\hat{I}^2$ and $f_2(\hat{t}) = B_1\left(\alpha_0 + \alpha_1\hat{I} + \alpha_2\hat{I}^2\right)\dfrac{d\hat{I}}{d\hat{t}} - \hat{I}\left(\beta_0 + \beta_1\hat{I} + \beta_2\hat{I}^2\right)$. The objective is to solve Eq. (10.21) using HAM. To that end, first, we construct the zeroth-order deformation equation as follows:

$$(1-q)\mathcal{L}\left[\Phi(\hat{t};q) - \hat{Q}_0(\hat{t})\right] = q\hbar H(\hat{t})\mathcal{N}\left[\Phi(\hat{t};q)\right] \tag{10.22}$$

subject to the initial condition:

$$\Phi(0;q) = \widehat{Q_0} \tag{10.23}$$

where q is the embedding parameter; $\Phi(\hat{t};q)$ is the representation of the solution across q; $\hat{Q}_0(\hat{t})$ is the initial approximation; $\hbar$ is the auxiliary parameter; $H(\hat{t})$ is the auxiliary function; and $\mathcal{L}$ and $\mathcal{N}$ are the linear and nonlinear operators, respectively. The nonlinear operator can be chosen as the operator of the governing equation:

$$\mathcal{N}\left[\Phi(\hat{t};q)\right] = f_1(\hat{t})\left[\alpha_0 + \alpha_1\Phi(\hat{t};q) + \alpha_2\Phi(\hat{t};q)^2\right]\frac{\partial\Phi(\hat{t};q)}{\partial\hat{t}}$$
$$+ A_1\left[\beta_0 f_2(\hat{t}) + \left(\beta_0 f_1(\hat{t}) + \beta_1 f_2(\hat{t})\right)\Phi(\hat{t};q) + \left(\beta_1 f_1(\hat{t}) + \beta_2 f_2(\hat{t})\right)\Phi(\hat{t};q)^2 + \beta_2 f_1(\hat{t})\Phi(\hat{t};q)^3\right] \tag{10.24}$$

The higher-order terms can be calculated from the higher-order deformation equation, given as follows:

$$\mathcal{L}\left[\hat{Q}_m(\hat{t}) - \chi_m\hat{Q}_{m-1}(\hat{t})\right] = \hbar H(\hat{t})R_m\left(\vec{\hat{Q}}_{m-1}\right),$$
$$m = 1,2,3,\ldots \text{ subject to } \hat{Q}_m(0) = 0 \tag{10.25}$$

where

$$\chi_m = \begin{cases} 0 & \text{when } m = 1, \\ 1 & \text{otherwise} \end{cases} \tag{10.26}$$

and:

$$R_m\left(\vec{\hat{Q}}_{m-1}\right) = \frac{1}{(m-1)!}\frac{\partial^{m-1}\mathcal{N}\left[\Phi(\hat{t};q)\right]}{\partial q^{m-1}}\bigg|_{q=0} \tag{10.27}$$

where $\hat{Q}_m$ for $m \geq 1$ are the higher-order terms. Using Eq. (10.24), $R_m\left(\vec{\hat{Q}}_{m-1}\right)$ can be obtained in closed form as follows:

$$R_m\left(\vec{\hat{Q}}_{m-1}\right) = f_1(\hat{t})\left[\alpha_0 \frac{d\hat{Q}_{m-1}}{d\hat{t}} + \alpha_1 \sum_{j=0}^{m-1} \hat{Q}_j \frac{d\hat{Q}_{m-1-j}}{d\hat{t}} + \alpha_2 \sum_{j=0}^{m-1} \hat{Q}_{m-1-j} \sum_{k=0}^{j} \hat{Q}_k \frac{d\hat{Q}_{j-k}}{d\hat{t}}\right]$$

$$+ A_1\left[\beta_0 f_2(\hat{t})(1-\chi_m) + (\beta_0 f_1(\hat{t}) + \beta_1 f_2(\hat{t}))\hat{Q}_{m-1} + (\beta_1 f_1(\hat{t}) + (\beta_2 f_2(\hat{t}))\sum_{j=0}^{m-1}\hat{Q}_j\hat{Q}_{m-1-j}\right.$$

$$\left. + \beta_2 f_1(\hat{t})\sum_{j=0}^{m-1}\hat{Q}_{m-1-j}\sum_{k=0}^{j}\hat{Q}_k\hat{Q}_{j-k}\right]$$

$$(10.28)$$

The higher-order deformation in Eq. (10.25) produces the higher-order terms with the help of Eq. (10.28). Then, the final solution is obtained as follows:

$$\hat{Q}(\hat{t}) = \hat{Q}_0(\hat{t}) + \sum_{m=1}^{\infty} \hat{Q}_m(\hat{t}) \qquad (10.29)$$

For solving the given equation using HAM, first, we need to choose a proper set of base functions to represent the solution. It can be seen that the equation contains the inflow rate $\hat{I}$, and the form of the base function will depend on the expression of $\hat{I}$. As an example, suppose the inflow rate follows a gamma distribution-type hydrograph. For that case, one can consider the following set of base functions:

$$\left\{\hat{t}^{an}exp\left(-bm\hat{t}\right) \mid n, m = 0, 1, 2, 3...\right\}, \; a \text{ and } b \text{ are constants} \qquad (10.30)$$

so that:

$$\hat{Q}(\hat{t}) = \sum_{n=0}^{\infty}\sum_{m=0}^{\infty} a_{n,m}\hat{t}^{an}exp\left(-bm\hat{t}\right) \qquad (10.31)$$

Constants a and b depend on the form of $\hat{I}$. Eq. (10.31) provides us with the so-called *rule of solution expression*. In accordance with the rule of solution expression, the linear operator can be of the following form:

$$\mathcal{L}\left[\Phi(\hat{t};q)\right] = \gamma_1(\hat{t})\frac{\partial\Phi(\hat{t};q)}{\partial\hat{t}} + \gamma_2(\hat{t})\Phi(\hat{t};q) \text{ where } \gamma_1(\hat{t}) \neq 0 \qquad (10.32)$$

As stated in the previous chapters, HAM provides a great choice in choosing the linear operator and functions. For the present problem, we choose the following linear operator:

$$\mathcal{L}\left[\Phi(\hat{t};q)\right] = \frac{\partial\Phi(\hat{t};q)}{\partial\hat{t}} + b\Phi(\hat{t};q) \qquad (10.33)$$

Following the rule of solution expression, the initial approximation is selected as:

$$\hat{Q}_0\left(\hat{t}\right)=\hat{I}\left(\hat{t}\right) \tag{10.34}$$

Interestingly, $\hat{I}\left(\hat{t}\right)$ always satisfies the initial condition, and hence the initial approximation $\hat{Q}_0\left(\hat{t}\right)$. It can be noted that within the framework of HAM, one has great freedom in choosing the linear operator and the initial approximation. Therefore, the selections made in this study given by Eqs. (10.33) and (10.34) are not unique. Using the inverse operator for Eq. (10.33), the higher-order terms can be calculated from Eq. (10.25) as follows:

$$\hat{Q}_m\left(\hat{t}\right)=\chi_m\hat{Q}_{m-1}\left(\hat{t}\right)+\hbar\mathcal{L}^{-1}\left[H\left(\hat{t}\right)R_m\left(\vec{\hat{Q}}_{m-1}\right)\right]+C_2=\chi_m\hat{Q}_{m-1}\left(\hat{t}\right)$$

$$+\hbar\exp\left(-b\hat{t}\right)\int_0^{\hat{t}}H\left(\tau\right)\exp\left(b\hat{t}\right)R_m\left(\vec{\hat{Q}}_{m-1}\right)d\tau+C_2\exp\left(-b\hat{t}\right),\ m=1,2,3,\dots \tag{10.35}$$

where R_m is given by Eq. (10.28) and the constant C_2 can be determined from the initial condition for the higher-order deformation equations, i.e., $\hat{Q}_m\left(0\right)=0$, which simply yields $C_2=0$ for all $m\geq 1$. Now, the auxiliary function $H\left(\hat{t}\right)$ can be determined from the *rule of coefficient ergodicity*. Based on the rule of solution expression and Eq. (10.35), the general form of $H\left(\hat{t}\right)$ should be:

$$H\left(\hat{t}\right)=\hat{t}^{an}exp\left(-bm\hat{t}\right) \tag{10.36}$$

where m and n are integers. It can be seen that if we consider $a<0$ or $b<0$ or both, then the terms appearing in the solution violate the *rule of solution expression*. This disobeys the *rule of coefficient ergodicity*, which tells us to include all the terms of the base function for the completeness of the solution. On the other hand, if we consider $a>1$ or $b\geq 0$ or both or $a=0$ and $b=0$, then certain terms do not appear in the solution, which violates the *rule of solution expression*. Therefore, we are left with the only choice: $a=1$ and $b=0$, i.e., $H\left(\hat{t}\right)=\hat{t}$. Eq. (10.35) now takes on the form:

$$\hat{Q}_m\left(\hat{t}\right)=\chi_m\hat{Q}_{m-1}\left(\hat{t}\right)+\hbar\exp\left(-b\hat{t}\right)\int_0^{\hat{t}}H\left(\tau\right)\exp\left(b\hat{t}\right)R_m\left(\vec{\hat{Q}}_{m-1}\right)d\tau,\ m=1,2,3,\dots \tag{10.37}$$

Proceeding in a like manner, the other terms can be calculated. Finally, the approximate solution can be obtained as:

$$\hat{Q}\left(\hat{t}\right)=\hat{Q}_0\left(\hat{t}\right)+\sum_{m=1}^M\hat{Q}_m\left(\hat{t}\right) \tag{10.38}$$

Based on the fundamental rules developed by Liao (2012), Eq. (10.38) was obtained. As mentioned in Chapter 2, the auxiliary parameter $\hbar$ plays the role of monitoring the rate and region of convergence of the series solution in Eq. (10.38).

10.3.3 HPM-Based Analytical Solution for Eq. (10.6)

Now, we aim to solve Eq. (10.6) using the homotopy perturbation method (HPM). For that purpose, one can use the same procedure as with HAM to convert the equation to a relatively weaker form using the Padé approximation. Therefore, we arrive at Eq. (10.21). For convenience, we now rewrite Eq. (10.21) as follows:

$$\left[\gamma_0 + \gamma_1 \hat{Q} + \gamma_2 \hat{Q}^2\right]\frac{d\hat{Q}}{d\hat{t}} + k_1 + k_2\hat{Q} + k_3\hat{Q}^2 + k_4\hat{Q}^3 = 0 \tag{10.39}$$

where $\gamma_0 = \alpha_0 f_1\left(\hat{t}\right), \gamma_1 = \alpha_1 f_1\left(\hat{t}\right), \gamma_2 = \alpha_2 f_1\left(\hat{t}\right), k_1 = A_1\beta_0 f_2\left(\hat{t}\right), k_2 = A_1\left(\beta_0 f_1\left(\hat{t}\right) + \beta_1 f_2\left(\hat{t}\right)\right),$ $k_3 = A_1\left(\beta_1 f_1\left(\hat{t}\right) + \beta_2 f_2\left(\hat{t}\right)\right),$ and $k_4 = A_1\beta_2 f_1.$ Eq. (10.39) can indeed be written in the form of:

$$\mathcal{N}\left(\hat{Q}\right) = f(t) \tag{10.40}$$

We construct a homotopy $\Phi\left(\hat{t};q\right)$, which satisfies:

$$(1-q)\left[\mathcal{L}\left(\Phi\left(\hat{t};q\right)\right) - \mathcal{L}\left(\hat{Q}_0\left(\hat{t}\right)\right)\right] + q\left[\mathcal{N}\left(\Phi\left(\hat{t};q\right)\right) - f\left(\hat{t}\right)\right] = 0 \tag{10.41}$$

where q is the embedding parameter that lies in $[0,1]$, $\mathcal{L}$ is the linear operator, and $\hat{Q}_0\left(\hat{t}\right)$ is the initial approximation to the final solution. Eq. (10.41) shows that as $q = 0$, $\Phi\left(\hat{t};0\right) = \hat{Q}_0\left(\hat{t}\right)$ and as $q = 1$, $\Phi\left(\hat{t};1\right) = \hat{Q}\left(\hat{t}\right)$. We now express $\Phi\left(\hat{t};q\right)$, as a series in terms of q, as follows:

$$\Phi\left(\hat{t};q\right) = \Phi_0 + q\Phi_1 + q^2\Phi_2 + q^3\Phi_3 + q^4\Phi_4 \ldots \tag{10.42}$$

As $q \to 1$, Eq. (10.42) produces the final solution as:

$$\hat{Q}\left(\hat{t}\right) = \lim_{q \to 1} \Phi\left(\hat{t};q\right) = \sum_{i=0}^{\infty}\Phi_i \tag{10.43}$$

For simplicity, we consider the single-term linear operator and the nonlinear operator as follows:

$$\mathcal{L}\left(\Phi\left(\hat{t};q\right)\right) = \frac{\partial\Phi\left(\hat{t};q\right)}{\partial\hat{t}} \tag{10.44}$$

$$\mathcal{N}\left(\Phi\left(\hat{t};q\right)\right) = \left[\gamma_0 + \gamma_1\Phi\left(\hat{t};q\right) + \gamma_2\Phi\left(\hat{t};q\right)^2\right]\frac{\partial\Phi\left(\hat{t};q\right)}{\partial\hat{t}}$$
$$+ k_1 + k_2\Phi\left(\hat{t};q\right) + k_3\Phi\left(\hat{t};q\right)^2 + k_4\Phi\left(\hat{t};q\right)^3 \tag{10.45}$$

Using these expressions, Eq. (10.41) takes on the form:

$$(1-q)\left[\frac{\partial\Phi(\hat{t};q)}{\partial\hat{t}}-\frac{d\hat{Q}_0}{d\hat{t}}\right]$$

$$+q\left[\left[\gamma_0+\gamma_1\Phi(\hat{t};q)+\gamma_2\Phi(\hat{t};q)^2\right]\frac{\partial\Phi(\hat{t};q)}{\partial\hat{t}}+k_1+k_2\Phi(\hat{t};q)+k_3\Phi(\hat{t};q)^2+k_4\Phi(\hat{t};q)^3\right]=0$$

$$(10.46)$$

Expanding the expressions, one can have the following relations:

$$\left(\Phi_0+q\Phi_1+q^2\Phi_2+q^3\Phi_3+q^4\Phi_4\cdots\right)^2=\left(\Phi_0\right)^2+q\left(2\Phi_0\Phi_1\right)$$

$$+q^2\left(\Phi_1^2+2\Phi_0\Phi_2\right)+q^3\left(2\Phi_0\Phi_3+2\Phi_1\Phi_2\right)+\cdots \qquad (10.47)$$

$$\left(\Phi_0+q\Phi_1+q^2\Phi_2+q^3\Phi_3+q^4\Phi_4\cdots\right)^3=\left(\Phi_0\right)^3+q\left(3\Phi_0^2\Phi_1\right)$$

$$+q^2\left(3\Phi_0^2\Phi_2+3\Phi_1^2\Phi_0\right)+q^3\left(3\Phi_0^2\Phi_3+6\Phi_0\Phi_1\Phi_2+\Phi_1^3\right)+\cdots \qquad (10.48)$$

Using Eqs. (10.47) and (10.48), we have:

$$\left[\gamma_0+\gamma_1\Phi(\hat{t};q)+\gamma_2\Phi(\hat{t};q)^2\right]\frac{\partial\Phi(\hat{t};q)}{\partial\hat{t}}$$

$$=\left[\gamma_0+\gamma_1\left\{\Phi_0+q\Phi_1+q^2\Phi_2+q^3\Phi_3+q^4\Phi_4\cdots\right\}\right.$$

$$+\gamma_2\left\{\left(\Phi_0\right)^2+q\left(2\Phi_0\Phi_1\right)+q^2\left(\Phi_1^2+2\Phi_0\Phi_2\right)+q^3\left(2\Phi_0\Phi_3+2\Phi_1\Phi_2\right)+\cdots\right\}\right]$$

$$\times\left[\frac{d\Phi_0}{d\hat{t}}+q\frac{d\Phi_1}{d\hat{t}}+q^2\frac{d\Phi_2}{d\hat{t}}+q^3\frac{d\Phi_3}{d\hat{t}}+q^4\frac{d\Phi_4}{d\hat{t}}+\cdots\right]$$

$$=\left[\left(\gamma_0+\gamma_1\Phi_0+\gamma_2\Phi_0^2\right)+q\left(\gamma_1\Phi_1+2\gamma_2\Phi_0\Phi_1\right)+q^2\left(\gamma_1\Phi_2+\gamma_2\Phi_1^2+2\gamma_2\Phi_0\Phi_2\right)\right.$$

$$+q^3\left(\gamma_1\Phi_3+2\gamma_2\Phi_0\Phi_3+2\gamma_2\Phi_1\Phi_2\right)+\cdots\right]$$

$$\times\left[\frac{d\Phi_0}{d\hat{t}}+q\frac{d\Phi_1}{d\hat{t}}+q^2\frac{d\Phi_2}{d\hat{t}}+q^3\frac{d\Phi_3}{d\hat{t}}+q^4\frac{d\Phi_4}{d\hat{t}}+\cdots\right]=\frac{d\Phi_0}{d\hat{t}}\left(\gamma_0+\gamma_1\Phi_0+\gamma_2\Phi_0^2\right)$$

$$+q\left\{\frac{d\Phi_0}{d\hat{t}}\left(\gamma_1\Phi_1+2\gamma_2\Phi_0\Phi_1\right)+\frac{d\Phi_1}{d\hat{t}}\left(\gamma_0+\gamma_1\Phi_0+\gamma_2\Phi_0^2\right)\right\}$$

$$+q^2\left\{\frac{d\Phi_0}{d\hat{t}}\left(\gamma_1\Phi_2+\gamma_2\Phi_1^2+2\gamma_2\Phi_0\Phi_2\right)+\frac{d\Phi_2}{d\hat{t}}\left(\gamma_0+\gamma_1\Phi_0+\gamma_2\Phi_0^2\right)+\frac{d\Phi_1}{d\hat{t}}\left(\gamma_1\Phi_1+2\gamma_2\Phi_0\Phi_1\right)\right\}$$

$$+q^3\left\{\frac{d\Phi_0}{d\hat{t}}\left(\gamma_1\Phi_3+2\gamma_2\Phi_0\Phi_3+2\gamma_2\Phi_1\Phi_2\right)+\frac{d\Phi_3}{d\hat{t}}\left(\gamma_0+\gamma_1\Phi_0+\gamma_2\Phi_0^2\right)\right.$$

$$+\frac{d\Phi_1}{d\hat{t}}\left(\gamma_1\Phi_2+\gamma_2\Phi_1^2+2\gamma_2\Phi_0\Phi_2\right)+\frac{d\Phi_2}{d\hat{t}}\left(\gamma_1\Phi_1+2\gamma_2\Phi_0\Phi_1\right)\right\}+\cdots \qquad (10.49)$$

$$k_1 + k_2 \Phi\left(\hat{t}; q\right) + k_3 \Phi\left(\hat{t}; q\right)^2 + k_4 \Phi\left(\hat{t}; q\right)^3 = k_1 + k_2 \left\{\Phi_0 + q\Phi_1 + q^2\Phi_2 + q^3\Phi_3 + q^4\Phi_4 \cdots\right\}$$

$$+k_3 \left\{\left(\Phi_0\right)^2 + q\left(2\Phi_0\Phi_1\right) + q^2\left(\Phi_1^2 + 2\Phi_0\Phi_2\right) + q^3\left(2\Phi_0\Phi_3 + 2\Phi_1\Phi_2\right) + \ldots\right\}$$

$$+k_4 \left\{\left(\Phi_0\right)^3 + q\left(3\Phi_0^2\Phi_1\right) + q^2\left(3\Phi_0^2\Phi_2 + 3\Phi_1^2\Phi_0\right) + q^3\left(3\Phi_0^2\Phi_3 + 6\Phi_0\Phi_1\Phi_2 + \Phi_1^3\right) + \cdots\right\}$$

$$= \left(k_1 + k_2\Phi_0 + k_3\Phi_0^2 + k_4\Phi_0^3\right) + q\left(k_2\Phi_1 + 2k_3\Phi_0\Phi_1 + 3k_4\Phi_0^2\Phi_1\right)$$

$$+q^2 \left\{k_2\Phi_2 + k_3\left(\Phi_1^2 + 2\Phi_0\Phi_2\right) + k_4\left(3\Phi_0^2\Phi_2 + 3\Phi_1^2\Phi_0\right)\right\}$$

$$+q^3 \left\{k_2\Phi_3 + k_3\left(2\Phi_0\Phi_3 + 2\Phi_1\Phi_2\right) + k_4\left(3\Phi_0^2\Phi_3 + 6\Phi_0\Phi_1\Phi_2 + \Phi_1^3\right)\right\} \quad (10.50)$$

Also, substituting Eq. (10.42) into Eq. (10.46) and using the initial condition $\hat{Q}(0) = \hat{Q}_0$, one can obtain:

$$\frac{\partial \Phi\left(\hat{t}; q\right)}{\partial \hat{t}} - \frac{d\hat{Q}_0}{d\hat{t}} = \left(\frac{d\Phi_0}{d\hat{t}} - \frac{d\hat{Q}_0}{d\hat{t}}\right) + q\frac{d\Phi_1}{d\hat{t}} + q^2\frac{d\Phi_2}{d\hat{t}} + q^3\frac{d\Phi_3}{d\hat{t}} + q^4\frac{d\Phi_4}{d\hat{t}} + \cdots \quad (10.51)$$

$$\Phi_0(0) = \hat{Q}_0, \ \Phi_1(0) = 0, \ \Phi_2(0) = 0, \ \ldots. \quad (10.52)$$

Using Eqs. (10.49)–(10.51) and equating the like powers of q in Eq. (10.46), the following system of differential equations is obtained:

$$\frac{d\Phi_0}{d\hat{t}} - \frac{d\hat{Q}_0}{d\hat{t}} = 0 \text{ subject to } \Phi_0(0) = \hat{Q}_0 \quad (10.53)$$

$$\frac{d\Phi_1}{d\hat{t}} - \left(\frac{d\Phi_0}{d\hat{t}} - \frac{d\hat{Q}_0}{d\hat{t}}\right) + \frac{d\Phi_0}{d\hat{t}}\left(\gamma_0 + \gamma_1\Phi_0 + \gamma_2\Phi_0^2\right) + \left(k_1 + k_2\Phi_0 + k_3\Phi_0^2 + k_4\Phi_0^3\right)$$

$$= 0 \text{ subject to } \Phi_1(0) = 0 \quad (10.54)$$

$$\frac{d\Phi_2}{d\hat{t}} - \frac{d\Phi_1}{d\hat{t}} + \frac{d\Phi_0}{d\hat{t}}\left(\gamma_1\Phi_1 + 2\gamma_2\Phi_0\Phi_1\right) + \frac{d\Phi_1}{d\hat{t}}\left(\gamma_0 + \gamma_1\Phi_0 + \gamma_2\Phi_0^2\right) + k_2\Phi_1$$

$$+2k_3\Phi_0\Phi_1 + 3k_4\Phi_0^2\Phi_1 = 0 \text{ subject to } \Phi_2(0) = 0 \quad (10.55)$$

$$\frac{d\Phi_3}{d\hat{t}} - \frac{d\Phi_2}{d\hat{t}} + \frac{d\Phi_0}{d\hat{t}}\left(\gamma_1\Phi_2 + \gamma_2\Phi_1^2 + 2\gamma_2\Phi_0\Phi_2\right) + \frac{d\Phi_2}{d\hat{t}}\left(\gamma_0 + \gamma_1\Phi_0 + \gamma_2\Phi_0^2\right)$$

$$+\frac{d\Phi_1}{d\hat{t}}\left(\gamma_1\Phi_1 + 2\gamma_2\Phi_0\Phi_1\right) + k_2\Phi_2 + k_3\left(\Phi_1^2 + 2\Phi_0\Phi_2\right)$$

$$+k_4\left(3\Phi_0^2\Phi_2 + 3\Phi_1^2\Phi_0\right) = 0 \text{ subject to } \Phi_3(0) = 0 \quad (10.56)$$

$$\frac{d\Phi_4}{d\hat{t}} - \frac{d\Phi_3}{d\hat{t}} + \frac{d\Phi_0}{d\hat{t}}\left(\gamma_1\Phi_3 + 2\gamma_2\Phi_0\Phi_3 + 2\gamma_2\Phi_1\Phi_2\right) + \frac{d\Phi_3}{d\hat{t}}\left(\gamma_0 + \gamma_1\Phi_0 + \gamma_2\Phi_0^2\right)$$

$$+\frac{d\Phi_1}{d\hat{t}}\left(\gamma_1\Phi_2 + \gamma_2\Phi_1^2 + 2\gamma_2\Phi_0\Phi_2\right) + \frac{d\Phi_2}{d\hat{t}}\left(\gamma_1\Phi_1 + 2\gamma_2\Phi_0\Phi_1\right) + k_2\Phi_3$$

$$+k_3\left(2\Phi_0\Phi_3 + 2\Phi_1\Phi_2\right) + k_4\left(3\Phi_0^2\Phi_3 + 6\Phi_0\Phi_1\Phi_2 + \Phi_1^3\right) = 0 \text{ subject to } \Phi_4(0) = 0 \quad (10.57)$$

Taking the initial approximation $\Phi_0 = \hat{I}$, we can solve the equations iteratively using symbolic software. Finally, the HPM-based solution can be approximated as follows:

$$\hat{Q}(\hat{t}) \approx \sum_{i=0}^{M} \Phi_i \qquad (10.58)$$

10.3.4   OHAM-BASED ANALYTICAL SOLUTION FOR EQ. (10.6)

When applying the optimal homotopy asymptotic method (OHAM), the approximate Eq. (10.21) can be written in the following form:

$$\mathcal{L}\left[\hat{Q}(\hat{t})\right] + \mathcal{N}\left[\hat{Q}(\hat{t})\right] + h(\hat{t}) = 0 \qquad (10.59)$$

where $\mathcal{L}\left[\hat{Q}(\hat{t})\right] = \dfrac{d\hat{Q}}{d\hat{t}}$, $\mathcal{N}\left[\hat{Q}(\hat{t})\right] = \left[\gamma_0 - 1 + \gamma_1\hat{Q} + \gamma_2\hat{Q}^2\right]\dfrac{d\hat{Q}}{d\hat{t}} + k_1 + k_2\hat{Q} + k_3\hat{Q}^2 + k_4\hat{Q}^3$, and $h(\hat{t}) = 0$. With these considerations, the zeroth-order problem becomes:

$$\mathcal{L}\left(\hat{Q}_0(\hat{t})\right) + h(\hat{t}) = 0 \text{ subject to } \hat{Q}_0(0) = \hat{Q}_0 \qquad (10.60)$$

Solving Eq. (10.60), one obtains:

$$\hat{Q}_0(\hat{t}) = \hat{Q}_0 \qquad (10.61)$$

It may be noted that one can use Eqs. (10.49) and (10.50) from the previous section to obtain the values $\mathcal{N}_0, \mathcal{N}_1, \mathcal{N}_2$, etc. The first-order problem reduces to:

$$\frac{d\hat{Q}_1}{d\hat{t}} = H_1(\hat{t}, C_i)\left\{\frac{d\hat{Q}_0}{d\hat{t}}\left(\gamma_0 - 1 + \gamma_1\hat{Q}_0 + \gamma_2\hat{Q}_0^2\right) + k_1 + k_2\hat{Q}_0 + k_3\hat{Q}_0^2 + k_4\hat{Q}_0^3\right\}$$

subject to $\hat{Q}_1(0) = 0$ \qquad (10.62)

We simply choose $H_1(\hat{t}, C_1) = C_1$. Using $k = 2$, the second-order problem becomes:

$$\mathcal{L}\left[\hat{Q}_2(\hat{t}, C_i) - \hat{Q}_1(\hat{t}, C_i)\right] = H_2(\hat{t}, C_i)\mathcal{N}_0\left(\hat{Q}_0(\hat{t})\right)$$
$$+ H_1(\hat{t}, C_i)\left[\mathcal{L}\left[\hat{Q}_1(\hat{t}, C_i)\right] + \mathcal{N}_1\left[\hat{Q}_0(\hat{t}), \hat{Q}_1(\hat{t}, C_i)\right]\right] \text{ subject to } \hat{Q}_2(0) = 0 \quad (10.63)$$

Further, we choose $H_2(t, C_i) = C_2$. Therefore, Eq. (10.63) becomes:

$$\frac{d\hat{Q}_2}{d\hat{t}} = \frac{d\hat{Q}_1}{d\hat{t}} + C_2\left\{\frac{d\hat{Q}_0}{d\hat{t}}\left(\gamma_0 - 1 + \gamma_1\hat{Q}_0 + \gamma_2\hat{Q}_0^2\right) + k_1 + k_2\hat{Q}_0 + k_3\hat{Q}_0^2 + k_4\hat{Q}_0^3\right\}$$
$$+ C_1\left[\frac{d\hat{Q}_1}{d\hat{t}} + \frac{d\hat{Q}_0}{d\hat{t}}\left(\gamma_1\hat{Q}_1 + 2\gamma_2\hat{Q}_0\hat{Q}_1\right) + \frac{d\hat{Q}_1}{d\hat{t}}\left(\gamma_0 - 1 + \gamma_1\hat{Q}_0 + \gamma_2\hat{Q}_0^2\right) + k_2\hat{Q}_1 + 2k_3\hat{Q}_0\hat{Q}_1 + 3k_4\hat{Q}_0^2\hat{Q}_1\right]$$

subject to $\hat{Q}_2(0) = 0$ \qquad (10.64)

It can be seen that we do not write the explicit forms of the terms here, as those are lengthy. Indeed, one can compute these terms without any difficulty using symbolic computation software, such as MATLAB. Further, following the discussion in Chapter 2, the higher-order terms can be computed in a similar manner. However, our aim is to produce an accurate solution with just two to three terms of the OHAM-based series. Therefore, we restrict our calculation up to $k = 2$. Finally, the approximate solution can be found as:

$$\hat{Q}(\hat{t}) \approx \hat{Q}_0(\hat{t}) + \hat{Q}_1(\hat{t}, C_1) + \hat{Q}_2(\hat{t}, C_1, C_2) \tag{10.65}$$

where the terms are given by Eqs. (10.61), (10.62), and (10.64).

10.4 CONVERGENCE THEOREMS

The convergence of the HAM-based and OHAM-based solutions are proved theoretically using the following theorems.

10.4.1 CONVERGENCE THEOREM OF THE HAM-BASED SOLUTION FOR EQ. (10.38)

The convergence theorems for the HAM-based solution for Eq. (10.38) can be proved using the following theorems.

Theorem 10.1: If the homotopy series $\Sigma_{m=0}^{\infty} \hat{Q}_m(\hat{t})$ and $\Sigma_{m=0}^{\infty} \hat{Q}_m{}'(\hat{t})$ converge, then $R_m\left(\vec{\hat{Q}}_{m-1}\right)$ given by Eq. (10.28) satisfies the relation $\Sigma_{m=1}^{\infty} R_m\left(\vec{\hat{Q}}_{m-1}\right) = 0$. [Here '''' denotes the derivative with respect to $\hat{t}$.]

Proof: The proof of this theorem follows exactly from the related theorem of the previous chapters.

Theorem 10.2: If $\hbar$ is so properly chosen that the series $\Sigma_{m=0}^{\infty} \hat{Q}_m(\hat{t})$ and $\Sigma_{m=0}^{\infty} \hat{Q}_m{}'(\hat{t})$ converge absolutely to $\hat{Q}(\hat{t})$ _and_ $\hat{Q}'(\hat{t})$, respectively, then the homotopy series $\Sigma_{m=0}^{\infty} \hat{Q}_m(\hat{t})$ satisfies the original governing Eq. (10.21).

Proof: Using the Cauchy product defined in Theorem 2.2 of Chapter 2 in relation to Eq. (10.28), we get:

$$\sum_{m=1}^{\infty}\sum_{j=0}^{m-1}\hat{Q}_j\hat{Q}_{m-1-j} = \left(\sum_{m=0}^{\infty}\hat{Q}_m\right)^2 \tag{10.66}$$

$$\sum_{m=1}^{\infty}\sum_{j=0}^{m-1}\hat{Q}_j\hat{Q}'_{m-1-j} = \left(\sum_{m=0}^{\infty}\hat{Q}_m\right)\left(\sum_{k=0}^{\infty}\hat{Q}'_k\right) \tag{10.67}$$

$$\sum_{m=1}^{\infty}\sum_{j=0}^{m-1}\hat{Q}_{m-1-j}\sum_{k=0}^{j}\hat{Q}_k\hat{Q}_{j-k} = \left(\sum_{m=0}^{\infty}\hat{Q}_m\right)^3 \qquad (10.68)$$

$$\sum_{m=1}^{\infty}\sum_{j=0}^{m-1}\hat{Q}_{m-1-j}\sum_{k=0}^{j}\hat{Q}_k\hat{Q}'_{j-k} = \left(\sum_{m=0}^{\infty}\hat{Q}_m\right)^2\left(\sum_{k=0}^{\infty}\hat{Q}'_k\right) \qquad (10.69)$$

Theorem 10.1 shows that if $\sum_{m=0}^{\infty}\hat{Q}_m\left(\hat{t}\right)$ and $\sum_{m=0}^{\infty}\hat{Q}_m'\left(\hat{t}\right)$ converge then $\sum_{m=1}^{\infty}R_m\left(\vec{\hat{Q}}_{m-1}\right)=0.$

Therefore, substituting the previous expressions in Eq. (10.28) and simplifying further lead to:

$$f_1\left(\hat{t}\right)\left[\alpha_0\sum_{m=0}^{\infty}\hat{Q}'_m+\alpha_1\left(\sum_{m=0}^{\infty}\hat{Q}'_m\right)\left(\sum_{k=0}^{\infty}\hat{Q}_k\right)+\alpha_2\left(\sum_{m=0}^{\infty}\hat{Q}'_m\right)\left(\sum_{k=0}^{\infty}\hat{Q}_k\right)^2\right]$$

$$+A_1\left[\beta_0 f_2\left(\hat{t}\right)\sum_{m=0}^{\infty}\left(1-\chi_{m+1}\right)+\left(\beta_0 f_1\left(\hat{t}\right)+\beta_1 f_2\left(\hat{t}\right)\right)\sum_{m=0}^{\infty}\hat{Q}_m\right.$$

$$\left.+\left(\beta_1 f_1\left(\hat{t}\right)+\beta_2 f_2\left(\hat{t}\right)\right)\left(\sum_{k=0}^{\infty}\hat{Q}_k\right)^2+\beta_2 f_1\left(\hat{t}\right)\left(\sum_{k=0}^{\infty}\hat{Q}_k\right)^3\right]=0 \qquad (10.70)$$

which is basically the original governing equation in Eq. (10.21). Furthermore, subject to the initial condition $\hat{Q}_0\left(0\right)=\hat{Q}_0$ and the conditions for the higher-order deformation equation $\hat{Q}_m\left(0\right)=0$, for $m\geq 1$, we easily obtain $\sum_{m=0}^{\infty}\hat{Q}_m\left(0\right)=\hat{Q}_0$. Hence, the convergence result follows.

10.4.2 CONVERGENCE THEOREM OF THE OHAM-BASED SOLUTION FOR EQ. (10.65)

Theorem 10.3: If the series $\hat{Q}_0\left(\hat{t}\right)+\sum_{j=1}^{\infty}\hat{Q}_j\left(\hat{t},C_i\right)$, $i=1,2,\ldots,s$ converges, where $\hat{Q}_j\left(\hat{t},C_i\right)$ is governed by Eqs. (10.61), (10.62), and (10.64), then Eq. (10.65) is a solution of the original Eq. (10.21).

Proof: The proof of this theorem follows exactly from Theorem 3 of Chapter 2.

10.5 RESULTS AND DISCUSSION

In this section, first, we validate the solutions obtained using HAM, HPM, and OHAM. Then, we compare them to get a comparative idea about the performances of the methods.

10.5.1 VALIDATION OF THE HAM-BASED SOLUTION

To deal with the high nonlinearity of the equation, we converted the original equation into a relatively weaker form using the Padé approximation technique. The coefficients of the approximation can be obtained using the MATHEMATICA command *PadeApproximant*. Here, we consider a test case where we select some random values for the required parameters and find the corresponding solution using HAM. The parameter values, including the Padé approximant coefficients, are shown in Table 10.1. It can be noted that to assess the solution, we need an expression for the inflow rate $\hat{I}(\hat{t})$. For this particular case, we consider $\hat{I}(\hat{t}) = \frac{1}{3} + 50\hat{t}^2\exp(-7\hat{t})$. The auxiliary parameter $\hbar$ is obtained by minimizing the following squared residual error:

$$\Delta_m = \int\limits_{\hat{t}\in\Omega} \left(\mathcal{N}\left[\hat{Q}(\hat{t})\right]\right)^2 d\hat{t} \qquad (10.71)$$

For the test case considered, Figure 10.2 shows the squared residual error for a different order of approximation. It can be seen from the figure that as the order of approximation increases, the corresponding residual error decreases, which shows the stability of the method. Also, the numerical results are reported in Table 10.2. For the test case considered, we compute HAM-based analytical solutions for a different order of approximations and compare them with the corresponding numerical solution. The MATLAB script *ode45* is used to obtain the numerical solution of the nonlinear equation, and the HAM-based approximation is achieved using the MATHEMATICA package BVPh 2.0. In Figure 10.3, the 10th-order HAM-based solution and the numerical solution are compared. Also, the numerical solution of the original equation without the application of the Padé approximation is compared in the figure to check the accuracy of the approximation. The numerical results are provided in Table 10.3. It can be seen from the table and the figure that the [2,2] order Padé approximant is a robust approximation for the nonlinear term, as it does not alter the solution of the original problem. Also, the HAM-based solution is seen to accurately agree with the corresponding numerical solution to the equation. A flowchart containing the steps of the method is provided in Figure 10.4.

TABLE 10.1

Required Parameters and Padé Approximant Coefficients for the Considered Cases

		k_1		t_{right}	I_{max}	Q_0			Padé Coefficients	
Case	m	(m³(1-m)/sec⁻ᵐ)	α_1	(days)	(m³/sec)	(m³/sec)	A_1	B_1	$(\alpha_0, \alpha_1, \alpha_2)$	$(\beta_0, \beta_1, \beta_2)$
Test case	0.50	1.50	0.25	6	150	50	130.64	0.38	(5,10,1)	(1,10,5)

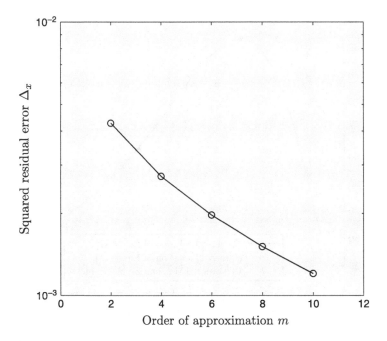

FIGURE 10.2   Squared residual error (Δ_m) versus different orders of approximation (m) of the HAM-based approximation for the test case.

10.5.2   VALIDATION OF THE HPM-BASED SOLUTION

The HPM-based analytical solution derived in Section 10.3.3 is validated here against a numerical solution obtained using *ode45* of MATLAB. Figure 10.5 shows the comparison between the HPM-based solution and the corresponding numerical solution. It is observed from the figure that HPM produces accurate solutions only for a restricted domain, specifically $t \in [0,1]$, and it starts deviating from the numerical solution for the remaining range. For a quantitative assessment, we

TABLE 10.2

Squared Residual Error (Δ_m) versus Different Orders of Approximation (m) for the Test Case

Order of Approximations (m)	Squared Residual Error (Δ_m)
2	4.3×10^{-3}
4	2.7×10^{-3}
6	2.0×10^{-3}
8	1.5×10^{-3}
10	1.2×10^{-3}

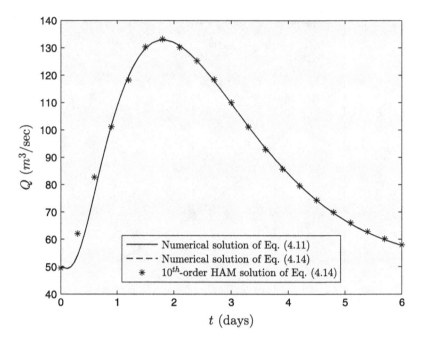

FIGURE 10.3 Comparison between the numerical solution and 10th-order HAM-based solution for the test case.

TABLE 10.3

Comparison between HAM-Based Approximation and Numerical Solution

$\hat{t}$	Numerical Solution		HAM-Based Approximation (Eq. (10.38))		
	Eq. (10.18)	Eq. (10.21)	6th Order	8th Order	10th Order
0.0	0.3300	0.3300	0.3300	0.3300	0.3300
0.1	0.5118	0.5119	0.5609	0.5559	0.5514
0.2	0.7982	0.7982	0.7858	0.7862	0.7884
0.3	0.8859	0.8858	0.8917	0.8910	0.8878
0.4	0.8352	0.8353	0.8319	0.8310	0.8350
0.5	0.7309	0.7310	0.7279	0.7356	0.7333
0.6	0.6229	0.6230	0.6282	0.6247	0.6190
0.7	0.5320	0.5321	0.5383	0.5285	0.5306
0.8	0.4641	0.4640	0.4669	0.4600	0.4659
0.9	0.4166	0.4166	0.4164	0.4144	0.4190
1.0	0.3849	0.3849	0.3832	0.3846	0.3864

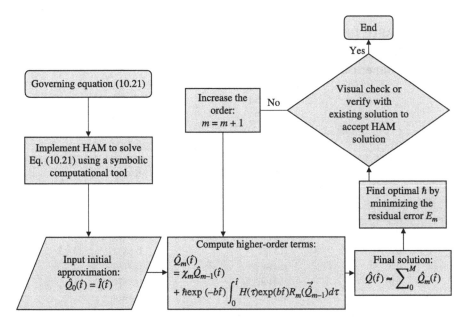

FIGURE 10.4   Flowchart for the HAM solution in Eq. (10.38).

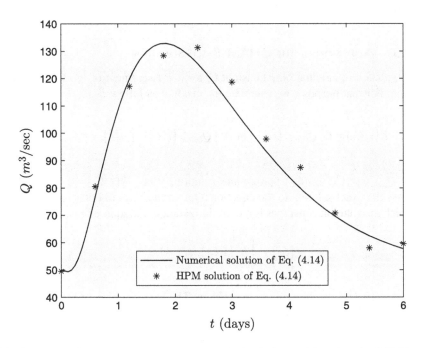

FIGURE 10.5   Comparison between the numerical solution and fourth-order HPM-based solution.

TABLE 10.4

Comparison between HPM-Based Approximation and Numerical Solution

$\hat{t}$	Numerical Solution Eq. (10.21)	HPM-Based Approximation (Eq. 10.21) 4th Order
0.0	0.3300	0.3300
0.1	0.5119	0.5367
0.2	0.7982	0.7812
0.3	0.8858	0.8557
0.4	0.8353	0.8754
0.5	0.7310	0.7911
0.6	0.6230	0.6524
0.7	0.5321	0.5832
0.8	0.4640	0.4719
0.9	0.4166	0.3871
1.0	0.3849	0.3974

consider a fourth-order approximation and report it in Table 10.4 by comparing it with a numerical solution. A flowchart containing the steps of the method is given in Figure 10.6.

10.5.3 VALIDATION OF THE OHAM-BASED SOLUTION

First, we need to calculate the constant C_i for the assessment of the OHAM-based solution. For that purpose, we calculate the residual as follows:

$$R(\hat{t},C_i) = \mathcal{L}\left[\hat{Q}_{OHAM}(\hat{t},C_i)\right] + \mathcal{N}\left[\hat{Q}_{OHAM}(\hat{t},C_i)\right] + h(\hat{t}), \; i = 1,2,\ldots,s \quad (10.72)$$

where $\hat{Q}_{OHAM}(\hat{t},C_i)$ is the approximate solution. When $R(\hat{t},C_i) = 0$, $\hat{Q}_{OHAM}(\hat{t},C_i)$ becomes the exact solution to the equation. One of the ways to obtain the optimal C_i for which the solution converges is the minimization of the squared residual error, i.e.,

$$J(C_i) = \int_{t \in D} R^2(\hat{t},C_i) d\hat{t}, \; i = 1,2,\ldots,s \quad (10.73)$$

where $D = [0,1]$ is the domain of the problem. The minimization of Eq. (10.73) leads to a system of algebraic equations as follows:

$$\frac{\partial J}{\partial C_1} = \frac{\partial J}{\partial C_2} = \cdots = \frac{\partial J}{\partial C_s} = 0 \quad (10.74)$$

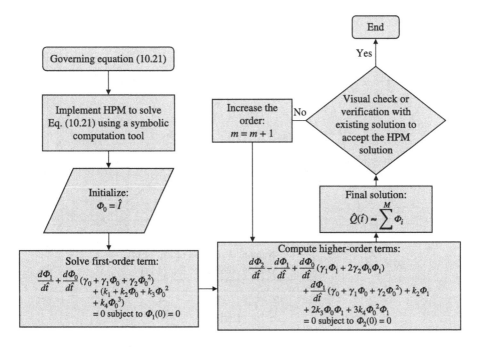

FIGURE 10.6   Flowchart for the HPM solution in Eq. (10.58).

Solving this system, one can obtain the optimal values of the parameters. We obtain the optimal values using the MATLAB routine *fminsearch*, which minimizes an unconstrained multivariable function. Using these values, the three-term OHAM solution is computed and compared in Figure 10.7 with a numerical solution to see the effectiveness of the proposed method. It can be seen that even with three terms of the OHAM-based series, the solution agrees well with the corresponding numerical solution to the equation. For a quantitative assessment, we also compare the numerical values of the solutions in Table 10.5. A flowchart containing the steps of the method is provided in Figure 10.8.

10.6   CONCLUDING REMARKS

The well-known Muskingum method for flood routing is governed by a first-order ODE. Based on the storage-discharge relationship, the equation can be linear or nonlinear. This chapter solves the highly nonlinear governing equation for flood routing using three different analytical methods, namely HAM, HPM, and OHAM. To deal with the highly nonlinear term present in the governing equation, first, the Padé approximation is applied to convert the equation into a relatively weaker nonlinear form. Then, the approximate equation is solved analytically using the methods mentioned earlier. It is seen that the HAM and OHAM produce accurate solutions to the problem over the entire domain, whereas the HPM fails to provide a good approximation over the full domain. Numerical examples are considered to compare

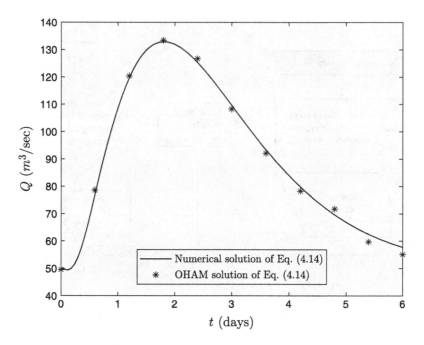

FIGURE 10.7 Comparison between the numerical solution and three-term OHAM-based solution.

TABLE 10.5

Comparison between OHAM-Based Approximation and Numerical Solution

$\hat{t}$	Numerical Solution Eq. (10.21)	OHAM-Based Approximation (Eq. 10.21) Three-Term Solution
0.0	0.3300	0.3300
0.1	0.5119	0.5241
0.2	0.7982	0.8027
0.3	0.8858	0.8891
0.4	0.8353	0.8446
0.5	0.7310	0.7219
0.6	0.6230	0.6143
0.7	0.5321	0.5220
0.8	0.4640	0.4783
0.9	0.4166	0.3980
1.0	0.3849	0.3672

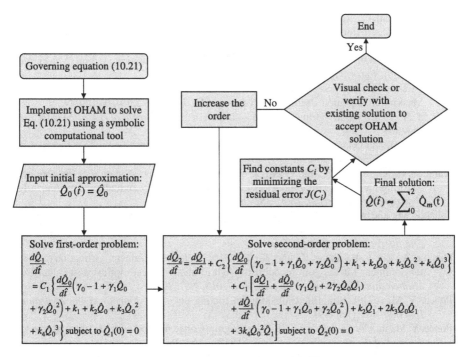

FIGURE 10.8   Flowchart for the OHAM solution in Eq. (10.65).

the solutions graphically. Also, quantitative assessment is carried out in support of the approximate results.

REFERENCES

Diskin, M. H. (1967). On the solution of the Muskingum flood routing equation. *Journal of Hydrology*, 5, 286–289.

Liao, S. (2012). *Homotopy analysis method in nonlinear differential equations*. Higher Education Press, Beijing.

Scarlatos, P. D. (1982). A pure finite-element method for the Saint-Venant equations. *Coastal Engineering*, 6(1), 27–45.

Singh, V. P., and McCann, R. C. (1980). Some notes on Muskingum method of flood routing. *Journal of Hydrology*, 48(3–4), 343–361.

Singh, V. P., and Scarlatos, P. D. (1987). Analysis of nonlinear Muskingum flood routing. *Journal of Hydraulic Engineering*, 113(1), 61–79.

Taylor, C., and Davis, J. (1973). Finite element numerical modelling of flow and dispersion in estuaries. River and estuary model analysis, Volume *III*.

FURTHER READING

Barati, R. (2011). Parameter estimation of nonlinear Muskingum models using Nelder-mead simplex algorithm. *Journal of Hydrologic Engineering*, 16(11), 946–954.

Bayrami, M., Vatankhah, A., and Nazi Ghameshlou, A. (2019). Flood routing using Muskingum model with fractional derivative. *Iranian Journal of Soil and Water Research*, 50(7), 1667–1676.

Bozorg-Haddad, O., Abdi-Dehkordi, M., Hamedi, F., Pazoki, M., and Loáiciga, H. A. (2019). Generalized storage equations for flood routing with nonlinear Muskingum models. *Water Resources Management, 33*(8), 2677–2691.

Cunge, J. A. (1969). On the subject of a flood propagation computation method (Muskingum method). *Journal of Hydraulic Research, 7*(2), 205–230.

Gąsiorowski, D., and Szymkiewicz, R. (2018). Dimensionally consistent nonlinear Muskingum equation. *Journal of Hydrologic Engineering, 23*(9), 04018039.

Gill, M. A. (1978). Flood routing by the Muskingum method. *Journal of Hydrology, 36*(3–4), 353–363.

Hirpurkar, P., and Ghare, A. D. (2015). Parameter estimation for the nonlinear forms of the Muskingum model. *Journal of Hydrologic Engineering, 20*(8), 04014085.

Karahan, H., Gurarslan, G., and Geem, Z. W. (2013). Parameter estimation of the nonlinear Muskingum flood-routing model using a hybrid harmony search algorithm. *Journal of Hydrologic Engineering, 18*(3), 352–360.

Niazkar, M., and Afzali, S. H. (2015). Assessment of modified honey bee mating optimization for parameter estimation of nonlinear Muskingum models. *Journal of Hydrologic Engineering, 20*(4), 04014055.

Ouyang, A., Liu, L. B., Sheng, Z., and Wu, F. (2015). A class of parameter estimation methods for nonlinear Muskingum model using hybrid invasive weed optimization algorithm. *Mathematical Problems in Engineering, 2015*, A573894.

Ponce, V. M. (1979). Simplified Muskingum routing equation. *Journal of the Hydraulics Division, 105*(1), 85–91.

Ponce, V. M., and Yevjevich, V. (1978). Muskingum-Cunge method with variable parameters. *Journal of the Hydraulics Division, 104*(12), 1663–1667.

11 Velocity and Sediment Concentration Distribution in Open-Channel Flow

11.1 INTRODUCTION

In channel or river flow, sediment and water are mixed, and the flow is sediment-laden. Therefore, for modeling sediment-laden flow, simultaneous treatment of water and sediment movement is needed. The fluid-sediment mixture density increases because of the presence of sediment particles, which stratify the flow. Fluid velocity and volumetric particle concentration are closely interrelated via the particle-turbulence interaction. The vertical distribution of streamwise velocity and sediment concentration in steady, uniform open-channel turbulent flow is described using the governing equations, which constitute a system of coupled nonlinear ordinary differential equations (ODEs). This chapter develops three kinds of analytical solutions for the governing equations, based on homotopy analysis method (HAM), homotopy perturbation method (HPM), and optimal homotopy asymptotic method (OHAM).

11.2 GOVERNING EQUATION AND ANALYTICAL SOLUTIONS

We consider a gravity-driven, sediment-laden open-channel flow of depth h with suspended sediment particles. The time-averaged velocity u and suspension sediment concertation c are assumed to vary only in the vertical direction, say z, from a certain reference level $z = a$ to the free surface $z = h$, as shown in Figure 11.1. The suspension concentration distribution under steady, uniform flow is governed by the following equation (Rouse 1937):

$$\omega_s c + \varepsilon_s \frac{dc}{dz} = 0 \tag{11.1}$$

where ε_s is the sediment diffusion coefficient and ω_s is the settling velocity of sediment particles. If the shear stress varies linearly with the distance from the channel bed, the time-averaged velocity profile in steady, uniform flow can be given by the following equation:

$$\varepsilon_t \frac{du}{dz} = u_*^2 \left(1 - \frac{z}{h}\right) \tag{11.2}$$

where u_* is the shear velocity, given as $u_* = \sqrt{\tau_0 / \rho}$, where τ_0 is the bed shear stress and ρ is the fluid density.

DOI: 10.1201/9781003368984-14

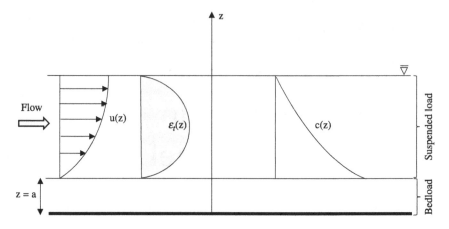

FIGURE 11.1   Schematic diagram of streamwise velocity, eddy viscosity, and sediment concentration along a vertical stream in open-channel, sediment-laden flow.

The whole flow region of sediment-laden flow behaves like a stably stratified flow region, as the entire flow domain can be classified into two regions: The heavy fluid zone near the channel bed having a high concentration of sediment and the light fluid zone for the rest of the domain. Due to this, the process of turbulent mixing of the fluid momentum and the sediment mass slows down along the vertical. Following Smith and McLean (1977), Herrmann and Madsen (2007) expressed sediment diffusivity and turbulent diffusivity, respectively, by incorporating the effect of stratification in the following forms:

$$\varepsilon_t = \varepsilon_{t0}\left(1 - \beta R_f\right) \tag{11.3}$$

$$\varepsilon_s = \varepsilon_{s0}\left(1 - \beta R_f\right) = \frac{\varepsilon_{t0}}{\alpha}\left(1 - \beta R_f\right) \tag{11.4}$$

where ε_{t0} and ε_{s0} are the neutral turbulent and sediment diffusivities, respectively; β is the stratification correction parameter; α is the Schmidt number; and R_f is the flux Richardson number. Parameter R_f can be given as (Monin and Yaglom 1971):

$$R_f = -\frac{g\varepsilon_s\left(s-1\right)\dfrac{dc}{dz}}{\varepsilon_t\left(\dfrac{du}{dz}\right)^2} \tag{11.5}$$

where g is the acceleration due to gravity and s is the ratio of sediment density to water density. The neutral turbulent diffusivity can be of different types, such as constant, linear, parabolic, etc. (van Rijn 1984). Here, we consider the parabolic profile given as:

$$\varepsilon_{t0} = \kappa u_* z\left(1 - \frac{z}{h}\right) \tag{11.6}$$

where κ is the von Karman constant taken as 0.41. It may be noted that without the stratification effect, i.e., $\varepsilon_t = \varepsilon_{t0}$, one can obtain the well-known logarithmic velocity profile and the Rouse equation for velocity and concentration from Eqs. (11.2) and (11.1), respectively. Using Eqs. (11.1)–(11.4) and (11.6) in Eq. (11.5), the flux Richardson number can be obtained as:

$$R_f = \frac{\gamma z c}{1 - \dfrac{z}{h} + \beta \gamma z c} \tag{11.7}$$

Inserting Eq. (11.7) in the governing Eq. (11.1), we get:

$$\frac{dc}{dz} + \frac{\omega_s \alpha}{\kappa u_*} \frac{c}{z\left(1 - \dfrac{z}{h}\right)} + \frac{\omega_s \alpha \beta \gamma}{\kappa u_*} \left(\frac{c}{1 - \dfrac{z}{h}}\right)^2 = 0 \tag{11.8}$$

Similarly, using Eq. (11.7) in Eq. (11.2), the governing equation for the velocity can be given as:

$$\frac{du}{dz} = \frac{u_*}{\kappa z}\left(1 + \beta \gamma \frac{z c}{1 - \dfrac{z}{h}}\right) \tag{11.9}$$

Equations (11.8) and (11.9) constitute a system of coupled differential equations governing the flow velocity and sediment concentration, respectively, proposed by Herrman and Madsen (2007). For solving the system, one needs to specify the initial conditions at a common elevation. Here, we consider the starting point to be the reference level $z = a$ and accordingly provide the conditions as follows:

$$c(z = a) = c_a, \ u(z = a) = u_a \tag{11.10}$$

where c_a and u_a are the reference concentration and reference velocity, respectively.

To simplify the model, we now nondimensionalize the equations using the following scales:

$$\xi = \frac{z}{h}, \ C = \frac{c}{c_a}, \ U = \frac{u}{u_*} \tag{11.11}$$

Accordingly, the governing Eqs. (11.8) and (11.9) and the initial conditions in Eq. (11.10) take on the following forms:

$$\frac{dC}{d\xi} + A_1 \frac{C}{\xi(1 - \xi)} + A_2 \left(\frac{C}{1 - \xi}\right)^2 = 0 \tag{11.12}$$

where $A_1 = \dfrac{\omega_s \alpha}{\kappa u_*}$ and $A_2 = \dfrac{\omega_s \alpha \beta \gamma c_a h}{\kappa u_*}$.

$$\frac{dU}{d\xi} = \frac{1}{\kappa\xi}\left(1 + A_3 \frac{\xi C}{1-\xi}\right) \tag{11.13}$$

where $A_3 = \beta\gamma c_a h$.

$$C\left(\xi = \xi_a\right) = 1, \; U\left(\xi = \xi_a\right) = \frac{u_a}{u_*} = U_a \tag{11.14}$$

11.2.1 HAM-BASED SOLUTION

To solve Eqs. (11.12) and (11.13) using HAM, we first rewrite the equations as follows:

$$\xi\left(1-\xi\right)^2 \frac{dC}{d\xi} + A_1\left(1-\xi\right)C + A_2\xi C^2 = 0 \tag{11.15}$$

$$\kappa\xi\left(1-\xi\right)\frac{dU}{d\xi} - A_3\xi C + \xi - 1 = 0 \tag{11.16}$$

According to Eqs. (11.15) and (11.16) and the given initial conditions in Eq. (11.14), the zeroth-order deformation equation can be constructed as follows:

$$\left(1-q\right)\mathcal{L}_i\left[\Phi_i\left(\xi;q\right) - y_{i,0}\left(\xi\right)\right] - q\hbar_i H_i\left(\xi\right)\mathcal{N}_i\left[\Phi_i\left(\xi;q\right)\right] = 0 \tag{11.17}$$

subject to the initial conditions:

$$\Phi_i\left(\xi_a;q\right) = y_i\left(\xi = \xi_a\right) \text{ for } i = 1,2 \tag{11.18}$$

where q is the embedding parameter, $\Phi_i\left(\xi;q\right)$ are the representation of solutions across q, $y_{i,0}\left(\xi\right)$ is the initial approximation, $\hbar_i$ is the auxiliary parameter, $H_i\left(\xi\right)$ is the auxiliary function, and $\mathcal{L}_i$ and $\mathcal{N}_i$ are the linear and nonlinear operators, respectively. As q increases from 0 to 1, $\Phi_i\left(\xi;q\right)$ varies from the initial approximation $y_{i,0}\left(\xi\right)$ to the exact solution $y_i\left(\xi\right)$, i.e., $\Phi_i\left(\xi;0\right) = y_{i,0}\left(\xi\right)$ and $\Phi_i\left(\xi;1\right) = y_i\left(\xi\right)$. It can be seen that here $y_1\left(\xi\right) = C\left(\xi\right)$ and $y_2\left(\xi\right) = U\left(\xi\right)$. The higher-order terms can be calculated from the higher-order deformation equation, given as follows:

$$\mathcal{L}_i\left[y_{i,m}\left(\xi\right) - \chi_m y_{i,m-1}\left(\xi\right)\right] = \hbar_i H_i\left(\xi\right)R_{i,m}\left(\vec{y}_{i,m-1}\right) \tag{11.19}$$

subject to

$$y_{i,m}\left(\xi = \xi_a\right) = 0 \tag{11.20}$$

where

$$\chi_m = \begin{cases} 0 & \text{when } m = 1, \\ 1 & \text{otherwise} \end{cases} \tag{11.21}$$

and

$$R_{i,m}\left(\vec{y}_{i,m-1}\right) = \frac{1}{(m-1)!} \frac{\partial^{m-1} \mathcal{N}_i\left[\Phi_i\left(\xi;q\right)\right]}{\partial q^{m-1}}\Bigg|_{q=0} \tag{11.22}$$

where $y_{i,m}$ for $m \geq 1$ is the higher-order term. Now, the final solution can be obtained as follows:

$$y_i\left(\xi\right) = y_{i,0}\left(\xi\right) + \sum_{m=1}^{\infty} y_{i,m}\left(\xi\right) \tag{11.23}$$

The nonlinear operators for the problem are chosen as the original governing equations, given as follows:

$$\mathcal{N}_1\left[\Phi_i\left(\xi;q\right)\right] = \xi\left(1-\xi\right)^2 \frac{\partial \Phi_1\left(\xi;q\right)}{\partial \xi} + A_1\left(1-\xi\right)\Phi_1\left(\xi;q\right) + A_2\xi\Phi_1\left(\xi;q\right)^2 \tag{11.24}$$

$$\mathcal{N}_2\left[\Phi_i\left(\xi;q\right)\right] = \kappa\xi\left(1-\xi\right)\frac{\partial \Phi_2\left(\xi;q\right)}{\partial \xi} - A_3\xi\Phi_1\left(\xi;q\right) + \xi - 1 \tag{11.25}$$

Therefore, using Eqs. (11.24) and (11.25), the term $R_{i,m}$ can be calculated from Eq. (11.22) as follows:

$$R_{1,m}\left(\vec{y}_{i,m-1}\right) = \xi\left(1-\xi\right)^2 \frac{dy_{1,m-1}}{d\xi} + A_1\left(1-\xi\right)y_{1,m-1} + A_2\xi\sum_{i=0}^{m-1} y_{1,m-i-1}y_{1,i} \tag{11.26}$$

$$R_{2,m}\left(\vec{y}_{i,m-1}\right) = \kappa\xi\left(1-\xi\right)\frac{dy_{2,m-1}}{d\xi} - A_3\xi y_{1,m-1} + \left(\xi-1\right)\left(1-\chi_m\right) \tag{11.27}$$

We consider the following set of base functions to represent the solutions $C\left(\xi\right)$ and $U\left(\xi\right)$ of the present problem:

$$\left\{\xi^n \mid n = 0,12, 3,...\right\} \tag{11.28}$$

so that

$$C\left(\xi\right) = \alpha_0 + \sum_{n=1}^{\infty} \alpha_n\left(\xi\right)^n \tag{11.29}$$

$$U\left(\xi\right) = \beta_0 + \sum_{n=1}^{\infty} \beta_n\left(\xi\right)^n \tag{11.30}$$

where α_n and β_n are the coefficients of the series. Equation (11.39) provides the so-called *rule of solution expression*. Following the rule of solution expression, the linear operators and the initial approximations are chosen, respectively, as follows:

$$\mathcal{L}_1\left[\Phi_1\left(\xi;q\right)\right] = \frac{\partial \Phi_1\left(\xi;q\right)}{\partial \xi} \tag{11.31}$$

$$\mathcal{L}_2\left[\Phi_2\left(\xi;q\right)\right] = \frac{\partial \Phi_2\left(\xi;q\right)}{\partial \xi} \tag{11.32}$$

$$y_{1,0}\left(\xi\right) = 1 \tag{11.33}$$

$$y_{2,0}\left(\xi\right) = U_a \tag{11.34}$$

Using Eqs. (11.31) and (11.32), the higher-order terms can be obtained from Eq. (11.19) as follows:

$$y_{1,m}\left(\xi\right) = \chi_m y_{1,m-1}\left(\xi\right) + \hbar_1 \int_{\xi_a}^{\xi} H_1\left(\xi\right) R_{1,m}\left(\vec{y}_{i,m-1}\right) d\xi + C_2, \ m = 1,2,3,\dots \tag{11.35}$$

$$y_{2,m}\left(\xi\right) = \chi_m y_{2,m-1}\left(\xi\right) + \hbar_2 \int_{\xi_a}^{\xi} H_2\left(\xi\right) R_{2,m}\left(\vec{y}_{i,m-1}\right) d\xi + C_3, \ m = 1,2,3,\dots \tag{11.36}$$

where $R_{i,m}$ is given by Eqs. (11.26) and (11.27) and constants C_2 and C_3 can be determined from the initial condition for the higher-order deformation equations, which simply yields $C_2 = C_3 = 0$ for all $m \geq 1$.

Now, the auxiliary functions $H_i\left(\xi\right)$ can be determined from the *rule of coefficient ergodicity*. Based on the rule of solution expression, the general form of $H_i\left(\xi\right)$ should be:

$$H_i\left(\xi\right) = \xi^{\alpha} \tag{11.37}$$

where α is an integer. Substituting Eq. (11.37) into Eqs. (11.35) and (11.36) and computing some of $y_{i,m}\left(\xi\right)$, we have

$$R_{1,1}\left(y_{i,0}\right) = A_1\left(1-\xi\right) + A_2\xi \tag{11.38}$$

$$R_{2,1}\left(y_{i,0}\right) = -A_3\xi + \left(\xi-1\right) \tag{11.39}$$

$$y_{1,1}\left(\xi\right) = \hbar_1 \int_{\xi_a}^{\xi} \xi^{\alpha}\left[A_1\left(1-\xi\right) + A_2\xi\right] d\xi$$

$$= \frac{\hbar_1 A_1}{\alpha+1}\left(\xi^{\alpha+1} - \xi_0^{\alpha+1}\right) + \frac{\hbar_1\left(A_2 - A_1\right)}{\alpha+2}\left(\xi^{\alpha+2} - \xi_0^{\alpha+2}\right) \tag{11.40}$$

$$y_{2,1}(\xi) = \hbar_2 \int_{\xi_a}^{\xi} \xi^\alpha \left[-A_3 \xi + (\xi - 1) \right] d\xi$$

$$= \frac{\hbar_2 (1 - A_3)}{\alpha + 2} \left(\xi^{\alpha+2} - \xi_0^{\alpha+2} \right) + \frac{\hbar_2}{\alpha + 1} \left(\xi^{\alpha+1} - \xi_0^{\alpha+1} \right) \quad (11.41)$$

Here, we can infer two observations. If we consider $\alpha \geq 1$, then some of the terms for ξ^α do not appear in the solution expression. This does not satisfy the *rule of coefficient ergodicity*, which allows us to include all the terms of the base function for the completeness of the solution. On the other hand, if we consider $\alpha \leq -2$, then the terms appearing in the solution violate the *rule of solution expression*. Therefore, we are left with the only choices $\alpha = 0$ and $\alpha = 1$, out of which we choose $H_i(\xi) = 1$. Finally, the approximate solutions can be obtained as follows:

$$y_i(\xi) \approx y_{i,0}(\xi) + \sum_{m=1}^{M} y_{i,m}(\xi) \quad (11.42)$$

11.2.2 HPM-BASED SOLUTION

For applying HPM, we rewrite Eqs. (11.15) and (11.16) in the following form:

$$\mathcal{N}_i(y_i) = f_i(\xi) \quad (11.43)$$

We construct a homotopy $\Phi_i(\xi; q)$ that satisfies:

$$(1 - q) \left[\mathcal{L}_i \left(\Phi_i(\xi; q) \right) - \mathcal{L}_i \left(y_{i,0}(\xi) \right) \right] + q \left[\mathcal{N}_i(y_i) - f_i(\xi) \right] = 0 \quad (11.44)$$

where q is the embedding parameter that lies on $[0,1]$, $\mathcal{L}_i$ is the linear operator, and $y_{i,0}(\xi)$ is the initial approximation to the final solutions. Equation (11.40) shows that as $q = 0$, $\Phi_i(\xi; 0) = y_{i,0}(\xi)$ and as $q = 1$, $\Phi(\xi; 1) = y_i(\xi)$. It may be noted that $y_1(\xi) = C(\xi)$ and $y_2(\xi) = U(\xi)$. We now express $\Phi_i(\xi; q)$ as a series in terms of q as follows:

$$\Phi_i(\xi; q) = \Phi_{i,0} + q\Phi_{i,1} + q^2\Phi_{i,2} + q^3\Phi_{i,3} + q^4\Phi_{i,4} \cdots \quad (11.45)$$

where $\Phi_{i,m}$ for $m \geq 1$ are the higher-order terms. As $q \to 1$, Eq. (11.41) produces the final solution as:

$$y_i(\xi) = \lim_{q \to 1} \Phi_i(\xi; q) = \sum_{k=0}^{\infty} \Phi_{i,k} \quad (11.46)$$

For simplicity, we consider the linear operators as follows:

$$\mathcal{L}_i \left[\Phi_i(\xi; q) \right] = \frac{\partial \Phi_i(\xi; q)}{\partial \xi} \quad (11.47)$$

The nonlinear operators for the governing equations are:

$$\mathcal{N}_1\left[\Phi_i(\xi;q)\right]=\xi(1-\xi)^2\frac{\partial\Phi_1(\xi;q)}{\partial\xi}+A_1(1-\xi)\Phi_1(\xi;q)+A_2\xi\Phi_1(\xi;q)^2 \quad (11.48)$$

$$\mathcal{N}_2\left[\Phi_i(\xi;q)\right]=\kappa\xi(1-\xi)\frac{\partial\Phi_2(\xi;q)}{\partial\xi}-A_3\xi\Phi_1(\xi;q)+\xi-1 \quad (11.49)$$

Using these expressions, Eq. (11.44) takes on the forms:

$$(1-q)\left[\frac{\partial\Phi_1(\xi;q)}{\partial\xi}-\frac{\partial y_{1,0}}{\partial\xi}\right]$$

$$+q\left[\xi(1-\xi)^2\frac{\partial\Phi_1(\xi;q)}{\partial\xi}+A_1(1-\xi)\Phi_1(\xi;q)+A_2\xi\Phi_1(\xi;q)^2\right]=0 \quad (11.50)$$

$$(1-q)\left[\frac{\partial\Phi_2(\xi;q)}{\partial\xi}-\frac{\partial y_{2,0}}{\partial\xi}\right]$$

$$+q\left[\kappa\xi(1-\xi)\frac{\partial\Phi_2(\xi;q)}{\partial\xi}-A_3\xi\Phi_1(\xi;q)+\xi-1\right]=0 \quad (11.51)$$

The following expression can be used for finding the terms of the series:

$$\left(\Phi_{i,0}+q\Phi_{i,1}+q^2\Phi_{i,2}+q^3\Phi_{i,3}+q^4\Phi_{i,4}\cdots\right)^2=\left(\Phi_{i,0}\right)^2+q\left(2\Phi_{i,0}\Phi_{i,1}\right)$$

$$+q^2\left(\Phi_{i,1}^2+2\Phi_{i,0}\Phi_{i,2}\right)+q^3\left(2\Phi_{i,0}\Phi_{i,3}+2\Phi_{i,1}\Phi_{i,2}\right)+\cdots \quad (11.52)$$

Using Eq. (11.45), we have

$$(1-q)\left[\frac{\partial\Phi_i(\xi;q)}{\partial\xi}-\frac{dy_{i,0}}{d\xi}\right]$$

$$=(1-q)\left[\left(\frac{d\Phi_{i,0}}{d\xi}-\frac{dy_{i,0}}{d\xi}\right)+q\frac{d\Phi_{i,1}}{d\xi}+q^2\frac{d\Phi_{i,2}}{d\xi}+q^3\frac{d\Phi_{i,3}}{d\xi}+q^4\frac{d\Phi_{i,4}}{d\xi}+\cdots\right]$$

$$=\left(\frac{d\Phi_{i,0}}{d\xi}-\frac{dy_{i,0}}{d\xi}\right)+q\left(\frac{d\Phi_{i,1}}{d\xi}-\left(\frac{d\Phi_{i,0}}{d\xi}-\frac{dy_{i,0}}{d\xi}\right)\right)+q^2\left(\frac{d\Phi_{i,2}}{d\xi}-\frac{d\Phi_{i,1}}{d\xi}\right)$$

$$+q^3\left(\frac{d\Phi_{i,3}}{d\xi}-\frac{d\Phi_{i,2}}{d\xi}\right)+q^4\left(\frac{d\Phi_{i,4}}{d\xi}-\frac{d\Phi_{i,3}}{d\xi}\right)+\cdots \quad (11.53)$$

$$\xi(1-\xi)^2 \frac{\partial \Phi_1(\xi;q)}{\partial \xi} + A_1(1-\xi)\Phi_1(\xi;q) + A_2\xi\Phi_1(\xi;q)^2$$

$$= \xi(1-\xi)^2 \left(\frac{d\Phi_{1,0}}{d\xi} + q\frac{d\Phi_{1,1}}{d\xi} + q^2\frac{d\Phi_{1,2}}{d\xi} + q^3\frac{d\Phi_{1,3}}{d\xi} + q^4\frac{d\Phi_{1,4}}{d\xi} + \cdots \right)$$

$$+ A_1(1-\xi)\left(\Phi_{1,0} + q\Phi_{1,1} + q^2\Phi_{1,2} + q^3\Phi_{1,3} + q^4\Phi_{1,4} + \cdots \right)$$

$$+ A_2\xi\left[(\Phi_{1,0})^2 + q(2\Phi_{1,0}\Phi_{1,1}) + q^2\left(\Phi_{1,1}^2 + 2\Phi_{1,0}\Phi_{1,2}\right) + q^3\left(2\Phi_{1,0}\Phi_{1,3} + 2\Phi_{1,1}\Phi_{1,2}\right) \right]$$

$$= \left[\xi(1-\xi)^2 \frac{d\Phi_{1,0}}{d\xi} + A_1(1-\xi)\Phi_{1,0} + A_2\xi(\Phi_{1,0})^2 \right]$$

$$+ q\left[\xi(1-\xi)^2 \frac{d\Phi_{1,1}}{d\xi} + A_1(1-\xi)\Phi_{1,1} + 2A_2\xi\Phi_{1,0}\Phi_{1,1} \right]$$

$$+ q^2\left[\xi(1-\xi)^2 \frac{d\Phi_{1,2}}{d\xi} + A_1(1-\xi)\Phi_{1,2} + A_2\xi\left(\Phi_{1,1}^2 + 2\Phi_{1,0}\Phi_{1,2}\right) \right]$$

$$+ q^3\left[\xi(1-\xi)^2 \frac{d\Phi_{1,3}}{d\xi} + A_1(1-\xi)\Phi_{1,3} + A_2\xi\left(2\Phi_{1,0}\Phi_{1,3} + 2\Phi_{1,1}\Phi_{1,2}\right) \right] + \cdots \qquad (11.54)$$

$$\kappa\xi(1-\xi)\frac{\partial \Phi_2(\xi;q)}{\partial \xi} - A_3\xi\Phi_1(\xi;q) + \xi - 1$$

$$= \kappa\xi(1-\xi)\left(\frac{d\Phi_{2,0}}{d\xi} + q\frac{d\Phi_{2,1}}{d\xi} + q^2\frac{d\Phi_{2,2}}{d\xi} + q^3\frac{d\Phi_{2,3}}{d\xi} + q^4\frac{d\Phi_{2,4}}{d\xi} + \cdots \right)$$

$$- A_3\xi\left(\Phi_{1,0} + q\Phi_{1,1} + q^2\Phi_{1,2} + q^3\Phi_{1,3} + q^4\Phi_{1,4} + \cdots \right) + \xi - 1$$

$$= \left[\kappa\xi(1-\xi)\frac{d\Phi_{2,0}}{d\xi} - A_3\xi\Phi_{1,0} + \xi - 1 \right] + q\left[\kappa\xi(1-\xi)\frac{d\Phi_{2,1}}{d\xi} - A_3\xi\Phi_{1,1} \right]$$

$$+ q^2\left[\kappa\xi(1-\xi)\frac{d\Phi_{2,2}}{d\xi} - A_3\xi\Phi_{1,2} \right] + q^3\left[\kappa\xi(1-\xi)\frac{d\Phi_{2,3}}{d\xi} - A_3\xi\Phi_{1,3} \right] + \cdots \qquad (11.55)$$

Using the boundary conditions given by Eq. (11.14), we have

$$\Phi_{1,0}(\xi_a) = 1, \; \Phi_{1,1}(\xi_a) = 0, \; \Phi_{1,2}(\xi_a) = 0, \; \dots \qquad (11.56)$$

$$\Phi_{2,0}(\xi_a) = U_a, \; \Phi_{2,1}(\xi_a) = 0, \; \Phi_{2,2}(\xi_a) = 0, \; \dots \qquad (11.57)$$

Using Eqs. (11.53)–(11.55) and equating the like powers of q in Eqs. (11.50) and (11.51), the following systems of differential equations are obtained:

$$\frac{d\Phi_{1,0}}{d\xi} - \frac{dy_{1,0}}{d\xi} = 0 \text{ subject to } \Phi_{1,0}(\xi_a) = 1 \qquad (11.58)$$

$$\frac{d\Phi_{2,0}}{d\xi} - \frac{dy_{2,0}}{d\xi} = 0 \text{ subject to } \Phi_{2,0}(\xi_a) = U_a \tag{11.59}$$

$$\frac{d\Phi_{1,1}}{d\xi} - \left(\frac{d\Phi_{1,0}}{d\xi} - \frac{dy_{1,0}}{d\xi}\right) + \xi(1-\xi)^2 \frac{d\Phi_{1,0}}{d\xi} + A_1(1-\xi)\Phi_{1,0} + A_2\xi(\Phi_{1,0})^2$$
$$= 0 \text{ subject to } \Phi_{1,1}(\xi_a) = 0 \tag{11.60}$$

$$\frac{d\Phi_{2,1}}{d\xi} - \left(\frac{d\Phi_{2,0}}{d\xi} - \frac{dy_{2,0}}{d\xi}\right) + \kappa\xi(1-\xi)\frac{d\Phi_{2,0}}{d\xi} - A_3\xi\Phi_{1,0} + \xi - 1$$
$$= 0 \text{ subject to } \Phi_{2,1}(\xi_a) = 0 \tag{11.61}$$

$$\frac{d\Phi_{1,2}}{d\xi} - \frac{d\Phi_{1,1}}{d\xi} + \xi(1-\xi)^2 \frac{d\Phi_{1,1}}{d\xi} + A_1(1-\xi)\Phi_{1,1} + 2A_2\xi\Phi_{1,0}\Phi_{1,1}$$
$$= 0 \text{ subject to } \Phi_{1,2}(\xi_a) = 0 \tag{11.62}$$

$$\frac{d\Phi_{2,2}}{d\xi} - \frac{d\Phi_{2,1}}{d\xi} + \kappa\xi(1-\xi)\frac{d\Phi_{2,1}}{d\xi} - A_3\xi\Phi_{1,1} = 0 \text{ subject to } \Phi_{2,2}(\xi_a) = 0 \tag{11.63}$$

$$\frac{d\Phi_{1,3}}{d\xi} - \frac{d\Phi_{1,2}}{d\xi} + \xi(1-\xi)^2 \frac{d\Phi_{1,2}}{d\xi} + A_1(1-\xi)\Phi_{1,2} + A_2\xi\left(\Phi_{1,1}^2 + 2\Phi_{1,0}\Phi_{1,2}\right)$$
$$= 0 \text{ subject to } \Phi_{1,3}(\xi_a) = 0 \tag{11.64}$$

$$\frac{d\Phi_{2,3}}{d\xi} - \frac{d\Phi_{2,2}}{d\xi} + \kappa\xi(1-\xi)\frac{d\Phi_{2,2}}{d\xi} - A_3\xi\Phi_{1,2} = 0 \text{ subject to } \Phi_{2,3}(\xi_a) = 0 \tag{11.65}$$

The initial approximations can be chosen as $\Phi_{1,0} = \left(\dfrac{1-\xi}{\xi}\dfrac{\xi_a}{1-\xi_a}\right)$ and $\Phi_{2,0} = U_a + \ln\left(\dfrac{\xi}{1-\xi}\dfrac{1-\xi_a}{\xi_a}\right)$. Using these initial approximations, we can solve the equations iteratively using symbolic software. We avoid those expressions here, as they are lengthy. Finally, the HPM-based solutions can be approximated as follows:

$$y_i(\xi) \approx \sum_{k=0}^{M} \Phi_{i,k} \tag{11.66}$$

11.2.3 OHAM-BASED SOLUTION

For applying OHAM, Eqs. (11.15) and (11.16) can be written in the following form:

$$\mathcal{L}_i\left[y_i(\xi)\right] + \mathcal{N}_i\left[y_i(\xi)\right] + h_i(\xi) = 0 \tag{11.67}$$

We select

$$\mathcal{L}_1\left[y_1(\xi)\right] = \frac{\partial y_1}{\partial \xi} + y_1 \tag{11.68}$$

$$\mathcal{L}_2\left[y_2(\xi)\right]=\frac{dy_2}{\partial\xi}-y_2 \tag{11.69}$$

$$\mathcal{N}_1\left[y_i(\xi)\right]=\left[\xi(1-\xi)^2-1\right]\frac{dy_1}{d\xi}+\left[A_1(1-\xi)-1\right]y_1+A_2\xi y_1^2 \tag{11.70}$$

$$\mathcal{N}_2\left[y_i(\xi)\right]=\left[\kappa\xi(1-\xi)-1\right]\frac{dy_2}{d\xi}+y_2-A_3\xi y_1 \tag{11.71}$$

$$h_1(\xi)=0 \tag{11.72}$$

$$h_2(\xi)=\xi-1 \tag{11.73}$$

With these considerations, the zeroth-order problems become:

$$\mathcal{L}_i\left[y_{i,0}(\xi)\right]+h_i(\xi)=0 \text{ subject to } y_{i,0}(\xi_a)=y_i(\xi=\xi_a) \tag{11.74}$$

Solving Eq. (11.74), one obtains:

$$y_{1,0}(\xi)=exp(\xi_a-\xi) \tag{11.75}$$

$$y_{2,0}(\xi)=\xi-(\xi_a-U_a)exp(\xi-\xi_a) \tag{11.76}$$

To obtain $\mathcal{N}_{i,0}$, $\mathcal{N}_{i,1}$, $\mathcal{N}_{i,2}$, etc., the following can be used:

$$\left[\xi(1-\xi)^2-1\right]\frac{dy_1}{d\xi}+\left[A_1(1-\xi)-1\right]y_1+A_2\xi y_1^2$$

$$=\left[\xi(1-\xi)^2-1\right]\left(\frac{dy_{1,0}}{d\xi}+q\frac{dy_{1,1}}{d\xi}+q^2\frac{dy_{1,2}}{d\xi}+q^3\frac{dy_{1,3}}{d\xi}+q^4\frac{dy_{1,4}}{d\xi}+\cdots\right)$$

$$+\left[A_1(1-\xi)-1\right]\left(y_{1,0}+qy_{1,1}+q^2y_{1,2}+q^3y_{1,3}+q^4y_{1,4}+\cdots\right)$$

$$+A_2\xi\left[\left(y_{1,0}\right)^2+q\left(2y_{1,0}y_{1,1}\right)+q^2\left(y_{1,1}^2+2y_{1,0}y_{1,2}\right)+q^3\left(2y_{1,0}y_{1,3}+2y_{1,1}y_{1,2}\right)+\cdots\right]$$

$$=\left[\left\{\xi(1-\xi)^2-1\right\}\frac{dy_{1,0}}{d\xi}+\left[A_1(1-\xi)-1\right]y_{1,0}+A_2\xi\left(y_{1,0}\right)^2\right]$$

$$+q\left[\left\{\xi(1-\xi)^2-1\right\}\frac{dy_{1,1}}{d\xi}+\left[A_1(1-\xi)-1\right]y_{1,1}+2A_2\xi y_{1,0}y_{1,1}\right]$$

$$+q^2\left[\left\{\xi(1-\xi)^2-1\right\}\frac{dy_{1,2}}{d\xi}+\left[A_1(1-\xi)-1\right]y_{1,2}+A_2\xi\left(y_{1,1}^2+2y_{1,0}y_{1,2}\right)\right]$$

$$+q^3\left[\left\{\xi(1-\xi)^2-1\right\}\frac{dy_{1,3}}{d\xi}+\left[A_1(1-\xi)-1\right]y_{1,3}+A_2\xi\left(2y_{1,0}y_{1,3}+2y_{1,1}y_{1,2}\right)\right]+\cdots \tag{11.77}$$

$$\left[\kappa\xi(1-\xi)-1\right]\frac{dy_2}{d\xi}+y_2-A_3\xi y_1$$

$$=\left[\kappa\xi(1-\xi)-1\right]\left(\frac{dy_{2,0}}{d\xi}+q\frac{dy_{2,1}}{d\xi}+q^2\frac{dy_{2,2}}{d\xi}+q^3\frac{dy_{2,3}}{d\xi}+q^4\frac{dy_{2,4}}{d\xi}+\cdots\right)$$

$$+\left(y_{2,0}+qy_{2,1}+q^2y_{2,2}+q^3y_{2,3}+q^4y_{2,4}+\cdots\right)$$

$$-A_3\xi\left(y_{1,0}+qy_{1,1}+q^2y_{1,2}+q^3y_{1,3}+q^4y_{1,4}+\cdots\right)$$

$$=\left[\left\{\kappa\xi(1-\xi)-1\right\}\frac{dy_{2,0}}{d\xi}+y_{2,0}-A_3\xi y_{1,0}\right]$$

$$+q\left[\left\{\kappa\xi(1-\xi)-1\right\}\frac{dy_{2,1}}{d\xi}+y_{2,1}-A_3\xi y_{1,1}\right]$$

$$+q^2\left[\left\{\kappa\xi(1-\xi)-1\right\}\frac{dy_{2,2}}{d\xi}+y_{2,2}-A_3\xi y_{1,2}\right]$$

$$+q^3\left[\left\{\kappa\xi(1-\xi)-1\right\}\frac{dy_{2,3}}{d\xi}+y_{2,3}-A_3\xi y_{1,3}\right]+\cdots \tag{11.78}$$

Using Eqs. (11.77) and (11.78), the first-order problems reduce to:

$$\frac{dy_{1,1}}{d\xi}+y_{1,1}=H_{1,1}(\xi,C_{1,i})\left\{\left\{\xi(1-\xi)^2-1\right\}\frac{dy_{1,0}}{d\xi}+\left[A_1(1-\xi)-1\right]y_{1,0}+A_2\xi(y_{1,0})^2\right\}$$

subject to $y_{1,1}(\xi=\xi_a)=0$ $\tag{11.79}$

$$\frac{dy_{2,1}}{d\xi}-y_{2,1}=H_{2,1}(\xi,C_{2,i})\left\{\left\{\kappa\xi(1-\xi)-1\right\}\frac{dy_{2,0}}{d\xi}+y_{2,0}-A_3\xi y_{1,0}\right\}$$

subject to $y_{2,1}(\xi=\xi_a)=0$ $\tag{11.80}$

The auxiliary functions can be chosen in many ways. Here, we select $H_{1,1}(\xi,C_{1,i})=C_{1,1}$ and $H_{2,1}(\xi,C_{2,i})=C_{2,1}$. Putting $k=2$, the second-order problems become:

$$\mathcal{L}_i\left[y_{i,2}(\xi)-y_{i,1}(\xi)\right]=H_{i,2}(\xi,C_{2,i})\mathcal{N}_{i,0}(y_{i,0}(\xi))$$

$$+H_{i,1}(\xi,C_i)\left[\mathcal{L}_i\left[y_{i,1}(\xi,C_i)\right]+\mathcal{N}_{i,1}\left[y_{i,0}(\xi),y_{i,1}(\xi)\right]\right]$$

subject to $y_{i,2}(\xi=\xi_a)=0$ $\tag{11.81}$

Further, we choose $H_{1,2}(\xi,C_{2,i})=C_{1,2}$ and $H_{2,2}(\xi,C_{2,i})=C_{2,2}$. Therefore, Eq. (11.81) becomes:

$$\frac{dy_{1,2}}{d\xi}+y_{1,2}=\frac{dy_{1,1}}{d\xi}+y_{1,1}+C_{1,2}\left\{\left\{\xi(1-\xi)^2-1\right\}\frac{dy_{1,0}}{d\xi}+\left[A_1(1-\xi)-1\right]y_{1,0}+A_2\xi(y_{1,0})^2\right\}$$

$$+C_{1,1}\left[\frac{dy_{1,1}}{d\xi}+y_{1,1}+\left\{\xi(1-\xi)^2-1\right\}\frac{dy_{1,1}}{d\xi}+\left[A_1(1-\xi)-1\right]y_{1,1}+2A_2\xi y_{1,0}y_{1,1}\right]$$

subject to $y_{1,2}(\xi=\xi_a)=0$ $\tag{11.82}$

$$\frac{dy_{2,2}}{d\xi} - y_{2,2} = \frac{dy_{2,1}}{d\xi} - y_{2,1} + C_{2,2}\left\{ \left\{ \kappa\xi(1-\xi) - 1 \right\}\frac{dy_{2,0}}{d\xi} + y_{2,0} - A_3\xi y_{1,0} \right\}$$

$$+ C_{2,1}\left[\frac{dy_{2,1}}{d\xi} - y_{2,2} + \left\{ \kappa\xi(1-\xi) - 1 \right\}\frac{dy_{2,1}}{d\xi} + y_{2,1} - A_3\xi y_{1,1} \right]$$

$$\text{subject to } y_{2,2}\left(\xi = \xi_a\right) = 0 \qquad (11.83)$$

The terms of the series can be computed using the equations developed here. One can compute these terms without any difficulty using symbolic computation software, such as MATLAB. Furthermore, following Chapter 2, the higher-order terms can be computed in a similar manner. However, our aim is to produce an accurate solution with just two or three terms of the OHAM-based series. Therefore, we restrict our calculation up to $k = 2$. Finally, the approximate solution can be found as:

$$y_i\left(\xi\right) \approx y_{i,0}\left(\xi\right) + y_{i,1}\left(\xi\right) + y_{i,2}\left(\xi\right) \qquad (11.84)$$

where the terms are given by Eqs. (11.75), (11.76), (11.79), (11.80), (11.82), and (11.83).

11.3   CONVERGENCE THEOREMS

The convergence of the HAM-based and OHAM-based solutions are proved theoretically using the following theorems.

11.3.1   CONVERGENCE THEOREM OF THE HAM-BASED SOLUTION

The convergence theorem for the HAM-based solution given by Eq. (11.42) can be proved using the following theorems.

Theorem 11.1: If the homotopy series $\Sigma_{m=0}^{\infty} y_{i,m}\left(\xi\right)$ and $\Sigma_{m=0}^{\infty} y_{i,m}{}'\left(\xi\right)$ converge, then $R_{i,m}\left(\vec{y}_{i,m-1}\right)$ given by Eqs. (11.26) and (11.27) satisfies the relation $\Sigma_{m=1}^{\infty} R_{i,m}\left(\vec{y}_{i,m-1}\right) = 0$.

Proof: The auxiliary linear operator was defined as follows:

$$\mathcal{L}_i\left[y_i\right] = \frac{dy_i}{d\xi} \qquad (11.85)$$

According to Eq. (11.19), we obtain:

$$\mathcal{L}_i\left[y_{i,1}\right] = \hbar_i R_{i,1}\left(\overrightarrow{y_{i,0}}\right) \qquad (11.86)$$

$$\mathcal{L}_i\left[y_{i,2} - y_{i,1}\right] = \hbar_i R_{i,2}\left(\overrightarrow{y_{i,1}}\right) \qquad (11.87)$$

$$\mathcal{L}_i\left[y_{i,3} - y_{i,2}\right] = \hbar_i R_{i,3}\left(\overrightarrow{y_{i,2}}\right) \qquad (11.88)$$

$$\mathcal{L}_i\left[y_{i,m} - y_{i,m-1}\right] = \hbar_i R_{i,m}\left(\overrightarrow{y_{i,m-1}}\right) \qquad (11.89)$$

Adding these terms, we get:

$$\mathcal{L}_i\left[y_{i,m}\right] = \hbar_i \sum_{k=1}^{m} R_{i,k}\left(\overrightarrow{y_{i,k-1}}\right) \tag{11.90}$$

As the series $\Sigma_{m=0}^{\infty} \, y_{i,m}\left(\xi\right)$ and $\Sigma_{m=0}^{\infty} \, y_{i,m}{}'\left(\xi\right)$ are convergent, $\lim_{m\to\infty} y_{i,m}\left(\xi\right) = 0$ and $\lim_{m\to\infty} y_{i,m}{}'\left(\xi\right) = 0$. Now, recalling the previous summand and taking the limit, the required result follows as:

$$\hbar_i \sum_{k=1}^{\infty} R_{i,k}\left(\overrightarrow{y_{i,k-1}}\right) = \hbar_i \lim_{m\to\infty} \sum_{k=1}^{m} R_{i,k}\left(\overrightarrow{y_{i,k-1}}\right)$$

$$= \lim_{m\to\infty} \mathcal{L}_i\left[y_{i,m}\right] = \lim_{m\to\infty} y_{i,m}{}' = 0 \tag{11.91}$$

Theorem 11.2: If $\hbar_i$ is so properly chosen that the series $\Sigma_{m=0}^{\infty} \, y_{i,m}\left(\xi\right)$ and $\Sigma_{m=0}^{\infty} \, y_{i,m}{}'\left(\xi\right)$ converge absolutely to $y_i\left(\xi\right)$ and $y_i{}'\left(\xi\right)$, respectively, then the homotopy series $\Sigma_{m=0}^{\infty} \, y_{i,m}\left(\xi\right)$ satisfies the original governing Eqs. (11.15) and (11.16).

Proof: From Theorem 11.1, we have

$$\sum_{m=1}^{\infty} R_{i,m}\left(\vec{y}_{i,m-1}\right) = 0 \tag{11.92}$$

According to the definition of $R_{i,m}\left(\vec{y}_{i,m-1}\right)$ given by Eq. (11.22):

$$\sum_{m=1}^{\infty} R_{i,m}\left(\vec{y}_{i,m-1}\right) = \sum_{m=1}^{\infty} \frac{1}{(m-1)!} \left.\frac{\partial^{m-1} \mathcal{N}_i\left[\Phi_i\left(\xi;q\right)\right]}{\partial q^{m-1}}\right|_{q=0}$$

$$= \sum_{m=0}^{\infty} \frac{1}{m!} \left.\frac{\partial^{m} \mathcal{N}_i\left[\Phi_i\left(\xi;q\right)\right]}{\partial q^{m}}\right|_{q=0} \tag{11.93}$$

Let $\varepsilon_i\left(\xi;q\right) = \mathcal{N}_i\left[\Phi_i\left(\xi;q\right)\right]$ denote the residual error of the system in Eqs. (11.15) and (11.16). Therefore, $\varepsilon_i\left(\xi;q\right) = 0$ corresponds to the exact solution to the system. Now, expanding $\varepsilon_i\left(\xi;q\right)$ as a Maclaurin series about q, we have

$$\varepsilon_i\left(\xi;q\right) = \sum_{m=0}^{\infty} \frac{q^m}{m!} \left.\frac{\partial^{m} \varepsilon_i\left(\xi;q\right)}{\partial q^{m}}\right|_{q=0} = \sum_{m=0}^{\infty} \frac{q^m}{m!} \left.\frac{\partial^{m} \mathcal{N}_i\left[\Phi_i\left(\xi;q\right)\right]}{\partial q^{m}}\right|_{q=0} \tag{11.94}$$

Using Eqs. (11.49)–(11.51), we obtain:

$$\varepsilon_i\left(\xi;1\right) = \sum_{m=0}^{\infty} \frac{1}{m!} \left.\frac{\partial^{m} \mathcal{N}_i\left[\Phi_i\left(\xi;q\right)\right]}{\partial q^{m}}\right|_{q=0} = 0 \tag{11.95}$$

which shows that the system achieves the exact solution when $q = 1$. Hence, the convergence result follows.

11.3.2 CONVERGENCE THEOREM OF THE OHAM-BASED SOLUTION

Theorem 11.3: If the series $y_{i,0}(\xi) + \Sigma_{j=1}^{\infty} y_{i,j}(\xi)$, $i = 1,2$, converges, where $y_{i,j}(\xi)$ is governed by Eqs. (11.75), (11.76), (11.79), (11.80), (11.82), and (11.83), then Eq. (11.84) is a solution of the original Eqs. (11.15) and (11.16).

Proof: The proof of this theorem follows exactly from Theorem 2.3 of Chapter 2. In place of a single equation, one needs to consider a system of equations.

11.4 RESULTS AND DISCUSSION

First, we discuss the expressions and the values of parameters needed for the assessment of the developed solution. Then, the numerical convergence of the HAM-based analytical solution is established for a specific test case, and thereafter the solution is validated against a numerical solution. Also, the derived analytical solution is tested under different physical conditions to evaluate the efficiency of the method. Finally, the HPM- and OHAM-based analytical solutions are validated by comparing them with the numerical solution.

11.4.1 SELECTION OF PARAMETERS

For the assessment of the solutions, it can be seen that several parameters need to be calculated. The objective here is to show the efficiencies of the analytical methods. Therefore, a detailed discussion of the physical parameters is avoided. The settling velocity ω_s is assumed to be the same as the clear water settling velocity ω_0. A simple expression for ω_0 is available in Cheng (1997). There are several studies available for the determination of the Schmidt number α (Pal and Ghoshal 2016). Also, the boundary conditions and the reference level are selected based on an available dataset, namely Coleman (1981). A detailed discussion can be found in Mohan et al. (2019). Here, for the considered test case, we mention the selected values for the parameters: $s = 2.65$, $g = 980$ cm/sec^2, $\beta = 6.25$, $u_* = 4.1$ cm/sec, $\kappa = 0.41$, $\alpha = 1.235$, $\omega_s = 0.6623$ cm/sec, $h = 17.1$ cm, $c_a = 0.014$, $\xi_a = 0.035$, and $U_a = 14.63$.

11.4.2 NUMERICAL CONVERGENCE AND VALIDATION
OF THE HAM-BASED SOLUTION

The convergence of the HAM-based series solutions depends on a suitable choice for the auxiliary parameter $\hbar_i$. For that purpose, the squared residual error with regard to Eq. (11.15) or Eq. (11.16) can be calculated as follows:

$$\Delta_m = \int_{\xi \in \Omega} \left[\left(\mathcal{N}_1 \left[C(\xi), U(\xi) \right] \right)^2 + \left(\mathcal{N}_2 \left[C(\xi), U(\xi) \right] \right)^2 \right] d\xi \qquad (11.96)$$

where $\Omega = [\xi_a, 1]$ is the solution domain. The HAM-based series solutions lead to the exact solutions when Δ_m approaches zero. Therefore, for a particular order of approximation m, it is sufficient to minimize the corresponding Δ_m, which then yields optimal values of parameters $\hbar_i$. Here, we consider a test case where the values of parameters are considered, as mentioned in the previous section. With these values, the HAM-based solutions are obtained. First, the squared residual errors are calculated from Eq. (11.96), and then the corresponding approximate solutions are obtained. Figure 11.2 shows the squared residual errors for a different order of approximation for the selected case. It can be seen from the figure that as the order of approximation increases, the corresponding residual error decreases. Thus, the adequacy of selecting linear operators, initial approximations, and auxiliary functions is established. Also, for a quantitative assessment, in Table 11.1, we provide the computational time taken by the computer to produce the corresponding order of approximations. All the computations reported here are performed using the BVPh 2.0 package developed by Zhao and Liao (2002).

Now, we obtain the HAM-based analytical solution for the system of equations and compare it with a numerical solution for the selected test case. For the numerical solution, the MATLAB routine *ode45* is used for the system. The 30th-order HAM-based solution is considered and is compared with the numerical solution in Figure 11.3. It can be observed that the HAM-based solution agrees well with the numerical solution throughout the water column. For a numerical assessment, considering some points in the domain, the exact solution and 30th-order HAM

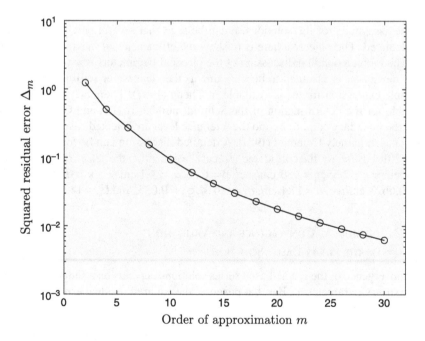

FIGURE 11.2 Squared residual error (Δ_m) versus different orders of approximation (m) of the HAM-based solution for the selected case.

TABLE 11.1

Squared Residual Error (Δ_m) and Computational Time versus Different Orders of Approximation (m) for the Selected Case

Order of Approximation (m)	Squared Residual Error (Δ_m)	Computational Time (sec)
2	1.23887	0.215
4	0.50483	0.455
6	0.27175	0.761
8	0.15391	1.045
10	0.09321	1.408
12	0.06063	1.817
14	0.04196	2.259
16	0.03046	2.733
18	0.02293	3.272
20	0.01773	3.834
24	0.01123	5.208
28	0.00753	6.797
30	0.00626	7.692

approximations are reported in Table 11.2. A flowchart containing the steps of the method is provided in Figure 11.4.

11.4.3 VALIDATION OF THE HPM-BASED SOLUTION

The HPM-based analytical solutions for the selected test case are validated over the numerical solution obtained using *ode45* of MATLAB. It is seen that the four-term HPM solutions produce satisfactory results over the entire flow depth. Figure 11.5 shows a comparison between the numerical solution and the HPM approximations. The HPM-based values and the numerical solution are compared numerically in Table 11.3. A flowchart containing the steps of the method is given in Figure 11.6.

11.4.4 VALIDATION OF THE OHAM-BASED SOLUTION

For the assessment of the OHAM-based analytical solution, one needs to calculate the constant C_i. For that purpose, we calculate the residual as follows:

$$R_i\left(\xi,C_i\right) = \mathcal{L}_i\left[y_{i,\text{OHAM}}\left(\xi,C_i\right)\right] + \mathcal{N}_i\left[y_{i,\text{OHAM}}\left(\xi,C_i\right)\right] + h_i\left(\xi\right), i = 1,2 \quad (11.97)$$

where $y_{i,\text{OHAM}}\left(\xi,C_i\right)$ is the approximate solution. When $R_i\left(\xi,C_i\right) = 0$, $y_{i,\text{OHAM}}\left(\xi,C_i\right)$ becomes the exact solution to the problem. One of the ways to obtain the optimal C_i for which the solution converges is the minimization of the squared residual error, i.e.,

$$J\left(C_i\right) = \int_{\xi \epsilon D} \left[R_1^{\ 2}\left(\xi,C_i\right) + R_2^{\ 2}\left(\xi,C_i\right)\right] d\xi, \ i = 1,2,\ldots,s \quad (11.98)$$

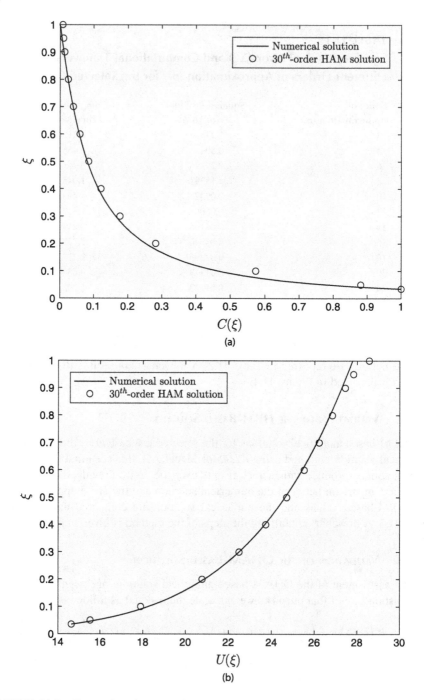

FIGURE 11.3 Comparison between the numerical solution and 30th-order HAM-based solution for the selected case: (a) concentration profile and (b) velocity profile.

TABLE 11.2

Comparison between the HAM-Based Approximation and Numerical Solution for the Selected Case

	Numerical Solution		30th-Order HAM-Based Approximation	
ξ	$C(\xi)$	$U(\xi)$	$C(\xi)$	$U(\xi)$
0.035	1.0000	14.6300	1.0000	14.6300
0.050	0.7803	15.8207	0.8807	15.5261
0.100	0.4627	18.2548	0.5733	17.8957
0.200	0.2517	20.8649	0.2819	20.7708
0.300	0.1640	22.4812	0.1767	22.4851
0.400	0.1139	23.6715	0.1215	23.7206
0.500	0.0807	24.6217	0.0852	24.7001
0.600	0.0567	25.4175	0.0589	25.5148
0.700	0.0382	26.1055	0.0390	26.2132
0.800	0.0234	26.7148	0.0240	26.8304
0.900	0.0110	27.2652	0.0135	27.4311
1.000	0.0000	27.7734	0.0080	28.5688

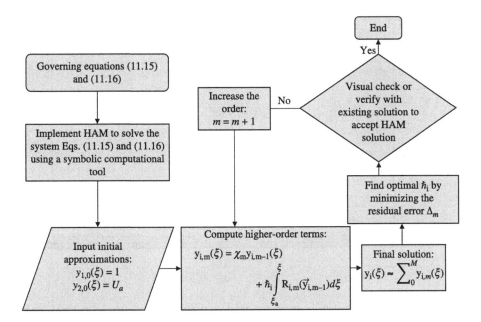

FIGURE 11.4 Flowchart for the HAM solution in Eq. (11.42).

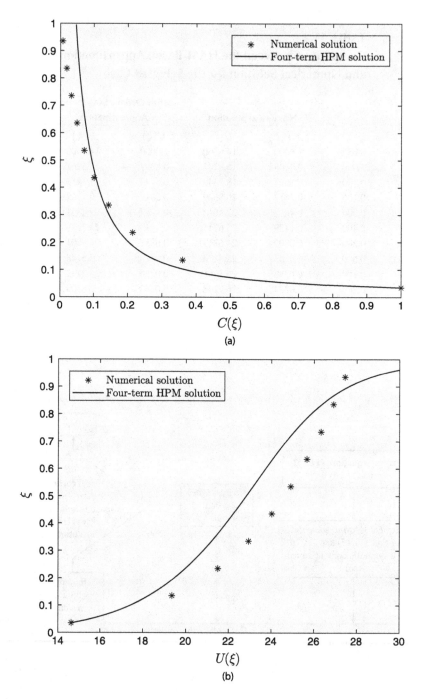

FIGURE 11.5 Comparison between the numerical solution and four-term HPM-based solution for the selected case: (a) concentration profile and (b) velocity profile.

TABLE 11.3

Comparison between the Four-Term HPM-Based Approximation and Numerical Solution for the Selected Case

	Numerical Solution		Four-Term HPM-Based Approximation	
ξ	$C(\xi)$	$U(\xi)$	$C(\xi)$	$U(\xi)$
0.035	1.0000	14.6300	1.0000	14.6300
0.050	0.7803	15.8207	0.7045	15.5551
0.100	0.4627	18.2548	0.3673	17.4263
0.200	0.2517	20.8649	0.2022	19.4694
0.300	0.1640	22.4812	0.1452	20.8110
0.400	0.1139	23.6715	0.1143	21.8797
0.500	0.0807	24.6217	0.0940	22.8245
0.600	0.0567	25.4175	0.0791	23.7330
0.700	0.0382	26.1055	0.0678	24.6913
0.800	0.0234	26.7148	0.0588	25.8440
0.900	0.0110	27.2652	0.0519	27.6117
1.000	0.0000	27.7734	0.0466	50.0673

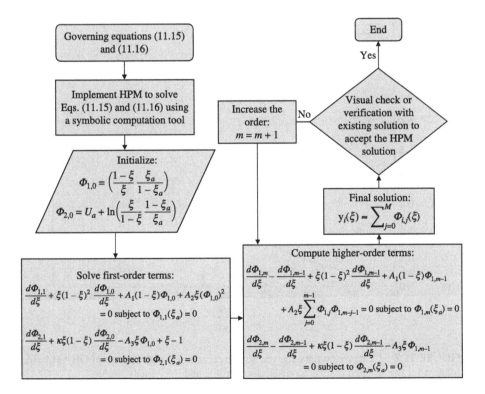

FIGURE 11.6   Flowchart for the HPM solution in Eq. (11.66).

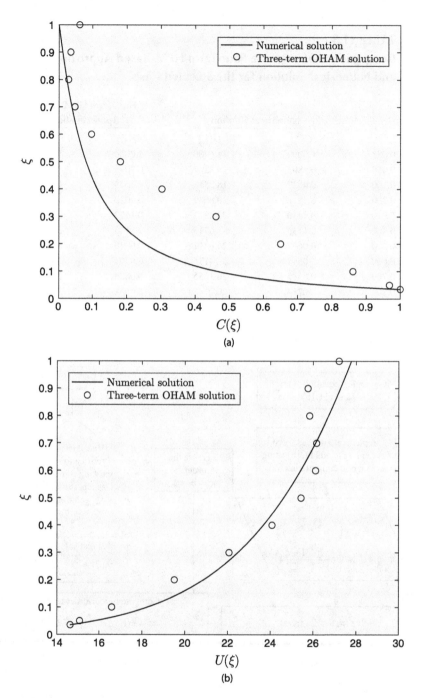

FIGURE 11.7 Comparison between the numerical solution and three-term OHAM-based solution for the selected case: (a) concentration profile and (b) velocity profile.

where D is the domain of the problem. The minimization of Eq. (11.98) leads to a system of algebraic equations as follows:

$$\frac{\partial J}{\partial C_1} = \frac{\partial J}{\partial C_2} = \cdots = \frac{\partial J}{\partial C_s} = 0 \tag{11.99}$$

Solving this system, one can obtain the optimal values of the parameters. We obtain the optimal values using the MATLAB routine *fminsearch*, which minimizes an unconstrained multivariable function. The three-term OHAM solution is computed and compared in Figure 11.7 with the numerical solution obtained using *ode45* to determine the effectiveness of the OHAM. It can be seen that the OHAM-based approximation agrees favorably well with the numerical solution for both cases. Indeed, one can expect a more accurate solution by considering other types of operators and auxiliary functions. For a quantitative assessment, we also compare the numerical values of the solutions in Table 11.4. A flowchart containing the steps of the method is provided in Figure 11.8.

11.5 CONCLUDING REMARKS

In this chapter, we derive HAM-, HPM-, and OHAM-based analytical solutions for the simultaneous distributions of flow velocity and suspended sediment concentration in sediment-laden, open-channel flow. Here, the classical model developed by Herrman and Madsen (2007) is selected, which accounts for the stratification effects in flow-carrying sediment. The governing equations constituted a system of coupled

TABLE 11.4

Comparison between Three-Term OHAM-Based Approximation and Numerical Solution for the Selected Case

	Numerical Solution		Three-Term OHAM-Based Approximation	
ξ	$c(\xi)$	$u(\xi)$	$c(\xi)$	$u(\xi)$
0.035	1.0000	14.6300	1.0000	14.6300
0.050	0.7803	15.8207	0.9682	15.0793
0.100	0.4627	18.2548	0.8609	16.5883
0.200	0.2517	20.8649	0.6501	19.5087
0.300	0.1640	22.4812	0.4602	22.0685
0.400	0.1139	23.6715	0.3023	24.0813
0.500	0.0807	24.6217	0.1811	25.4399
0.600	0.0567	25.4175	0.0970	26.1093
0.700	0.0382	26.1055	0.0475	26.1558
0.800	0.0234	26.7148	0.0285	25.8334
0.900	0.0110	27.2652	0.0349	25.7571
1.000	0.0000	27.7734	0.0614	27.2009

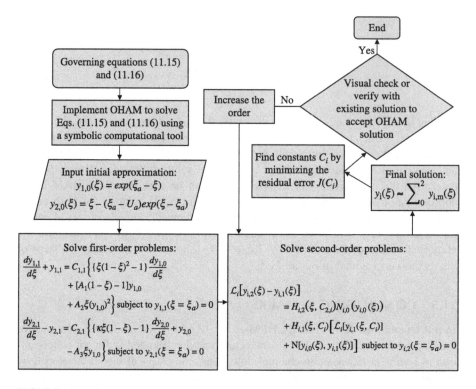

FIGURE 11.8 Flowchart for the OHAM solution in Eq. (11.84).

nonlinear ODEs, which is solved using three different analytical methods, and solutions are provided in terms of a series. It is shown that all three methods perform well to predict the flow variables throughout the entire flow depth. The numerical assessment is also carried out for the selected case considered in the study.

REFERENCES

Cheng, N. S. (1997). Simplified settling velocity formula for sediment particle. *Journal of Hydraulic Engineering, 123*(2), 149–152.

Coleman, N. L. (1981). Velocity profiles with suspended sediment. *Journal of Hydraulic Research, 19*(3), 211–229.

Herrmann, M. J., and Madsen, O. S. (2007). Effect of stratification due to suspended sand on velocity and concentration distribution in unidirectional flows. *Journal of Geophysical Research: Oceans, 112*, C02006, doi:10.1029/2006JC003569.

Mohan, S., Kumbhakar, M., Ghoshal, K., and Kumar, J. (2019). Semianalytical solution for simultaneous distribution of fluid velocity and sediment concentration in open-channel flow. *Journal of Engineering Mechanics, 145*(11), 04019090.

Monin, A., and Yaglom, A. (1971). *Vol. I of statistical fluid dynamics*. MIT Press, Cambridge, MA.

Pal, D., and Ghoshal, K. (2016). Effect of particle concentration on sediment and turbulent diffusion coefficients in open-channel turbulent flow. *Environmental Earth Sciences, 75*(18), 1–11.

Rijn, L. C. V. (1984). Sediment transport, part II: Suspended load transport. *Journal of Hydraulic Engineering*, *110*(11), 1613–1641.

Rouse, H. (1937). Modern conceptions of the mechanics of fluid turbulence. *Transactions of the American Society of Civil Engineers*, *102*(1), 463–505.

Smith, J. D., and McLean, S. R. (1977). Spatially averaged flow over a wavy surface. *Journal of Geophysical Research*, *82*(12), 1735–1746.

Zhao, Y., and Liao, S. (2002). User guide to BVPh 2.0. School of naval architecture, ocean and civil engineering, Shanghai, *40*.

FURTHER READING

Aminikhah, H. (2012). The combined Laplace transform and new homotopy perturbation methods for stiff systems of ODEs. *Applied Mathematical Modelling*, *36*(8), 3638–3644.

Aminikhah, H., and Hemmatnezhad, M. (2011). An effective modification of the homotopy perturbation method for stiff systems of ordinary differential equations. *Applied Mathematics Letters*, *24*(9), 1502–1508.

Anakira, N. (2019, April). Solution of system of ordinary differential equations by optimal homotopy asymptotic method. In *AIP Conference Proceedings* (Vol. 2096, No. 1, p. 020023). AIP Publishing LLC.

Bataineh, A. S., Noorani, M. S. M., and Hashim, I. (2008). Solving systems of ODEs by homotopy analysis method. *Communications in Nonlinear Science and Numerical Simulation*, *13*(10), 2060–2070.

Bataineh, A. S., Noorani, M. S. M., and Hashim, I. (2009). Modified homotopy analysis method for solving systems of second-order BVPs. *Communications in Nonlinear Science and Numerical Simulation*, *14*(2), 430–442.

Biazar, J., Asadi, M. A., and Salehi, F. (2015). Rational homotopy perturbation method for solving stiff systems of ordinary differential equations. *Applied Mathematical Modelling*, *39*(3–4), 1291–1299.

Chien, N., and Wan, Z. (1999, June). *Mechanics of sediment transport*. American Society of Civil Engineers.

Darvishi, M. T., and Khani, F. (2008). Application of He's homotopy perturbation method to stiff systems of ordinary differential equations. *Zeitschrift Fuer Naturforschung A*, *63*(1–2), 19–23.

Dey, S. (2014). *Fluvial hydrodynamics*. Springer, Berlin.

Einstein, H., and Chien, N. (1955). *Effects of heavy sediment concentration near the bed on velocity and sediment distribution*. MRD Sediment Series No. 8. University of California, Berkeley, CA.

Ganji, D. D., Mirgolbabaei, H., Miansari, M., and Miansari, M. (2008). Application of homotopy perturbation method to solve linear and non-linear systems of ordinary differential equations and differential equation of order three. *Journal of Applied Sciences*, *8*(7), 1256–1261.

Ghoshal, K., and Mazumder, B. S. (2005). Sediment-induced stratification in turbulent open-channel flow. *Environmetrics: The Official Journal of the International Environmetrics Society*, *16*(7), 673–686.

Hashim, I., and Chowdhury, M. S. H. (2008). Adaptation of homotopy-perturbation method for numeric–analytic solution of system of ODEs. *Physics Letters A*, *372*(4), 470–481.

Haq, S., and Ishaq, M. (2014). Solution of coupled Whitham–Broer–Kaup equations using optimal homotopy asymptotic method. *Ocean Engineering*, *84*, 81–88.

Huang, S. H., Sun, Z. L., Xu, D., and Xia, S. S. (2008). Vertical distribution of sediment concentration. *Journal of Zhejiang University-Science A*, *9*(11), 1560–1566.

Hunt, J. N. (1954). The turbulent transport of suspended sediment in open channels. *Proceedings of the Royal Society of London. Series A. Mathematical and Physical Sciences*, *224*(1158), 322–335.

Jobson, H. E., and Sayre, W. W. (1970). Vertical transfer in open channel flow. *Journal of the Hydraulics Division*, *96*(3), 703–724.

Lane, E. W., and Kalinske, A. A. (1941). Engineering calculations of suspended sediment. *Eos, Transactions American Geophysical Union*, *22*(3), 603–607.

Maude, A. D., and Whitmore, R. L. (1958). A generalized theory of sedimentation. *British Journal of Applied Physics*, *9*(12), 477.

Mazumder, B. S., and Ghoshal, K. (2002). Velocity and suspension concentration in sediment-mixed fluid. 国际泥沙研究: 英文版, *17*(3), 220–232.

Nandeppanavar, M. M., Vajravelu, K., and Datti, P. S. (2016). Optimal homotopy asymptotic solutions for nonlinear ordinary differential equations arising in flow and heat transfer due to nonlinear stretching sheet. *Heat Transfer—Asian Research*, *45*(1), 15–29.

Pal, D., and Ghoshal, K. (2016). Vertical distribution of fluid velocity and suspended sediment in open channel turbulent flow. *Fluid Dynamics Research*, *48*(3), 035501.

Pal, D., and Ghoshal, K. (2017). Hydrodynamic interaction in suspended sediment distribution of open channel turbulent flow. *Applied Mathematical Modelling*, *49*, 630–646.

Rashidi, M. M., Pour, S. M., Hayat, T., and Obaidat, S. (2012). Analytic approximate solutions for steady flow over a rotating disk in porous medium with heat transfer by homotopy analysis method. *Computers & Fluids*, *54*, 1–9.

Saadatmandi, A., Dehghan, M., and Eftekhari, A. (2009). Application of He's homotopy perturbation method for non-linear system of second-order boundary value problems. *Nonlinear Analysis: Real World Applications*, *10*(3), 1912–1922.

Tsai, C. H., and Tsai, C. T. (2000). Velocity and concentration distributions of sediment-laden open channel flow 1. *JAWRA Journal of the American Water Resources Association*, *36*(5), 1075–1086.

Villaret, C., and Trowbridge, J. H. (1991). Effects of stratification by suspended sediments on turbulent shear flows. *Journal of Geophysical Research: Oceans*, *96*(C6), 10659–10680.

Wright, S., and Parker, G. (2004). Flow resistance and suspended load in sand-bed rivers: Simplified stratification model. *Journal of Hydraulic Engineering*, *130*(8), 796–805.

Part IV

---

*Partial Differential Equations
(Single and System)*

12 Unsteady Confined Radial Ground-Water Flow Equation

12.1 INTRODUCTION

Groundwater flow in confined aquifers is described by coupling the continuity equation and the Darcy flux law. The resulting equation is a partial differential equation of diffusion type. Under simplified assumptions, approximate analytical solutions have been derived. One common assumption is that the confined aquifer is of uniform thickness and is homogeneous and isotropic. Using the Boltzmann transformation, Theis (1935) transformed the partial differential equation into an ordinary differential equation and then derived a solution in terms of an exponential integral, which came to be known as a well function, and gave tabular values of this function for different values of the Boltzmann variable or the argument of the well function. Expressing the exponential integral as a power series and considering only two terms of the series, an approximate solution has also been proposed. This chapter derives a series solution to the Theis equation using homotopy-based methods: homotopy analysis method (HAM), homotopy perturbation method (HPM), and optimal homotopy asymptotic method (OHAM).

12.2 GOVERNING EQUATION

We consider a confined aquifer's simple configuration, which is horizontal, infinite in a horizontal extent, of constant thickness, and homogeneous and isotropic. It is further assumed that the aquifer has a single pumping well, the rate of pumping is constant with respect to time, the diameter of the well is infinitesimally small, the well penetrates the entire aquifer, and the hydraulic head in the aquifer before pumping is uniform throughout the aquifer.

For transient flow in a saturated porous medium, the conservation of mass requires that the net rate of fluid mass flow into an arbitrary control volume be equal to the net rate of fluid mass out of the control volume and the time rate of change of fluid mass storage within that control volume. Combining the continuity equation with Darcy's law, the partial differential equation representing saturated flow in a horizontal confined aquifer with transmittivity T and storativity S can be given as (Walton 1970):

$$\frac{\partial^2 h}{\partial x^2} + \frac{\partial^2 h}{\partial y^2} = \frac{S}{T}\frac{\partial h}{\partial t} \tag{12.1}$$

where x and y are the spatial variables, t is the temporal variable, and h is the hydraulic head. Due to the radial symmetry of the hydraulic-head drawdown around

DOI: 10.1201/9781003368984-16

a well, it is convenient to convert Eq. (12.1) into radial coordinates. For that, we use $r = \sqrt{x^2 + y^2}$, where r is the radial coordinate, and obtain:

$$\frac{\partial h}{\partial x} = \frac{\partial h}{\partial r}\frac{\partial r}{\partial x} = \frac{x}{r}\frac{\partial h}{\partial r} \tag{12.2}$$

$$\frac{\partial^2 h}{\partial x^2} = \frac{\partial}{\partial x}\left[\frac{x}{r}\frac{\partial h}{\partial r}\right] = \frac{x^2}{r^2}\frac{\partial^2 h}{\partial r^2} + \left(\frac{1}{r} - \frac{x^2}{r^3}\right)\frac{\partial h}{\partial r} \tag{12.3}$$

$$\frac{\partial h}{\partial y} = \frac{\partial h}{\partial r}\frac{\partial r}{\partial y} = \frac{y}{r}\frac{\partial h}{\partial r} \tag{12.4}$$

$$\frac{\partial^2 h}{\partial y^2} = \frac{\partial}{\partial y}\left[\frac{y}{r}\frac{\partial h}{\partial r}\right] = \frac{y^2}{r^2}\frac{\partial^2 h}{\partial r^2} + \left(\frac{1}{r} - \frac{y^2}{r^3}\right)\frac{\partial h}{\partial r} \tag{12.5}$$

Using Eqs. (12.2)–(12.5) in Eq. (12.1), the governing equation becomes:

$$\frac{\partial^2 h}{\partial r^2} + \frac{1}{r}\frac{\partial h}{\partial r} = \frac{S}{T}\frac{\partial h}{\partial t} \tag{12.6}$$

Using the radial coordinate, the space dimensions are converted into a one-dimensional line starting from the well center at $r = 0$ to the infinite extremity at $r = \infty$. The schematic is given in Figure 12.1.

The initial condition can be given as:

$$h(r,0) = h_0 \text{ for all } r \tag{12.7}$$

where h_0 is the constant initial hydraulic head.

The boundary conditions can be prescribed as follows. It is assumed that no hydraulic-head drawdown is there at the infinite extremity, i.e.,

$$h(\infty,t) = h_0 \text{ for all } t \tag{12.8}$$

Again, a constant pumping rate is assumed at the wall:

$$\lim_{r \to 0}\left(r\frac{\partial h}{\partial r}\right) = \frac{Q}{2\pi T} \text{ for all } t \tag{12.9}$$

The condition given by Eq. (12.9) can be attributed to the application of Darcy's law at the well face. The solution $h(r,t)$ represents the hydraulic head at any radial distance r and time t after the start of pumping. The solutions are often presented in terms of the drawdown expressed in head $h_0 - h(r,t)$, which can also be seen from Figure 12.1.

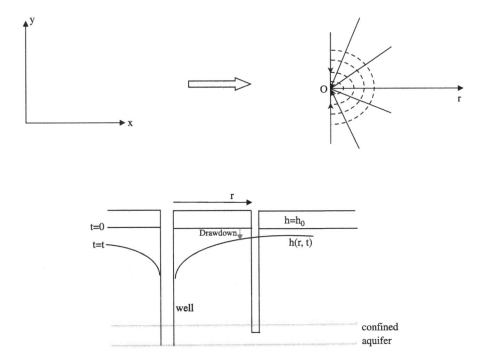

FIGURE 12.1   Radial flow to a well in a horizontal confined aquifer.

12.3   THEIS SOLUTION

Theis (1935) provided an analytical solution to Eq. (12.6) together with the initial and boundary conditions in Eqs. (12.7)–(12.9). Theis used an analogy to heat conduction in solids to arrive at the solution, which is expressed as:

$$h_0 - h(r,t) = \frac{Q}{4\pi T} \int_u^\infty \frac{\exp(-u)}{u} du \tag{12.10}$$

where $u = \dfrac{r^2 S}{4Tt}$. Equation (12.10) is known as the *Theis solution*. In general, one can obtain the solution for Eq. (12.10) using different techniques, such as the Laplace transform, Fourier transform, similarity transformation, etc. The integral in the right side of Eq. (12.10) is known as the *exponential integral*. In relation to the present context, it is also known as the *well function* and is denoted by $W(u)$. Equation (12.10) is therefore written as:

$$h_0 - h(r,t) = \frac{Q}{4\pi T} W(u) \tag{12.11}$$

The function $W(u)$ exhibits many interesting properties. For example:

$$W(-\infty) = -\infty, \; W(0) = +\infty, \; W(+\infty) = 0, \; W(u) = \Gamma(0,u) \tag{12.12}$$

where $\Gamma(s,u)$ is the incomplete gamma function, defined as:

$$\Gamma(s,u) = \int_{u}^{\infty} t^{s-1} \exp(-t) \, dt \qquad (12.13)$$

There are two convergent series for the integral $W(u)$. One of them can be expressed as follows (Jacob 1940; Abramowitz and Stegun 1970, p. 229):

$$W(u) = -\gamma - \ln u + \sum_{k=1}^{\infty} \frac{(-1)^{k+1} u^k}{k \, k!} \quad |Arg(u)| < \pi \qquad (12.14)$$

where γ is the Euler-Mascheroni constant, whose numerical value up to four decimal places is 0.5772. Equation (12.14) can be used to compute the value of the integral. However, if we consider a few terms, the series gives inaccurate results for $u > 2.5$ due to cancellation. In Figure 12.2, we test the performance of the series in Eq. (12.14) by comparing it with the numerical solution of the exponential integral. The numerical values of $W(u)$ are calculated using the MATLAB function *integral*, which uses the global adaptive quadrature rule (Shampine 2008). It can be seen from the figure that even if the series is convergent, its convergence rate is too slow. Specifically, to get an accurate approximation for u up to 6.5, we need 40 terms of the series. Therefore, it may not be practically efficient for large values of u.

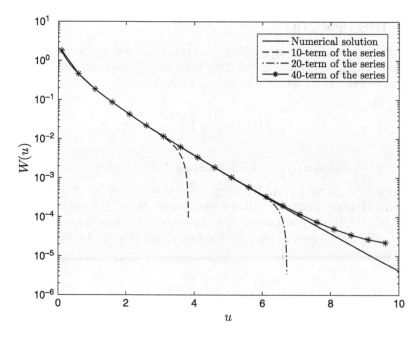

FIGURE 12.2 Comparison between the numerical solution and 10, 20, and 40 terms of the series in Eq. (12.14) for $W(u)$.

A more rapidly convergent series, credited to S. Ramanujan, is given by (Berndt 1994):

$$Ei(u) = \gamma + \ln u + \exp(u/2) \sum_{n=1}^{\infty} \frac{(-1)^{n-1} u^n}{n! 2^{n-1}} \sum_{k=0}^{[(n-1)/2]} \frac{1}{2k+1} \qquad (12.15)$$

where $[(n-1)/2]$ denotes the *floor function*. The integral $W(u)$ is related to Eq. (12.15) as follows:

$$W(u) = -Ei(-u) \qquad (12.16)$$

Equation (12.15) gives good estimates for the integral for small values of u. However, both Eqs. (12.14) and (12.15) are not useful for larger values of u. For large values of u, there is an asymptotic series approximation as follows (Masina 2019):

$$W(u) = \frac{\exp(-u)}{u} \left(\sum_{n=0}^{N-1} \frac{n!}{(-z)^n} + \mathcal{O}\left(|u|^{-N}\right) \right) \qquad (12.17)$$

where $\mathcal{O}$ denotes the 'Big-O' notation. Equation (12.17) can be obtained by expanding $W(u)$ in parts. Figure 12.3 compares the numerical solution and the series approximation in Eq. (12.17) of $W(u)$ for several terms of the series. As expected, the series performs well for large values of u. However, the series produces infeasible

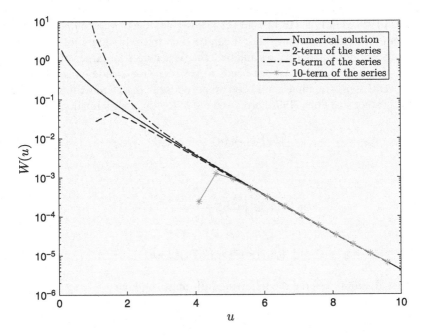

FIGURE 12.3 Comparison between the numerical solution and 2, 5, and 10 terms of the series in Eq. (12.17) for $W(u)$.

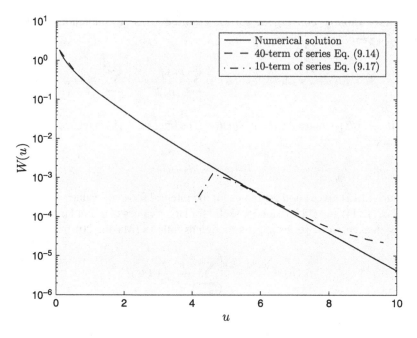

FIGURE 12.4 Comparison between the numerical solution, 40 terms of the series in Eq. (12.14), and 10 terms of the series in Eq. (12.17) for $W(u)$.

(or inaccurate) results for small values of u. To get a comparative idea of the performances of series in Eqs. (12.14) and (12.17), we plot together the numerical solution and both the series in Figure 12.4. It can be seen from the figure that one of them performs well for smaller values and the other is accurate for large values of u.

For practical applications, one needs to get $W(u)$ for a wide range of u. To that end, several approximations have been proposed for $W(u)$. Here we mention some of them. Swamee and Ojha (1990) proposed the following approximation:

$$W(u) = \left[A_1(u)^{-7.7} + B_1(u) \right]^{-0.13} \tag{12.18}$$

where

$$A_1(u) = \ln\left[\left(0.65 + \frac{0.56146}{u} \right)(1+u) \right] \tag{12.19}$$

$$B_1(u) = u^4 exp(7.7u)(2+u)^{3.7} \tag{12.20}$$

Barry et al. (2000) offered the following full-range solution:

$$W(u) = \frac{exp(-u)}{B_1 + (1-B_1)exp\left(-\dfrac{u}{1-B_1} \right)} \ln\left[1 + \frac{B_1}{u} - \frac{1-B_1}{(B_2(u)+B_3u)^2} \right] \tag{12.21}$$

where

$$B_1 = \exp(-\gamma), \; B_3 = \sqrt{\frac{2(1-B_1)}{B_1(2-B_1)}}, \; \gamma \text{ is the Euler} - \text{Mascheroni constant} \quad (12.22)$$

$$B_2(u) = \frac{1}{1+u\sqrt{u}} + \frac{\widehat{B}_2 \widetilde{B}_2}{1+\widetilde{B}_2}, \; \widehat{B}_2 = \frac{(1-B_1)(B_1^2 - 6B_1 + 12)}{3B_1(2-B_1)^2 B_3}, \; \widetilde{B}_2 = \frac{20}{47} u^{\sqrt{\frac{31}{26}}} \quad (12.23)$$

In a recent study, Vatankhah (2014) proposed the following approximation:

$$W(u) = \left\{ \left[\left(1 + c_1 u^{c_2}\right) \ln\left(\frac{c_3}{u} + c_4\right) \right]^{-p} + \left[\frac{1}{u \exp(u)} \left(\frac{u+c_5}{u+c_6}\right) \right]^{-p} \right\}^{-\frac{1}{p}} \quad (12.24)$$

where $p = 2$, $c_1 = -0.19$, $c_2 = 0.7$, $c_3 = 0.565$, $c_4 = 4$, $c_5 = 0.444$, and $c_6 = 1.384$. To get a comparative idea of the performances of the approximations given by Swamee and Ojha (1990), Barry et al. (2000), and Vatankhah (2014), we calculate the percentage error as PE $(\%) = 100 \times \dfrac{(W_{\text{num}} - W_{\text{apprx}})}{W_{\text{num}}}$ where W_{num} and W_{apprx} are the values of $W(u)$ obtained from the numerical method and the corresponding approximation, respectively. The percentage errors are calculated for the approximations and compared in Figure 12.5 (a), (b), and (c). It can be seen that the approximation given by Vatankhah (2014) performs better as compared to the others.

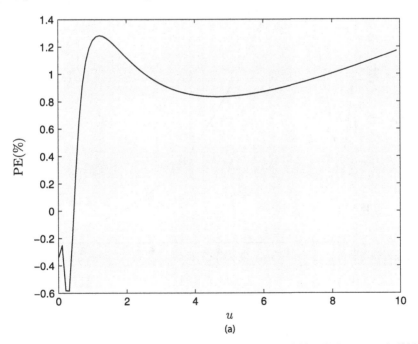

FIGURE 12.5   Percentage errors for (a) Swamee and Ojha (1990), (b) Barry et al. (2000), and (c) Vatankhah (2014) for $u \in [0.01, 10]$. (*Continued*)

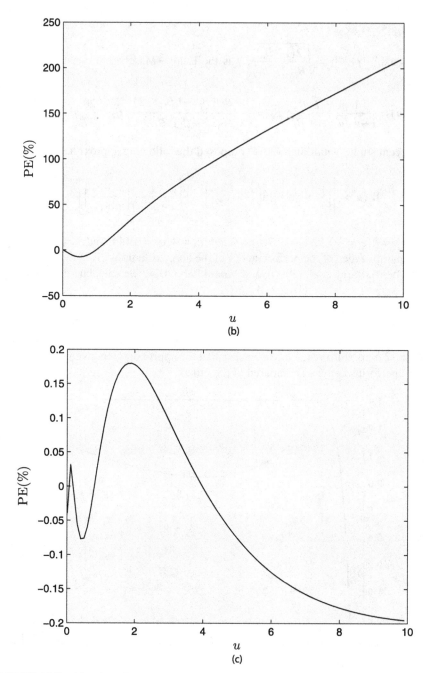

FIGURE 12.5 (*Continued*)

12.4   HAM SOLUTION

Here, we use HAM to obtain a convergent series solution for the governing equation. It may be noted that it is always advantageous to solve an ordinary differential equation (ODE) using HAM rather than a partial differential equation (PDE). PDE involves more than one independent variable, which makes it difficult to choose the linear operator and carry out the initial approximations. For the present problem, it is possible to convert the original governing Eq. (12.6) into an ODE using the following similarity transformation:

$$u = \frac{r^2 S}{4tT} \tag{12.25}$$

Then

$$\frac{\partial h}{\partial r} = \frac{dh}{du}\frac{\partial u}{\partial r} = \frac{2rS}{4tT}\frac{dh}{du} \Rightarrow \frac{1}{r}\frac{\partial h}{\partial r} = \frac{2S}{4tT}\frac{dh}{du} \tag{12.26}$$

$$\frac{\partial^2 h}{\partial r^2} = \frac{\partial}{\partial r}\left[\frac{2rS}{4tT}\frac{dh}{du}\right] = \frac{2S}{4tT}\frac{dh}{du} + \frac{4r^2 S^2}{16t^2 T^2}\frac{d^2 h}{du^2} \tag{12.27}$$

$$\frac{\partial h}{\partial t} = \frac{dh}{du}\frac{\partial u}{\partial t} = -\frac{r^2 S}{4t^2 T}\frac{dh}{du} \tag{12.28}$$

Using Eqs. (12.26)–(12.28), the governing Eq. (12.6) becomes:

$$\frac{4r^2 S^2}{16t^2 T^2}\frac{d^2 h}{du^2} + \left[\frac{2S}{4tT} + \frac{2S}{4tT}\right]\frac{dh}{du} + \frac{S}{T}\frac{r^2 S}{4t^2 T}\frac{dh}{du} = 0$$

$$\Rightarrow \frac{S}{tT}\left[\frac{r^2 S}{4tT}\frac{d^2 h}{du^2} + \left(1 + \frac{r^2 S}{4tT}\right)\frac{dh}{du}\right] = 0 \Rightarrow u\frac{d^2 h}{du^2} + (1+u)\frac{dh}{du} = 0 \tag{12.29}$$

Following Eq. (12.25), the conditions in Eqs. (12.7) and (12.8) change to:

$$h(u \to \infty) = h_0 \tag{12.30}$$

and Eq. (12.9) becomes:

$$\lim_{r \to 0}\left(r\frac{\partial h}{\partial r}\right) = \frac{Q}{2\pi T} \Rightarrow \lim_{u \to 0}\left(u\frac{dh}{du}\right) = \frac{Q}{4\pi T} \tag{12.31}$$

To be in line with Theis's method, we intend to obtain the solution as $h_0 - h(u)$. For that, we use the transformation:

$$h(u) = h_0 + \overline{h}(u) \tag{12.32}$$

Accordingly, Eq. (12.29) becomes:

$$u\frac{d^2\overline{h}}{du^2}+(1+u)\frac{d\overline{h}}{du}=0 \qquad (12.33)$$

The boundary conditions in Eqs. (12.30) and (12.31) become:

$$\overline{h}(u\rightarrow\infty)=0 \qquad (12.34)$$

$$\lim_{u\to0}\left(u\frac{d\overline{h}}{du}\right)=-\frac{Q}{4\pi T} \qquad (12.35)$$

To solve Eq. (12.33) using HAM subject to the boundary conditions in Eqs. (12.34) and (12.35), the zeroth-order deformation equation can be constructed as follows:

$$(1-q)\mathcal{L}\left[\Phi(u;q)-\overline{h}_0(u)\right]=q\hbar H(u)\mathcal{N}\left[\Phi(u;q)\right] \qquad (12.36)$$

subject to the boundary conditions:

$$\Phi(\infty;q)=0 \qquad (12.37)$$

$$\lim_{u\to0}\left(u\frac{d\Phi}{du}\right)=-\frac{Q}{4\pi T} \qquad (12.38)$$

where q is the embedding parameter, $\Phi(u;q)$ is the representation of the solution across q, $\overline{h}_0(u)$ is the initial approximation, $\hbar$ is the auxiliary parameter, $H(u)$ is the auxiliary function, and $\mathcal{L}$ and $\mathcal{N}$ are the linear and nonlinear operators, respectively. The higher-order terms can be calculated from the higher-order deformation equation, given as follows:

$$\mathcal{L}\left[\overline{h}_m(u)-\chi_m\overline{h}_{m-1}(u)\right]=\hbar H(u)R_m\left(\vec{\overline{h}}_{m-1}\right),\ m=1,2,3,\ldots$$

subject to:

$$\overline{h}_m(\infty)=0,\ \lim_{u\to0}\left(u\frac{d\,\overline{h}_m}{du}\right)=0 \qquad (12.39)$$

where

$$\chi_m=\begin{cases}0 & \text{when } m=1,\\ 1 & \text{otherwise}\end{cases} \qquad (12.40)$$

and:

$$R_m\left(\vec{\overline{h}}_{m-1}\right)=\frac{1}{(m-1)!}\frac{\partial^{m-1}\mathcal{N}\left[\Phi(u;q)\right]}{\partial q^{m-1}}\Bigg|_{q=0} \tag{12.41}$$

where $\overline{h}_m$ for $m\geq1$ is the higher-order term. Now, the final solution can be obtained as follows:

$$\overline{h}(u)=\overline{h}_0(u)+\sum_{m=1}^{\infty}\overline{h}_m(u) \tag{12.42}$$

The nonlinear operator for the problem is selected as:

$$\mathcal{N}\left[\Phi(u;q)\right]=u\frac{\partial^2\Phi(u;q)}{\partial u^2}+(1+u)\frac{\partial\Phi(u;q)}{\partial u} \tag{12.43}$$

Therefore, using Eq. (12.43), the term R_m can be calculated from Eq. (12.41) as follows:

$$R_m\left(\vec{\overline{h}}_{m-1}\right)=u\frac{d^2\overline{h}_{m-1}}{du^2}+(1+u)\frac{d\overline{h}_{m-1}}{du} \tag{12.44}$$

As discussed in Chapter 2, it may not always be possible to choose base functions based on the physical background of the problem and other factors. In those cases, one needs to test the solution using different sets of operators and auxiliary functions. To that end, for the present equation, the following set of base functions is chosen to represent the solution:

$$\left\{\exp(-u)\ln(u)+\left[W(u)+\exp(-u)\right]u^m \mid m=0,1,2,...\right\} \tag{12.45}$$

so that:

$$\overline{h}(u)=a_0\exp(-u)\ln(u)+\left[W(u)+\exp(-u)\right]\sum_{m=0}^{\infty}b_m(u)^m \tag{12.46}$$

where a_0 and b_m are the coefficients of the series. Equation (12.46) provides the so-called *rule of solution expression*. Following the rule of solution expression, the linear operator and the initial approximation are chosen, respectively, as follows:

$$\mathcal{L}\left[\Phi(u;q)\right]=\frac{\partial^2\Phi(u;q)}{\partial u^2} \text{ with the property } \mathcal{L}\left[C_2u+C_3\right]=0 \tag{12.47}$$

$$\overline{h}_0(u)=-\frac{Q}{4\pi T}\exp(-u)\ln(u) \tag{12.48}$$

where C_2 and C_3 are integral constants. Using Eq. (12.47), the higher-order terms can be obtained from Eq. (12.39) as follows:

$$\bar{h}_m(u) = \chi_m \bar{h}_{m-1}(u) + \hbar \int_0^u \int_0^u H(x) R_m\left(\vec{\bar{h}}_{m-1}\right) dx dy + C_2 u + C_3, \ m = 1,2,3,\ldots \quad (12.49)$$

where R_m is given by Eq. (12.44) and constants C_2 and C_3 can be determined from the boundary conditions for the higher-order deformation equations given by Eq. (12.39).

Now, the auxiliary function $H(u)$ can be determined from the *rule of coefficient ergodicity*. Based on the rule of solution expression in Eq. (12.46), the function $H(u)$ can take on many values. However, for simplicity, we take $H(u)$ as simply 1. This can always be done because of the freedom that HAM gives, as discussed by Vajravelu and van Gorder (2013). Also, this consideration simplifies the computation. Finally, the approximate solution can be obtained as follows:

$$\bar{h}(u) = h_0 - h(u) \approx \bar{h}_0(u) + \sum_{m=1}^{M} \bar{h}_m(u) \quad (12.50)$$

12.5   HPM-BASED SOLUTION

When applying HPM, we rewrite Eq. (12.33) in the following form:

$$\mathcal{N}\left(\bar{h}\right) = f(u) \quad (12.51)$$

We construct a homotopy $\Phi(u;q)$ that satisfies:

$$(1-q)\left[\mathcal{L}\left(\Phi(u;q)\right) - \mathcal{L}\left(\bar{h}_0(u)\right)\right] + q\left[\mathcal{N}\left(\Phi(u;q)\right) - f(u)\right] = 0 \quad (12.52)$$

where q is the embedding parameter that lies in $[0,1]$, $\mathcal{L}$ is the linear operator, and $\bar{h}_0(u)$ is the initial approximation to the final solution. Equation (12.52) shows that as $q = 0$, $\Phi(u;0) = \bar{h}_0(u)$ and as $q = 1$, $\Phi(u;1) = \bar{h}(u)$. Let us now express $\Phi(u;q)$ as a series in terms of q as follows:

$$\Phi(u;q) = \Phi_0 + q\Phi_1 + q^2\Phi_2 + q^3\Phi_3 + q^4\Phi_4 \ldots \quad (12.53)$$

where Φ_m for $m \geq 1$ is the higher-order term. As $q \to 1$, Eq. (12.53) produces the final solution as:

$$\bar{h}(u) = \lim_{q \to 1} \Phi(u;q) = \sum_{i=0}^{\infty} \Phi_i \quad (12.54)$$

For simplicity, we consider the linear and nonlinear operators as follows:

$$\mathcal{L}\left(\Phi(u;q)\right) = \frac{\partial^2 \Phi(u;q)}{\partial u^2} \quad (12.55)$$

$$\mathcal{N}\big(\Phi(u;q)\big) = u\frac{\partial^2 \Phi(u;q)}{\partial u^2} + (1+u)\frac{\partial \Phi(u;q)}{\partial u} \tag{12.56}$$

Using these expressions, Eq. (12.52) takes on the form:

$$(1-q)\left[\frac{\partial^2 \Phi(u;q)}{\partial u^2} - \frac{d^2\bar{h}_0}{du^2}\right] + q\left[u\frac{\partial^2 \Phi(u;q)}{\partial u^2} + (1+u)\frac{\partial \Phi(u;q)}{\partial u}\right] = 0 \tag{12.57}$$

Using Eq. (12.53), we have:

$$(1-q)\left[\frac{\partial^2 \Phi(u;q)}{\partial u^2} - \frac{d^2\bar{h}_0}{du^2}\right]$$

$$= (1-q)\left[\left(\frac{d^2\Phi_0}{du^2} - \frac{d^2\bar{h}_0}{du^2}\right) + q\frac{d^2\Phi_1}{du^2} + q^2\frac{d^2\Phi_2}{du^2} + q^3\frac{d^2\Phi_3}{du^2} + q^4\frac{d^2\Phi_4}{du^2} + \cdots\right]$$

$$= \left(\frac{d^2\Phi_0}{du^2} - \frac{d^2\bar{h}_0}{du^2}\right) + q\left(\frac{d^2\Phi_1}{du^2} - \left(\frac{d^2\Phi_0}{du^2} - \frac{d^2\bar{h}_0}{du^2}\right)\right) + q^2\left(\frac{d^2\Phi_2}{du^2} - \frac{d^2\Phi_1}{du^2}\right)$$

$$+ q^3\left(\frac{d^2\Phi_3}{du^2} - \frac{d^2\Phi_2}{du^2}\right) + q^4\left(\frac{d^2\Phi_4}{du^2} - \frac{d^2\Phi_3}{du^2}\right) + \cdots \tag{12.58}$$

$$u\frac{\partial^2 \Phi(u;q)}{\partial u^2} + (1+u)\frac{\partial \Phi(u;q)}{\partial u}$$

$$= u\left(\frac{d^2\Phi_0}{du^2} + q\frac{d^2\Phi_1}{du^2} + q^2\frac{d^2\Phi_2}{du^2} + q^3\frac{d^2\Phi_3}{du^2} + q^4\frac{d^2\Phi_4}{du^2} + \cdots\right)$$

$$+ (1+u)\left(\frac{d\Phi_0}{du} + q\frac{d\Phi_1}{du} + q^2\frac{d\Phi_2}{du} + q^3\frac{d\Phi_3}{du} + q^4\frac{d\Phi_4}{du} + \cdots\right)$$

$$= \left[u\frac{d^2\Phi_0}{du^2} + (1+u)\frac{d\Phi_0}{du}\right] + q\left[u\frac{d^2\Phi_1}{du^2} + (1+u)\frac{d\Phi_1}{du}\right]$$

$$+ q^2\left[u\frac{d^2\Phi_2}{du^2} + (1+u)\frac{d\Phi_2}{du}\right] + q^3\left[u\frac{d^2\Phi_3}{du^2} + (1+u)\frac{d\Phi_3}{du}\right] + \cdots \tag{12.59}$$

Using the boundary conditions $\bar{h}(u \to \infty) = 0$ and $\lim\limits_{u \to 0}\left(u\frac{d\bar{h}}{du}\right) = -\frac{Q}{4\pi T}$, we have:

$$\Phi_0(u \to \infty) = 0,\ \Phi_1(u \to \infty) = 0,\ \Phi_2(u \to \infty) = 0,\ \dots. \tag{12.60}$$

$$\lim\limits_{u \to 0}\left(u\frac{d\Phi_0}{du}\right) = -\frac{Q}{4\pi T},\ \lim\limits_{u \to 0}\left(u\frac{d\Phi_1}{du}\right) = 0,\ \lim\limits_{u \to 0}\left(u\frac{d\Phi_2}{du}\right)$$

$$= 0,\ \lim\limits_{u \to 0}\left(u\frac{d\Phi_3}{du}\right) = 0,\dots \tag{12.61}$$

Using Eqs. (12.58) and (12.59) and equating the like powers of q in Eq. (12.57), the following system of differential equations is obtained:

$$\frac{d^2\Phi_0}{du^2} - \frac{d^2\bar{h}_0}{du^2} = 0 \text{ subject to } \Phi_0(u \to \infty) = 0, \lim_{u \to 0}\left(u\frac{d\Phi_0}{du}\right) = -\frac{Q}{4\pi T} \quad (12.62)$$

$$\frac{d^2\Phi_1}{du^2} - \left(\frac{d^2\Phi_0}{du^2} - \frac{d^2\bar{h}_0}{du^2}\right) + u\frac{d^2\Phi_0}{du^2} + (1+u)\frac{d\Phi_0}{du} = 0$$

$$\text{subject to } \Phi_1(u \to \infty) = 0, \lim_{u \to 0}\left(u\frac{d\Phi_1}{du}\right) = 0 \quad (12.63)$$

$$\frac{d^2\Phi_2}{du^2} - \frac{d^2\Phi_1}{du^2} + u\frac{d^2\Phi_1}{du^2} + (1+u)\frac{d\Phi_1}{du} = 0 \text{ subject to } \Phi_2(u \to \infty)$$

$$= 0, \lim_{u \to 0}\left(u\frac{d\Phi_2}{du}\right) = 0 \quad (12.64)$$

$$\frac{d^2\Phi_3}{du^2} - \frac{d^2\Phi_2}{du^2} + u\frac{d^2\Phi_2}{du^2} + (1+u)\frac{d\Phi_2}{du} = 0 \text{ subject to } \Phi_3(u \to \infty) = 0,$$

$$\lim_{u \to 0}\left(u\frac{d\Phi_3}{du}\right) = 0 \quad (12.65)$$

$$\frac{d^2\Phi_4}{du^2} - \frac{d^2\Phi_3}{du^2} + u\frac{d^2\Phi_3}{du^2} + (1+u)\frac{d\Phi_3}{du} = 0 \text{ subject to } \Phi_4(u \to \infty) = 0,$$

$$\lim_{u \to 0}\left(u\frac{d\Phi_4}{du}\right) = 0 \quad (12.66)$$

Proceeding in a like manner, one can arrive at the following recurrence relation:

$$\frac{d^2\Phi_m}{du^2} = (1-u)\frac{d^2\Phi_{m-1}}{du^2} - (1+u)\frac{d\Phi_{m-1}}{du} \text{ subject to } \Phi_m(u \to \infty) = 0,$$

$$\lim_{u \to 0}\left(u\frac{d\Phi_m}{du}\right) = 0 \text{ for } m \geq 2 \quad (12.67)$$

The initial approximation can be chosen as $\Phi_0 = -\dfrac{Q}{4\pi T}\exp(-u)\ln u$. Using this initial approximation, we can solve the equations iteratively using symbolic software. We avoid those expressions here, as they are lengthy. Finally, the HPM-based solution can be approximated as follows:

$$\bar{h}(u) \approx \sum_{i=0}^{M}\Phi_i \quad (12.68)$$

12.6   OHAM-BASED SOLUTION

It is concluded that due to the boundary conditions in Eqs. (12.34) and (12.35), the governing Eq. (12.33) may not be possible to solve using OHAM. Therefore, here we first convert Eq. (12.33) to a first-order ODE as follows. Equation (12.33) can be rewritten as:

$$u\frac{d^2\bar{h}}{du^2}+(1+u)\frac{d\bar{h}}{du}=0 \Rightarrow \frac{d}{du}\left(u\frac{d\bar{h}}{du}\right)+u\frac{d\bar{h}}{du}=0 \Rightarrow \frac{dv}{du}+v=0 \quad (12.69)$$

subject to:

$$\lim_{u\to 0}(v)=-\frac{Q}{4\pi T} \quad (12.70)$$

Solving Eq. (12.69) together with Eq. (12.70), one obtains:

$$u\frac{d\bar{h}}{du}+\frac{Q}{4\pi T}exp(-u)=0 \ subject \ to \ \bar{h}(\infty)=0 \quad (12.71)$$

When applying OHAM, Eq. (12.71) can be written in the following form:

$$\mathcal{L}\left[\bar{h}(u)\right]+\mathcal{N}\left[\bar{h}(u)\right]+h(u)=0 \quad (12.72)$$

Here, we select $\mathcal{L}\left[\bar{h}(u)\right]=\dfrac{d\bar{h}}{du}$, $\mathcal{N}\left[\bar{h}(u)\right]=(u-1)\dfrac{d\bar{h}}{du}$, and $h(u)=Aexp(-u)$. With these considerations, the zeroth-order representation becomes:

$$\mathcal{L}\left(\bar{h}_0(u)\right)+h(u)=0 \ subject \ to \ \bar{h}_0(u\to\infty)=0 \quad (12.73)$$

Solving Eq. (12.73), one obtains:

$$\bar{h}_0(u)=\frac{Q}{4\pi T}exp(-u) \quad (12.74)$$

To obtain $\mathcal{N}_0$, $\mathcal{N}_1$, $\mathcal{N}_2$, etc., the following can be used:

$$(u-1)\frac{d\bar{h}}{du}=(u-1)\left(\frac{d\bar{h}_0}{du}+q\frac{d\bar{h}_1}{du}+q^2\frac{d\bar{h}_2}{du}+q^3\frac{d\bar{h}_3}{du}+q^4\frac{d\bar{h}_4}{du}+\cdots\right)$$

$$=\left[(u-1)\frac{d\bar{h}_0}{du}\right]+q\left[(u-1)\frac{d\bar{h}_1}{du}\right]+q^2\left[(u-1)\frac{d\bar{h}_2}{du}\right]$$

$$+q^3\left[(u-1)\frac{d\bar{h}_3}{du}\right]+q^4\left[(u-1)\frac{d\bar{h}_4}{du}\right]+\cdots \quad (12.75)$$

Using Eq. (12.76), the first-order representation reduces to:

$$\frac{d\overline{h_1}}{du} = H_1(u, C_i) \left\{ (u-1)\frac{d\overline{h_0}}{du} \right\} \text{ subject to } \overline{h_1}(u \to \infty) = 0 \qquad (12.76)$$

We simply choose $H_1(u, C_i) = C_1 + C_2 exp(-u)$. Putting $k = 2$, the second-order representation becomes:

$$\mathcal{L}\left[\overline{h_2}(u, C_i) - \overline{h_1}(u, C_i)\right] = H_2(u, C_i)\mathcal{N}_0\left(\overline{h_0}(u)\right)$$
$$+ H_1(u, C_i)\left[\mathcal{L}\left[\overline{h_1}(u, C_i)\right] + \mathcal{N}_1\left[\overline{h_0}(u), \overline{h_1}(u, C_i)\right]\right]$$
$$\text{subject to } \overline{h_2}(u \to \infty) = 0 \qquad (12.77)$$

Further, we choose $H_2(u, C_i) = C_3 + C_4 exp(-u)$. Therefore, Eq. (12.77) becomes:

$$\frac{d\overline{h_2}}{du} = \frac{d\overline{h_1}}{du} + (C_3 + C_4 exp(-u))\left\{ (u-1)\frac{d\overline{h_0}}{du} \right\}$$
$$+ (C_1 + C_2 exp(-u))\left[\frac{d\overline{h_1}}{du} + (u-1)\frac{d\overline{h_1}}{du} \right] \text{ subject to } \overline{h_2}(u \to \infty) = 0 \quad (12.78)$$

The terms of the series can be computed using the equations developed here. One can compute these terms without any difficulty using symbolic computation software, such as MATLAB. Further, following Chapter 2, the higher-order terms can be computed in a similar manner. However, our aim is to produce an accurate solution with just two to three terms of OHAM-based series. Therefore, we restrict our calculation up to $k = 2$. Finally, the approximate solution can be found as:

$$\overline{h}(u) \approx \overline{h_0}(u) + \overline{h_1}(u, C_1, C_2) + \overline{h_2}(u, C_1, C_2, C_3, C_4) \qquad (12.79)$$

where the terms are given by Eqs. (12.74), (12.76), and (12.78).

12.7   CONVERGENCE THEOREMS

The convergence of the HAM-based and OHAM-based solutions are proved theoretically using the following theorems.

12.7.1   CONVERGENCE THEOREM OF THE HAM-BASED SOLUTION

The convergence theorems for the HAM-based solution given by Eq. (12.50) can be proved using the following theorems.

Theorem 12.1: If the homotopy series $\sum_{m=0}^{\infty} \bar{h}_m(u)$, $\sum_{m=0}^{\infty} \bar{h}_m'(u)$ and $\sum_{m=0}^{\infty} \bar{h}_m''(u)$ converge, then $R_m\left(\overrightarrow{\bar{h}}_{m-1}\right)$ given by Eq. (12.44) satisfies the relation $\sum_{m=1}^{\infty} R_m\left(\overrightarrow{\bar{h}}_{m-1}\right) = 0$. [Here "'" and "''" denote the first and second derivatives with respect to u.]

Proof: The proof of this theorem follows exactly from the previous chapters.

Theorem 12.2: If $\hbar$ is so properly chosen that the series $\sum_{m=0}^{\infty} \bar{h}_m(u)$, $\sum_{m=0}^{\infty} \bar{h}_m'(u)$ and $\sum_{m=0}^{\infty} \bar{h}_m''(u)$ converge absolutely to $\bar{h}(u)$, $\bar{h}'(u)$, and $\bar{h}''(u)$, respectively, then the homotopy series $\sum_{m=0}^{\infty} \bar{h}_m(u)$ satisfies the original governing Eq. (12.33).

Proof: Theorem 12.1 shows that if $\sum_{m=0}^{\infty} \bar{h}_m(u)$, $\sum_{m=0}^{\infty} \bar{h}_m'(u)$, and $\sum_{m=0}^{\infty} \bar{h}_m''(u)$ converge, then $\sum_{m=1}^{\infty} R_m\left(\overrightarrow{\bar{h}}_{m-1}\right) = 0$.

Therefore, substituting the previous expressions in Eq. (12.44) and simplifying further lead to:

$$u\sum_{m=0}^{\infty} \bar{h}_m'' + (1+u)\sum_{m=0}^{\infty} \bar{h}_m' = 0 \qquad (12.80)$$

which is basically the original governing Eq. (12.33). Furthermore, subject to the boundary conditions $\bar{h}_0(\infty) = 0$, $\lim_{u \to 0}\left(u\dfrac{d\bar{h}_0}{du}\right) = -\dfrac{Q}{4\pi T}$, and the conditions for the higher-order deformation equation $\bar{h}_m(\infty) = 0$, $\lim_{u \to 0}\left(u\dfrac{d\bar{h}_m}{du}\right) = 0$, for $m \geq 1$, we easily obtain $\sum_{m=0}^{\infty} \bar{h}_m(\infty) = 0$ and $\lim_{u \to 0}\left(u\sum_{m=0}^{\infty} \bar{h}_m'\right) = -\dfrac{Q}{4\pi T}$. Hence, the convergence result follows.

12.7.2 CONVERGENCE THEOREM OF THE OHAM-BASED SOLUTION

Theorem 12.3: If the series $\bar{h}_0(u) + \sum_{j=1}^{\infty} \bar{h}_j(u, C_i)$, $i = 1, 2, \ldots, s$ converges, where $\bar{h}_j(u, C_i)$ are governed by Eqs. (12.74), (12.76), and (12.78), then Eq. (12.79) is a solution of the original Eq. (12.71).

Proof: The proof of this theorem follows exactly from Theorem 2.3 of Chapter 2.

12.8 RESULTS AND DISCUSSION

First, the numerical convergence of the HAM-based analytical solutions is shown for a specific test case, and then the solution is validated over the existing analytical solution proposed by Theis (1935). Then, the derived analytical solution is tested under different physical conditions to check the efficiency of the method. Finally, the HPM and OHAM-based analytical solutions are validated by comparing them with the solution of Theis (1935).

12.8.1 NUMERICAL CONVERGENCE AND VALIDATION OF THE HAM-BASED SOLUTION

A suitable choice for the auxiliary parameter $\hbar$ determines the convergence of the HAM-based series solution. For that purpose, the squared residual error of Eq. (12.33) can be calculated as follows:

$$\Delta_m = \int_{u \in \Omega} \left(\mathcal{N} \left[\overline{h}(u) \right] \right)^2 du \tag{12.81}$$

where Ω is the domain of the equation. The HAM-based series solution leads to the exact solution of the problem when Δ_m tends to zero. Therefore, for a particular order of approximation m, it is sufficient to minimize the corresponding Δ_m, which then yields an optimal value of parameter $\hbar$. Here, we consider a test case, where the parameters are chosen as $Q = 4 \times 10^{-3}$ m³s^{-1}, $T = 0.0023$ m²s^{-1}, and $S = 7.5 \times 10^{-4}$. It may be noted that HAM can indeed provide a convergent solution for the whole domain of u. However, as an example, we consider the domain of u as $\Omega = (0,10]$. Using these parameters, the HAM solution is assessed and the squared residual errors are calculated. Figure 12.6 shows the squared residual errors for a different order of approximation for the selected case. It can be seen from the figure that as the order of approximation increases, the corresponding residual error decreases. Thus, the adequacy of selecting the linear operator, initial approximation, and auxiliary

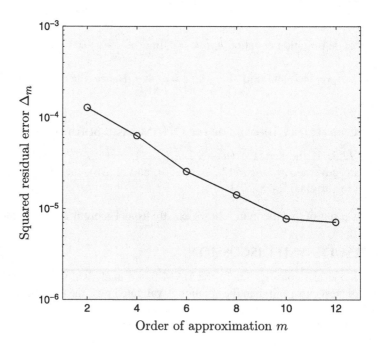

FIGURE 12.6 Squared residual error (Δ_m) versus different orders of approximation (m) of the HAM-based solution for the selected case.

TABLE 12.1

Squared Residual Error (Δ_m) and Computational Time versus Different Orders of Approximation (m) for the Selected Case

Order of Approximation (m)	Squared Residual Error (Δ_m)	Computational Time (sec)
2	1.28×10^{-4}	0.295
4	6.31×10^{-5}	1.032
6	2.55×10^{-5}	2.255
8	1.42×10^{-5}	5.374
10	7.74×10^{-6}	6.694
12	7.08×10^{-6}	10.058

function is established. Also, for a quantitative assessment, the numerical results are reported in Table 12.1 along with the computational time taken by the computer to produce the corresponding order of approximations. For the selected case, the comparison between Theis's analytical solution, where the integral was computed using numerical technique, and the 10th-order HAM-based approximate solution is shown in Figure 12.7. An excellent agreement between the computed and observed values can be seen from both figures for $u \in [0.1, 10]$. In addition, we choose the 4th, 7th, and 10th orders of approximation and compare them with the Theis solution in Table 12.2.

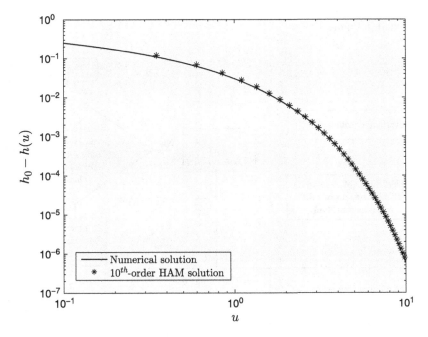

FIGURE 12.7 Comparison between the Theis solution and 10th-order HAM-based solution for the selected case.

TABLE 12.2

Comparison between HAM-Based Approximation and Numerical Solution for the Selected Case in the *Logarithm* Scale

u	Numerical Solution (Log-scale)	HAM-Based Approximation (Log-scale)		
		4th Order	7th Order	10th Order
0.1	−1.38	−1.20	−1.27	−1.30
1	−3.49	−3.22	−3.34	−3.41
2	−4.99	−4.68	−4.83	−4.93
3	−6.32	−5.97	−6.14	−6.27
4	−7.56	−7.18	−7.38	−7.53
5	−8.75	−8.35	−8.58	−8.72
6	−9.91	−9.49	−9.74	−9.87
7	−11.04	−10.62	−10.89	−10.99
8	−12.16	−11.72	−12.01	−12.08
9	−13.27	−12.82	−13.13	−13.14
10	−14.37	−13.90	−14.23	−14.20

To get a clear comparative idea, the numerical values are calculated taking the logarithm. It can be observed from the table that the higher the order of approximation, the better the accuracy. All computations are performed using the BVPh 2.0 package developed by Zhao and Liao (2002). A flowchart containing the steps of the method is provided in Figure 12.8.

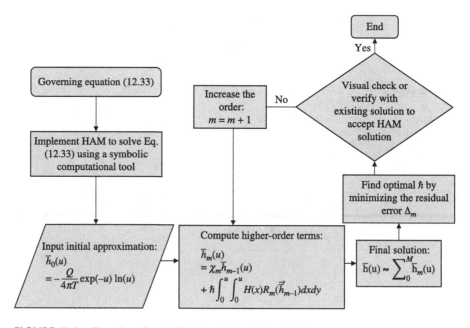

FIGURE 12.8 Flowchart for the HAM solution in Eq. (12.50).

12.8.2 COMPARISON OF THE HAM SOLUTION WITH SERIES APPROXIMATIONS

Here we test the performance of the HAM-based analytical solution for the present equation by comparing it with the Theis solution, where the well function is approximated by the series in Eqs. (12.14) and (12.17) and the expression of Vatankhah (2014). For that purpose, we consider the same test parameters, i.e., $Q = 4 \times 10^{-3}$ m^{3}s^{-1}, $T = 0.0023$ m^{2}s^{-1}, and $S = 7.5 \times 10^{-4}$. A comparison is provided in Figure 12.9 with four different subfigures. For each of the cases, we computed the well function numerically using *integral* of MATLAB to get a main solution. The series in Eq. (12.14) with 40 terms, Eq. (12.17) with 5 terms, the Vatankhah (2014) model, and the 10th-order HAM-based solution are plotted in Figure 12.9 (a), (b), (c), and (d), respectively. As already discussed in the previous section, the figures show that the series in Eq. (12.14) performs poorly for larger values of u, and Eq. (12.17) produces infeasible results for smaller u-values. The model of Vatankhah (2014) and the 10th-order HAM solution well approximate the solution, as can be seen from the figures. However, Vatankhah's (2014) method is an empirical method, while the HAM proposed here solves the original governing equation efficiently, producing an accurate solution with just a few terms of the series solution.

12.8.3 VALIDATION OF THE HPM-BASED SOLUTION

The HPM-based analytical solution for the selected test case is validated over the analytical solution given by Theis (1935). After assessing the solution, it was found that only the two-term HPM solution provides feasible values. Further, this solution produces accurate results only for a restricted domain, specifically $u \in [0.1, 1]$. Figure 12.10 shows a comparison of the initial approximation and two-term HPM solution with the existing numerical solution for a restricted domain. However, the HPM solution with more terms produces infeasible values, and therefore is not applicable. One can indeed try with some other initial approximations, which might work in producing accurate solutions to the problem. The HPM-based values and the existing analytical solution are compared in Table 12.3. A flowchart containing the steps of HPM is given in Figure 12.11.

12.8.4 VALIDATION OF THE OHAM-BASED SOLUTION

For the assessment of the OHAM-based analytical solution, one needs to calculate the constant C_i. For that purpose, we calculate the residual as follows:

$$R(u, C_i) = \mathcal{L}\left[\bar{h}_{\text{OHAM}}(u, C_i)\right] + \mathcal{N}\left[\bar{h}_{\text{OHAM}}(u, C_i)\right] + h(u), i = 1, 2, \ldots, s \quad (12.82)$$

where $\bar{h}_{\text{OHAM}}(u, C_i)$ is the approximate solution. When $R(u, C_i) = 0$, $\bar{h}_{\text{OHAM}}(u, C_i)$ becomes the exact solution to the equation. One of the ways to obtain the optimal C_i for which the solution converges is the minimization of the squared residual error, i.e.,

$$J(C_i) = \int_{u \in D} R^2(u, C_i) du, \ i = 1, 2, \ldots, s \quad (12.83)$$

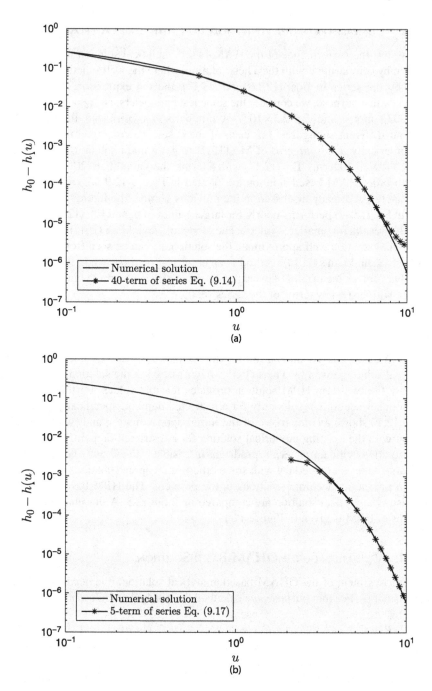

FIGURE 12.9 Comparison between (a) series in Eq. (12.14) with 40 terms, (b) Eq. (12.17) with five terms, (b) Vatankhah's (2014) approximation, and (d) 10th-order HAM solution for the selected case. (*Continued*)

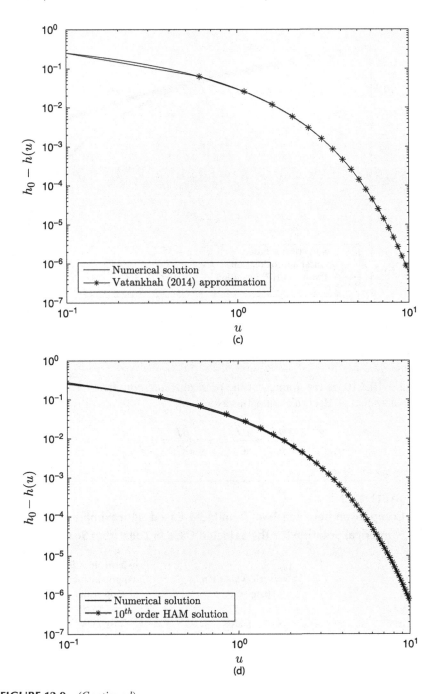

FIGURE 12.9 *(Continued)*

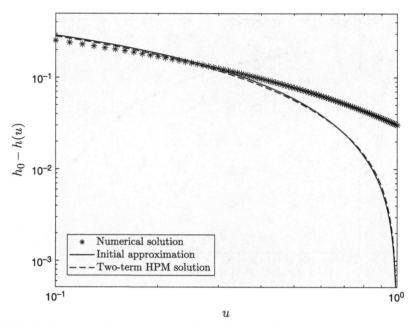

FIGURE 12.10 Comparison between the Theis solution, initial approximation, and two-term HPM-based solution for the selected case.

where $D = [0.1, 10]$ is the domain of the problem. The minimization of Eq. (12.83) leads to a system of algebraic equations as follows:

$$\frac{\partial J}{\partial C_1} = \frac{\partial J}{\partial C_2} = \ldots = \frac{\partial J}{\partial C_s} = 0 \qquad (12.84)$$

TABLE 12.3
Comparison between Two-Term HPM-Based Approximation and Numerical Solution for the Selected Case in *Logarithm* Scale

u	Numerical Solution (Log-scale)	Two-Term HPM-Based Approximation (Log-scale)
0.1	−1.3772	−1.2685
0.2	1.7766	−1.7378
0.3	−2.0767	−2.1343
0.4	−2.3309	−2.5075
0.5	−2.5579	−2.8796
0.6	−2.7665	−3.2694
0.7	−2.9618	−3.7037
0.8	−3.1469	−4.2374
0.9	−3.3240	−5.0416
1.0	−3.4946	−39.1039

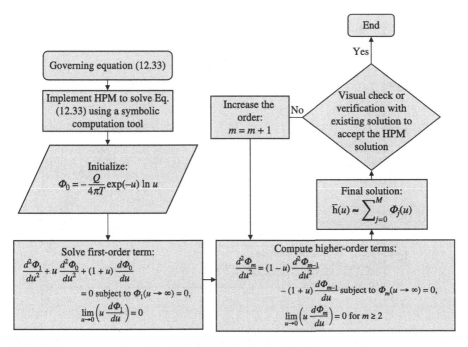

FIGURE 12.11   Flowchart for the HPM solution in Eq. (12.68).

Solving this system, one can obtain the optimal values of parameters. We obtain the optimal values using the MATLAB routine *fminsearch*, which minimizes an unconstrained multivariable function. It is found that only a three-term solution produces accurate values. Therefore, the three-term OHAM solution is computed and compared in Figure 12.12 with the existing analytical solution of Theis (1935) to see the effectiveness of the proposed method. It can be seen that just three terms of the OHAM-based series agree well with the corresponding analytical solution. For a quantitative assessment, we also compare the numerical values of the solutions in Table 12.4. A flowchart containing the steps of OHAM is provided in Figure 12.13.

12.9   CONCLUDING REMARKS

Three kinds of analytical methods are applied to the governing equation of unsteady confined radial ground-water flow. The governing equation is a PDE, which is first converted to a second-order ODE using a similarity transformation, which simplifies the methods further. Theis provided the classical solution to the equation, which expresses the solution in terms of the exponential integral. Several series and closed-form approximations have been proposed for the integral in the last few decades. This work summarizes them and carries out a detailed analysis in terms of their performance. One of the major drawbacks of the proposed approximation is that it is applicable only to a restricted domain of the equation. On the other hand, the HAM- and OHAM-based solutions seem to overcome this drawback and provide a convergent solution. OHAM provides an accurate solution with just three terms of the

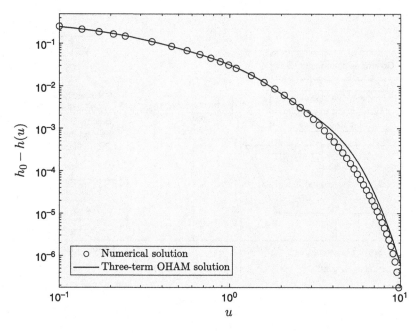

FIGURE 12.12 Comparison between the Theis solution and three-term OHAM-based solution for the selected case.

TABLE 12.4

Comparison between Three-Term OHAM-Based Approximation and Numerical Solution for the Selected Case in *Logarithm* Scale

u	Numerical Solution (Log-scale)	Three-Term OHAM-Based Approximation (Log-scale)
0.1	−1.38	−1.40
1	−3.49	−3.46
2	−4.99	−5.03
3	−6.32	−6.16
4	−7.56	−7.07
5	−8.75	−8.09
6	−9.91	−9.21
7	−11.04	−10.41
8	−12.16	−11.68
9	−13.27	−12.99
10	−14.37	−14.33

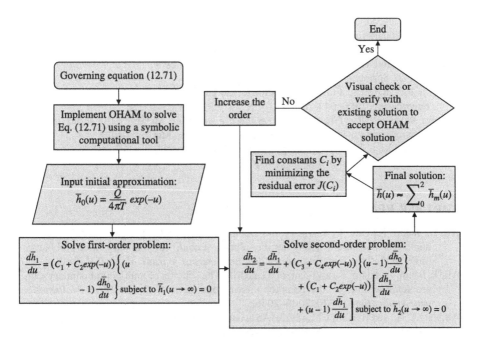

FIGURE 12.13   Flowchart for the OHAM solution in Eq. (12.79).

series solution. It is also found that HPM can produce an accurate solution only for a restricted domain of the equation. A better agreement between the derived HAM and OHAM solutions and the Theis solution with a numerical method for the exponential integral is found. Different test cases are used to justify the applicability of the proposed method further. This work shows the potential of HAM and OHAM for obtaining a series solution that is valid over a wide range of the domain. Theoretical and numerical convergences of the series solutions are provided.

REFERENCES

Abramowitz, M., and Stegun, I. A. (Eds.). (1970). *Handbook of mathematical functions with formulas, graphs, and mathematical tables* (Vol. 55). US Government Printing Office.

Barry, D. A., Parlange, J. Y., and Li, L. (2000). Approximation for the exponential integral (Theis well function). *Journal of Hydrology*, 227(1–4), 287–291.

Berndt, B. C. (1994). *Ramanujan's notebook part IV*. Springer-Verlag, New York.

Jacob, C. E. (1940). On the flow of water in an elastic artesian aquifer. *Eos, Transactions American Geophysical Union*, 21(2), 574–586.

Masina, E. (2019). Useful review on the Exponential-Integral special function. *arXiv preprint arXiv:1907.12373*.

Shampine, L. F. (2008). Vectorized adaptive quadrature in MATLAB. *Journal of Computational and Applied Mathematics*, 211(2), 131–140.

Swamee, P. K., and Ojha, C. S. P. (1990). Pump test analysis of confined aquifer. *Journal of Irrigation and Drainage Engineering*, 116(1), 99–106.

Theis, C. V. (1935). The relation between the lowering of the piezometric surface and the rate and duration of discharge of a well using ground-water storage. *Eos, Transactions American Geophysical Union, 16*(2), 519–524.

Vajravelu, K., and Van Gorder, R. (2013). *Nonlinear flow phenomena and homotopy analysis* (Vol. 2). Springer, Berlin.

Vatankhah, A. R. (2014). Full-range solution for the Theis well function. *Journal of Hydrologic Engineering, 19*(3), 649–653.

Walton, W. C. (1970). *Groundwater resource evaluation.* McGraw-Hill series in water resources and environmental engineering (USA) Eng. McGraw-Hill.

Zhao, Y., and Liao, S. (2002). User guide to BVPh 2.0. School of naval architecture, ocean and civil engineering, Shanghai, *40.*

FURTHER READING

Ali, L., Islam, S., Gul, T., Khan, I., and Dennis, L. C. C. (2016). New version of optimal homotopy asymptotic method for the solution of nonlinear boundary value problems in finite and infinite intervals. *Alexandria Engineering Journal, 55*(3), 2811–2819.

Ali, J., Islam, S., Islam, S., and Zaman, G. (2010). The solution of multipoint boundary value problems by the optimal homotopy asymptotic method. *Computers & Mathematics with Applications, 59*(6), 2000–2006.

Alkheir, A. A., and Ibnkahla, M. (2013). An accurate approximation of the exponential integral function using a sum of exponentials. *IEEE Communications Letters, 17*(7), 1364–1367.

Baalousha, H. M. (2015). Approximation of the exponential integral (well function) using sampling methods. *Journal of Hydrology, 523*, 278–282.

Chun, C., and Sakthivel, R. (2010). Homotopy perturbation technique for solving two-point boundary value problems–comparison with other methods. *Computer Physics Communications, 181*(6), 1021–1024.

Cody, W. J., and Thacher, H. C. (1968). Rational Chebyshev approximations for the exponential integral E_1 (x). *Mathematics of Computation, 22*(103), 641–649.

Cooper Jr, H. H., and Jacob, C. E. (1946). A generalized graphical method for evaluating formation constants and summarizing well-field history. *Eos, Transactions American Geophysical Union, 27*(4), 526–534.

Das, S., Kumar, S., and Singh, O. P. (2010). Solutions of nonlinear second order multi-point boundary value problems by homotopy perturbation method. *Applied Mathematics: An International Journal, 5*(10), 1592–1600.

Geng, F. Z., and Cui, M. G. (2009). Solving nonlinear multi-point boundary value problems by combining homotopy perturbation and variational iteration methods. *International Journal of Nonlinear Sciences and Numerical Simulation, 10*(5), 597–600.

Hassan, H. N., and El-Tawil, M. A. (2011). An efficient analytic approach for solving two-point nonlinear boundary value problems by homotopy analysis method. *Mathematical Methods in the Applied Sciences, 34*(8), 977–989.

Hassan, H. N., and El-Tawil, M. A. (2012). A new technique of using homotopy analysis method for second order nonlinear differential equations. *Applied Mathematics and Computation, 219*(2), 708–728.

Manafian, J. A. L. I. L. (2021). An optimal Galerkin-homotopy asymptotic method applied to the nonlinear second-order BVPs. *Proceedings of the Institute of Mathematics and Mechanics, 47*, 156–182.

McWhorter, D. B., and Sunada, D. K. (1977). *Ground-water hydrology and hydraulics.* Water Resources Publication, Fort Collins, Colorado.

Morel-Seytoux, H. J. (1975). A simple case of conjunctive surface-ground-water management. *Groundwater, 13*(6), 506–515.

Motsa, S. S., Sibanda, P., and Shateyi, S. (2010). A new spectral-homotopy analysis method for solving a nonlinear second order BVP. *Communications in Nonlinear Science and Numerical Simulation*, *15*(9), 2293–2302.

Prodanoff, J. H. A., Mansur, W. J., and Mascarenhas, F. C. B. (2006). Numerical evaluation of Theis and Hantush-Jacob well functions. *Journal of Hydrology*, *318*(1–4), 173–183.

Shivanian, E., and Abbasbandy, S. (2014). Predictor homotopy analysis method: Two points second order boundary value problems. *Nonlinear Analysis: Real World Applications*, *15*, 89–99.

Srivastava, R. (1995). Implications of using approximate expressions for well function. *Journal of Irrigation and Drainage Engineering*, *121*(6), 459–462.

Tseng, P. H., and Lee, T. C. (1998). Numerical evaluation of exponential integral: Theis well function approximation. *Journal of Hydrology*, *205*(1–2), 38–51.

Wang, Y. G., Song, H. F., and Li, D. (2009). Solving two-point boundary value problems using combined homotopy perturbation method and green's function method. *Applied Mathematics and Computation*, *212*(2), 366–376.

13 Series Solutions for Burger's Equation

13.1 INTRODUCTION

Burger's equation is a nonlinear partial differential equation (PDE) and well known in fluid dynamics and has received extensive treatment in applied mathematics and fluid mechanics. In water engineering, it has been applied to drainage and flow routing, but in general it has received relatively little attention (Warrick and Parkin 1995). Under certain conditions, it is possible to derive Burger's equation by combining the continuity equation with a flux law. The flux law expresses the relation between flux and concentration and the gradient of the concentration. Thus, this flux law is a mixture of a nonlinear flux law and linear gradient law. This chapter presents homotopy-based methods for solving Burger's equation.

13.2 GOVERNING EQUATION

We consider the governing equation for a one-dimensional velocity field given by

$$\frac{\partial w}{\partial t} + w \frac{\partial w}{\partial x} = \frac{\partial^2 w}{\partial x^2} \tag{13.1}$$

subject to the initial and boundary conditions given as follows:

$$w(x,0) = 0 \tag{13.2}$$

$$w(0,t) = \lambda \alpha(t),\ t > 0 \text{ and } \lambda \neq 0 \tag{13.3}$$

$$\lim_{x \to \infty} w(x,t) = 0 \tag{13.4}$$

where x and t are spatial and temporal coordinates, respectively; $w(x,t)$ is the velocity field; $\alpha(t)$ is a function of t; and λ is a parameter. Abd-el-Malek and El-Mansi (2000) considered the equation and the conditions given by Eqs. (13.1)–(13.4) and solved it using the group-theoretic approach. In this approach, a one-parameter group was used to transform Burger's equation into an ordinary differential equation (ODE). Before applying the group-theoretic approach, the equations were converted using the transformation $w(x,t) = u(x,t)\alpha(t)$, as follows:

$$\alpha \frac{\partial u}{\partial t} + u \frac{\partial \alpha}{\partial t} + \alpha^2 u \frac{\partial u}{\partial x} = \alpha \frac{\partial^2 u}{\partial x^2} \tag{13.5}$$

DOI: 10.1201/9781003368984-17

subject to the conditions:

$$u(x,0)=0 \tag{13.6}$$

$$u(0,t)=\lambda, \; t>0 \text{ and } \lambda \neq 0 \tag{13.7}$$

$$\lim_{x \to \infty} u(x,t)=0 \tag{13.8}$$

The one-parameter group transformation can be used now to convert the PDE to an ODE. The group-theoretic approach adopted here needs mathematical details of abstract algebra, which may not fit into the scope of the monograph. However, one can find the details in Abd-el-Malek and El-Mansi (2000). Here, we consider the transformed equation directly, which is given as:

$$2\frac{d^2U}{d\eta^2}+(\eta-2U)\frac{dU}{d\eta}+U=0 \tag{13.9}$$

Subject to:

$$U(0)=\lambda \text{ and } U(\infty)=0 \tag{13.10}$$

where $\eta=\dfrac{x}{\sqrt{t}}$ and $u(x,t)=U(\eta)$.

13.3   HAM-BASED SOLUTION

To solve Eq. (13.9) using HAM subject to the boundary conditions in Eq. (13.10), the zeroth-order deformation equation can be constructed as follows:

$$(1-q)\mathcal{L}\left[\Phi(\eta;q)-U_0(\eta)\right]=q\hbar H(\eta)\mathcal{N}\left[\Phi(\eta;q)\right] \tag{13.11}$$

subject to the boundary conditions:

$$\Phi(0;q)=\lambda \tag{13.12}$$

$$\Phi(\infty;q)=0 \tag{13.13}$$

where q is the embedding parameter, $\Phi(\eta;q)$ is the representation of the solution across q, $U_0(\eta)$ is the initial approximation, $\hbar$ is the auxiliary parameter, $H(\eta)$ is the auxiliary function, and $\mathcal{L}$ and $\mathcal{N}$ are the linear and nonlinear operators, respectively. The higher-order terms can be calculated from the higher-order deformation equation, given as follows:

$$\mathcal{L}\left[U_m(\eta)-\chi_m U_{m-1}(\eta)\right]=\hbar H(\eta)R_m\left(\vec{U}_{m-1}\right), \; m=1,2,3,\ldots$$

subject to:

$$U_m(0) = 0, \; U_m(\infty) = 0 \tag{13.14}$$

where:

$$\chi_m = \begin{cases} 0 & \text{when } m = 1, \\ 1 & \text{otherwise} \end{cases} \tag{13.15}$$

and:

$$R_m\left(\vec{U}_{m-1}\right) = \frac{1}{(m-1)!} \frac{\partial^{m-1} \mathcal{N}\left[\Phi(\eta; q)\right]}{\partial q^{m-1}} \Bigg|_{q=0} \tag{13.16}$$

where U_m for $m \geq 1$ are the higher-order terms. Now, the final solution can be obtained as follows:

$$U(\eta) = U_0(\eta) + \sum_{m=1}^{\infty} U_m(\eta) \tag{13.17}$$

The nonlinear operator for the problem is selected as the original governing equation, i.e.,

$$\mathcal{N}\left[\Phi(\eta; q)\right] = 2\frac{\partial^2 \Phi(\eta; q)}{\partial \eta^2} + \left(\eta - 2\Phi(\eta; q)\right)\frac{\partial \Phi(\eta; q)}{\partial \eta} + \Phi(\eta; q) \tag{13.18}$$

Therefore, using Eq. (13.18), the term R_m can be calculated from Eq. (13.16) as follows:

$$R_m\left(\vec{U}_{m-1}\right) = 2\frac{d^2 U_{m-1}}{d\eta^2} + \eta\frac{dU_{m-1}}{d\eta} - 2\sum_{j=0}^{m-1} U_j \frac{dU_{m-j-1}}{d\eta} + U_{m-1} \tag{13.19}$$

Based on the analytical solution given by Abd-el-Malek and El-Mansi (2000) and the boundedness of the solution at infinity (boundary condition in Eq. 13.10), the following set of base functions is chosen to represent the solution:

$$\left\{\eta^m \exp(-n\eta) \mid m, n = 0, 1, 2, \ldots\right\} \tag{13.20}$$

so that:

$$U(\eta) = \sum_{m=0}^{\infty} \sum_{n=0}^{\infty} b_{m,n} \eta^m \exp(-n\eta) \tag{13.21}$$

where $b_{m,n}$ is the coefficient of the series. Eq. (13.21) provides the so-called *rule of solution expression*. Following the rule of solution expression, the linear operator and the initial approximation are chosen, respectively, as follows:

$$\mathcal{L}\left[\Phi(\eta;q)\right] = \frac{\partial^2 \Phi(\eta;q)}{\partial \eta^2} + \frac{\partial \Phi(\eta;q)}{\partial \eta} \text{ with the property } \mathcal{L}\left[C_2 + C_3 \exp(-\eta)\right] = 0$$

$$(13.22)$$

$$U_0(\eta) = \lambda \exp(-\eta) \tag{13.23}$$

where C_2 and C_3 are integral constants. Using Eq. (13.22), the higher-order terms can be obtained from Eq. (13.14) as follows:

$$U_m(\eta) = \chi_m U_{m-1}(\eta) + \hbar \int_0^\eta \int_0^\eta H(x) R_m\left(\vec{U}_{m-1}\right) dx\, dy + C_2 exp(-\eta) + C_3, \ m = 1,2,3,\ldots$$

$$(13.24)$$

where R_m is given by Eq. (13.19) and constants C_2 and C_3 can be determined from the boundary conditions for the higher-order deformation equations given by Eq. (13.14).

Now, the auxiliary function $H(\eta)$ can be determined from the *rule of coefficient ergodicity*. Based on the rule of solution expression in Eq. (13.20), function $H(\eta)$ can take on the form:

$$H(\eta) = \eta^a \exp(-b\eta) \tag{13.25}$$

where a and b are integers. However, to be in accordance with the *rule of coefficient ergodicity*, we found $H(\eta)$ to be simply 1, which corresponds to the case $a = b = 0$. Finally, the approximate solution can be obtained as follows:

$$U(\eta) \approx U_0(\eta) + \sum_{m=1}^{M} U_m(\eta) \tag{13.26}$$

13.4   HPM-BASED SOLUTION

When applying HPM, we rewrite Eq. (13.9) in the following form:

$$\mathcal{N}(U) = f(\eta) \tag{13.27}$$

We construct a homotopy $\Phi(\eta;q)$ that satisfies:

$$(1-q)\left[\mathcal{L}(\Phi(\eta;q)) - \mathcal{L}(U_0(\eta))\right] + q\left[\mathcal{N}(\Phi(\eta;q)) - f(\eta)\right] = 0 \tag{13.28}$$

where q is the embedding parameter that lies on $[0,1]$, $\mathcal{L}$ is the linear operator, and $U_0(\eta)$ is the initial approximation to the final solution. Eq. (13.28) shows that as $q = 0$, $\Phi(\eta;0) = U_0(\eta)$ and as $q = 1$, $\Phi(\eta;1) = U(\eta)$. Let us now express $\Phi(\eta;q)$ as a series in terms of q as:

$$\Phi(\eta;q) = \Phi_0 + q\Phi_1 + q^2\Phi_2 + q^3\Phi_3 + q^4\Phi_4 \ldots \tag{13.29}$$

where Φ_m for $m \geq 1$ are the higher-order terms. As $q \to 1$, Eq. (13.29) produces the final solution as:

$$U(\eta) = \lim_{q \to 1} \Phi(\eta;q) = \sum_{i=0}^{\infty} \Phi_i \tag{13.30}$$

For simplicity, we consider the linear and nonlinear operators as:

$$\mathcal{L}(\Phi(\eta;q)) = \frac{\partial^2 \Phi(\eta;q)}{\partial \eta^2} + \frac{\partial \Phi(\eta;q)}{\partial \eta} \tag{13.31}$$

$$\mathcal{N}(\Phi(\eta;q)) = 2\frac{\partial^2 \Phi(\eta;q)}{\partial \eta^2} + (\eta - 2\Phi(\eta;q))\frac{\partial \Phi(\eta;q)}{\partial \eta} + \Phi(\eta;q) \tag{13.32}$$

Using these expressions, Eq. (13.28) takes on the form:

$$(1-q)\left[\frac{\partial^2 \Phi(\eta;q)}{\partial \eta^2} + \frac{\partial \Phi(\eta;q)}{\partial \eta} - \frac{d^2 U_0}{d\eta^2} - \frac{dU_0}{d\eta}\right]$$
$$+ q\left[2\frac{\partial^2 \Phi(\eta;q)}{\partial \eta^2} + (\eta - 2\Phi(\eta;q))\frac{\partial \Phi(\eta;q)}{\partial \eta} + \Phi(\eta;q)\right] = 0 \tag{13.33}$$

The nonlinear term can be expanded as follows:

$$\left(\Phi_0 + q\Phi_1 + q^2\Phi_2 + q^3\Phi_3 + q^4\Phi_4 + \cdots\right)$$

$$\times \left(\frac{d\Phi_0}{d\eta} + q\frac{d\Phi_1}{d\eta} + q^2\frac{d\Phi_2}{d\eta} + q^3\frac{d\Phi_3}{d\eta} + q^4\frac{d\Phi_4}{d\eta} + \cdots\right)$$

$$= \Phi_0 \frac{d\Phi_0}{d\eta} + q\left(\Phi_0 \frac{d\Phi_1}{d\eta} + \Phi_1 \frac{d\Phi_0}{d\eta}\right) + q^2\left(\Phi_0 \frac{d\Phi_2}{d\eta} + \Phi_1 \frac{d\Phi_1}{d\eta} + \Phi_2 \frac{d\Phi_0}{d\eta}\right)$$

$$+ q^3\left(\Phi_0 \frac{d\Phi_3}{d\eta} + \Phi_3 \frac{d\Phi_0}{d\eta} + \Phi_2 \frac{d\Phi_1}{d\eta} + \Phi_1 \frac{d\Phi_2}{d\eta}\right)$$

$$+ q^4\left(\Phi_0 \frac{d\Phi_4}{d\eta} + \Phi_4 \frac{d\Phi_0}{d\eta} + \Phi_1 \frac{d\Phi_3}{d\eta} + \Phi_3 \frac{d\Phi_1}{d\eta} + \Phi_2 \frac{d\Phi_2}{d\eta}\right) \tag{13.34}$$

Using Eq. (13.33), we have:

$$(1-q)\left[\frac{\partial^2\Phi(\eta;q)}{\partial\eta^2}+\frac{\partial\Phi(\eta;q)}{\partial\eta}-\frac{d^2U_0}{d\eta^2}-\frac{dU_0}{d\eta}\right]$$

$$=(1-q)\left[\left(\frac{d^2\Phi_0}{d\eta^2}+\frac{d\Phi_0}{d\eta}-\frac{d^2U_0}{d\eta^2}-\frac{dU_0}{d\eta}\right)+q\left(\frac{d^2\Phi_1}{d\eta^2}+\frac{d\Phi_1}{d\eta}\right)+q^2\left(\frac{d^2\Phi_2}{d\eta^2}+\frac{d\Phi_2}{d\eta}\right)\right.$$

$$\left.+q^3\left(\frac{d^2\Phi_3}{d\eta^2}+\frac{d\Phi_3}{d\eta}\right)+q^4\left(\frac{d^2\Phi_4}{d\eta^2}+\frac{d\Phi_4}{d\eta}\right)+\cdots\right]$$

$$=\left(\frac{d^2\Phi_0}{d\eta^2}+\frac{d\Phi_0}{d\eta}-\frac{d^2U_0}{d\eta^2}-\frac{dU_0}{d\eta}\right)+q\left(\frac{d^2\Phi_1}{d\eta^2}+\frac{d\Phi_1}{d\eta}-\left(\frac{d^2\Phi_0}{d\eta^2}+\frac{d\Phi_0}{d\eta}-\frac{d^2U_0}{d\eta^2}-\frac{dU_0}{d\eta}\right)\right)$$

$$+q^2\left(\frac{d^2\Phi_2}{d\eta^2}+\frac{d\Phi_2}{d\eta}-\frac{d^2\Phi_1}{d\eta^2}-\frac{d\Phi_1}{d\eta}\right)+q^3\left(\frac{d^2\Phi_3}{d\eta^2}+\frac{d\Phi_3}{d\eta}-\frac{d^2\Phi_2}{d\eta^2}-\frac{d\Phi_2}{d\eta}\right)$$

$$+q^4\left(\frac{d^2\Phi_4}{d\eta^2}+\frac{d\Phi_4}{d\eta}-\frac{d^2\Phi_3}{d\eta^2}-\frac{d\Phi_3}{d\eta}\right)+\cdots \tag{13.35}$$

$$2\frac{\partial^2\Phi(\eta;q)}{\partial\eta^2}+\left(\eta-2\Phi(\eta;q)\right)\frac{\partial\Phi(\eta;q)}{\partial\eta}+\Phi(\eta;q)$$

$$=2\left(\frac{d^2\Phi_0}{d\eta^2}+q\frac{d^2\Phi_1}{d\eta^2}+q^2\frac{d^2\Phi_2}{d\eta^2}+q^3\frac{d^2\Phi_3}{d\eta^2}+q^4\frac{d^2\Phi_4}{d\eta^2}+\cdots\right)$$

$$+\eta\left(\frac{d\Phi_0}{d\eta}+q\frac{d\Phi_1}{d\eta}+q^2\frac{d\Phi_2}{d\eta}+q^3\frac{d\Phi_3}{d\eta}+q^4\frac{d\Phi_4}{d\eta}+\cdots\right)$$

$$-2\left[\Phi_0\frac{\partial\Phi_0}{\partial\Phi}+q\left(\Phi_0\frac{\partial\Phi_1}{\partial\eta}+\Phi_1\frac{\partial\Phi_0}{\partial\eta}\right)+q^2\left(\Phi_0\frac{\partial\Phi_2}{\partial\eta}+\Phi_1\frac{\partial\Phi_1}{\partial\eta}+\Phi_2\frac{\partial\Phi_0}{\partial\eta}\right)\right.$$

$$+q^3\left(\Phi_0\frac{\partial\Phi_3}{\partial\eta}+\Phi_3\frac{\partial\Phi_0}{\partial\eta}+\Phi_2\frac{\partial\Phi_1}{\partial\eta}+\Phi_1\frac{\partial\Phi_2}{\partial\eta}\right)$$

$$\left.+q^4\left(\Phi_0\frac{\partial\Phi_4}{\partial\eta}+\Phi_4\frac{\partial\Phi_0}{\partial\eta}+\Phi_1\frac{\partial\Phi_3}{\partial\eta}+\Phi_3\frac{\partial\Phi_1}{\partial\eta}+\Phi_2\frac{\partial\Phi_2}{\partial\eta}\right)\right]$$

$$+\left(\Phi_0+q\Phi_1+q^2\Phi_2+q^3\Phi_3+q^4\Phi_4+\cdots\right)=\left[2\frac{d^2\Phi_0}{d\eta^2}+\eta\frac{d\Phi_0}{d\eta}-2\Phi_0\frac{\partial\Phi_0}{\partial\eta}+\Phi_0\right]$$

$$+q\left[2\frac{d^2\Phi_1}{d\eta^2}+\eta\frac{d\Phi_1}{d\eta}-2\left(\Phi_0\frac{\partial\Phi_1}{\partial\eta}+\Phi_1\frac{\partial\Phi_0}{\partial\eta}\right)+\Phi_1\right]$$

$$+q^2\left[2\frac{d^2\Phi_2}{d\eta^2}+\eta\frac{d\Phi_2}{d\eta}-2\left(\Phi_0\frac{\partial\Phi_2}{\partial\eta}+\Phi_1\frac{\partial\Phi_1}{\partial\eta}+\Phi_2\frac{\partial\Phi_0}{\partial\eta}\right)+\Phi_2\right]$$

$$+q^3\left[2\frac{d^2\Phi_3}{d\eta^2}+\eta\frac{d\Phi_3}{d\eta}-2\left(\Phi_0\frac{\partial\Phi_3}{\partial\eta}+\Phi_3\frac{\partial\Phi_0}{\partial\eta}+\Phi_2\frac{\partial\Phi_1}{\partial\eta}+\Phi_1\frac{\partial\Phi_2}{\partial\eta}\right)+\Phi_3\right]+\cdots \tag{13.36}$$

Using the boundary conditions $U(0) = \lambda$ and $U(\infty) = 0$, we have:

$$\Phi_0(0) = \lambda, \ \Phi_1(0) = 0, \ \Phi_2(0) = 0, \ \dots \qquad (13.37)$$

$$\Phi_0(\infty) = 0, \ \Phi_1(\infty) = 0, \ \Phi_2(\infty) = 0, \ \dots \qquad (13.38)$$

Using Eqs. (13.34)–(13.36) and equating the like powers of q in Eq. (13.33), the following system of differential equations is obtained:

$$\frac{d^2\Phi_0}{d\eta^2} + \frac{d\Phi_0}{d\eta} - \frac{d^2 U_0}{d\eta^2} - \frac{dU_0}{d\eta} = 0 \text{ subject to } \Phi_0(0) = \lambda, \Phi_0(\infty) = 0 \quad (13.39)$$

$$\frac{d^2\Phi_1}{d\eta^2} + \frac{d\Phi_1}{d\eta} - \left(\frac{d^2\Phi_0}{d\eta^2} + \frac{d\Phi_0}{d\eta} - \frac{d^2 U_0}{d\eta^2} - \frac{dU_0}{d\eta} \right)$$

$$+ 2\frac{d^2\Phi_0}{d\eta^2} + \eta\frac{d\Phi_0}{d\eta} - 2\Phi_0\frac{\partial\Phi_0}{\partial\eta} + \Phi_0 = 0 \text{ subject to } \Phi_1(0) = 0, \Phi_1(\infty) = 0$$

$$(13.40)$$

$$\frac{d^2\Phi_2}{d\eta^2} + \frac{d\Phi_2}{d\eta} - \frac{d^2\Phi_1}{d\eta^2} - \frac{d\Phi_1}{d\eta} + 2\frac{d^2\Phi_1}{d\eta^2} + \eta\frac{d\Phi_1}{d\eta}$$

$$-2\left(\Phi_0\frac{\partial\Phi_1}{\partial\eta} + \Phi_1\frac{\partial\Phi_0}{\partial\eta} \right) + \Phi_1 = 0 \text{ subject to } \Phi_2(0) = 0, \Phi_2(\infty) = 0 \quad (13.41)$$

$$\frac{d^2\Phi_3}{d\eta^2} + \frac{d\Phi_3}{d\eta} - \frac{d^2\Phi_2}{d\eta^2} - \frac{d\Phi_2}{d\eta} + 2\frac{d^2\Phi_2}{d\eta^2} + \eta\frac{d\Phi_2}{d\eta} - 2\left(\Phi_0\frac{\partial\Phi_2}{\partial\eta} + \Phi_1\frac{\partial\Phi_1}{\partial\eta} + \Phi_2\frac{\partial\Phi_0}{\partial\eta} \right)$$

$$+ \Phi_2 = 0 \text{ subject to } \Phi_3(0) = 0, \Phi_3(\infty) = 0 \quad (13.42)$$

$$\frac{d^2\Phi_4}{d\eta^2} + \frac{d\Phi_4}{d\eta} - \frac{d^2\Phi_3}{d\eta^2} - \frac{d\Phi_3}{d\eta} + 2\frac{d^2\Phi_3}{d\eta^2} + \eta\frac{d\Phi_3}{d\eta}$$

$$-2\left(\Phi_0\frac{\partial\Phi_3}{\partial\eta} + \Phi_3\frac{\partial\Phi_0}{\partial\eta} + \Phi_2\frac{\partial\Phi_1}{\partial\eta} + \Phi_1\frac{\partial\Phi_2}{\partial\eta} \right) + \Phi_3$$

$$= 0 \text{ subject to } \Phi_4(0) = 0, \Phi_4(\infty) = 0 \quad (13.43)$$

The initial approximation can be chosen as $\Phi_0 = \lambda\exp(-\eta)$. This is chosen following the homotopy analysis method (HAM), and also it is a solution of the selected linear operator. Using this initial approximation, we can solve the equations iteratively using symbolic software. We avoid those expressions here, as they are lengthy.

Finally, the homotopy perturbation method (HPM)-based solution can be approximated as follows:

$$U(\eta) \approx \sum_{i=0}^{M} \Phi_i \tag{13.44}$$

13.5   OHAM-BASED SOLUTION

When applying the optimal homotopy asymptotic method (OHAM), Eq. (13.9) can be written in the following form:

$$\mathcal{L}\big[U(\eta)\big] + \mathcal{N}\big[U(\eta)\big] + h(\eta) = 0 \tag{13.45}$$

Here, we select $\mathcal{L}\big[U(\eta)\big] = 2\dfrac{d^2U}{d\eta^2} + \dfrac{dU}{d\eta}$, $\mathcal{N}\big[U(\eta)\big] = (\eta - 2U - 1)\dfrac{dU}{d\eta} + U$, and $h(\eta) = 0$. With these considerations, the zeroth-order problem becomes:

$$\mathcal{L}\big(U_0(\eta)\big) + h(\eta) = 0 \text{ subject to } U_0(0) = \lambda,\ U_0(\infty) = 0 \tag{13.46}$$

Solving Eq. (13.46), one obtains:

$$U_0(\eta) = \lambda \exp\left(-\frac{\eta}{2}\right) \tag{13.47}$$

To obtain $\mathcal{N}_0$, $\mathcal{N}_1$, $\mathcal{N}_2$, etc., the following can be used:

$$(\eta - 2U - 1)\frac{dU}{d\eta} + U = (\eta - 1)\left(\frac{dU_0}{d\eta} + q\frac{dU_1}{d\eta} + q^2\frac{dU_2}{d\eta} + q^3\frac{dU_3}{d\eta} + q^4\frac{dU_4}{d\eta} + \cdots\right)$$

$$-2\left(U_0 + qU_1 + q^2U_2 + q^3U_3 + q^4U_4 + \cdots\right)$$

$$\times\left(\frac{dU_0}{d\eta} + q\frac{dU_1}{d\eta} + q^2\frac{dU_2}{d\eta} + q^3\frac{dU_3}{d\eta} + q^4\frac{dU_4}{d\eta} + \cdots\right)$$

$$+\left(U_0 + qU_1 + q^2U_2 + q^3U_3 + q^4U_4 + \cdots\right) = \left[(\eta - 1)\frac{dU_0}{d\eta} - 2U_0\frac{dU_0}{d\eta} + U_0\right]$$

$$+ q\left[(\eta - 1)\frac{dU_1}{d\eta} - 2\left(U_0\frac{dU_1}{d\eta} + U_1\frac{dU_0}{d\eta}\right) + U_1\right]$$

$$+ q^2\left[(\eta - 1)\frac{dU_2}{d\eta} - 2\left(U_2\frac{dU_0}{d\eta} + U_0\frac{dU_2}{d\eta} + U_1\frac{dU_1}{d\eta}\right) + U_2\right]$$

$$+ q^3\left[(\eta - 1)\frac{dU_3}{d\eta} - 2\left(U_0\frac{dU_3}{d\eta} + U_3\frac{dU_0}{d\eta} + U_1\frac{dU_2}{d\eta} + U_2\frac{dU_1}{d\eta}\right) + U_3\right] + \cdots \tag{13.48}$$

Using Eq. (13.48), the first-order problem reduces to:

$$2\frac{d^2U_1}{d\eta^2} + \frac{dU_1}{d\eta}$$

$$= H_1(\eta, C_i)\left\{(\eta-1)\frac{dU_0}{d\eta} - 2U_0\frac{dU_0}{d\eta} + U_0\right\} \text{ subject to } U_1(0) = 0, \ U_1(\infty) = 0$$

$$\text{(13.49)}$$

We simply choose $H_1(\eta, C_i) = C_1 + C_2 exp\left(-\frac{\eta}{2}\right) + C_3 exp(-\eta)$. Putting $k = 2$, the second-order problem becomes:

$$\mathcal{L}[U_2(\eta, C_i) - U_1(\eta, C_i)] = H_2(\eta, C_i)\mathcal{N}_0(U_0(\eta)) + H_1(\eta, C_i)$$

$$\times\left[\mathcal{L}[U_1(\eta, C_i)] + \mathcal{N}_1[U_0(\eta), U_1(\eta, C_i)]\right] \text{ subject to } U_2(0) = 0, \ U_2(\infty) = 0$$

$$\text{(13.50)}$$

Further, we choose $H_2(\eta, C_i) = C_4 + C_5 exp\left(-\frac{\eta}{2}\right) + C_6 exp(-\eta)$. Therefore, Eq. (13.50) becomes:

$$2\frac{d^2U_2}{d\eta^2} + \frac{dU_2}{d\eta} = 2\frac{d^2U_1}{d\eta^2} + \frac{dU_1}{d\eta} + \left(C_4 + C_5 exp\left(-\frac{\eta}{2}\right) + C_6 exp(-\eta)\right)$$

$$\times\left\{(\eta-1)\frac{dU_0}{d\eta} - 2U_0\frac{dU_0}{d\eta} + U_0\right\} + \left(C_1 + C_2 exp\left(-\frac{\eta}{2}\right) + C_3 exp(-\eta)\right)$$

$$\times\left[2\frac{d^2U_1}{d\eta^2} + \frac{dU_1}{d\eta} + (\eta-1)\frac{dU_1}{d\eta} - 2\left(U_0\frac{dU_1}{d\eta} + U_1\frac{dU_0}{d\eta}\right) + U_1\right]$$

$$\text{subject to } U_2(0) = 0, \ U_2(\infty) = 0 \qquad \text{(13.51)}$$

The terms of the series can be computed using the equations developed here. One can compute these terms without any difficulty using symbolic computation software, such as MATLAB. Further, following Chapter 2, the higher-order terms can be computed in a similar manner. However, our aim is to produce an accurate solution with just two to three terms of the OHAM-based series. Therefore, we restrict our calculation up to $k = 2$. Finally, the approximate solution can be found as:

$$U(\eta) \approx U_0(\eta) + U_1(\eta, C_i) + U_2(\eta, C_i) \qquad \text{(13.52)}$$

where the terms are given by Eqs. (13.47), (13.49), and (13.51).

13.6 CONVERGENCE THEOREMS

The convergence of the HAM-based and OHAM-based solutions are proved theoretically using the following theorems.

13.6.1 CONVERGENCE THEOREM OF THE HAM-BASED SOLUTION

The convergence theorems for the HAM-based solution given by Eq. (13.26) can be proved using the following theorems.

Theorem 13.1: If the homotopy series $\Sigma_{m=0}^{\infty} U_m(\eta)$, $\Sigma_{m=0}^{\infty} U_m{}'(\eta)$ and $\Sigma_{m=0}^{\infty} U_m{}''(\eta)$ converge, then $R_m\left(\vec{U}_{m-1}\right)$ given by Eq. (13.19) satisfies the relation $\Sigma_{m=1}^{\infty} R_m\left(\vec{U}_{m-1}\right) = 0$. [Here ' $'$ ' and ' $''$ ' denote the first and second derivatives with respect to η.]

Proof: The proof of this theorem follows exactly from the previous chapters.

Theorem 13.2: If $\hbar$ is so properly chosen that the series $\Sigma_{m=0}^{\infty} U_m(\eta)$, $\Sigma_{m=0}^{\infty} U_m{}'(\eta)$ and $\Sigma_{m=0}^{\infty} U_m{}''(\eta)$ converge absolutely to $U(\eta), U'(\eta)$, and $U''(\eta)$, respectively, then the homotopy series $\Sigma_{m=0}^{\infty} U_m(\eta)$ satisfies the original governing Eq. (13.9).

Proof: Theorem 13.1 shows that if $\Sigma_{m=0}^{\infty} U_m(\eta)$, $\Sigma_{m=0}^{\infty} U_m{}'(\eta)$, and $\Sigma_{m=0}^{\infty} U_m{}''(\eta)$ converge then $\Sigma_{m=1}^{\infty} R_m\left(\vec{U}_{m-1}\right) = 0$.

Using the Cauchy product defined in Theorem 2.2 of Chapter 2, we get

$$\sum_{m=1}^{\infty}\sum_{j=0}^{m-1} U_j U'_{m-1-j} = \left(\sum_{m=0}^{\infty} U_m\right)\left(\sum_{k=0}^{\infty} U'_k\right) \tag{13.53}$$

Therefore, substituting the previous expressions in Eq. (13.19) and simplifying further lead to:

$$2\sum_{m=0}^{\infty} U_m{}'' + \eta\sum_{m=0}^{\infty} U_m{}' - 2\left(\sum_{m=0}^{\infty} U_m\right)\left(\sum_{k=0}^{\infty} U'_k\right) + \sum_{m=0}^{\infty} U_m = 0 \tag{13.54}$$

which is basically the original governing Eq. (13.9). Furthermore, subject to the boundary conditions $U_0(0) = \lambda$, $U_0(\infty) = 0$ and the conditions for the higher-order deformation equation $U_m(0) = 0$, $U_m(\infty) = 0$, for $m \geq 1$, we easily obtain $\Sigma_{m=0}^{\infty} U_m(0) = \lambda$ and $\Sigma_{m=0}^{\infty} U_m(\infty) = 0$. Hence, the convergence result follows.

13.6.2 CONVERGENCE THEOREM OF THE OHAM-BASED SOLUTION

Theorem 13.3: If the series $U_0(\eta) + \Sigma_{j=1}^{\infty} U_j(\eta, C_i)$, $i = 1, 2, \ldots, s$ converges, where $U_j(\eta, C_i)$ are governed by Eqs. (13.47), (13.49), and (13.51), then Eq. (13.52) is a solution of the original Eq. (13.9).

Proof: The proof of this theorem follows exactly from Theorem 2.3 of Chapter 2.

13.7   RESULTS AND DISCUSSION

First, the numerical convergence of the HAM-based analytical solution is established for a specific test case. Then the solution is validated against the existing analytical solution proposed by Abd-el-Malek and El-Mansi (2000). Thereafter, the derived analytical solution is tested under different physical conditions to determine the efficiency of the method. Finally, the HPM- and OHAM-based analytical solutions are validated by comparing them with the solution of Abd-el-Malek and El-Mansi (2000).

13.7.1   NUMERICAL CONVERGENCE AND VALIDATION OF THE HAM-BASED SOLUTION

A suitable choice for the auxiliary parameter $\hbar$ determines the convergence of the HAM-based series solution. For that purpose, the squared residual error with regard to Eq. (13.9) can be calculated as:

$$\Delta_m = \int\limits_{\eta \in \Omega} \left(\mathcal{N}\left[U(\eta) \right] \right)^2 d\eta \tag{13.55}$$

where Ω is the domain of the problem. The HAM-based series solution leads to the exact solution of the problem when Δ_m tends to zero. Therefore, for a particular order of approximation m, it is sufficient to minimize the corresponding Δ_m, which then yields an optimal value of parameter $\hbar$. Here, we consider a test case, where parameter λ is chosen as $\lambda = 1$. It may be noted that the HAM can indeed provide a convergent solution for the whole domain of η. However, as an example, we consider the domain of η as $\Omega = [0,10]$. Using these parameters, the HAM solution is assessed, and the squared residual errors are calculated. Figure 13.1 shows the squared residual errors for a different order of approximation for the selected case. It can be seen from the figure that as the order of approximation increases, the corresponding residual error decreases. Thus, the adequacy of selecting the linear operator, initial approximation, and auxiliary function is established. Also, for a quantitative assessment, the numerical results are reported in Table 13.1 along with the computational time taken by the computer to produce the corresponding order of approximations. For the selected case, a comparison between the analytical solution where the integral is computed using a numerical technique and the 25th-order HAM-based approximate solution is shown in Figure 13.2. An excellent agreement between the computed and observed values can be found from both figures for $u \in [0,10]$. In addition, we choose the 25th-order approximation and compare it with the analytical solution in Table 13.2. It can be observed from the table that the HAM-based approximation provides a good estimated solution. All the computations are performed using the BVPh 2.0 package developed by Zhao and Liao (2002). A flowchart containing the steps of the method is provided in Figure 13.3.

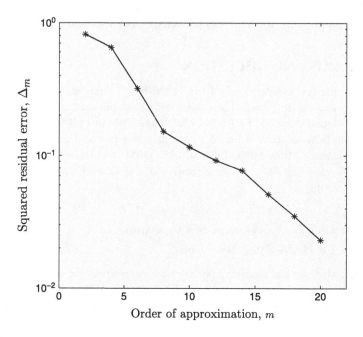

FIGURE 13.1   Squared residual error (Δ_m) versus different orders of approximation (m) of the HAM-based solution for the selected case.

13.7.2   VALIDATION OF THE HPM-BASED SOLUTION

The HPM-based analytical solution for the selected test case is validated over the analytical solution given by Abd-el-Malek and El-Mansi (2000). After assessing the solution, it is found that the four-term HPM solution provides satisfactory values. Figure 13.4 shows a comparison of the four-term HPM solution with the existing

TABLE 13.1

Squared Residual Error (Δ_m) and Computational Time versus Different Orders of Approximation (m) for the Selected Case

Order of Approximation (m)	Squared Residual Error (Δ_m)	Computational Time (sec)
2	8.18×10^{-1}	0.695
4	6.53×10^{-1}	2.032
6	3.21×10^{-1}	4.255
8	1.52×10^{-1}	8.374
10	1.16×10^{-1}	13.694
12	0.92×10^{-1}	37.058
14	0.77×10^{-1}	54.321
16	0.51×10^{-1}	82.178
18	0.35×10^{-1}	102.331
20	0.23×10^{-1}	145.021

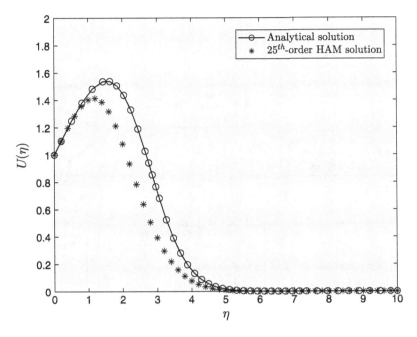

FIGURE 13.2 Comparison between the analytical solution and 25th-order HAM-based solution for the selected case.

analytical solution for a restricted domain, namely [0,5]. One can indeed try with some other initial approximations and/or linear operators, which might work in producing more accurate solutions. The HPM-based values and the existing analytical solution are compared in Table 13.3. A flowchart containing the steps of HPM is given in Figure 13.5.

TABLE 13.2

Comparison between 25th-Order HAM-Based Approximation and Analytical Solution for the Selected Case

η	Analytical Solution	HAM-Based Approximation 25th Order
0	1.0000	1.0000
1	1.4457	1.4017
2	1.4531	1.0800
3	0.7329	0.3924
4	0.1553	0.0772
5	0.0169	0.0091
6	0.0011	0.0007
7	0.0000	0.0001
8	0.0000	0.0001
9	0.0000	0.0000
10	0.0000	0.0000

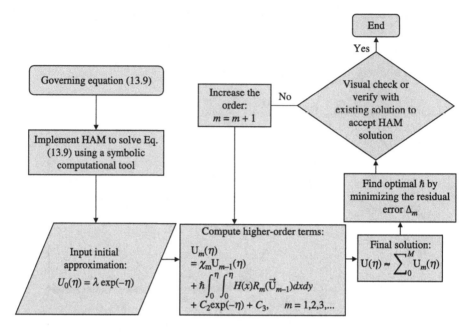

FIGURE 13.3   Flowchart for the HAM solution in Eq. (13.26).

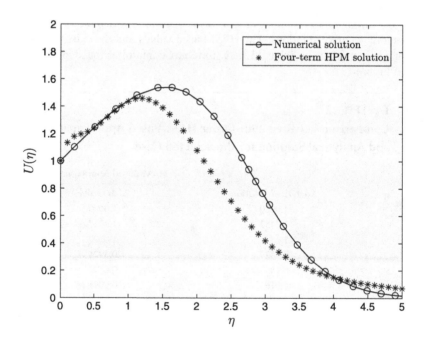

FIGURE 13.4   Comparison between the analytical solution and four-term HPM-based solution for the selected case.

TABLE 13.3

Comparison between Four-Term HPM-Based Approximation and Analytical Solution for the Selected Case

η	Analytical Solution	Four-Term HPM-Based Approximation
0.0	1.0000	1.0000
0.5	1.2441	1.2102
1.0	1.4457	1.3920
1.5	1.5410	1.3382
2.0	1.4531	1.0182
2.5	1.1511	0.6409
3.0	0.7329	0.3463
3.5	0.3724	0.1659
4.0	0.1553	0.0735
4.5	0.0550	0.0302
5.0	0.0169	0.0067

13.7.3 VALIDATION OF THE OHAM-BASED SOLUTION

For the assessment of the OHAM-based analytical solution, one needs to calculate the constant C_i. For that purpose, we calculate the residual as follows:

$$R(\eta, C_i) = \mathcal{L}\left[U_{OHAM}(\eta, C_i)\right] + \mathcal{N}\left[U_{OHAM}(\eta, C_i)\right] + h(\eta), i = 1, 2, \ldots, s \quad (13.56)$$

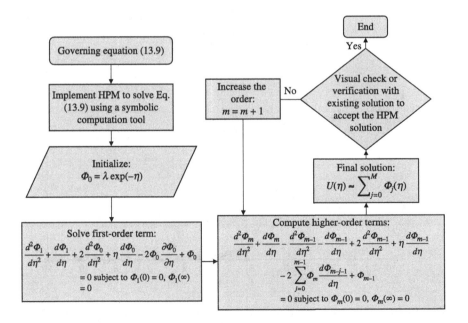

FIGURE 13.5   Flowchart for the HPM solution in Eq. (13.44).

where $U_{OHAM}(\eta, C_i)$ is the approximate solution. When $R(\eta, C_i) = 0$, $U_{OHAM}(\eta, C_i)$ becomes the exact solution. One of the ways to obtain the optimal C_i for which the solution converges is the minimization of the squared residual error, i.e.,

$$J(C_i) = \int_{\eta \in D} R^2(\eta, C_i)\, d\eta, \ i = 1, 2, \ldots, s \tag{13.57}$$

where $\eta = [0, 10]$ is the domain of the problem. The minimization of Eq. (13.57) leads to a system of algebraic equations as follows:

$$\frac{\partial J}{\partial C_1} = \frac{\partial J}{\partial C_2} = \cdots = \frac{\partial J}{\partial C_s} = 0 \tag{13.58}$$

Solving this system, one can obtain the optimal values of the parameters. We obtain the optimal values using the MATLAB routine *fminsearch*, which minimizes an unconstrained multivariable function. It is found that only the three-term solution produces feasible values. Therefore, the three-term OHAM solution is computed and compared in Figure 13.6 with the existing analytical solution of Abd-el-Malek and El-Mansi (2000) to see the effectiveness of the proposed approach. It can be seen that the OHAM-based series does not produce accurate results against the corresponding analytical solution. One might expect to get accurate results by modifying the operators and the auxiliary functions. For a quantitative assessment, we also compare the numerical values of the solutions in Table 13.4. A flowchart containing the steps of OHAM is provided in Figure 13.7.

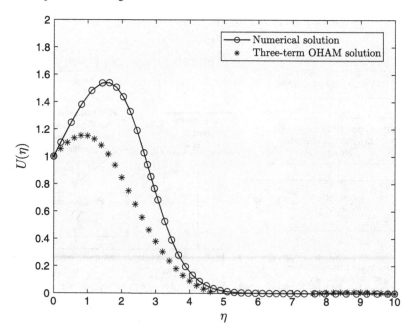

FIGURE 13.6 Comparison between the analytical solution and three-term OHAM-based solution for the selected case.

TABLE 13.4

Comparison between Three-Term OHAM-Based Approximation and Analytical Solution for the Selected Case

η	Analytical Solution	Three-Term OHAM-Based Approximation
0.0	1.0000	1.0000
0.5	1.2441	1.1187
1.0	1.4457	1.1500
1.5	1.5410	1.0508
2.0	1.4531	0.8454
2.5	1.1511	0.6023
3.0	0.7329	0.3799
3.5	0.3724	0.2075
4.0	0.1553	0.0903
4.5	0.0550	0.0203
5.0	0.0169	−0.0148

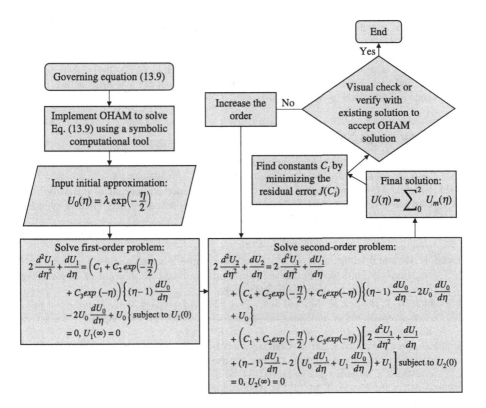

FIGURE 13.7 Flowchart for the OHAM solution in Eq. (13.52).

13.8 CONCLUDING REMARKS

Three kinds of analytical methods are applied to determine the series solution for Burger's equation. A relatively generalized form for Burger's equation is considered together with the initial and boundary conditions given in Abd-el-Malek and El-Mansi (2000), who provided a one-dimensional group-theoretic approach to convert the original PDE into a second-order ODE, which is easier to solve. This chapter follows the same approach proposed by them and derives three kinds of analytical solutions of the ODE using HAM, HPM, and OHAM. The existing group-theoretic approach can be potentially applied to solve Burger's equation arising in water engineering. Moreover, the HAM- and HPM-based solutions provide a good approximation to the problem when compared with the existing analytical solution. The OHAM provides a feasible solution with just three terms of the series solution. However, one might obtain a more accurate solution using OHAM if different forms of operators and auxiliary functions are chosen. This chapter shows the potential of HAM and HPM in the context of obtaining a series solution that is valid over a wide range of the problem. The theoretical and numerical convergences of the series solutions are also provided.

REFERENCES

Abd-el-Malek, M. B., and El-Mansi, S. M. (2000). Group theoretic methods applied to Burgers' equation. *Journal of Computational and Applied Mathematics*, *115*(1–2), 1–12.

Warrick, A. W., and Parkin, G. W. (1995). Analytical solution for one-dimensional drainage: Burgers' and simplified forms. *Water Resources Research*, *31*(11), 2891–2894.

Zhao, Y., and Liao, S. (2002). User guide to BVPh 2.0. School of naval architecture, ocean and civil engineering, Shanghai, *40*.

FURTHER READING

Abd-El-Malek, M. B. (1998). Application of the group-theoretical method to physical problems. *Journal of Nonlinear Mathematical Physics*, *5*(3), 314–330.

Abd-el-Malek, M. B., and Badran, N. A. (1990). Group method analysis of unsteady free-convective laminar boundary-layer flow on a nonisothermal vertical circular cylinder. *Acta Mechanica*, *85*(3), 193–206.

Basha, H. A. (2002). Burgers' equation: A general nonlinear solution of infiltration and redistribution. *Water Resources Research*, *38*(11), 29–1.

Desai, K. R., and Pradhan, V. H. (2012). Solution of Burger's equation and coupled Burger's equations by homotopy perturbation method. *International Journal of Engineering Research and Applications*, *2*(3), 2033–2040.

Fushchich, W. I., Shtelen, W. M., and Serov, N. I. (2013). *Symmetry analysis and exact solutions of equations of nonlinear mathematical physics* (Vol. 246). Springer Science & Business Media.

Gupta, A. K., and Ray, S. S. (2014). Comparison between homotopy perturbation method and optimal homotopy asymptotic method for the soliton solutions of Boussinesq–Burger equations. *Computers & Fluids*, *103*, 34–41.

Hamza, A. E., and Elzaki, T. M. (2015). Application of homotopy perturbation and Sumudu transform method for solving Burgers equations. *American Journal of Theoretical and Applied Statistics*, *4*(6), 480–483.

Inc, M. (2008). On numerical solution of Burgers' equation by homotopy analysis method. *Physics Letters A*, *372*(4), 356–360.

Islam, S., Nawaz, R., Arif, M., and Shah, I. A. (2012). Application of optimal homotopy asymptotic method to the equal width wave and Burger equations. *Life Science Journal*, *9*(4), 2380–2386.

Joshi, M. S., Desai, N. B., and Mehta, M. N. (2012). Solution of the burger's equation for longitudinal dispersion phenomena occurring in miscible phase flow through porous media. *ITB Journal of Engineering Sciences*, *44*(1), 61–76.

Molabahrami, A., and Khani, F. (2009). The homotopy analysis method to solve the Burgers–Huxley equation. *Nonlinear Analysis: Real World Applications*, *10*(2), 589–600.

Nawaz, R., Ullah, H., Islam, S., and Idrees, M. (2013). Application of optimal homotopy asymptotic method to Burger equations. *Journal of Applied Mathematics*, *2013* https://doi.org/10.1155/2013/387478.

Rashidi, M. M., Domairry, G., and Dinarvand, S. (2009). Approximate solutions for the Burger and regularized long wave equations by means of the homotopy analysis method. *Communications in Nonlinear Science and Numerical Simulation*, *14*(3), 708–717.

Ryu, S., Lyu, G., Do, Y., and Lee, G. (2020). Improved rainfall nowcasting using Burgers' equation. *Journal of Hydrology*, *581*, 124140.

Schwarz, F. (1988). Symmetries of differential equations: From Sophus lie to computer algebra. *Siam Review*, *30*(3), 450–481.

Shah, K. (2016). Solution of Burger's equation in a one-dimensional groundwater recharge by spreading using q-homotopy analysis method. *European Journal of Pure and Applied Mathematics*, *9*(1), 114–124.

Singh, R., Maurya, D. K., and Rajoria, Y. K. (2020). A mathematical model to solve the nonlinear Burger's equation by homotopy perturbation method. *Mathematics in Engineering, Science and Aerospace*, *11*(1), 115–125.

Singh, V. P., and Vimal, S. (2022). A unified framework for governing equations of hydrologic flows. *Journal of Hydrologic Engineering*, *27*(1), 04021044.

Smith, R. E., Smettem, K. R., and Broadbridge, P. (2002). *Infiltration theory for hydrologic applications*. American Geophysical Union.

14 Diffusive Wave Flood Routing Problem with Lateral Inflow

14.1 INTRODUCTION

Flow routing in open channels is done for a variety of hydraulic and hydrologic applications, such as flood forecasting and warning, evacuation measures, design of levees, design of drainage structures, assessment of inundation, damage assessment, flood protection measures, navigation, and river basin management. The hydraulic method of flow routing employs the continuity equation and momentum equation either in full form or in approximate form. In the momentum equation, the convective and local acceleration terms are of the same order of magnitude but of opposite signs, and they therefore tend to cancel out each other. With these acceleration terms deleted, the momentum equation then contains terms pertaining to the depth differential, bed slope, and frictional slope and is referred to as the diffusion wave approximation. When combined with the continuity equation, the diffusion wave model is obtained.

In flow routing, the channel may be subject to lateral inflow and outflow. In long river reaches, lateral flow terms are often neglected, but if a river is not leveed, then there is lateral inflow, and in arid and semi-arid areas, a river may be subject to significant leakage. Therefore, it is important to consider flow routing with the incorporation of lateral inflow or outflow. This is done in this chapter.

14.2 GOVERNING EQUATION

The St. Venant equations are derived from the depth integration of the Navier-Stokes equations, which are a result of the combined application of conservation of mass and momentum at a cross-section within the flow. For practical applications, the acceleration terms are small compared to the channel bed slope, and hence can be neglected (Henderson 1963; 1966; Dooge and Harley 1967; Vieira 1983). The diffusive wave equation can then be written as (Moussa 1996):

$$\frac{\partial Q}{\partial t} + C\left(\frac{\partial Q}{\partial x} - q\right) - D\left(\frac{\partial^2 Q}{\partial x^2} - \frac{\partial q}{\partial x}\right) = 0 \tag{14.1}$$

where t is the time; x is the downstream distance; C and D are commonly known as the celerity and diffusivity, respectively; and $q(x,t)$ is the distribution of lateral inflow. The total lateral inflow hydrograph can be given as:

$$Q_a(t) = \int_0^L q(x,t)dx \tag{14.2}$$

where L is the length of the river reach. The boundary and initial conditions are $Q(0,t)$ and $Q(x,0)$, where:

$$Q(x,0)=Q(0,0)+\int_0^x q(s,0)ds \qquad (14.3)$$

and for $x = L$:

$$Q(L,0)=Q(0,0)+Q_a(0) \qquad (14.4)$$

The simplified St. Venant equations using some hypotheses can be generally met in practical situations, though the validity may be questionable for some cases. However, the diffusive model can be applied successfully for most of the practical applications (Hayami 1951; Price 1974; Singh 1994).

14.2.1 DIFFUSIVE WAVE EQUATION WITHOUT LATERAL INFLOW

For a river reach with lateral inflow or outflow, we have $q(x,t)=0 \Rightarrow Q_a(t)=0$ and $Q(L,0)=Q(0,0)$. Accordingly, the diffusive wave in Eq. (14.1) becomes:

$$\frac{\partial Q}{\partial t}+C\frac{\partial Q}{\partial x}-D\frac{\partial^2 Q}{\partial x^2}=0 \qquad (14.5)$$

For a particular case $D=0$, the diffusive wave in Eq. (14.5) becomes the kinematic wave equation, whose solution requires only an upstream boundary condition (Li et al. 1975; Beven 1979; Weinmann and Laurenson 1979; Hromadka and DeVries 1988). However, this equation has a drawback, as it cannot predict the backwater effects or any dispersion or attenuation of the flood peak.

On the other hand, the diffusive wave equation allows us to compute a backwater profile. The analytical solutions were derived for Eq. (14.5) with constant C and D using the perturbation method (Hayami 1951) or Laplace transform method (Dooge 1973; Natale and Todini 1977). For a semi-infinite channel, no physical downstream boundary condition exists. Then, the diffusive wave equation has an analytical solution given by Hayami (1951) as follows:

$$Q(x,t)=Q(0,0)+\frac{x}{2(\pi D)^{0.5}}exp\left(\frac{Cx}{2D}\right)\int_0^t [Q(0,t-\tau)-Q(0,0)]$$

$$\times exp\left[\frac{-\frac{Cx}{4D}\left(\frac{x}{C\tau}+\frac{C\tau}{x}\right)}{\tau^{3/2}}\right]d\tau \text{ for } 0 \le x \le L \qquad (14.6)$$

Let us now define:

$$I(t)=Q(0,t)-Q(0,0) \qquad (14.7)$$

$$O(t) = Q(L,t) - Q(L,0) \tag{14.8}$$

$$K(t) = \frac{L}{2(\pi D)^{0.5}} \frac{exp\left[\dfrac{CL}{4D}\left(2 - \dfrac{L}{Ct} - \dfrac{Ct}{L}\right)\right]}{t^{3/2}}$$

$$= \left(\frac{\theta z}{\pi}\right)^{0.5} \frac{exp\left[z\left(2 - \dfrac{\theta}{t} - \dfrac{t}{\theta}\right)\right]}{t^{3/2}} = K_{\theta,z}(t) \tag{14.9}$$

$$\text{with } \theta = \frac{L}{C} \quad z = \frac{CL}{4D} \tag{14.10}$$

Here, $I(t)$ and $O(t)$ denote the upstream inflow minus the baseflow and the downstream outflow minus baseflow, respectively, and $K(t)$ is the Hayami kernel. Using Eqs. (14.7)–(14.10), one can obtain from Eq. (14.6):

$$O(t) = \int_0^t I(t-\tau)K(\tau)d\tau = I(t) * K(t) \tag{14.11}$$

Here * represents the convolution relation. In practical situations, attenuation and dispersion of the flood wave may be masked by significant lateral inflows or outflows. Finite difference methods are used when the lateral inflow is incorporated. However, finite difference methods may lead to numerical instabilities and some dispersion and attenuation.

14.2.2   DIFFUSIVE WAVE EQUATION WITH LATERAL INFLOW

Moussa (1996) proposed an analytical solution to the diffusive wave equation with lateral inflow (or outflow) uniformly distributed over a channel reach using the Hayami hypotheses, i.e., C and D are constant and there is no physical downstream boundary condition.

Larinier and Saucerotte (1970) proposed the lateral inflow $q(x,t)$ as a multiplication of two functions:

$$q(x,t) = f(x)g(t) \tag{14.12}$$

where $f(x)$ is a function of space that describes the lateral subcatchment shape and $g(t)$ is a function of time that describes the distribution of effective rainfall. Here, $q(x,t) > 0$ or < 0 represents the lateral inflow and outflow (i.e., seepage), respectively. Using Eq. (14.12), Eq. (14.1) reduces to:

$$\frac{\partial Q}{\partial t} + C\frac{\partial Q}{\partial x} - D\frac{\partial^2 Q}{\partial x^2} = \left[Cf(x) - Df'(x)\right]g(t) \tag{14.13}$$

When the lateral inflow is uniformly distributed along a river reach, then:

$$q(x,t) = \frac{Q_a(t)}{L} \qquad (14.14)$$

with:

$$f(x) = \frac{1}{L} \text{ and } g(t) = Q_a(t) \qquad (14.15)$$

Using Eq. (14.15) in Eq. (14.13), we have:

$$\frac{\partial Q}{\partial t} + C\frac{\partial Q}{\partial x} - D\frac{\partial^2 Q}{\partial x^2} = C\frac{Q_a(t)}{L} \qquad (14.16)$$

with the initial condition:

$$Q(x,0) = Q(0,0) + \frac{x}{L}Q_a(0) \qquad (14.17)$$

Moussa (1996) proposed an analytical solution to Eq. (14.16) together with the initial condition in Eq. (14.17) using the Laplace transform method. Here, we aim to solve the equation using three kinds of series solution techniques.

14.3 HAM-BASED SOLUTION

To solve Eq. (14.16) using the homotopy analysis method (HAM) subject to the initial condition in Eq. (14.17), the zeroth-order deformation equation can be constructed as follows:

$$(1-q)\mathcal{L}\big[\Phi(x,t;q) - Q_0(x,t)\big] = q\hbar H(x,t)\mathcal{N}\big[\Phi(x,t;q)\big] \qquad (14.18)$$

subject to the initial condition:

$$\Phi(x,0;q) = Q(0,0) + \frac{x}{L}Q_a(0) \qquad (14.19)$$

where q is the embedding parameter, $\Phi(x,t;q)$ is the representation of the solution across q, $Q_0(x,t)$ is the initial approximation, $\hbar$ is the auxiliary parameter, $H(x,t)$ is the auxiliary function, and $\mathcal{L}$ and $\mathcal{N}$ are the linear and nonlinear operators, respectively. The higher-order terms can be calculated from the higher-order deformation equation, given as follows:

$$\mathcal{L}\big[Q_m(x,t) - \chi_m Q_{m-1}(x,t)\big] = \hbar H(x,t)R_m\big(\vec{Q}_{m-1}\big), \ m = 1,2,3,\ldots \qquad (14.20)$$

subject to:

$$Q_m(x,0) = 0 \qquad (14.21)$$

where:

$$\chi_m = \begin{cases} 0 & \text{when } m = 1, \\ 1 & \text{otherwise} \end{cases} \qquad (14.22)$$

and:

$$R_m \left(\vec{Q}_{m-1} \right) = \frac{1}{(m-1)!} \frac{\partial^{m-1} \mathcal{N} \left[\Phi(x,t;q) \right]}{\partial q^{m-1}} \Bigg|_{q=0} \qquad (14.23)$$

where Q_m for $m \geq 1$ is the higher-order term. Now, the final solution can be obtained as follows:

$$Q(x,t) = Q_0(x,t) + \sum_{m=1}^{\infty} Q_m(x,t) \qquad (14.24)$$

The nonlinear operator for the problem is selected as:

$$\mathcal{N} \left[\Phi(x,t;q) \right] = \frac{\partial \Phi(x,t;q)}{\partial t} + C \frac{\partial \Phi(x,t;q)}{\partial x} - D \frac{\partial^2 \Phi(x,t;q)}{\partial x^2} - C \frac{Q_a(t)}{L} \qquad (14.25)$$

Therefore, using Eq. (14.25), the term R_m can be calculated from Eq. (14.23) as follows:

$$R_m \left(\vec{Q}_{m-1} \right) = \frac{\partial Q_{m-1}}{\partial t} + C \frac{\partial Q_{m-1}}{\partial x} - D \frac{\partial^2 Q_{m-1}}{\partial x^2} - C \frac{Q_a(t)}{L} (1 - \chi_m) \qquad (14.26)$$

For the present problem, the linear operator and the initial approximation are chosen, respectively, as follows:

$$\mathcal{L} \left[\Phi(x,t;q) \right] = \frac{\partial \Phi(x,t;q)}{\partial t} \text{ with the property } \mathcal{L}[C_2] = 0 \qquad (14.27)$$

$$Q_0(x,t) = Q(0,0) + \frac{x}{L} Q_a(0) \qquad (14.28)$$

where C_2 is an integral constant. Using Eq. (14.27), the higher-order terms can be obtained from Eq. (14.20) as follows:

$$Q_m(x,t) = \chi_m Q_{m-1}(x,t) + \hbar \int_0^t H(x,t) R_m \left(\vec{Q}_{m-1} \right) dt + C_2, \ m = 1,2,3,\ldots \qquad (14.29)$$

where R_m is given by Eq. (14.26) and the constant C_2 can be determined from the given initial condition for the higher-order deformation equations given by Eq. (14.20).

Now, the auxiliary function $H(x,t)$ can be determined from the *rule of coefficient ergodicity*. However, for simplicity one can assume $H(x,t)=1$ (Vajravelu and Van Gorder 2013). Finally, the approximate solution can be obtained as follows:

$$Q(x,t) \approx Q_0(x,t) + \sum_{m=1}^{M} Q_m(x,t) \qquad (14.30)$$

14.4   HPM-BASED SOLUTION

When applying homotopy perturbation method (HPM), we rewrite Eq. (14.16) in the following form:

$$\mathcal{N}(Q) = f(x,t) \qquad (14.31)$$

We construct a homotopy $\Phi(x,t;q)$ that satisfies:

$$(1-q)\left[\mathcal{L}(\Phi(x,t;q)) - \mathcal{L}(Q_0(x,t))\right] + q\left[\mathcal{N}(\Phi(x,t;q)) - f(x,t)\right] = 0 \qquad (14.32)$$

where q is the embedding parameter that lies on $[0,1]$, $\mathcal{L}$ is the linear operator, and $Q_0(x,t)$ is the initial approximation to the final solution. Eq. (14.32) shows that as $q = 0$, $\Phi(x,t;0) = Q_0(x,t)$ and as $q = 1$, $\Phi(x,t;1) = Q(x,t)$. We now express $\Phi(x,t;q)$ as a series in terms of q, as follows:

$$\Phi(x,t;q) = \Phi_0 + q\Phi_1 + q^2\Phi_2 + q^3\Phi_3 + q^4\Phi_4 \ldots \qquad (14.33)$$

where Φ_m for $m \geq 1$ is the higher-order term. As $q \to 1$, Eq. (14.33) produces the final solution as:

$$Q(x,t) = \lim_{q \to 1} \Phi(x,t;q) = \sum_{i=0}^{\infty} \Phi_i \qquad (14.34)$$

For simplicity, let us consider the linear and nonlinear operators as follows:

$$\mathcal{L}(\Phi(x,t;q)) = \frac{\partial \Phi(x,t;q)}{\partial t} \qquad (14.35)$$

$$\mathcal{N}(\Phi(x,t;q)) = \frac{\partial \Phi(x,t;q)}{\partial t} + C\frac{\partial \Phi(x,t;q)}{\partial x} - D\frac{\partial^2 \Phi(x,t;q)}{\partial x^2} - C\frac{Q_a(t)}{L} \qquad (14.36)$$

Using these expressions, Eq. (14.32) takes on the form:

$$(1-q)\left[\frac{\partial \Phi(x,t;q)}{\partial t} - \frac{\partial Q_0}{\partial t}\right]$$
$$+ q\left[\frac{\partial \Phi(x,t;q)}{\partial t} + C\frac{\partial \Phi(x,t;q)}{\partial x} - D\frac{\partial^2 \Phi(x,t;q)}{\partial x^2} - C\frac{Q_a(t)}{L}\right] = 0 \quad (14.37)$$

Using Eq. (14.33), we have:

$$(1-q)\left[\frac{\partial \Phi(x,t;q)}{\partial t} - \frac{\partial Q_0}{\partial t}\right]$$

$$= (1-q)\left[\left(\frac{\partial \Phi_0}{\partial t} - \frac{\partial Q_0}{\partial t}\right) + q\frac{\partial \Phi_1}{\partial t} + q^2\frac{\partial \Phi_2}{\partial t} + q^3\frac{\partial \Phi_3}{\partial t} + q^4\frac{\partial \Phi_4}{\partial t} + \cdots\right]$$

$$= \left(\frac{\partial \Phi_0}{\partial t} - \frac{\partial Q_0}{\partial t}\right) + q\left(\frac{\partial \Phi_1}{\partial t} - \left(\frac{\partial \Phi_0}{\partial t} - \frac{\partial Q_0}{\partial t}\right)\right) + q^2\left(\frac{\partial \Phi_2}{\partial t} - \frac{\partial \Phi_1}{\partial t}\right)$$

$$+ q^3\left(\frac{\partial \Phi_3}{\partial t} - \frac{\partial \Phi_2}{\partial t}\right) + q^4\left(\frac{\partial \Phi_4}{\partial t} - \frac{\partial \Phi_3}{\partial t}\right) + \cdots \qquad (14.38)$$

$$\frac{\partial \Phi(x,t;q)}{\partial t} + C\frac{\partial \Phi(x,t;q)}{\partial x} - D\frac{\partial^2 \Phi(x,t;q)}{\partial x^2} - C\frac{Q_a(t)}{L}$$

$$= \left(\frac{\partial \Phi_0}{\partial t} + q\frac{\partial \Phi_1}{\partial t} + q^2\frac{\partial \Phi_2}{\partial t} + q^3\frac{\partial \Phi_3}{\partial t} + q^4\frac{\partial \Phi_4}{\partial t} + \cdots\right)$$

$$+ C\left(\frac{\partial \Phi_0}{\partial x} + q\frac{\partial \Phi_1}{\partial x} + q^2\frac{\partial \Phi_2}{\partial x} + q^3\frac{\partial \Phi_3}{\partial x} + q^4\frac{\partial \Phi_4}{\partial x} + \cdots\right)$$

$$- D\left(\frac{\partial^2 \Phi_0}{\partial x^2} + q\frac{\partial^2 \Phi_1}{\partial x^2} + q^2\frac{\partial^2 \Phi_2}{\partial x^2} + q^3\frac{\partial^2 \Phi_3}{\partial x^2} + q^4\frac{\partial^2 \Phi_4}{\partial x^2} + \cdots\right)$$

$$- C\frac{Q_a(t)}{L} = \left[\frac{\partial \Phi_0}{\partial t} - C\frac{\partial \Phi_0}{\partial x} - D\frac{\partial^2 \Phi_0}{\partial x^2} - C\frac{Q_a(t)}{L}\right]$$

$$+ q\left[\frac{\partial \Phi_1}{\partial t} - C\frac{\partial \Phi_1}{\partial x} - D\frac{\partial^2 \Phi_1}{\partial x^2}\right] + q^2\left[\frac{\partial \Phi_2}{\partial t} - C\frac{\partial \Phi_2}{\partial x} - D\frac{\partial^2 \Phi_2}{\partial x^2}\right]$$

$$+ q^3\left[\frac{\partial \Phi_3}{\partial t} - C\frac{\partial \Phi_3}{\partial x} - D\frac{\partial^2 \Phi_3}{\partial x^2}\right] + \cdots \qquad (14.39)$$

Using the initial condition given by Eq. (14.17), we have:

$$\Phi_0(x,0) = Q(0,0) + \frac{x}{L}Q_a(0), \quad \Phi_1(x,0) = 0, \quad \Phi_2(x,0) = 0,\dots \qquad (14.40)$$

Using Eqs. (14.38)–(14.40) and equating the like powers of q in Eq. (14.37), the following system of differential equations is obtained:

$$\frac{\partial \Phi_0}{\partial t} - \frac{\partial Q_0}{\partial t} = 0 \text{ subject to } \Phi_0(x,0) = Q(0,0) + \frac{x}{L}Q_a(0) \qquad (14.41)$$

$$\frac{\partial \Phi_1}{\partial t} - \left(\frac{\partial \Phi_0}{\partial t} - \frac{\partial Q_0}{\partial t}\right) + \frac{\partial \Phi_0}{\partial t} - C\frac{\partial \Phi_0}{\partial x} - D\frac{\partial^2 \Phi_0}{\partial x^2} - C\frac{Q_a(t)}{L}$$

$$= 0 \text{ subject to } \Phi_1(x,0) = 0 \qquad (14.42)$$

$$\frac{\partial \Phi_2}{\partial t} - \frac{\partial \Phi_1}{\partial t} + \frac{\partial \Phi_1}{\partial t} - C\frac{\partial \Phi_1}{\partial x} - D\frac{\partial^2 \Phi_1}{\partial x^2} = 0 \text{ subject to } \Phi_2(x,0) = 0 \quad (14.43)$$

$$\frac{\partial \Phi_3}{\partial t} - \frac{\partial \Phi_2}{\partial t} + \frac{\partial \Phi_2}{\partial t} - C\frac{\partial \Phi_2}{\partial x} - D\frac{\partial^2 \Phi_2}{\partial x^2} = 0 \text{ subject to } \Phi_3(x,0) = 0 \quad (14.44)$$

$$\frac{\partial \Phi_4}{\partial t} - \frac{\partial \Phi_3}{\partial t} + \frac{\partial \Phi_3}{\partial t} - C\frac{\partial \Phi_3}{\partial x} - D\frac{\partial^2 \Phi_3}{\partial x^2} = 0 \text{ subject to } \Phi_4(x,0) = 0 \quad (14.45)$$

Proceeding in a like manner, one can arrive at the following recurrence relation:

$$\frac{\partial \Phi_m}{\partial t} = C\frac{\partial \Phi_{m-1}}{\partial x} - D\frac{\partial^2 \Phi_{m-1}}{\partial x^2} \text{ subject to } \Phi_m(x,0) = 0 \quad (14.46)$$

The initial approximation can be chosen as $\Phi_0 = Q(0,0) + \frac{x}{L}Q_a(0)$. Using this initial approximation, we can solve the equations iteratively using symbolic software. We do not include those expressions here, as they are lengthy. Finally, the HPM-based solution can be approximated as follows:

$$Q(x,t) \approx \sum_{i=0}^{M} \Phi_i \quad (14.47)$$

14.5 OHAM-BASED SOLUTION

When applying the optimal homotopy asymptotic method (OHAM), Eq. (14.16) can be written in the following form:

$$\mathcal{L}[Q(x,t)] + \mathcal{N}[Q(x,t)] + h(x,t) = 0 \quad (14.48)$$

We select $\mathcal{L}[Q(x,t)] = \frac{\partial Q}{\partial t}$, $\mathcal{N}[Q(x,t)] = C\frac{\partial Q}{\partial x} - D\frac{\partial^2 Q}{\partial x^2} - C\frac{Q_a(t)}{L}$, and $h(x,t) = 0$. With these considerations, the zeroth-order problem becomes:

$$\mathcal{L}(Q_0(x,t)) + h(x,t) = 0 \text{ subject to } Q_0(x,0) = Q(0,0) + \frac{x}{L}Q_a(0) \quad (14.49)$$

Solving Eq. (14.49), one obtains:

$$Q_0(x,t) = Q(0,0) + \frac{x}{L}Q_a(0) \quad (14.50)$$

To obtain $\mathcal{N}_0, \mathcal{N}_1, \mathcal{N}_2$, etc., the following can be used:

$$C\frac{\partial Q}{\partial x} - D\frac{\partial^2 Q}{\partial x^2} - C\frac{Q_a(t)}{L} = C\left(\frac{\partial Q_0}{\partial x} + q\frac{\partial Q_1}{\partial x} + q^2\frac{\partial Q_2}{\partial x} + q^3\frac{\partial Q_3}{\partial x} + q^4\frac{\partial Q_4}{\partial x} + \cdots\right)$$

$$-D\left(\frac{\partial^2 Q_0}{\partial x^2} + q\frac{\partial^2 Q_1}{\partial x^2} + q^2\frac{\partial^2 Q_2}{\partial x^2} + q^3\frac{\partial^2 Q_3}{\partial x^2} + q^4\frac{\partial^2 Q_4}{\partial x^2} + \cdots\right)$$

$$-C\frac{Q_a(t)}{L} = \left[C\frac{\partial Q_0}{\partial x} - D\frac{\partial^2 Q_0}{\partial x^2} - C\frac{Q_a(t)}{L}\right] + q\left[C\frac{\partial Q_1}{\partial x} - D\frac{\partial^2 Q_1}{\partial x^2}\right]$$

$$+q^2\left[C\frac{\partial Q_2}{\partial x} - D\frac{\partial^2 Q_2}{\partial x^2}\right] + q^3\left[C\frac{\partial Q_3}{\partial x} - D\frac{\partial^2 Q_3}{\partial x^2}\right] + q^4\left[C\frac{\partial Q_4}{\partial x} - D\frac{\partial^2 Q_4}{\partial x^2}\right] + \cdots$$

$$(14.51)$$

Using Eq. (14.51), the first-order problem reduces to:

$$\frac{\partial Q_1}{\partial t} = H_1(x,t,C_i)\left\{C\frac{\partial Q_0}{\partial x} - D\frac{\partial^2 Q_0}{\partial x^2} - C\frac{Q_a(t)}{L}\right\} \text{ subject to } Q_1(x,0)=0 \qquad (14.52)$$

The auxiliary functions can be chosen in many ways. Here, we select $H_1(x,t,C_i)=C_1$. Putting $k=2$, the second-order problem becomes:

$$\mathcal{L}\left[Q_2(x,t,C_i)-Q_1(x,t,C_i)\right] = H_2(x,t,C_i)\mathcal{N}_0\left(Q_0(x,t)\right)$$
$$+ H_1(x,t,C_i)\left[\mathcal{L}\left[Q_1(x,t,C_i)\right]+\mathcal{N}_1\left[Q_0(x,t),Q_1(x,t,C_i)\right]\right]$$
$$\text{subject to } Q_2(x,0)=0 \qquad (14.53)$$

Further, we choose $H_2(x,t,C_i)=C_2$. Therefore, Eq. (14.53) becomes:

$$\frac{\partial Q_2}{\partial t} = \frac{\partial Q_1}{\partial t} + C_2\left\{C\frac{\partial Q_0}{\partial x} - D\frac{\partial^2 Q_0}{\partial x^2} - C\frac{Q_a(t)}{L}\right\}$$
$$+ C_1\left[\frac{\partial Q_1}{\partial t} + C\frac{\partial Q_1}{\partial x} - D\frac{\partial^2 Q_1}{\partial x^2}\right] \text{ subject to } Q_2(x,0)=0 \qquad (14.54)$$

The terms of the series can be computed using the equations developed here. One can compute these terms without any difficulty using symbolic computation software, such as MATLAB. Further, following Chapter 2, the higher-order terms can be computed in a similar manner. However, our aim is to produce an accurate solution with just two to three terms of the OHAM-based series. Therefore, we restrict our calculation up to $k=2$. Finally, the approximate solution can be found as:

$$Q(x,t)\approx Q_0(x,t)+Q_1(x,t,C_i)+Q_2(x,t,C_i) \qquad (14.55)$$

where the terms are given by Eqs. (14.50), (14.52), and (14.54).

14.6 CONVERGENCE THEOREMS

The convergence of the HAM-based and OHAM-based solutions are proved theoretically using the following theorems.

14.6.1 CONVERGENCE THEOREM OF THE HAM-BASED SOLUTION

The convergence theorem for the HAM-based solution given by Eq. (14.30) can be proved using the following theorems.

Theorem 14.1: If the homotopy series $\sum_{m=0}^{\infty} Q_m(x,t), \sum_{m=0}^{\infty}\frac{\partial Q_m(x,t)}{\partial t}, \sum_{m=0}^{\infty}\frac{\partial Q_m(x,t)}{\partial x}$
and $\sum_{m=0}^{\infty}\frac{\partial^2 Q_m(x,t)}{\partial x^2}$ converge, then $R_m\left(\vec{Q}_{m-1}\right)$ given by Eq. (14.26) satisfies the relation $\sum_{m=1}^{\infty} R_m\left(\vec{Q}_{m-1}\right)=0$.

Proof: The proof of this theorem follows exactly from the previous chapters.

Theorem 14.2: If $\hbar$ is so properly chosen that the series $\Sigma_{m=0}^{\infty} Q_m(x,t)$, $\Sigma_{m=0}^{\infty}\dfrac{\partial Q_m(x,t)}{\partial t}$, $\Sigma_{m=0}^{\infty}\dfrac{\partial Q_m(x,t)}{\partial x}$ and $\Sigma_{m=0}^{\infty}\dfrac{\partial^2 Q_m(x,t)}{\partial x^2}$ converge absolutely to $Q(x,t)$, $\dfrac{\partial Q(x,t)}{\partial t}$, $\dfrac{\partial Q(x,t)}{\partial x}$, and $\dfrac{\partial^2 Q(x,t)}{\partial x^2}$, respectively, then the homotopy series $\Sigma_{m=0}^{\infty} Q_m(x,t)$ satisfies the original governing Eq. (14.16).

Proof: Theorem 14.1 shows that if $\Sigma_{m=0}^{\infty} Q_m(x,t)$, $\Sigma_{m=0}^{\infty}\dfrac{\partial Q_m(x,t)}{\partial t}$, $\Sigma_{m=0}^{\infty}\dfrac{\partial Q_m(x,t)}{\partial x}$ and $\Sigma_{m=0}^{\infty}\dfrac{\partial^2 Q_m(x,t)}{\partial x^2}$ converge, then $\Sigma_{m=1}^{\infty} R_m\left(\vec{Q}_{m-1}\right)=0$.

Therefore, using Eq. (14.26) and simplifying further lead to:

$$\sum_{m=0}^{\infty}\frac{\partial Q_m}{\partial t}+C\sum_{m=0}^{\infty}\frac{\partial Q_m}{\partial x}-D\sum_{m=0}^{\infty}\frac{\partial^2 Q_m}{\partial x^2}-C\frac{Q_a(t)}{L}\sum_{m=0}^{\infty}(1-\chi_m)=0 \quad (14.56)$$

which is basically the original governing Eq. (14.16). Furthermore, subject to the initial condition $Q(x,0)=Q(0,0)+\dfrac{x}{L}Q_a(0)$ and the conditions for the higher-order deformation equation $Q_m(x,0)=0$, for $m\geq 1$, we easily obtain $\Sigma_{m=0}^{\infty} Q_m(x,0)=Q(0,0)+\dfrac{x}{L}Q_a(0)$. Hence, the convergence result follows.

14.6.2 Convergence Theorem of the OHAM-Based Solution

Theorem 14.3: If the series $Q_0(x,t)+\Sigma_{j=1}^{\infty} Q_j(x,t,C_i)$, $i=1,2,\ldots,s$ converges, where $Q_j(x,t,C_i)$ are governed by Eqs. (14.50), (14.52), and (14.54), then Eq. (14.55) is a solution of the original Eq. (14.16).

Proof: The proof of this theorem follows exactly from Theorem 2.3 of Chapter 2.

14.7 RESULTS AND DISCUSSION

First, we discuss the expressions and the values of the parameters needed for the assessment of the developed solution. Then, the numerical convergence of the HAM-based analytical solution is established for a specific test case, and then the solution is validated over the existing analytical solution given by Moussa (1996). Finally, the HPM- and OHAM-based analytical solutions are validated by comparing them with the solution given by Moussa (1996).

14.7.1 Selection of Expressions and Parameters

The identification of lateral inflow or outflow is given in Moussa (1996), who carried out an inverse problem analysis to identify the function $Q_a(t)$. Functions C and D can be taken as constants. The values considered herein are $C=1.33$ ms^{-1} and $D=4734$ m^{2}s^{-1}.

14.7.2 NUMERICAL CONVERGENCE AND VALIDATION OF THE HAM SOLUTION

Here, we have calculated the so-called $\hbar-$curves to find a suitable choice for the auxiliary parameter $\hbar$, which determines the convergence of the HAM-based series solution. For a particular order of approximation, we plot the approximate solution $Q(x,t)$ (or its derivatives) at some point within the domain. The flatness of the $\hbar-$curve determines a suitable choice for the auxiliary parameter $\hbar$. In Figure 14.1, we plot the $\hbar-$curve for the 15th-order HAM-based approximation for the values $Q(0.5,0)$ and $Q(0,0.5)$, respectively. From the exact solution, those quantities can be calculated, and it is observed from the figures that the curves become flat for a specific range of $\hbar$. Any choice of $\hbar$ within this range determines an optimal value for which the series solution converges (Abbasbandy et al. 2011).

Now, we test the performance of the HAM-based analytical solution for the present problem by comparing it with the analytical solution of Moussa (1996). For that purpose, we consider the 15th-order HAM-based solution and comparison, as shown in Figure 14.2 at two different locations along the river reach. It may be noted that for better visualization, we have nondimensionalized the variables. The nondimensionalization is done following Cimorelli et al. (2014). It can be observed from Figure 14.2 that the homotopy based solutions accurately predict the analytical solution considered here. For a quantitative assessment, a numerical comparison between the HAM-based solution and analytical solution is shown in Table 14.1, where we consider some selected values of the domain. A flowchart containing the steps of the method is provided in Figure 14.3. All the computations are performed following the Mathematica package developed by Zhao and Liao (2002).

14.7.3 VALIDATION OF THE HPM-BASED SOLUTION

The HPM-based analytical solution for the selected test case is validated over the analytical solution of Moussa (1996). It is seen that the three-term HPM produces accurate results over the restricted range of the domain. Figure 14.4 compares the numerical solution and the HPM approximation. The HPM-based values and the numerical solution are compared numerically in Table 14.2. A flowchart containing the steps of the method is given in Figure 14.5.

14.7.4 VALIDATION OF THE OHAM-BASED SOLUTION

For the assessment of the OHAM-based analytical solution, one needs to calculate the constant C_i. For that purpose, we calculate the residual as follows:

$$R(x,t,C_i) = \mathcal{L}\left[Q_{OHAM}(x,t,C_i)\right] + \mathcal{N}\left[Q_{OHAM}(x,t,C_i)\right] + h(x,t),\, i = 1,2,\ldots,s \quad (14.57)$$

where $Q_{OHAM}(x,t,C_i)$ is the approximate solution. When $R(x,t,C_i) = 0$, $Q_{OHAM}(x,t,C_i)$ becomes the exact solution to the equation. One of the ways to obtain the optimal C_i for which the solution converges is the minimization of the squared residual error, i.e.,

$$J(C_i) = \int_{(x,t)\in D} R^2(x,t,C_i)\,dxdt,\, i = 1,2,\ldots,s \quad (14.58)$$

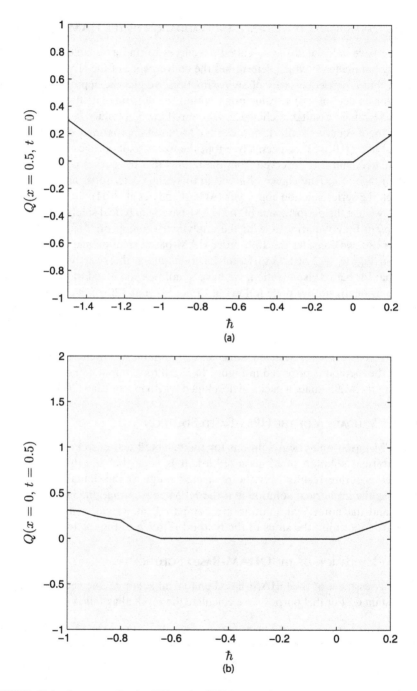

FIGURE 14.1 $\hbar$–curves for the 15th-order HAM approximation considering (A) $Q(0.5,0)$ and (B) $Q(0,0.5)$.

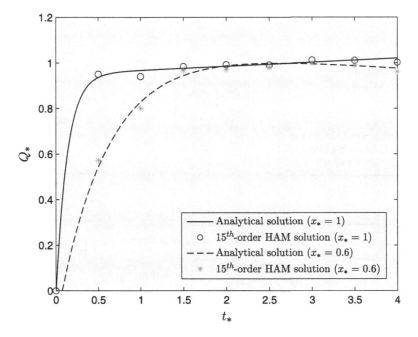

FIGURE 14.2 Comparison between the 15th-order HAM-based solution and analytical solution (Moussa 1996) for two locations.

TABLE 14.1
Comparison between HAM-Based
Approximation and Analytical Solution

t_*	Analytical Solution		15th-Order HAM-Based Approximation	
	$x_* = 1$	$x_* = 0.6$	$x_* = 1$	$x_* = 0.6$
0.0	0.00	0.00	0.00	0.00
0.5	0.94	0.54	0.95	0.57
1.0	0.97	0.83	0.94	0.80
1.5	0.98	0.94	0.98	0.97
2.0	0.99	0.99	0.99	0.97
2.5	0.99	1.00	0.99	0.98
3.0	1.00	1.00	1.01	1.00
3.5	1.01	0.99	1.01	1.00
4.0	1.02	0.98	1.00	0.96

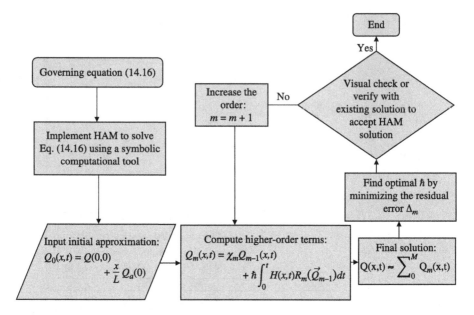

FIGURE 14.3   Flowchart for the HAM solution in Eq. (14.30).

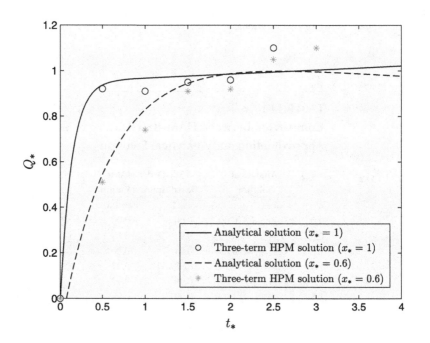

FIGURE 14.4   Comparison between the three-term HPM-based solution and analytical solution of Moussa (1996).

TABLE 14.2

Comparison between HPM-Based Approximation and Analytical Solution

t_*	Analytical Solution		Three-Term HPM-Based Approximation	
	$x_* = 1$	$x_* = 0.6$	$x_* = 1$	$x_* = 0.6$
0.0	0.00	0.00	0.00	0.00
0.5	0.94	0.54	0.92	0.51
1.0	0.97	0.83	0.91	0.74
1.5	0.98	0.94	0.95	0.91
2.0	0.99	0.99	0.96	0.92
2.5	0.99	1.00	1.1	1.05
3.0	1.00	1.00	1.2	1.1
3.5	1.01	0.99	1.3	1.3
4.0	1.02	0.98	1.8	1.9

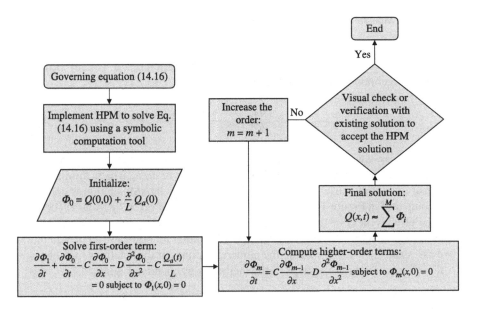

FIGURE 14.5 Flowchart for the HPM solution in Eq. (14.47).

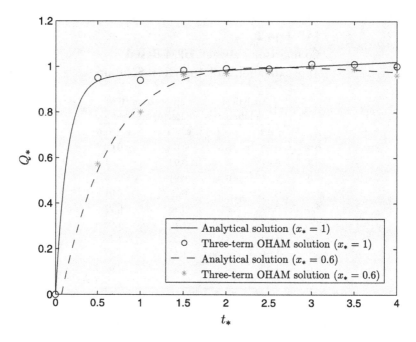

FIGURE 14.6 Comparison between the three-term OHAM-based solution and the analytical solution (Moussa 1996).

where D is the domain of the problem. The minimization of Eq. (14.58) leads to a system of algebraic equations as follows:

$$\frac{\partial J}{\partial C_1} = \frac{\partial J}{\partial C_2} = \cdots = \frac{\partial J}{\partial C_s} = 0 \qquad (14.59)$$

Solving this system, one can obtain the optimal values of the parameters. We obtain the optimal values using the MATLAB routine *fminsearch*, which minimizes an unconstrained multivariable function. The three-term OHAM solution is computed and compared in Figure 14.6 with the analytical solution given by Moussa (1996) to see the effectiveness of the HAM approach. It can be seen that the OHAM-based approximation agrees well with the numerical solution for both cases. For a quantitative assessment, we also compare the numerical values of the solutions in Table 14.3. A flowchart containing the steps of the method is provided in Figure 14.7.

14.8 CONCLUDING REMARKS

Three kinds of analytical methods are applied to derive analytical solutions for the diffusive wave equation used in flood routing in rivers following the work of Moussa (1996). The derived solution incorporates the lateral inflow or outflow uniformly distributed along a river reach. It is seen that the HAM- and OHAM-based solutions provide a good approximation to the equation when compared with the existing

TABLE 14.3

Comparison between OHAM-Based Approximation and Analytical Solution

t_*	Analytical Solution		Three-Term OHAM-Based Approximation	
	$x_* = 1$	$x_* = 0.6$	$x_* = 1$	$x_* = 0.6$
0.0	0.00	0.00	0.00	0.00
0.5	0.94	0.54	0.95	0.57
1.0	0.97	0.83	0.94	0.80
1.5	0.98	0.94	0.98	0.97
2.0	0.99	0.99	0.99	0.97
2.5	0.99	1.00	0.99	0.98
3.0	1.00	1.00	1.01	1.00
3.5	1.01	0.99	1.01	1.00
4.0	1.02	0.98	1.00	0.96

analytical solution. The OHAM provides an accurate solution with just three terms of the series solution. It is also found that the HPM can produce an accurate solution only for a restricted domain of the problem. However, one might obtain a more accurate solution using HPM if different forms of the initial approximation and linear operator are chosen. The theoretical and numerical convergences of the series solutions are provided.

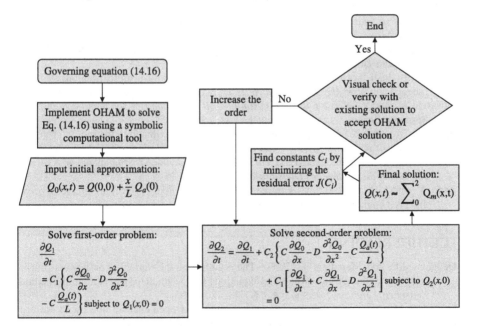

FIGURE 14.7 Flowchart for the OHAM solution in Eq. (14.55).

REFERENCES

Abbasbandy, S., Shivanian, E., and Vajravelu, K. (2011). Mathematical properties of ℏ-curve in the framework of the homotopy analysis method. *Communications in Nonlinear Science and Numerical Simulation*, *16*(11), 4268–4275.

Beven, K. (1979). On the generalized kinematic routing method. *Water Resources Research*, *15*(5), 1238–1242.

Cimorelli, L., Cozzolino, L., Della Morte, R., and Pianese, D. (2014). Analytical solutions of the linearized parabolic wave accounting for downstream boundary condition and uniform lateral inflows. *Advances in Water Resources*, *63*, 57–76.

Dooge, J. (1973). *Linear theory of hydrologic systems* (No. 1468). Agricultural Research Service, US Department of Agriculture, Washington, D.C.

Dooge, J. C. I. and Harley, B. M. 1967. 'Linear routing in uniform channels', *Proc. Int. Hydrol. Symp.*, 1, Colorado State University, Fort Collins. pp. 57–63.

Hayami, S. (1951). On the propagation of flood waves. *Bulletins-Disaster Prevention Research Institute, Kyoto University*, *1*, 1–16.

Henderson, F. M. (1963). Flood waves in prismatic channels. *Journal of the Hydraulics Division*, *89*(4), 39–67.

Henderson, F. M. (1966). *Open channel flow*. Macmillan, New York, NY.

Hromadka, T. V., and DeVries, J. J. (1988). Kinematic wave routing and computational error. *Journal of Hydraulic Engineering*, *114*(2), 207–217.

Larinier, M. and Saucerotte, H. 1970. 'La propagation des crues a partir des hypotheses d'Hayami', Rapport du Laboratoire dHydrologie Mathimatique, 16/70, Universite Montpellier 11. 45 pp.

Li, R. M., Simons, D. B., and Stevens, M. A. (1975). Nonlinear kinematic wave approximation for water routing. *Water Resources Research*, *11*(2), 245–252.

Moussa, R. (1996). Analytical Hayami solution for the diffusive wave flood routing problem with lateral inflow. *Hydrological Processes*, *10*(9), 1209–1227.

Natale, L. and Todini, E. (1977). A constrained parameter estimation technique for linear models in hydrology. In: *Mathematical models for surface water hydrology*. Wiley, Chichester.

Price, R. K. (1974). Comparison of four numerical methods for flood routing. *Journal of the Hydraulics Division*, *100*(7), 879–899.

Singh, V. P. (1994). Accuracy of kinematic wave and diffusion wave approximations for space independent flows. *Hydrological Processes*, *8*(1), 45–62.

Vajravelu, K., and Van Gorder, R. (2013). *Nonlinear flow phenomena and homotopy analysis*. Springer, Berlin.

Vieira, J. D. (1983). Conditions governing the use of approximations for the Saint-Venant equations for shallow surface water flow. *Journal of Hydrology*, *60*(1–4), 43–58.

Weinmann, P. E., and Laurenson, E. M. (1979). Approximate flood routing methods: A review. *Journal of the Hydraulics Division*, *105*(12), 1521–1536.

Zhao, Y., and Liao, S. (2002). User guide to BVPh 2.0. School of naval architecture, ocean and civil engineering, Shanghai, *40*.

FURTHER READING

Dehghan, M., and Manafian, J. (2009). The solution of the variable coefficients fourth-order parabolic partial differential equations by the homotopy perturbation method. *Zeitschrift für Naturforschung A*, *64*(7–8), 420–430.

Fan, P., and Li, J. C. (2006). Diffusive wave solutions for open channel flows with uniform and concentrated lateral inflow. *Advances in Water Resources*, *29*(7), 1000–1019.

Matinfar, M., and Saeidy, M. (2010). Application of homotopy analysis method to fourth-order parabolic partial differential equations. *Applications and Applied Mathematics: An International Journal (AAM)*, 5(1), 6.

Moramarco, T., Fan, Y., and Bras, R. L. (1999). Analytical solution for channel routing with uniform lateral inflow. *Journal of Hydraulic Engineering*, 125(7), 707–713.

Onyejekwe, O. N. (2014). Determination of two-time dependent coefficients in a parabolic partial differential equation by homotopy analysis method. *International Journal of Applied Mathematics Research*, 3(2), 161.

Wang, L., Wu, J. Q., Elliot, W. J., Fiedler, F. R., and Lapin, S. (2014). Linear diffusion-wave channel routing using a discrete Hayami convolution method. *Journal of Hydrology*, 509, 282–294.

Yıldırım, A. (2010). Application of the homotopy perturbation method for the Fokker–Planck equation. *International Journal for Numerical Methods in Biomedical Engineering*, 26(9), 1144–1154.

15 Kinematic Wave Equation

15.1 INTRODUCTION

The kinematic wave theory is the mainstay in physically based modeling of overland flow. The theory comprises the continuity equation and a nonlinear equation of the discharge-stage relationship. When interpreted using the momentum equation, this relationship is derived from the balance between the bed slope and friction or energy slope. Coupling these equations, a first-order nonlinear partial differential equation results, which is also referred to as the kinematic wave equation. This equation can be analytically solved if the lateral inflow and lateral outflow are assumed constant in space and time; for variable lateral inflow and outflow, numerical solutions are the only resort. This chapter presents homotopy analysis–based methods for deriving series solutions.

15.2 GOVERNING EQUATION

We consider flow over a sloping plane of length L and slope $S \equiv \sin\theta$, as shown in Figure 15.1. It is assumed that rainfall begins at a constant rate I per unit area.

The continuity equation takes on the form (Henderson and Wooding 1964):

$$\frac{\partial h}{\partial t} + \frac{\partial Q}{\partial x} = I(t) \tag{15.1}$$

where t is the time, x is the longitudinal distance from the origin, h is the water depth, Q is the flow discharge, and $I(t)$ is the rainfall rate given as:

$$I(t) = \begin{cases} I & \text{for } 0 \le t < D \\ 0 & \text{for } t \ge D \end{cases} \tag{15.2}$$

where I is constant. If the water depth is sufficiently small, then the depth of the water surface relative to the slope S can be neglected, and the flow is related to the depth as follows:

$$Q = \alpha h^n \tag{15.3}$$

where parameters α and n take on values depending on the nature of the flow. For example, in laminar flow, $\alpha = gS/3v$ and $n = 3$, where g is the acceleration due to gravity and v is the kinematic viscosity of fluid. In the case of turbulent flow, $\alpha = CS^{1/2}$

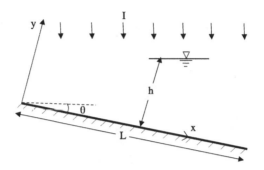

FIGURE 15.1 Schematic diagram for flow over a sloping plane.

and $n = 3/2$, where C is the Chezy coefficient. Eqs. (15.1) and (15.3) constitute a kinematic wave formulation. Using Eq. (15.3), the governing Eq. (15.1) becomes:

$$\frac{\partial h}{\partial t} + \alpha n h^{n-1} \frac{\partial h}{\partial x} = I(t) \tag{15.4}$$

where $I(t)$ is given by Eq. (15.2). The initial and boundary conditions can be provided as follows:

$$h(x = 0, t) = 0 \tag{15.5}$$

$$h(x, t = 0) = 0 \tag{15.6}$$

Different solution methodologies for Eq. (15.4), together with the conditions given by Eqs. (15.5) and (15.6), are described in what follows.

15.3 SOLUTION METHODOLOGIES

15.3.1 NUMERICAL SOLUTION

For convenience, we first nondimensionalize the independent variables as follows:

$$\hat{t} = \frac{t}{D}, \ \hat{x} = \frac{x}{L} \tag{15.7}$$

Using Eq. (15.7), the governing Eq. (15.4) and the conditions in Eqs. (15.5) and (15.6) become:

$$\frac{1}{D} \frac{\partial h}{\partial \hat{t}} + \frac{\alpha n}{L} h^{n-1} \frac{\partial h}{\partial \hat{x}} = I(\hat{t}) \tag{15.8}$$

$$h(\hat{x} = 0, \hat{t}) = 0 \tag{15.9}$$

$$h(\hat{x}, \hat{t} = 0) = 0 \tag{15.10}$$

Several authors have provided analytical solutions to the kinematic wave problem using the method of characteristics when the rainfall rate is constant (Woolhiser 1969; Singh 1996, 2001). Here, we can apply the method of lines (MOL) to convert the partial differential equation (PDE) to a system of ordinary differential equation (ODEs). MOL discretizes the equation in one direction, say the x-direction, while keeping the other direction (i.e., time) fixed. Thus, the governing equation is converted into a system of ODEs. For discretization with respect to the x-coordinate, we have used the following:

$$\left(\frac{\partial h}{\partial \hat{x}}\right)_i = \frac{h_i - h_{i-1}}{\Delta \hat{x}} \tag{15.11}$$

$$\left(h^{n-1}\right)_i = \left(\frac{h_i + h_{i-1}}{2}\right)^{n-1} \tag{15.12}$$

where $h_i = h_i\left(\hat{t}\right) = h\left(\hat{x} = \hat{x}_i, \hat{t}\right)$, the domain of $\hat{x}$, i.e., $[0,1]$, is divided into $x_0 \,(=0)$, $x_1, x_2, \ldots, x_N \,(=1)$. Using Eq. (15.9), we have:

$$h_0 = h_0\left(\hat{t}\right) = h\left(\hat{x} = 0, \hat{t}\right) = 0 \tag{15.13}$$

Using Eqs. (15.11) and (15.12), Eq. (15.8) takes on the form:

$$\frac{1}{D}\frac{dh_i}{d\hat{t}} + \frac{\alpha n}{\Delta \hat{x} L}\left(h_i - h_{i-1}\right)\left(\frac{h_i + h_{i-1}}{2}\right)^{n-1} = I\left(\hat{t}\right) \tag{15.14}$$

subject to:

$$h_i\left(0\right) = 0 \tag{15.15}$$

for $i = 1, 2, 3, \ldots, N$. Eqs. (15.14) and (15.15) constitute a system of the initial value problem (IVP), which can be solved numerically to obtain the unknowns.

The MOL-based approach discussed earlier converts the governing equation into a nonlinear system whose analytical solutions are not tractable because of the term having noninteger power, i.e., h^{n-1}, but it can be solved numerically. To that end, we use the Padé approximation technique to transform the term into a relatively weaker nonlinear form. The $[j,k]$ order Padé approximation of the term about the point a can be obtained as follows:

$$\left[h^{n-1}\right]_{j,k} = \frac{\displaystyle\sum_{i=0}^{j} a_i\left(h-a\right)^j}{\displaystyle\sum_{i=0}^{k} b_i\left(h-a\right)^k} \tag{15.16}$$

where a_i and b_i are the coefficients involving n. It can be seen from Eq. (15.16) that the Padé approximation technique converts the nonlinear term into a rational

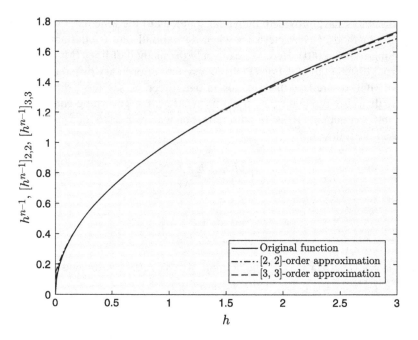

FIGURE 15.2 Padé approximations.

function, which is easier to deal with. Here, we test the efficiency of the approxima-tion. For that, we calculate the [2,2] and [3,3] order approximations and compare them with the original function. Taking $n = 3/2$ and $a = 0.5$, the [2,2] and [3,3] order approximations are obtained as:

$$\left[h^{n-1}\right]_{2,2} = \frac{0.177 + 3.536h + 3.536h^2}{1.250 + 5h + h^2} \tag{15.17}$$

$$\left[h^{n-1}\right]_{3,3} = \frac{0.088 + 3.712h + 12.374h^2 + 4.950h^3}{0.875 + 8.750h + 10.500h^2 + h^3} \tag{15.18}$$

Figure 15.2 compares approximations in Eqs. (15.17) and (15.18) with the original function. It can be observed from the figure that both approximations match the orig-inal function well. However, the higher the order, the better the accuracy. For sim-plicity, in this chapter, we have considered the [2,2] order approximation. Therefore, using Eq. (15.17), Eq. (15.14) takes on the form:

$$\frac{1}{D}\frac{dh_i}{d\hat{t}} + \frac{\alpha n}{\Delta\hat{x}L}\left(h_i - h_{i-1}\right)\frac{0.177 + 3.536h_i + 3.536h_i^2}{1.250 + 5h_i + h_i^2} = I\left(\hat{t}\right) \tag{15.19}$$

To check how the approximation works in solving the system of equations, we com-pare the solution of Eqs. (15.14) and (15.19). For that purpose, the MATLAB rou-tine *ode45* efficiently solves a system of ODEs using the Runge-Kutta higher-order

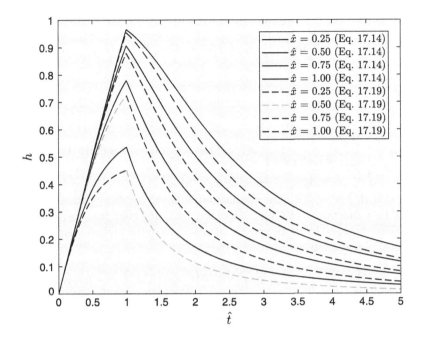

FIGURE 15.3 Comparison between the numerical solutions of Eqs. (15.14) and (15.19).

method. It may be noted that based on the form of $I\left(\hat{t}\right)$ given by Eq. (15.2), we solve Eq. (15.19) in two parts by splitting the domain into $[0,1]$ *and* $[1,\infty)$. The second part of the problem considers the initial conditions given by the solution of the first part evaluated at $\hat{t}=1$. We consider $N=4$, which produces the unknowns h_i, $i=1,\dots,4$, with $\widehat{x}_i=\{0.25,0.50,0.75,1.00\}$. The parameter values are $n=1.5$, $\alpha=1$ unit, $I=1$ unit, $D=1$ unit, and $L=2$ unit. Figure 15.3 compares numerical solutions of Eqs. (15.14) and (15.19). It is observed that the Padé approximation can provide an accurate solution to the equations.

15.3.2 HAM-BASED SOLUTION

As discussed in the previous section, the [2,2] order Padé approximation is an accurate formulation of the problem. Therefore, when applying the homotopy analysis method (HAM), we use Eq. (15.19) subject to the condition in Eq. (15.15). Before proceeding further, we rearrange Eq. (15.19) as follows:

$$\frac{1}{D}\left(1.250+5h_i+h_i^{\,2}\right)\frac{dh_i}{d\hat{t}}$$

$$+\frac{\alpha n}{\Delta \hat{x}L}\left(0.177h_i+3.536h_i^{\,2}+3.536h_i^{\,3}-0.177h_{i-1}-3.536h_ih_{i-1}-3.536h_i^{\,2}h_{i-1}\right)$$

$$=I\left(\hat{t}\right)\left(1.250+5h_i+h_i^{\,2}\right) \tag{15.20}$$

The zeroth-order deformation equation can be constructed as follows:

$$(1-q)\mathcal{L}_i\left[\Phi_i\left(\hat{t};q\right)-h_{i,0}\left(\hat{t}\right)\right]-q\hbar_i H_i\left(\hat{t}\right)\mathcal{N}_i\left[\Phi_i\left(\hat{t};q\right)\right]=0 \qquad (15.21)$$

subject to the condition:

$$\Phi_i\left(0;q\right)=\beta_i \qquad (15.22)$$

where $\beta_i = \begin{cases} 0 \ for \ 0\leq\hat{t}\leq1 \\ h_i\left(\hat{t}=1\right) \ for \ \hat{t}\geq1 \end{cases}$, $h_i\left(\hat{t}=1\right)$ are the solutions of the first part evaluated

at $\hat{t}=1$, q is the embedding parameter, $\Phi_i\left(\hat{t};q\right)$ is the representation of the solution across q, $h_{i,0}\left(\hat{t}\right)$ is the initial approximation, $\hbar_i$ is the auxiliary parameter, $H_i\left(\hat{t}\right)$ is the auxiliary function, and $\mathcal{L}_i$ and $\mathcal{N}_i$ are the linear and nonlinear operators, respectively. The higher-order terms can be calculated from the higher-order deformation equations, given as follows:

$$\mathcal{L}_i\left[h_{i,m}\left(\hat{t}\right)-\chi_{i,m}h_{i,m-1}\left(\hat{t}\right)\right]=\hbar_i H_i\left(\hat{t}\right)R_{i,m}\left(\vec{h}_{i,m-1}\right), \ m=1,2,3,\dots$$

subject to:

$$h_{i,m}\left(0\right)=0 \qquad (15.23)$$

where:

$$\chi_{i,m} = \begin{cases} 0 \quad \text{when } m=1, \\ 1 \quad \text{otherwise} \end{cases} \qquad (15.24)$$

and:

$$R_{i,m}\left(\vec{h}_{i,m-1}\right)=\frac{1}{(m-1)!}\frac{\partial^{m-1}\mathcal{N}_i\left[\Phi_i\left(\hat{t};q\right)\right]}{\partial q^{m-1}}\Bigg|_{q=0} \qquad (15.25)$$

where $h_{i,m}$ for $m\geq1$ is the higher-order term. Now, the final solution can be obtained as follows:

$$h_i\left(\hat{t}\right)=h_{i,0}\left(\hat{t}\right)+\sum_{m=1}^{\infty}h_{i,m}\left(\hat{t}\right) \qquad (15.26)$$

The nonlinear operators for the system are selected as:

$$\mathcal{N}_i\left[\Phi_i\left(\hat{t};q\right)\right]=\frac{1}{D}\left(1.250+5\Phi_i+\Phi_i^2\right)\frac{\partial\Phi_i\left(\hat{t};q\right)}{\partial\hat{t}}$$

$$+\frac{\alpha n}{\Delta\hat{x}L}\left(0.177\Phi_i+3.536\Phi_i^2+3.536\Phi_i^3-0.177\Phi_{i-1}-3.536\Phi_i\Phi_{i-1}-3.536\Phi_i^2\Phi_{i-1}\right)$$

$$-I\left(\hat{t}\right)\left(1.250+5\Phi_i+\Phi_i^2\right) \qquad (15.27)$$

Therefore, using Eq. (15.27), the term $R_{i,m}$ can be calculated from Eq. (15.25) as follows:

$$R_{i,m}\left(\vec{h}_{i,m-1}\right) = \frac{1.250}{D}\frac{dh_{i,m-1}}{d\hat{t}} + \frac{5}{D}\sum_{j=0}^{m-1}h_{i,j}\frac{dh_{i,m-1-j}}{d\hat{t}} + \frac{1}{D}\sum_{j=0}^{m-1}h_{i,m-1-j}\sum_{k=0}^{j}h_{i,k}\frac{dh_{i,j-k}}{d\hat{t}}$$

$$+\frac{\alpha n}{\Delta\hat{x}L}\left(0.177h_{i,m-1} + 3.536\sum_{j=0}^{m-1}h_{i,j}h_{i,m-1-j} + 3.536\sum_{j=0}^{m-1}h_{i,m-1-j}\sum_{k=0}^{j}h_{i,k}h_{i,j-k}\right.$$

$$-0.177h_{i-1,m-1} - 3.536\sum_{j=0}^{m-1}h_{i,j}h_{i-1,m-1-j} - 3.536\sum_{j=0}^{m-1}h_{i,m-1-j}\sum_{k=0}^{j}h_{i,k}h_{i-1,j-k}\right)$$

$$-I\left(\hat{t}\right)\left(1.250\left(1-\chi_{i,m}\right) + 5h_{i,m-1} + \sum_{j=0}^{m-1}h_{i,j}h_{i,m-1-j}\right) \tag{15.28}$$

The following set of base functions is chosen to represent the solution:

$$\left\{\hat{t}^{m}exp\left(-n\hat{t}\right) \mid m,n = 0,1,2,...\right\} \text{ for } 0\le\hat{t}\le1 \tag{15.29}$$

$$\left\{\hat{t}^{p} \mid p = 0,1,2,...\right\} \text{ for } \hat{t}\ge1 \tag{15.30}$$

so that:

$$h_{i}\left(\hat{t}\right) = \begin{cases} \displaystyle\sum_{m=0}^{\infty}\sum_{n=0}^{\infty}a_{m,n}\hat{t}^{m}exp\left(-n\hat{t}\right) \text{ for } 0\le\hat{t}\le1 \\ \displaystyle\sum_{p=0}^{\infty}b_{p}\hat{t}^{p} \text{ for } \hat{t}\ge1 \end{cases} \tag{15.31}$$

where $a_{m,n}$ and b_p are the coefficients of the series. Eq. (15.31) provides us with the so-called *rule of solution expression*. Following the rule of solution expression, the linear operators and the initial approximations are chosen, respectively, as follows:

$$\mathcal{L}_{i}\left[\Phi_{i}\left(\hat{t};q\right)\right] = \frac{\partial\Phi_{i}\left(\hat{t};q\right)}{\partial\hat{t}} \text{ with the property } \mathcal{L}_{i}\left[C_{i,2}\right] = 0 \tag{15.32}$$

$$h_{i,0}\left(\hat{t}\right) = \begin{cases} 1-exp\left(-\hat{t}\right) \\ \beta_{i} \end{cases} \tag{15.33}$$

where $C_{i,2}$ is the integral constant. Using Eq. (15.32), the higher-order terms can be obtained from Eq. (15.23) as follows:

$$h_{i,m}\left(\hat{t}\right) = \chi_{i,m}h_{i,m-1}\left(\hat{t}\right) + \hbar_i \int_0^{\hat{t}} H_i\left(x\right)R_{i,m}\left(\vec{h}_{i,m-1}\right)dx + C_{i,2}, \ m = 1,2,3,\ldots \qquad (15.34)$$

where $R_{i,m}$ is given by Eq. (15.28) and the constant $C_{i,2}$ can be determined from the initial conditions for the higher-order deformation equations given by Eq. (15.23).

Now, the auxiliary function $H_i\left(\hat{t}\right)$ can be determined from the *rule of coefficient ergodicity*. Based on the rule of solution expressions in Eq. (15.29) and (15.30), function $H_i\left(\hat{t}\right)$ can take on the form:

$$H_i\left(\hat{t}\right) = \begin{cases} \hat{t}^a \exp\left(-b\hat{t}\right) \ for \ 0 \le \hat{t} \le 1 \\ \hat{t}^c \ for \ \hat{t} \ge 1 \end{cases} \qquad (15.35)$$

where a, b, and c are integers. Substituting Eq. (15.35) into Eq. (15.34) and calculating some of $h_{i,m}$, we can infer two observations. For $0 \le \hat{t} \le 1$, if we consider $a,b \ge 1$ or $a = 0$ or $b = 0$, then term $\hat{t}^a$ will be missing in the solution expression. This does not satisfy the *rule of coefficient ergodicity*, which allows us to include all the terms of the base function for the completeness of the solution. On the other hand, if we consider $a,b \le -1$, then the terms appearing in the solution violate the *rule of solution expression*. Therefore, we are left with two choices $a,b = 0$, i.e., $H\left(\hat{t}\right) = 1$. For $\hat{t} \ge 1$, if we consider $c \ge 1$, then term $\hat{t}^c$ will be missing in the solution expression. This does not satisfy the *rule of coefficient ergodicity*, which allows us to include all the terms of the base function for the completeness of the solution. On the other hand, if we consider $c \le -1$, then the terms appearing in the solution violate the *rule of solution expression*. Therefore, we are left with two choices $c = 0$, i.e., $H\left(\hat{t}\right) = 1$. Finally, the approximate solution can be obtained as follows:

$$h_i\left(\hat{t}\right) \approx h_{i,0}\left(\hat{t}\right) + \sum_{m=1}^{M} h_{i,m}\left(\hat{t}\right) \qquad (15.36)$$

15.3.3 HPM-BASED SOLUTION

When applying the homotopy perturbation method (HPM), we rewrite Eq. (15.20) in the following form:

$$\mathcal{N}_i\left(h_i\right) = f_i\left(\hat{t}\right) \qquad (15.37)$$

We construct a homotopy $\Phi_i\left(\hat{t};q\right)$ that satisfies:

$$(1-q)\mathcal{L}_i\left[\Phi_i\left(\hat{t};q\right) - h_{i,0}\left(\hat{t}\right)\right] + q\left[\mathcal{N}_i\left[\Phi_i\left(\hat{t};q\right)\right] - f_i\left(\hat{t}\right)\right] = 0 \qquad (15.38)$$

where q is the embedding parameter that lies on $[0,1]$, $\mathcal{L}_i$ is the linear operator, and $h_{i,0}\left(\hat{t}\right)$ is the initial approximation to the final solutions. Eq. (15.38) shows that as $q = 0$, $\Phi_i\left(\hat{t};0\right) = h_{i,0}\left(\hat{t}\right)$ and as $q = 1$, $\Phi_i\left(\hat{t};1\right) = h_i\left(\hat{t}\right)$. We now express $\Phi_i\left(\hat{t};q\right)$ as a series in terms of q, as follows:

$$\Phi_i\left(\hat{t};q\right) = \Phi_{i,0} + q\Phi_{i,1} + q^2\Phi_{i,2} + q^3\Phi_{i,3} + q^4\Phi_{i,4} + \cdots \tag{15.39}$$

where $\Phi_{i,m}$ for $m \geq 1$ is the higher-order term. As $q \to 1$, Eq. (15.39) produces the final solution as:

$$h_i\left(\hat{t}\right) = \lim_{q \to 1} \Phi_i\left(\hat{t};q\right) = \sum_{j=0}^{\infty} \Phi_{i,j} \tag{15.40}$$

For simplicity, we consider the linear and nonlinear operators as follows:

$$\mathcal{L}_i\left(\Phi_i\left(\hat{t};q\right)\right) = \frac{1.25}{D}\frac{\partial \Phi_i\left(\hat{t};q\right)}{\partial \hat{t}} + \frac{0.177\alpha n}{\Delta \hat{x} L}\Phi_i \tag{15.41}$$

$$\mathcal{N}_i\left[\Phi_i\left(\hat{t};q\right)\right] = \frac{1}{D}\left(1.250 + 5\Phi_i + \Phi_i^2\right)\frac{\partial \Phi_i\left(\hat{t};q\right)}{\partial \hat{t}}$$

$$+ \frac{\alpha n}{\Delta \hat{x} L}\left(0.177\Phi_i + 3.536\Phi_i^2 + 3.536\Phi_i^3 - 0.177\Phi_{i-1} - 3.536\Phi_i\Phi_{i-1} - 3.536\Phi_i^2\Phi_{i-1}\right)$$

$$- I\left(\hat{t}\right)\left(1.250 + 5\Phi_i + \Phi_i^2\right) \tag{15.42}$$

Using these expressions, Eq. (15.38) takes on the form:

$$(1-q)\left[\frac{1.25}{D}\frac{\partial \Phi_i\left(\hat{t};q\right)}{\partial \hat{t}} + \frac{0.177\alpha n}{\Delta \hat{x} L}\Phi_i - \frac{1.25}{D}\frac{dh_{i,0}\left(\hat{t}\right)}{d\hat{t}} - \frac{0.177\alpha n}{\Delta \hat{x} L}h_{i,0}\left(\hat{t}\right)\right]$$

$$+ q\left[\frac{1}{D}\left(1.250 + 5\Phi_i + \Phi_i^2\right)\frac{\partial \Phi_i\left(\hat{t};q\right)}{\partial \hat{t}}\right.$$

$$+ \frac{\alpha n}{\Delta \hat{x} L}\left(0.177\Phi_i + 3.536\Phi_i^2 + 3.536\Phi_i^3 - 0.177\Phi_{i-1} - 3.536\Phi_i\Phi_{i-1} - 3.536\Phi_i^2\Phi_{i-1}\right)$$

$$\left. - I\left(\hat{t}\right)\left(1.250 + 5\Phi_i + \Phi_i^2\right)\right] = 0 \tag{15.43}$$

Using Eq. (15.39), we have:

$$
(1-q)\left[\frac{1.25}{D}\frac{\partial \Phi_i\left(\hat{t};q\right)}{\partial \hat{t}}+\frac{0.177\alpha n}{\Delta \hat{x}L}\Phi_i-\frac{1.25}{D}\frac{dh_{i,0}\left(\hat{t}\right)}{d\hat{t}}-\frac{0.177\alpha n}{\Delta \hat{x}L}h_{i,0}\left(\hat{t}\right)\right]
$$

$$
=(1-q)\left[\left(\frac{1.25}{D}\frac{d\Phi_{i,0}}{d\hat{t}}+\frac{0.177\alpha n}{\Delta \hat{x}L}\Phi_{i,0}-\frac{1.25}{D}\frac{dh_{i,0}\left(\hat{t}\right)}{d\hat{t}}-\frac{0.177\alpha n}{\Delta \hat{x}L}h_{i,0}\left(\hat{t}\right)\right)\right.
$$

$$
+q\left(\frac{1.25}{D}\frac{d\Phi_{i,1}}{d\hat{t}}+\frac{0.177\alpha n}{\Delta \hat{x}L}\Phi_{i,1}\right)+q^2\left(\frac{1.25}{D}\frac{d\Phi_{i,2}}{d\hat{t}}+\frac{0.177\alpha n}{\Delta \hat{x}L}\Phi_{i,2}\right)
$$

$$
\left.+q^3\left(\frac{1.25}{D}\frac{d\Phi_{i,3}}{d\hat{t}}+\frac{0.177\alpha n}{\Delta \hat{x}L}\Phi_{i,3}\right)+q^4\left(\frac{1.25}{D}\frac{d\Phi_{i,4}}{d\hat{t}}+\frac{0.177\alpha n}{\Delta \hat{x}L}\Phi_{i,4}\right)+\cdots\right]
$$

$$
=\left(\frac{1.25}{D}\frac{d\Phi_{i,0}}{d\hat{t}}+\frac{0.177\alpha n}{\Delta \hat{x}L}\Phi_{i,0}-\frac{1.25}{D}\frac{dh_{i,0}\left(\hat{t}\right)}{d\hat{t}}-\frac{0.177\alpha n}{\Delta \hat{x}L}h_{i,0}\left(\hat{t}\right)\right)
$$

$$
+q\left(\left(\frac{1.25}{D}\frac{d\Phi_{i,1}}{d\hat{t}}+\frac{0.177\alpha n}{\Delta \hat{x}L}\Phi_{i,1}\right)\right.
$$

$$
\left.-\left(\frac{1.25}{D}\frac{d\Phi_{i,0}}{d\hat{t}}+\frac{0.177\alpha n}{\Delta \hat{x}L}\Phi_{i,0}-\frac{1.25}{D}\frac{dh_{i,0}\left(\hat{t}\right)}{d\hat{t}}-\frac{0.177\alpha n}{\Delta \hat{x}L}h_{i,0}\left(\hat{t}\right)\right)\right)
$$

$$
+q^2\left(\left(\frac{1.25}{D}\frac{d\Phi_{i,2}}{d\hat{t}}+\frac{0.177\alpha n}{\Delta \hat{x}L}\Phi_{i,2}\right)-\left(\frac{1.25}{D}\frac{d\Phi_{i,1}}{d\hat{t}}+\frac{0.177\alpha n}{\Delta \hat{x}L}\Phi_{i,1}\right)\right)
$$

$$
+q^3\left(\left(\frac{1.25}{D}\frac{d\Phi_{i,3}}{d\hat{t}}+\frac{0.177\alpha n}{\Delta \hat{x}L}\Phi_{i,3}\right)-\left(\frac{1.25}{D}\frac{d\Phi_{i,2}}{d\hat{t}}+\frac{0.177\alpha n}{\Delta \hat{x}L}\Phi_{i,2}\right)\right)
$$

$$
+q^4\left(\left(\frac{1.25}{D}\frac{d\Phi_{i,4}}{d\hat{t}}+\frac{0.177\alpha n}{\Delta \hat{x}L}\Phi_{i,4}\right)-\left(\frac{1.25}{D}\frac{d\Phi_{i,3}}{d\hat{t}}+\frac{0.177\alpha n}{\Delta \hat{x}L}\Phi_{i,3}\right)\right)+\cdots
$$

$$
(15.44)
$$

$$
\Phi_i\frac{\partial \Phi_i\left(\hat{t};q\right)}{\partial \hat{t}}=\left(\Phi_{i,0}+q\Phi_{i,1}+q^2\Phi_{i,2}+q^3\Phi_{i,3}+q^4\Phi_{i,4}+\cdots\right)
$$

$$
\times\left(\frac{d\Phi_{i,0}}{d\hat{t}}+q\frac{d\Phi_{i,1}}{d\hat{t}}+q^2\frac{d\Phi_{i,2}}{d\hat{t}}+q^3\frac{d\Phi_{i,3}}{d\hat{t}}+q^4\frac{d\Phi_{i,4}}{d\hat{t}}+\cdots\right)
$$

$$
=\Phi_{i,0}\frac{d\Phi_{i,0}}{d\hat{t}}+q\left(\phi_{i,1}\frac{d\phi_{i,0}}{d\hat{t}}+\phi_{i,0}\frac{d\phi_{i,1}}{d\hat{t}}\right)+q^2\left(\phi_{i,0}\frac{d\phi_{i,2}}{d\hat{t}}+\phi_{i,1}\frac{d\phi_{i,1}}{d\hat{t}}+\phi_{i,2}\frac{d\phi_{i,0}}{d\hat{t}}\right)
$$

$$
+q^3\left(\phi_{i,1}\frac{d\phi_{i,2}}{d\hat{t}}+\phi_{i,2}\frac{d\phi_{i,1}}{d\hat{t}}+\phi_{i,3}\frac{d\phi_{i,0}}{d\hat{t}}+\phi_{i,0}\frac{d\phi_{i,3}}{d\hat{t}}\right)+\cdots \qquad (15.45)
$$

$$\phi_i{}^2 \frac{\partial \phi_i\left(\hat{t};q\right)}{\partial \hat{t}}$$

$$= \left[\left(\phi_{i,0}\right)^2 + q\left(2\phi_{i,0}\phi_{i,1}\right) + q^2\left(\phi_{i,1}{}^2 + 2\phi_{i,0}\phi_{i,2}\right) + q^3\left(2\phi_{i,0}\phi_{i,3} + 2\phi_{i,1}\phi_{i,2}\right) + \dots \right]$$

$$\left(\frac{d\phi_{i,0}}{d\hat{t}} + q\frac{d\phi_{i,1}}{d\hat{t}} + q^2\frac{d\phi_{i,2}}{d\hat{t}} + q^3\frac{d\phi_{i,3}}{d\hat{t}} + q^4\frac{d\phi_{i,4}}{d\hat{t}} + \dots\right)$$

$$= \left(\phi_{i,0}\right)^2 \frac{d\phi_{i,0}}{d\hat{t}} + q\left[\left(\phi_{i,0}\right)^2 \frac{d\phi_{i,1}}{d\hat{t}} + 2\phi_{i,0}\phi_{i,1}\frac{d\phi_{i,0}}{d\hat{t}}\right]$$

$$+ q^2\left[\left(\phi_{i,0}\right)^2 \frac{d\phi_{i,2}}{d\hat{t}} + \frac{d\phi_{i,0}}{d\hat{t}}\left(\phi_{i,1}{}^2 + 2\phi_{i,0}\phi_{i,2}\right) + 2\phi_{i,0}\phi_{i,1}\frac{d\phi_{i,1}}{d\hat{t}}\right]$$

$$+ q^3\left[\left(\phi_{i,0}\right)^2 \frac{d\phi_{i,3}}{d\hat{t}} + \frac{d\phi_{i,0}}{d\hat{t}}\left(2\phi_{i,0}\phi_{i,3} + 2\phi_{i,1}\phi_{i,2}\right)\right.$$

$$\left. + 2\phi_{i,0}\phi_{i,1}\frac{d\phi_{i,2}}{d\hat{t}} + \frac{d\phi_{i,1}}{d\hat{t}}\left(\phi_{i,1}{}^2 + 2\phi_{i,0}\phi_{i,2}\right)\right] + \dots \tag{15.46}$$

$$\left(\phi_i\right)^2 = \left(\phi_{i,0} + q\phi_{i,1} + q^2\phi_{i,2} + q^3\phi_{i,3} + q^4\phi_{i,4} + \dots\right)^2$$

$$= \left(\phi_{i,0}\right)^2 + q\left(2\phi_{i,0}\phi_{i,1}\right) + q^2\left(\phi_{i,1}{}^2 + 2\phi_{i,0}\phi_{i,2}\right) + q^3\left(2\phi_{i,0}\phi_{i,3} + 2\phi_{i,1}\phi_{i,2}\right) + \dots \tag{15.47}$$

$$\left(\phi_i\right)^3 = \left(\phi_{i,0} + q\phi_{i,1} + q^2\phi_{i,2} + q^3\phi_{i,3} + q^4\phi_{i,4} + \dots\right)^3$$

$$= \left(\phi_{i,0} + q\phi_{i,1} + q^2\phi_{i,2} + q^3\phi_{i,3} + q^4\phi_{i,4} + \dots\right)$$

$$\times \left[\left(\phi_{i,0}\right)^2 + q\left(2\phi_{i,0}\phi_{i,1}\right) + q^2\left(\phi_{i,1}{}^2 + 2\phi_{i,0}\phi_{i,2}\right) + q^3\left(2\phi_{i,0}\phi_{i,3} + 2\phi_{i,1}\phi_{i,2}\right) + \dots \right]$$

$$= \left(\phi_{i,0}\right)^3 + q\left[3\phi_{i,0}{}^2\phi_{i,1}\right] + q^2\left[\phi_{i,0}\left(\phi_{i,1}{}^2 + 2\phi_{i,0}\phi_{i,2}\right) + \phi_{i,2}\phi_{i,0}{}^2 + 2\phi_{i,0}\phi_{i,1}{}^2\right]$$

$$+ q^3\left[\phi_{i,0}\left(2\phi_{i,0}\phi_{i,3} + 2\phi_{i,1}\phi_{i,2}\right) + \phi_{i,3}\left(\phi_{i,0}\right)^2 + \phi_{i,1}\left(\phi_{i,1}{}^2 + 2\phi_{i,0}\phi_{i,2}\right) + 2\phi_{i,0}\phi_{i,1}\phi_{i,2}\right] + \dots \tag{15.48}$$

$$\phi_i\phi_{i-1} = \left(\phi_{i,0} + q\phi_{i,1} + q^2\phi_{i,2} + q^3\phi_{i,3} + q^4\phi_{i,4} + \dots\right)$$

$$\times \left(\phi_{i-1,0} + q\phi_{i-1,1} + q^2\phi_{i-1,2} + q^3\phi_{i-1,3} + q^4\phi_{i-1,4} + \dots\right)$$

$$= \phi_{i,0}\phi_{i-1,0} + q\left(\phi_{i,0}\phi_{i-1,1} + \phi_{i-1,0}\phi_{i,1}\right)q^2\left(\phi_{i,0}\phi_{i-1,2} + \phi_{i-1,0}\phi_{i,2} + \phi_{i,1}\phi_{i-1,1}\right)$$

$$+ q^3\left(\phi_{i,0}\phi_{i-1,3} + \phi_{i-1,0}\phi_{i,3} + \phi_{i,1}\phi_{i-1,2} + \phi_{i,2}\phi_{i-1,1} + \dots\right) \tag{15.49}$$

$$\phi_i^2 \phi_{i-1} = \left[\left(\phi_{i,0}\right)^2 + q\left(2\phi_{i,0}\phi_{i,1}\right) + q^2\left(\phi_{i,1}^2 + 2\phi_{i,0}\phi_{i,2}\right) + q^3\left(2\phi_{i,0}\phi_{i,3} + 2\phi_{i,1}\phi_{i,2}\right) + \cdots \right]$$

$$\times \left(\phi_{i-1,0} + q\phi_{i-1,1} + q^2\phi_{i-1,2} + q^3\phi_{i-1,3} + q^4\phi_{i-1,4} + \cdots\right)$$

$$= \phi_{i-1,0}\left(\phi_{i,0}\right)^2 + q\left[2\phi_{i,0}\phi_{i,1}\phi_{i-1,0} + \phi_{i-1,1}\left(\phi_{i,0}\right)^2\right]$$

$$+ q^2 \left[\phi_{i-1,2}\left(\phi_{i,0}\right)^2 + \phi_{i-1,0}\left(\phi_{i,1}^2 + 2\phi_{i,0}\phi_{i,2}\right) + 2\phi_{i,0}\phi_{i,1}\phi_{i-1,1}\right]$$

$$+ q^3 \left[\phi_{i-1,3}\left(\phi_{i,0}\right)^2 + \phi_{i-1,0}\left(2\phi_{i,0}\phi_{i,3} + 2\phi_{i,1}\phi_{i,2}\right)\right.$$

$$\left. + 2\phi_{i,0}\phi_{i,1}\phi_{i-1,2} + \phi_{i-1,1}\left(\phi_{i,1}^2 + 2\phi_{i,0}\phi_{i,2}\right)\right] + \cdots \quad (15.50)$$

Using Eqs. (15.45–15.50), we have:

$$\frac{1}{D}\left(1.250 + 5\phi_i + \phi_i^2\right)\frac{\partial \phi_i\left(\hat{t};q\right)}{\partial \hat{t}}$$

$$+ \frac{\alpha n}{\Delta \hat{x} L}\left(0.177\phi_i + 3.536\phi_i^2 + 3.536\phi_i^3 - 0.177\phi_{i-1} - 3.536\phi_i\phi_{i-1} - 3.536\phi_i^2\phi_{i-1}\right)$$

$$-I\left(\hat{t}\right)\left(1.250 + 5\phi_i + \phi_i^2\right) = \frac{1.250}{D}\frac{d\phi_{i,0}}{d\hat{t}} + \frac{5}{D}\phi_{i,0}\frac{d\phi_{i,0}}{d\hat{t}} + \frac{1}{D}\left(\phi_{i,0}\right)^2\frac{d\phi_{i,0}}{d\hat{t}} + \frac{\alpha n}{\Delta \hat{x} L}$$

$$\times\left(0.177\phi_{i,0} + 3.536\phi_{i,0}^2 + 3.536\phi_{i,0}^3 - 0.177\phi_{i-1,0} - 3.536\phi_{i,0}\phi_{i-1,0} - 3.536\phi_{i,0}^2\phi_{i-1,0}\right)$$

$$-I\left(\hat{t}\right)\left(1.250 + 5\phi_{i,0} + \phi_{i,0}^2\right) + q\left[\frac{1.250}{D}\frac{d\phi_{i,1}}{d\hat{t}} + \frac{5}{D}\left(\phi_{i,1}\frac{d\phi_{i,0}}{d\hat{t}} + \phi_{i,0}\frac{d\phi_{i,1}}{d\hat{t}}\right)\right.$$

$$+ \frac{1}{D}\left[\left(\phi_{i,0}\right)^2\frac{d\phi_{i,1}}{d\hat{t}} + 2\phi_{i,0}\phi_{i,1}\frac{d\phi_{i,0}}{d\hat{t}}\right] + \frac{\alpha n}{\Delta \hat{x} L}\left(0.177\phi_{i,1} + 3.536\left(2\phi_{i,0}\phi_{i,1}\right) + 3.536\left[3\phi_{i,0}^2\phi_{i,1}\right]\right.$$

$$\left. - 0.177\phi_{i-1,1} - 3.536\left(\phi_{i,0}\phi_{i-1,1} + \phi_{i-1,0}\phi_{i,1}\right) - 3.536\left[2\phi_{i,0}\phi_{i,1}\phi_{i-1,0} + \phi_{i-1,1}\left(\phi_{i,0}\right)^2\right]\right)$$

$$\left. -I\left(\hat{t}\right)\left(5\phi_{i,1} + 2\phi_{i,0}\phi_{i,1}\right)\right] + q^2\left[\frac{1.250}{D}\frac{d\phi_{i,2}}{d\hat{t}} + \frac{5}{D}\left(\phi_{i,0}\frac{d\phi_{i,2}}{d\hat{t}} + \phi_{i,1}\frac{d\phi_{i,1}}{d\hat{t}} + \phi_{i,2}\frac{d\phi_{i,0}}{d\hat{t}}\right)\right.$$

$$+ \frac{1}{D}\left[\left(\phi_{i,0}\right)^2\frac{d\phi_{i,2}}{d\hat{t}} + \frac{d\phi_{i,0}}{d\hat{t}}\left(\phi_{i,1}^2 + 2\phi_{i,0}\phi_{i,2}\right) + 2\phi_{i,0}\phi_{i,1}\frac{d\phi_{i,1}}{d\hat{t}}\right]$$

$$+ \frac{\alpha n}{\Delta \hat{x} L}\left(0.177\phi_{i,2} + 3.536\left(\phi_{i,1}^2 + 2\phi_{i,0}\phi_{i,2}\right) + 3.536\left[\phi_{i,0}\left(\phi_{i,1}^2 + 2\phi_{i,0}\phi_{i,2}\right) + \phi_{i,2}\phi_{i,0}^2 + 2\phi_{i,0}\phi_{i,1}^2\right]\right.$$

$$- 0.177\phi_{i-1,2} - 3.536\left(\phi_{i,0}\phi_{i-1,2} + \phi_{i-1,0}\phi_{i,2} + \phi_{i,1}\phi_{i-1,1}\right)$$

$$\left.\left. - 3.536\left[\phi_{i-1,2}\left(\phi_{i,0}\right)^2 + \phi_{i-1,0}\left(\phi_{i,1}^2 + 2\phi_{i,0}\phi_{i,2}\right) + 2\phi_{i,0}\phi_{i,1}\phi_{i-1,1}\right]\right) - I\left(\hat{t}\right)\left(5\phi_{i,2} + \left(\phi_{i,1}^2 + 2\phi_{i,0}\phi_{i,2}\right)\right)\right]$$

$$\quad (15.51)$$

Using the initial conditions $h_i(0) = 0$, we have:

$$h_{i,0}(0) = 0, \; h_{i,1}(0) = 0, \; h_{i,2}(0) = 0, \; \ldots \tag{15.52}$$

Using Eqs. (15.44)–(15.50) and equating the like powers of q in Eq. (15.43), the following systems of differential equations are obtained:

$$\frac{1.25}{D}\frac{d\Phi_{i,0}}{d\hat{t}} + \frac{0.177\alpha n}{\Delta\hat{x}L}\Phi_{i,0} - \frac{1.25}{D}\frac{dh_{i,0}(\hat{t})}{d\hat{t}} - \frac{0.177\alpha n}{\Delta\hat{x}L}h_{i,0}(\hat{t})$$

$$= 0 \text{ subject to } \phi_{i,0}(0) = 0 \tag{15.53}$$

$$\left(\frac{1.25}{D}\frac{d\Phi_{i,1}}{d\hat{t}} + \frac{0.177\alpha n}{\Delta\hat{x}L}\Phi_{i,1}\right)$$

$$-\left(\frac{1.25}{D}\frac{d\Phi_{i,0}}{d\hat{t}} + \frac{0.177\alpha n}{\Delta\hat{x}L}\Phi_{i,0} - \frac{1.25}{D}\frac{dh_{i,0}(\hat{t})}{d\hat{t}} - \frac{0.177\alpha n}{\Delta\hat{x}L}h_{i,0}(\hat{t})\right)$$

$$+\frac{1.250}{D}\frac{d\phi_{i,0}}{d\hat{t}} + \frac{5}{D}\phi_{i,0}\frac{d\phi_{i,0}}{d\hat{t}} + \frac{1}{D}(\phi_{i,0})^2\frac{d\phi_{i,0}}{d\hat{t}}$$

$$+\frac{\alpha n}{\Delta\hat{x}L}\left(0.177\phi_{i,0} + 3.536\phi_{i,0}{}^2 + 3.536\phi_{i,0}{}^3 - 0.177\phi_{i-1,0}\right)$$

$$-3.536\phi_{i,0}\phi_{i-1,0} - 3.536\phi_{i,0}{}^2\phi_{i-1,0}\Big) - I(\hat{t})\left(1.250 + 5\phi_{i,0} + \phi_{i,0}{}^2\right)$$

$$= 0 \text{ subject to } \phi_{i,1}(0) = 0 \tag{15.54}$$

$$\left(\frac{1.25}{D}\frac{d\Phi_{i,2}}{d\hat{t}} + \frac{0.177\alpha n}{\Delta\hat{x}L}\Phi_{i,2}\right) - \left(\frac{1.25}{D}\frac{d\Phi_{i,1}}{d\hat{t}} + \frac{0.177\alpha n}{\Delta\hat{x}L}\Phi_{i,1}\right)$$

$$+\frac{1.250}{D}\frac{d\phi_{i,1}}{d\hat{t}} + \frac{5}{D}\left(\phi_{i,1}\frac{d\phi_{i,0}}{d\hat{t}} + \phi_{i,0}\frac{d\phi_{i,1}}{d\hat{t}}\right) + \frac{1}{D}\left[(\phi_{i,0})^2\frac{d\phi_{i,1}}{d\hat{t}} + 2\phi_{i,0}\phi_{i,1}\frac{d\phi_{i,0}}{d\hat{t}}\right]$$

$$+\frac{\alpha n}{\Delta\hat{x}L}\left(0.177\phi_{i,1} + 3.536(2\phi_{i,0}\phi_{i,1}) + 3.536\left[3\phi_{i,0}{}^2\phi_{i,1}\right] - 0.177\phi_{i-1,1}\right.$$

$$\left. -3.536\left(\phi_{i,0}\phi_{i-1,1} + \phi_{i-1,0}\phi_{i,1}\right) - 3.536\left[2\phi_{i,0}\phi_{i,1}\phi_{i-1,0} + \phi_{i-1,1}(\phi_{i,0})^2\right]\right)$$

$$-I(\hat{t})\left(5\phi_{i,1} + 2\phi_{i,0}\phi_{i,1}\right) = 0 \text{ subject to } \phi_{i,2}(0) = 0 \tag{15.55}$$

$$\left(\frac{1.25}{D} \frac{d\Phi_{i,3}}{d\hat{t}} + \frac{0.177\alpha n}{\Delta \hat{x} L} \Phi_{i,3} \right) - \left(\frac{1.25}{D} \frac{d\Phi_{i,2}}{d\hat{t}} + \frac{0.177\alpha n}{\Delta \hat{x} L} \Phi_{i,2} \right)$$

$$+ \frac{1.250}{D} \frac{d\phi_{i,2}}{d\hat{t}} + \frac{5}{D} \left(\phi_{i,0} \frac{d\phi_{i,2}}{d\hat{t}} + \phi_{i,1} \frac{d\phi_{i,1}}{d\hat{t}} + \phi_{i,2} \frac{d\phi_{i,0}}{d\hat{t}} \right)$$

$$+ \frac{1}{D} \left[\left(\phi_{i,0} \right)^2 \frac{d\phi_{i,2}}{d\hat{t}} + \frac{d\phi_{i,0}}{d\hat{t}} \left(\phi_{i,1}{}^2 + 2\phi_{i,0}\phi_{i,2} \right) + 2\phi_{i,0}\phi_{i,1} \frac{d\phi_{i,1}}{d\hat{t}} \right]$$

$$+ \frac{\alpha n}{\Delta \hat{x} L} \left(0.177\phi_{i,2} + 3.536 \left(\phi_{i,1}{}^2 + 2\phi_{i,0}\phi_{i,2} \right) \right)$$

$$+ 3.536 \left[\phi_{i,0} \left(\phi_{i,1}{}^2 + 2\phi_{i,0}\phi_{i,2} \right) + \phi_{i,2}\phi_{i,0}{}^2 + 2\phi_{i,0}\phi_{i,1}{}^2 \right]$$

$$- 0.177\phi_{i-1,2} - 3.536 \left(\phi_{i,0}\phi_{i-1,2} + \phi_{i-1,0}\phi_{i,2} + \phi_{i,1}\phi_{i-1,1} \right)$$

$$- 3.536 \left[\phi_{i-1,2} \left(\phi_{i,0} \right)^2 + \phi_{i-1,0} \left(\phi_{i,1}{}^2 + 2\phi_{i,0}\phi_{i,2} \right) + 2\phi_{i,0}\phi_{i,1}\phi_{i-1,1} \right] \right)$$

$$- I\left(\hat{t} \right) \left(5\phi_{i,2} + \left(\phi_{i,1}{}^2 + 2\phi_{i,0}\phi_{i,2} \right) \right) = 0 \text{ subject to } \phi_{i,3}\left(0 \right) = 0 \qquad (15.56)$$

The initial approximations can be chosen as $\Phi_{1,0} = 1 - \exp\left(-\hat{t} \right)$ and $\Phi_{2,0} = \Phi_{3,0} = \Phi_{4,0} = \hat{t}$. Using these initial approximations, we can solve the equations iteratively using symbolic software. Finally, the HPM-based solutions can be approximated as follows:

$$h_i\left(\hat{t} \right) \approx \sum_{k=0}^{M} \Phi_{i,k} \qquad (15.57)$$

15.3.4 OHAM-BASED SOLUTION

When applying the optimal homotopy asymptotic method (OHAM), Eq. (15.20) can be written in the following form:

$$\mathcal{L}_i\left[h_i\left(\hat{t} \right) \right] + \mathcal{N}_i\left[h_i\left(\hat{t} \right) \right] + \bar{h}_i\left(\hat{t} \right) = 0 \qquad (15.58)$$

Here, we select:

$$\mathcal{L}_i\left[h_i\left(\hat{t} \right) \right] = \frac{1.250}{D} \frac{dh_i}{d\hat{t}} + \frac{0.177\alpha n}{\Delta \hat{x} L} h_i \qquad (15.59)$$

$$\mathcal{N}_i\left[h_i\left(\hat{t} \right) \right] = \frac{1}{D} \left(5h_i + h_i{}^2 \right) \frac{dh_i}{d\hat{t}}$$

$$+ \frac{\alpha n}{\Delta \hat{x} L} \left(3.536h_i{}^2 + 3.536h_i{}^3 - 0.177h_{i-1} - 3.536h_i h_{i-1} - 3.536h_i{}^2 h_{i-1} \right)$$

$$- I\left(\hat{t} \right) \left(5h_i + h_i{}^2 \right) \qquad (15.60)$$

and:

$$\bar{h}_i\left(\hat{t}\right) = -1.250I\left(\hat{t}\right) \tag{15.61}$$

With these considerations, the zeroth-order problem becomes:

$$\mathcal{L}_i\left[h_{i,0}\left(\hat{t}\right)\right] + \bar{h}_i\left(\hat{t}\right) = 0 \text{ subject to } h_{i,0}\left(0\right) = 0 \tag{15.62}$$

Solving Eq. (15.62), one obtains:

$$\frac{1.250}{D}\frac{dh_{i,0}}{d\hat{t}} + \frac{0.177\alpha n}{\Delta \hat{x} L}h_{i,0} = 1.250I\left(\hat{t}\right) \text{ subject to } h_{i,0}\left(0\right) = 0 \tag{15.63}$$

To obtain $\mathcal{N}_{i,0}$, $\mathcal{N}_{i,1}$, $\mathcal{N}_{i,2}$, etc., like in Eqs. (15.45)–(15.51), the following expressions are used:

$$h_i\frac{dh_i}{d\hat{t}} = h_{i,0}\frac{dh_{i,0}}{d\hat{t}} + q\left(h_{i,1}\frac{dh_{i,0}}{d\hat{t}} + h_{i,0}\frac{dh_{i,1}}{d\hat{t}}\right) + q^2\left(h_{i,0}\frac{dh_{i,2}}{d\hat{t}} + h_{i,1}\frac{dh_{i,1}}{d\hat{t}} + h_{i,2}\frac{dh_{i,0}}{d\hat{t}}\right) + \cdots \tag{15.64}$$

$$h_i^2\frac{dh_i}{d\hat{t}} = \left(h_{i,0}\right)^2\frac{dh_{i,0}}{d\hat{t}} + q\left[\left(h_{i,0}\right)^2\frac{dh_{i,1}}{d\hat{t}} + 2h_{i,0}h_{i,1}\frac{dh_{i,0}}{d\hat{t}}\right]$$

$$+ q^2\left[\left(h_{i,0}\right)^2\frac{dh_{i,2}}{d\hat{t}} + \frac{dh_{i,0}}{d\hat{t}}\left(h_{i,1}^2 + 2h_{i,0}h_{i,2}\right) + 2h_{i,0}h_{i,1}\frac{dh_{i,1}}{d\hat{t}}\right] + \cdots \tag{15.65}$$

$$\left(h_i\right)^2 = \left(h_{i,0}\right)^2 + q\left(2h_{i,0}h_{i,1}\right) + q^2\left(h_{i,1}^2 + 2h_{i,0}h_{i,2}\right) + \cdots \tag{15.66}$$

$$\left(h_i\right)^3 = \left(h_{i,0}\right)^3 + q\left[3h_{i,0}^2 h_{i,1}\right] + q^2\left[h_{i,0}\left(h_{i,1}^2 + 2h_{i,0}h_{i,2}\right) + h_{i,2}h_{i,0}^2 + 2h_{i,0}h_{i,1}^2\right] + \cdots \tag{15.67}$$

$$h_i h_{i-1} = h_{i,0}h_{i-1,0} + q\left(h_{i,0}h_{i-1,1} + h_{i-1,0}h_{i,1}\right) + q^2\left(h_{i,0}h_{i-1,2} + h_{i-1,0}h_{i,2} + h_{i,1}h_{i-1,1}\right) + \cdots \tag{15.68}$$

$$h_i^2 h_{i-1} = h_{i-1,0}\left(h_{i,0}\right)^2 + q\left[2h_{i,0}h_{i,1}h_{i-1,0} + h_{i-1,1}\left(h_{i,0}\right)^2\right]$$

$$+ q^2\left[h_{i-1,2}\left(h_{i,0}\right)^2 + h_{i-1,0}\left(h_{i,1}^2 + 2h_{i,0}h_{i,2}\right) + 2h_{i,0}h_{i,1}h_{i-1,1}\right] + \cdots \tag{15.69}$$

Using Eqs. (15.64–15.69), we have:

$$\frac{1}{D}\left(5h_i + h_i^2\right)\frac{dh_i}{d\hat{t}} + \frac{\alpha n}{\Delta\hat{x}L}\left(3.536h_i^2 + 3.536h_i^3 - 0.177h_{i-1} - 3.536h_ih_{i-1} - 3.536h_i^2h_{i-1}\right)$$

$$-I\left(\hat{t}\right)\left(5h_i + h_i^2\right) = \frac{5}{D}h_{i,0}\frac{dh_{i,0}}{d\hat{t}} + \frac{1}{D}\left(h_{i,0}\right)^2\frac{dh_{i,0}}{d\hat{t}} + \frac{\alpha n}{\Delta\hat{x}L}$$

$$\times\left(3.536h_{i,0}^2 + 3.536h_{i,0}^3 - 0.177h_{i-1,0} - 3.536h_{i,0}h_{i-1,0} - 3.536h_{i,0}^2h_{i-1,0}\right)$$

$$-I\left(\hat{t}\right)\left(5h_{i,0} + h_{i,0}^2\right) + q\left[\frac{5}{D}\left(h_{i,1}\frac{dh_{i,0}}{d\hat{t}} + h_{i,0}\frac{dh_{i,1}}{d\hat{t}}\right) + \frac{1}{D}\left[\left(h_{i,0}\right)^2\frac{dh_{i,1}}{d\hat{t}} + 2h_{i,0}h_{i,1}\frac{dh_{i,0}}{d\hat{t}}\right]\right.$$

$$+\frac{\alpha n}{\Delta\hat{x}L}\left(3.536\left(2h_{i,0}h_{i,1}\right) + 3.536\left[3h_{i,0}^2h_{i,1}\right] - 0.177h_{i-1,1} - 3.536\left(h_{i,0}h_{i-1,1} + h_{i-1,0}h_{i,1}\right)\right.$$

$$\left.-3.536\left[2h_{i,0}h_{i,1}h_{i-1,0} + h_{i-1,1}\left(h_{i,0}\right)^2\right]\right) - I\left(\hat{t}\right)\left(5h_{i,1} + 2h_{i,0}h_{i,1}\right)\right]$$

$$+q^2\left[\frac{5}{D}\left(h_{i,0}\frac{dh_{i,2}}{d\hat{t}} + h_{i,1}\frac{dh_{i,1}}{d\hat{t}} + h_{i,2}\frac{dh_{i,0}}{d\hat{t}}\right) + \frac{1}{D}\left[\left(h_{i,0}\right)^2\frac{dh_{i,2}}{d\hat{t}} + \frac{dh_{i,0}}{d\hat{t}}\left(h_{i,1}^2 + 2h_{i,0}h_{i,2}\right) + 2h_{i,0}h_{i,1}\frac{dh_{i,1}}{d\hat{t}}\right]\right.$$

$$+\frac{\alpha n}{\Delta\hat{x}L}\left(3.536\left(h_{i,1}^2 + 2h_{i,0}h_{i,2}\right) + 3.536\left[h_{i,0}\left(h_{i,1}^2 + 2h_{i,0}h_{i,2}\right) + h_{i,2}h_{i,0}^2 + 2h_{i,0}h_{i,1}^2\right] - 0.177h_{i-1,2}\right.$$

$$\left.-3.536\left(h_{i,0}h_{i-1,2} + h_{i-1,0}h_{i,2} + h_{i,1}h_{i-1,1}\right) - 3.536\left[h_{i-1,2}\left(h_{i,0}\right)^2 + h_{i-1,0}\left(h_{i,1}^2 + 2h_{i,0}h_{i,2}\right) + 2h_{i,0}h_{i,1}h_{i-1,1}\right]\right)$$

$$\left.-I\left(\hat{t}\right)\left(5h_{i,2} + \left(h_{i,1}^2 + 2h_{i,0}h_{i,2}\right)\right)\right] \tag{15.70}$$

Using Eq. (15.70), the first-order problem reduces to:

$$\frac{1.250}{D}\frac{dh_{i,1}}{d\hat{t}} + \frac{0.177\alpha n}{\Delta\hat{x}L}h_{i,1} = H_{i,1}\left(\hat{t},C_i\right)\left\{\frac{5}{D}h_{i,0}\frac{dh_{i,0}}{d\hat{t}} + \frac{1}{D}\left(h_{i,0}\right)^2\frac{dh_{i,0}}{d\hat{t}}\right.$$

$$+\frac{\alpha n}{\Delta\hat{x}L}\left(3.536h_{i,0}^2 + 3.536h_{i,0}^3 - 0.177h_{i-1,0} - 3.536h_{i,0}h_{i-1,0} - 3.536h_{i,0}^2h_{i-1,0}\right)$$

$$\left.-I\left(\hat{t}\right)\left(5h_{i,0} + h_{i,0}^2\right)\right\} \text{ subject to } h_{i,1}\left(0\right) = 0 \tag{15.71}$$

Let us simply choose $H_{i,1}\left(\hat{t},C_{i,j}\right) = C_{i,1}$. Using $k = 2$, the second-order problem becomes:

$$\mathcal{L}_i\left[h_{i,2}\left(\hat{t},C_{i,j}\right) - h_{i,1}\left(\hat{t},C_{i,j}\right)\right] = H_{i,2}\left(\hat{t},C_{i,j}\right)\mathcal{N}_{i,0}\left(h_{i,0}\left(\hat{t}\right)\right) + H_{i,1}\left(\hat{t},C_{i,j}\right)$$

$$\times\left[\mathcal{L}_i\left[h_{i,1}\left(\hat{t},C_{i,j}\right)\right] + \mathcal{N}_{i,1}\left[h_{i,0}\left(\hat{t}\right),h_{i,1}\left(\hat{t},C_{i,j}\right)\right]\right] \text{ subject to } h_{i,2}\left(0\right) = 0 \tag{15.72}$$

Further, we choose $H_{i,2}\left(\hat{t}, C_{i,j}\right) = C_{i,2}$. Therefore, Eq. (15.72) becomes:

$$\frac{1.250}{D}\frac{dh_{i,2}}{d\hat{t}} + \frac{0.177\alpha n}{\Delta\hat{x}L}h_{i,2} = \frac{1.250}{D}\frac{dh_{i,1}}{d\hat{t}} + \frac{0.177\alpha n}{\Delta\hat{x}L}h_{i,1}$$

$$+ C_{i,2}\left\{\frac{5}{D}h_{i,0}\frac{dh_{i,0}}{d\hat{t}} + \frac{1}{D}\left(h_{i,0}\right)^2\frac{dh_{i,0}}{d\hat{t}}\right.$$

$$+ \frac{\alpha n}{\Delta\hat{x}L}\left(3.536h_{i,0}^2 + 3.536h_{i,0}^3 - 0.177h_{i-1,0} - 3.536h_{i,0}h_{i-1,0} - 3.536h_{i,0}^2h_{i-1,0}\right)$$

$$- I\left(\hat{t}\right)\left(5h_{i,0} + h_{i,0}^2\right)\right\} + C_{i,1}\left[\frac{1.250}{D}\frac{dh_{i,1}}{d\hat{t}} + \frac{0.177\alpha n}{\Delta\hat{x}L}h_{i,1} + \frac{5}{D}\left(h_{i,1}\frac{dh_{i,0}}{d\hat{t}} + h_{i,0}\frac{dh_{i,1}}{d\hat{t}}\right)\right.$$

$$+ \frac{1}{D}\left[\left(h_{i,0}\right)^2\frac{dh_{i,1}}{d\hat{t}} + 2h_{i,0}h_{i,1}\frac{dh_{i,0}}{d\hat{t}}\right] + \frac{\alpha n}{\Delta\hat{x}L}\left(3.536\left(2h_{i,0}h_{i,1}\right) + 3.536\left[3h_{i,0}^2h_{i,1}\right]\right.$$

$$\left. - 0.177h_{i-1,1} - 3.536\left(h_{i,0}h_{i-1,1} + h_{i-1,0}h_{i,1}\right) - 3.536\left[2h_{i,0}h_{i,1}h_{i-1,0} + h_{i-1,1}\left(h_{i,0}\right)^2\right]\right)$$

$$\left. - I\left(\hat{t}\right)\left(5h_{i,1} + 2h_{i,0}h_{i,1}\right)\right] \text{ subject to } h_{i,2}(0) = 0 \qquad (15.73)$$

The terms of the series can be computed using the equations developed here. One can compute these terms without any difficulty using symbolic computation software, such as MATLAB. Further, following Chapter 2, the higher-order terms can be computed in a similar manner. However, our aim is to produce an accurate solution with just two to three terms of the OHAM-based series. Therefore, we restrict our calculation up to $k = 2$. Finally, the approximate solutions can be found as:

$$h_i\left(\hat{t}\right) \approx h_{i,0}\left(\hat{t}\right) + h_{i,1}\left(\hat{t}, C_{i,1}\right) + h_{i,2}\left(\hat{t}, C_{i,1}, C_{i,2}\right) \qquad (15.74)$$

where the terms are given by Eqs. (15.63), (15.71), and (15.73).

15.4 CONVERGENCE THEOREMS

The convergence of the HAM-based and OHAM-based solutions are proved theoretically using the following theorems.

15.4.1 CONVERGENCE THEOREM OF THE HAM-BASED SOLUTION

The convergence theorems for the HAM-based solutions given by Eq. (15.36) can be proved using the following theorems.

Theorem 15.1: If the homotopy series $\Sigma_{m=0}^{\infty} h_{i,m}\left(\hat{t}\right)$ and $\Sigma_{m=0}^{\infty} h_{i,m}'\left(\hat{t}\right)$ converge, then $R_{i,m}\left(\vec{h}_{i,m-1}\right)$ given by Eq. (15.28) satisfies the relation $\Sigma_{m=1}^{\infty} R_{i,m}\left(\vec{h}_{i,m-1}\right) = 0$. [Here " ' " denotes the first derivative with respect to $\hat{t}$.]

Proof: The proof of this theorem follows exactly from the previous chapters.

Theorem 15.2: If $\hbar$ is so properly chosen that the series $\sum_{m=0}^{\infty} h_{i,m}\left(\hat{t}\right)$ and $\sum_{m=0}^{\infty} h_{i,m}'\left(\hat{t}\right)$ converge absolutely to $h_i\left(\hat{t}\right)$ *and* $h_i'\left(\hat{t}\right)$, respectively, then the homotopy series $\sum_{m=0}^{\infty} h_{i,m}\left(\hat{t}\right)$ satisfies the original governing Eq. (15.20).

Proof: Using the Cauchy product defined in Theorem 2.2 of Chapter 2 in relation to Eq. (15.28), we have:

$$\sum_{m=1}^{\infty}\sum_{j=0}^{m-1} h_{i,j} h_{i,m-1-j}' = \left(\sum_{m=0}^{\infty} h_{i,m}\right)\left(\sum_{k=0}^{\infty} h_{i,k}'\right) \tag{15.75}$$

$$\sum_{m=1}^{\infty}\sum_{j=0}^{m-1} h_{i,m-1-j} \sum_{k=0}^{j} h_{i,k} h_{i,j-k}' = \left(\sum_{m=0}^{\infty} h_{i,m}\right)^2\left(\sum_{k=0}^{\infty} h_{i,k}'\right) \tag{15.76}$$

$$\sum_{m=1}^{\infty}\sum_{j=0}^{m-1} h_{i,j} h_{i,m-1-j} = \left(\sum_{m=0}^{\infty} h_{i,m}\right)^2 \tag{15.77}$$

$$\sum_{m=1}^{\infty}\sum_{j=0}^{m-1} h_{i,m-1-j} \sum_{k=0}^{j} h_{i,k} h_{i,j-k} = \left(\sum_{m=0}^{\infty} h_{i,m}\right)^3 \tag{15.78}$$

$$\sum_{m=1}^{\infty}\sum_{j=0}^{m-1} h_{i,j} h_{i-1,m-1-j} = \left(\sum_{m=0}^{\infty} h_{i,m}\right)\left(\sum_{k=0}^{\infty} h_{i-1,k}\right) \tag{15.79}$$

$$\sum_{m=1}^{\infty}\sum_{j=0}^{m-1} h_{i,m-1-j} \sum_{k=0}^{j} h_{i,k} h_{i-1,j-k} = \left(\sum_{m=0}^{\infty} h_{i,m}\right)^2\left(\sum_{k=0}^{\infty} h_{i-1,k}\right) \tag{15.80}$$

Theorem 15.1 shows that if $\sum_{m=0}^{\infty} h_{i,m}\left(\hat{t}\right)$ and $\sum_{m=0}^{\infty} h_{i,m}'\left(\hat{t}\right)$ converge, then $\sum_{m=0}^{\infty} R_m\left(\vec{h}_{m-1}\right)=0$.

Therefore, substituting the earlier expressions in Eq. (15.28) and simplifying further lead to:

$$\frac{1.250}{D}\sum_{m=0}^{\infty} h_{i,m}' + \frac{5}{D}\left(\sum_{m=0}^{\infty} h_{i,m}\right)\left(\sum_{k=0}^{\infty} h_{i,k}'\right)$$

$$+\frac{1}{D}\left(\sum_{m=0}^{\infty} h_{i,m}\right)^2\left(\sum_{k=0}^{\infty} h_{i,k}'\right) + \frac{\alpha n}{\Delta \hat{x} L}\left(0.177\sum_{m=0}^{\infty} h_{i,m} + 3.536\left(\sum_{m=0}^{\infty} h_{i,m}\right)^2\right.$$

$$+3.536\left(\sum_{m=0}^{\infty} h_{i,m}\right)^3 - 0.177\sum_{m=0}^{\infty} h_{i-1,m} - 3.536\left(\sum_{m=0}^{\infty} h_{i,m}\right)\left(\sum_{k=0}^{\infty} h_{i-1,k}\right)$$

$$-3.536\left(\sum_{m=0}^{\infty} h_{i,m}\right)^2\left(\sum_{k=0}^{\infty} h_{i-1,k}\right)\right) - I\left(\hat{t}\right)\left(1.250\sum_{m=0}^{\infty}\left(1-\chi_{i,m}\right) + 5\sum_{m=0}^{\infty} h_{i,m} + \left(\sum_{m=0}^{\infty} h_{i,m}\right)^2\right) = 0$$

$$\tag{15.81}$$

which is basically the original governing equation in Eq. (15.20). Furthermore, subject to the boundary conditions $h_{i,0}(0) = 0$ and the conditions for the higher-order deformation equations $h_{i,m}(0) = 0$ for $m \geq 1$, we easily obtain $\Sigma_{m=0}^{\infty} h_{i,m}(0) = 0$. Hence, the convergence result follows.

15.4.2 CONVERGENCE THEOREM OF THE OHAM-BASED SOLUTION

Theorem 15.3: If the series $h_{i,0}(\hat{t}) + \Sigma_{j=1}^{\infty} h_{i,j}(\hat{t}, C_{i,j})$, $j = 1, 2, \ldots, s$ converges, where $h_{i,j}(\hat{t}, C_{i,j})$ is governed by Eqs. (15.63), (15.71), and (15.73), then Eq. (15.74) is a solution of the original Eq. (15.20).

Proof: The proof of this theorem follows exactly from Theorem 2.3 of Chapter 2.

15.5 RESULTS AND DISCUSSION

First, the numerical convergence of the HAM-based analytical solutions is established for a specific test case, and then the solution is validated over a numerical solution discussed in this chapter. Finally, the HPM- and OHAM-based analytical solutions are validated by comparing them with the MOL-based numerical solution.

15.5.1 NUMERICAL CONVERGENCE AND VALIDATION
OF THE HAM-BASED SOLUTION

For validating the homotopy-based solutions, we consider $N = 4$, which produces the unknowns h_i, $i = 1, \ldots, 4$, with $\widehat{x}_i = \{0.25, 0.50, 0.75, 1.00\}$. The parameter values are taken as $n = 1.5$, $\alpha = 1$ unit, $I = 1$ unit, $D = 1$ unit, and $L = 2$ unit. It can be noted that suitable choices for the auxiliary parameters $\hbar_i$ determine the convergence of the HAM-based series solutions. For $0 \leq \hat{t} \leq 1$, we consider $\hbar_i = \hbar$, and for $\hat{t} \geq 1$, $\hbar_1 = \hbar_2$, $\hbar_3 = \hbar_4$. To that end, the squared residual error of Eq. (15.20) can be calculated as follows:

$$\Delta_m = \int_{\hat{t} \in \Omega} \left(\sum_{i=1}^{N} \mathcal{N}_i \left[h_i(\hat{t}) \right] \right)^2 d\hat{t} \tag{15.82}$$

where Ω is the domain of the problem. The HAM-based series solutions lead to the exact solutions of the problem when Δ_m tends to zero. Therefore, for a particular order of approximation m, it is sufficient to minimize the corresponding Δ_m, which yields an optimal value of parameter $\hbar$. Considering the parameters chosen earlier, the HAM solutions are assessed and the squared residual errors are calculated. Figure 15.4 shows the squared residual errors for a different order of approximation for the selected case. It can be seen from the figure that as the order of approximation increases, the corresponding residual error decreases. Thus, the adequacy of selecting the linear operators, initial approximations, and auxiliary functions is established. Also, for a quantitative assessment, the numerical results are reported in Table 15.1 along with the computational time taken by the computer to produce the

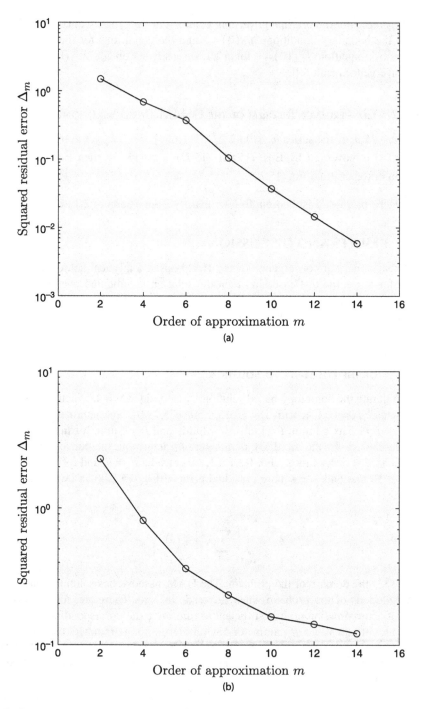

FIGURE 15.4   Squared residual error (Δ_m) versus different order of approximations (m) of the HAM-based solution for the selected case: (A) $0 \leq \hat{t} \leq 1$ and (B) $\hat{t} \geq 1$.

TABLE 15.1

Squared Residual Error (Δ_m) and Computational Time versus Different Orders of Approximation (m) for the Selected Case

Order of Approximation (m)	$0 \leq \hat{t} \leq 1$		$\hat{t} \geq 1$	
	Squared Residual Error (Δ_m)	Computational Time (sec)	Squared Residual Error (Δ_m)	Computational Time (sec)
2	7.51×10^{-2}	0.156	2.28×10^{0}	0.052
4	3.24×10^{-3}	0.412	8.12×10^{-1}	0.154
6	1.40×10^{-4}	1.005	3.62×10^{-1}	0.344
8	8.02×10^{-5}	3.289	2.32×10^{-1}	0.653
10	1.15×10^{-5}	7.996	1.61×10^{-1}	1.116
12	7.77×10^{-6}	16.481	1.42×10^{-1}	1.616
14	8.50×10^{-7}	35.330	1.21×10^{-1}	2.324

corresponding order of approximations. For the selected case, a comparison between the MOL-based numerical solution and 15th-order HAM-based approximate solution is shown in Figure 15.5. An excellent agreement between the computed and observed values is found from the figure. To get a clear comparative idea, the numerical values are calculated for both the solutions and reported in Table 15.2. All the computations are performed using the BVPh 2.0 package developed by Zhao and Liao (2002). A flowchart containing the steps of the method is provided in Figure 15.6.

15.5.2 VALIDATION OF THE HPM-BASED SOLUTION

The HPM-based analytical solutions for the selected test case are validated over the MOL-based numerical solution discussed earlier. Here, we consider the solution for the part $0 \leq \hat{t} \leq 1$. One can derive the solution for the other part following the same steps. After assessing the solution, it was found that the three-term HPM solution provides the best approximation in comparison with the other HPM approximations. However, it can be seen that the solution is more accurate for the first half of the domain of the independent variable. Figure 15.7 shows the comparison of the three-term HPM solution with the numerical solution. It can be noted that one can indeed try with some other initial approximations, which might work in producing more accurate solutions to the problem. The HPM-based values and the numerical solution are compared in Table 15.3. A flowchart containing the steps of the method is given in Figure 15.8.

15.5.3 VALIDATION OF THE OHAM-BASED SOLUTION

For the assessment of the OHAM-based analytical solution, one needs to calculate the constant $C_{i,j}$. For that purpose, we calculate the residual as follows:

$$R_i\left(\hat{t}, C_{i,j}\right) = \mathcal{L}_i\left[h_{i,OHAM}\left(\hat{t}, C_{i,j}\right)\right] + \mathcal{N}_i\left[h_{i,OHAM}\left(\hat{t}, C_{i,j}\right)\right] + \overline{h}_i\left(\hat{t}\right), \quad j = 1, 2, \ldots, s \quad (15.83)$$

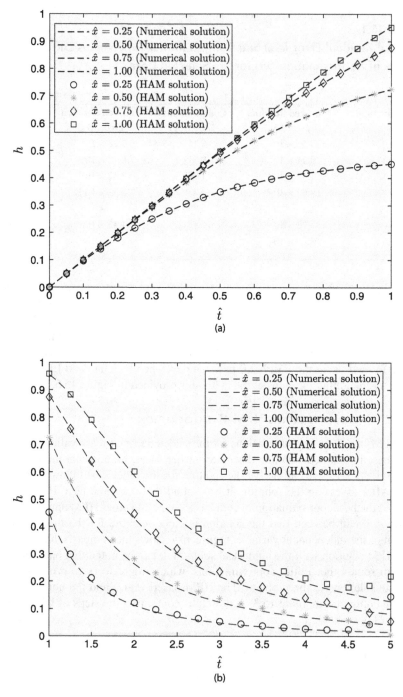

FIGURE 15.5 Comparison between the MOL-based numerical solution and 14th-order HAM-based solution for the selected case: (A) $0 \leq \hat{t} \leq 1$ and (B) $\hat{t} \geq 1$.

TABLE 15.2
Comparison between HAM-Based Approximation and Numerical Solution for the Selected Case

(a) $0 \leq \hat{t} \leq 1$

$\hat{t}$	Numerical Solution		14th-Order HAM-Based Approximations	
	$\hat{x}_i = 0.25$	$\hat{x}_i = 1.00$	$\hat{x}_i = 0.25$	$\hat{x}_i = 1.00$
0.0	0.0000	0.0000	0.0000	0.0000
0.1	0.0961	0.1000	0.0960	0.0992
0.2	0.1802	0.2000	0.1803	0.1981
0.3	0.2499	0.3000	0.2503	0.2973
0.4	0.3057	0.3998	0.3061	0.3971
0.5	0.3492	0.4990	0.3496	0.4972
0.6	0.3826	0.5971	0.3828	0.5967
0.7	0.4079	0.6930	0.4080	0.6942
0.8	0.4268	0.7854	0.4269	0.7874
0.9	0.4410	0.8730	0.4409	0.8737
1.0	0.4515	0.9544	0.4513	0.95067

(b) $\hat{t} \geq 1$

$\hat{t}$	Numerical Solution		14th-Order HAM-Based Approximations	
	$\hat{x}_i = 0.25$	$\hat{x}_i = 1.00$	$\hat{x}_i = 0.25$	$\hat{x}_i = 1.00$
1.0	0.4515	0.9544	0.4512	0.9575
1.4	0.2294	0.8251	0.2414	0.8283
1.8	0.1382	0.6616	0.1497	0.6740
2.2	0.0920	0.5174	0.1012	0.5360
2.6	0.0652	0.4056	0.0721	0.4269
3.0	0.0482	0.3222	0.0533	0.3437
3.4	0.0367	0.2601	0.0402	0.2806
3.8	0.0286	0.2132	0.0309	0.2321
4.2	0.0226	0.1773	0.0243	0.1949
4.6	0.0181	0.1493	0.0280	0.1782
5.0	0.0146	0.1270	0.1447	0.2191

where $h_{i,OHAM}\left(\hat{t}, C_{i,j}\right)$ is the approximate solution. When $R_i\left(\hat{t}, C_{i,j}\right) = 0$, $h_{i,OHAM}\left(\hat{t}, C_{i,j}\right)$ becomes the exact solution to the problem. One of the ways to obtain the optimal $C_{i,j}$ for which the solution converges is the minimization of the squared residual error, i.e.,

$$J\left(C_{i,j}\right) = \int_{\hat{t} \in D}\left(\sum_{i=1}^{N} R_i\left(\hat{t}, C_{i,j}\right)\right)^2 d\hat{t}, \ i = 1, 2, \ldots, s \qquad (15.84)$$

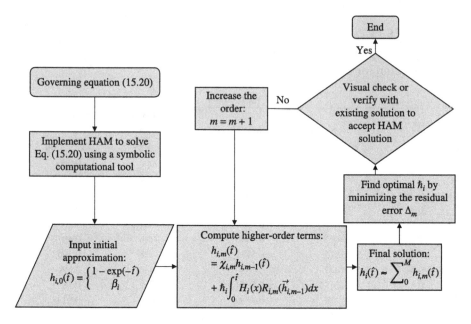

FIGURE 15.6   Flowchart for the HAM solution in Eq. (15.36).

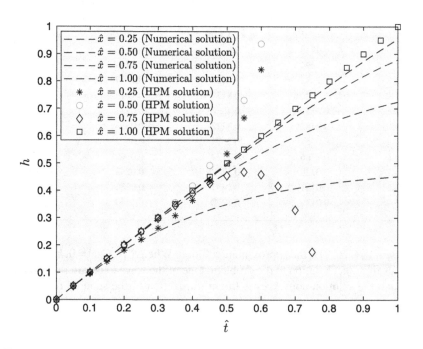

FIGURE 15.7   Comparison between the numerical solution and three-term HPM-based approximation for the selected case.

TABLE 15.3

Comparison between Three-Term HPM-Based Approximation and Numerical Solution for the Selected Case for $0 \le \hat{t} \le 1$

$\hat{t}$	Numerical Solution		Three-Term HPM-Based Approximations	
	$\hat{x}_i = 0.25$	$\hat{x}_i = 1.00$	$\hat{x}_i = 0.25$	$\hat{x}_i = 1.00$
0.0	0.0000	0.0000	0.0000	0.0000
0.1	0.0961	0.1000	0.0961	0.1000
0.2	0.1802	0.2000	0.1810	0.2000
0.3	0.2499	0.3000	0.2617	0.3000
0.4	0.3057	0.3998	0.3638	0.4000
0.5	0.3492	0.4990	0.5344	0.5000
0.6	0.3826	0.5971	0.8423	0.6000
0.7	0.4079	0.6930	1.3766	0.7000
0.8	0.4268	0.7854	2.2433	0.8000
0.9	0.4410	0.8730	3.5608	0.9000
1.0	0.4515	0.9544	5.4554	1.0000

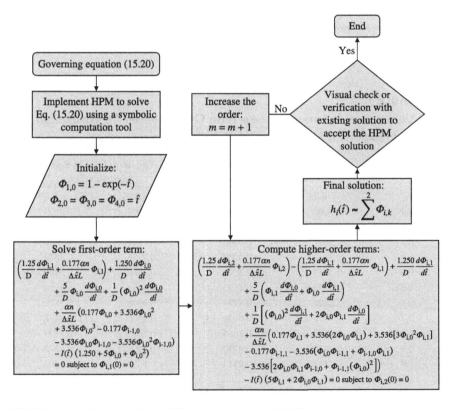

FIGURE 15.8 Flowchart for the HPM solution in Eq. (15.57).

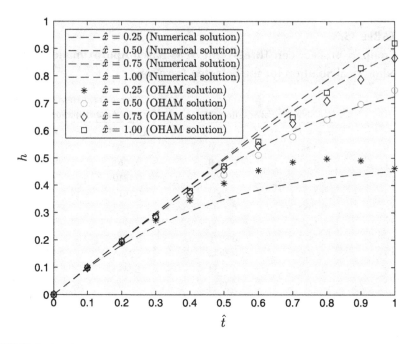

FIGURE 15.9 Comparison between the numerical solution and two-term OHAM-based solution for the selected case for $0 \leq \hat{t} \leq 1$.

TABLE 15.4

Comparison between the Two-Term OHAM-Based Approximation and Numerical Solution for the Selected Case for $0 \leq \hat{t} \leq 1$

$\hat{t}$	Numerical Solution		Two-Term OHAM-Based Approximations	
	$\hat{x}_i = 0.25$	$\hat{x}_i = 1.00$	$\hat{x}_i = 0.25$	$\hat{x}_i = 1.00$
0.0	0.0000	0.0000	0.0000	0.0000
0.1	0.0961	0.1000	0.0976	0.0984
0.2	0.1802	0.2000	0.1891	0.1940
0.3	0.2499	0.3000	0.2726	0.2874
0.4	0.3057	0.3998	0.3460	0.3792
0.5	0.3492	0.4990	0.4071	0.4699
0.6	0.3826	0.5971	0.4540	0.5600
0.7	0.4079	0.6930	0.4846	0.6498
0.8	0.4268	0.7854	0.4973	0.7396
0.9	0.4410	0.8730	0.4903	0.8296
1.0	0.4515	0.9544	0.4624	0.9203

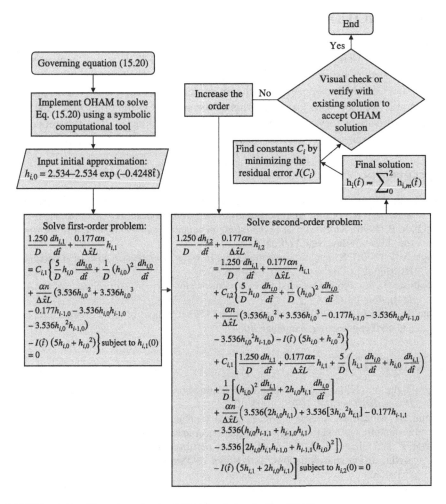

FIGURE 15.10 Flowchart for the OHAM solution in Eq. (15.74).

where D is the domain of the problem. The minimization of Eq. (15.82) leads to a system of algebraic equations as follows:

$$\frac{\partial J}{\partial C_{1,1}} = \frac{\partial J}{\partial C_{1,2}} = \cdots = \frac{\partial J}{\partial C_{1,s}} = 0 \tag{15.85}$$

Solving this system, one can obtain the optimal values of parameters. We obtain the optimal values using the MATLAB routine *fminsearch*, which minimizes an unconstrained multivariable function. Here, we consider the solution for the part $0 \leq \hat{t} \leq 1$. One can derive the solution for the other part following the same steps. It is found that only the two-term solution produces favorable results. Therefore, the two-term OHAM solution is computed and compared in Figure 15.9 with the MOL-based numerical solution to see the effectiveness of the proposed approach. For a quantitative assessment, we also compare the numerical values of the solutions in Table 15.4. A flowchart containing the steps of the method is provided in Figure 15.10.

15.6 CONCLUDING REMARKS

The HAM, HPM, and OHAM are applied to the governing equation of the kinematic wave solution. Using the method of lines, the original PDE is first converted to a nonlinear system using the finite difference method along the space direction. The HAM and OHAM provide good results for a selected test case. The OHAM provides a favorable solution with just two terms of the series. On the other hand, the HPM produces a satisfactory result only for the restricted domain. The theoretical and numerical convergences of the series solutions are provided.

REFERENCES

Henderson, F. M., and Wooding, R. A. (1964). Overland flow and groundwater flow from a steady rainfall of finite duration. *Journal of Geophysical Research*, *69*(8), 1531–1540.

Singh, V. P. (1996). *Kinematic wave modelling in water resources: Surface water hydrology* (pp. 1399). Wiley, New York, NY.

Singh, V. P. (2001). Kinematic wave modelling in water resources: A historical perspective. *Hydrological Processes*, *15*(4), 671–706.

Woolhiser, D. A. (1969). Overland flow on a converging surface. *Transactions of the ASAE*, *12*(4), 460–0462.

Zhao, Y., and Liao, S. (2002). User guide to BVPh 2.0. School of naval architecture, ocean and civil engineering, Shanghai, *40*.

FURTHER READINGS

Agiralioglu, N. (1981). Water routing on diverging-converging watersheds. *Journal of the Hydraulics Division*, *107*(8), 1003–1017.

Azimi, M., Mozaffari, A., and Ommi, F. (2014). Application of Galerkin optimal homotopy asymptotic method to shock wave equation. *Journal of Advanced Physics*, *3*(1), 35–38.

Biazar, J., and Ghazvini, H. (2008). Homotopy perturbation method for solving hyperbolic partial differential equations. *Computers & Mathematics with Applications*, *56*(2), 453–458.

Brakensiek, D. L. (1966). Hydrodynamics of overland flow and nonprismatic channels. *Transactions of the ASAE*, *9*(1), 119–0122.

Eagleson, P. S. (1972). Dynamics of flood frequency. *Water Resources Research*, *8*(4), 878–898.

Kibler, D. F., and Woolhiser, D. A. (1970). *Kinematic cascade as a hydrologic model, The* (Doctoral dissertation, Colorado State University. Libraries).

Kibler, D. F., and Woolhiser, D. A. (1972). Mathematical properties of the kinematic cascade. *Journal of Hydrology*, *15*(2), 131–147.

Li, R. M., Simons, D. B., and Stevens, M. A. (1975). Nonlinear kinematic wave approximation for water routing. *Water Resources Research*, *11*(2), 245–252.

Liggett, J. A., and Woolhiser, D. A. (1967). Difference solutions of the shallow-water equation. *Journal of the Engineering Mechanics Division*, *93*(2), 39–72.

Matinfar, M., Saeidy, M., Yasir, K. H. A. N., and Gharahsuflu, B. (2014). Finding the exact solution of special nonlinear partial differential equations by homotopy analysis method. *Walailak Journal of Science and Technology (WJST)*, *11*(3), 171–178.

Mohyud-Din, S. T., Yıldırım, A., and Kaplan, Y. (2010). Homotopy perturbation method for one-dimensional hyperbolic equation with integral conditions. *Zeitschrift für Naturforschung A*, *65*(12), 1077–1080.

Raftari, B., Khosravi, H., and Yildirim, A. (2013). Homotopy analysis method for the one-dimensional hyperbolic telegraph equation with initial conditions. *International Journal of Numerical Methods for Heat & Fluid Flow, 23* (2), 355–372.

Sherman, B., and Singh, V. P. (1976a). A distributed converging overland flow model: 1. Mathematical solutions. *Water Resources Research, 12*(5), 889–896.

Sherman, B., and Singh, V. P. (1976b). A distributed converging overland flow model: 2. Effect of infiltration. *Water Resources Research, 12*(5), 897–901.

Singh V. P. 1974. Kinematic wave model of surface runoff. PhD dissertation, Department of Civil Engineering, Colorado State University, Fort Collins, CO.

Singh, V. P. (1976a). A distributed converging overland flow model: 3. Application to natural watersheds. *Water Resources Research, 12*(5), 902–908.

Singh, V. P. (1976b). A note on the step error of some finite-difference schemes used to solve kinematic wave equations. *Journal of Hydrology, 30*(3), 247–255.

Singh, V. P. (1977). Criterion to choose step length for some numerical methods used in hydrology. *Journal of Hydrology, 33*(3–4), 287–299.

Singh, V. P., and Agiralioglu, N. (1981a). Diverging overland flow: 1. Analytical solutions. *Hydrology Research, 12*(2), 81–98.

Singh, V., and Agiralioglu, N. (1981b). Diverging overland flow: 2. Application to natural watersheds. *Hydrology Research, 12*(2), 99–110.

Singh, V. P., and Woolhiser, D. A. (1976). A nonlinear kinematic wave model for watershed surface runoff. *Journal of Hydrology, 31*(3–4), 221–243.

Wooding, R. A. (1965a). A hydraulic model for the catchment-stream problem: I. Kinematic-wave theory. *Journal of Hydrology, 3*(3–4), 254–267.

Wooding, R. A. (1965b). A hydraulic model for the catchment-stream problem: II. Numerical solutions. *Journal of Hydrology, 3*(3–4), 268–282.

Woolhiser, D. A., Hanson, C. L., and Kuhlman, A. R. (1970). Overland flow on rangeland watersheds. *Journal of Hydrology (New Zealand), 9*, 336–356.

16 Multispecies Convection-Dispersion Transport Equation with Variable Parameters

16.1 INTRODUCTION

Pollution is caused by a number of sources, such as agricultural fertilizers, erosion, industries, energy generation, waste disposal, hospital waste, vehicular transport, nuclear waste, and domestic waste. A significant portion of pollution generated by these sources finds its way either directly or indirectly into groundwater. Often pollutants are different species, and their chemical characteristics are different. In general, pollutant transport in groundwater is modeled using the conservation of mass of water, flux law for flow of water, conservation of mass of pollutant, and flux law for pollutant transport. Depending on the type, a pollutant can be physical, chemical, or biological. A chemical pollutant can be conservative or nonconservative, for which production and decay functions need to be specified. When these equations are combined, the resulting equation is the advection-dispersion equation. This equation is for a specific species. This chapter discusses the case of two-species pollutant transport.

16.2 GOVERNING EQUATION

In a one-dimensional flow field, the general form for a reactive transport system that describes the phenomenon of a multispecies migration having spatially and temporally varying parameters can be modeled as follows (Chaudhary and Singh 2020):

$$r_i \frac{\partial C_i}{\partial t} = \frac{\partial}{\partial x}\left(D(x,t)\frac{\partial C_i}{\partial x} - v(x,t)C_i\right) + k_{i-1}C_{i-1} - k_iC_i, \ i = 1, 2, \ldots, n \quad (16.1)$$

Here, x and t denote the spatial and temporal coordinates, respectively; r_i is the retardation factor for the i-th species; C_i is the concentration strength for the i-th species; k_i is the decay rate coefficient for the i-th species; and $D(x,t)$ and $v(x,t)$ are the dispersion coefficient and seepage velocity, respectively.

The present work considers a two-species system (i.e., $i = 1, 2$) and includes the effect of spatially and temporally dependent transport parameters on pollutant migration. To that end, the model becomes (Chaudhary and Singh 2020):

$$r_1 \frac{\partial C_1}{\partial t} = \frac{\partial}{\partial x}\left(D(x,t)\frac{\partial C_1}{\partial x} - v(x,t)C_1\right) - k_1C_1 \quad (16.2)$$

DOI: 10.1201/9781003368984-20

$$r_2 \frac{\partial C_2}{\partial t} = \frac{\partial}{\partial x}\left(D(x,t)\frac{\partial C_2}{\partial x} - v(x,t)C_2 \right) - k_2 C_2 + k_1 C_1 \qquad (16.3)$$

where C_1 and C_2 represent the pollutant concentration level for the parent and daughter species, respectively.

The initial and boundary conditions for the model can be given as follows:

$$C_1(x,0) = f_1(x), \; C_1(0,t) = f_1(0), \; \lim_{x \to \infty} C_1(x,t) = 0 \qquad (16.4)$$

$$C_2(x,0) = f_2(x), \; C_2(0,t) = f_2(0), \; \lim_{x \to \infty} C_2(x,t) = 0 \qquad (16.5)$$

where $f_1(x)$ and $f_2(x)$ are considered as (Simpson and Ellery 2014):

$$f_1(x) = a_1 x exp(-a_2 x), \; f_2(x) = 0 \qquad (16.6)$$

Eqs. (16.4)–(16.6) show that spatially dependent pollution exists initially for the parent species. However, for the daughter species, the initial concentration is zero.

Based on the forms of $D(x,t)$ and $v(x,t)$, several scenarios can be considered. Here, we consider a specific case, specifically the steady migration phenomenon in the case of steady groundwater flow. Accordingly, we assume:

$$D(x,t) = D_0(\alpha_1 + \alpha_2 x)^2, \; v(x,t) = v_0(\alpha_1 + \alpha_2 x) \qquad (16.7)$$

where D_0 and v_0 are the initial dispersion and velocity, respectively, and α_1 and α_2 are the parameters, where α_2 is known as the heterogeneity parameter. Eq. (16.7) shows that dispersion is directly proportional to the square of the velocity (Batu 2005). For $\alpha_1 = 1$ and $\alpha_2 \to 0$, i.e., α_2 is very small, the system converts into a homogeneous one, where the effect of space on dispersion and seepage velocity is absent. As α_2 increases, the system becomes heterogeneous.

Before applying the homotopy-based methods, let us rewrite the governing Eqs. (16.2) and Eqs. (16.3) in the following form:

$$r_1 \frac{\partial C_1}{\partial t} = D \frac{\partial^2 C_1}{\partial x^2} + \left(\frac{\partial D}{\partial x} - v \right)\frac{\partial C_1}{\partial x} - \left(k_1 + \frac{\partial v}{\partial x} \right)C_1 \qquad (16.8)$$

$$r_2 \frac{\partial C_2}{\partial t} = D \frac{\partial^2 C_2}{\partial x^2} + \left(\frac{\partial D}{\partial x} - v \right)\frac{\partial C_2}{\partial x} - \left(k_2 + \frac{\partial v}{\partial x} \right)C_2 + k_1 C_1 \qquad (16.9)$$

16.3   HAM-BASED SOLUTION

To solve Eqs. (16.8) and (16.9) using the homotopy analysis method (HAM) subject to the initial and boundary conditions in Eqs. (16.4) and (16.5), the zeroth-order deformation equation can be constructed as follows:

$$(1-q)\mathcal{L}_i\left[\Phi_i(x,t;q) - C_{i,0}(x,t) \right] = q\hbar_i H_i(x,t)\mathcal{N}_i\left[\Phi_i(x,t;q) \right], \; i = 1,2 \qquad (16.10)$$

subject to the initial and boundary conditions:

$$\Phi_i(x,0;q) = f_i(x), \ \Phi_i(0,t;q) = f_i(0), \ \lim_{x \to \infty} \Phi_i(x,t;q) = 0 \qquad (16.11)$$

where q is the embedding parameter, $\Phi_i(x,t;q)$ is the representation of solutions across q, $C_{i,0}(x,t)$ is the initial approximation, $\hbar_i$ is the auxiliary parameter, $H_i(x,t)$ is the auxiliary function, and $\mathcal{L}_i$ and $\mathcal{N}_i$ are the linear and nonlinear operators, respectively. The higher-order terms can be calculated from the higher-order deformation equation given as follows:

$$\mathcal{L}_i\left[C_{i,m}(x,t) - \chi_m C_{i,m-1}(x,t)\right] = \hbar_i H_i(x,t) R_{i,m}\left(\vec{C}_{i,m-1}\right), \ m = 1,2,3,\dots \qquad (16.12)$$

subject to:

$$C_{i,m}(x,0) = 0, \ C_{i,m}(0,t) = 0, \ \lim_{x \to \infty} C_{i,m}(x,t) = 0 \qquad (16.13)$$

where:

$$\chi_m = \begin{cases} 0 & \text{when } m = 1, \\ 1 & \text{otherwise} \end{cases} \qquad (16.14)$$

and:

$$R_{i,m}\left(\vec{C}_{i,m-1}\right) = \frac{1}{(m-1)!} \left. \frac{\partial^{m-1} \mathcal{N}_i\left[\Phi_i(x,t;q)\right]}{\partial q^{m-1}} \right|_{q=0} \qquad (16.15)$$

where $C_{i,m}$ for $m \geq 1$ is the higher-order term. Now, the final solutions can be obtained as follows:

$$C_i(x,t) = C_{i,0}(x,t) + \sum_{m=1}^{\infty} C_{i,m}(x,t) \qquad (16.16)$$

The nonlinear operators for the problem are selected as:

$$\mathcal{N}_1\left[\Phi_i(x,t;q)\right] = r_1 \frac{\partial \Phi_1(x,t;q)}{\partial t} - D \frac{\partial^2 \Phi_1(x,t;q)}{\partial x^2}$$
$$- \left(\frac{\partial D}{\partial x} - v\right) \frac{\partial \Phi_1(x,t;q)}{\partial x} + \left(k_1 + \frac{\partial v}{\partial x}\right) \Phi_1(x,t;q) \qquad (16.17)$$

$$\mathcal{N}_2\left[\Phi_i(x,t;q)\right] = r_2 \frac{\partial \Phi_2(x,t;q)}{\partial t} - D \frac{\partial^2 \Phi_2(x,t;q)}{\partial x^2}$$
$$- \left(\frac{\partial D}{\partial x} - v\right) \frac{\partial \Phi_2(x,t;q)}{\partial x} + \left(k_2 + \frac{\partial v}{\partial x}\right) \Phi_2(x,t;q) - k_1 \Phi_1(x,t;q) \qquad (16.18)$$

Therefore, using Eqs. (16.17) and (16.18), the term $R_{i,m}$ can be calculated from Eq. (16.15) as follows:

$$R_{1,m}\left(\vec{C}_{i,m-1}\right) = r_1 \frac{\partial C_{1,m-1}}{\partial t} - D \frac{\partial^2 C_{1,m-1}}{\partial x^2} - \left(\frac{\partial D}{\partial x} - v\right) \frac{\partial C_{1,m-1}}{\partial x} + \left(k_1 + \frac{\partial v}{\partial x}\right) C_{1,m-1} \quad (16.19)$$

$$R_{2,m}\left(\vec{C}_{i,m-1}\right) = r_2 \frac{\partial C_{2,m-1}}{\partial t} - D \frac{\partial^2 C_{2,m-1}}{\partial x^2} - \left(\frac{\partial D}{\partial x} - v\right) \frac{\partial C_{2,m-1}}{\partial x}$$

$$+ \left(k_2 + \frac{\partial v}{\partial x}\right) C_{2,m-1} - k_1 C_{1,m-1} \quad (16.20)$$

We consider the following set of base functions to represent the solution $C_i(x,t)$ of the present problem:

$$\{x^m exp(-nbx)t^p \mid m,n,p = 0,1,2,\,3,...\} \quad (16.21)$$

so that:

$$C_i(x,t) = \sum_{p=0}^{\infty}\sum_{m=0}^{\infty}\sum_{n=1}^{\infty} \beta_{m,n,p} x^m exp(-nbx) t^p \quad (16.22)$$

where $\beta_{m,n,p}$ are the coefficients of the series. Eq. (16.22) provides the so-called *rule of solution expression*. Following the rule of solution expression, the linear operators and the initial approximations are chosen, respectively, as follows:

$$\mathcal{L}_i\left[\Phi_i(x,t;q)\right] = r_i \frac{\partial \Phi_i(x,t;q)}{\partial t} \text{ with the property } \mathcal{L}_i[E_2] = 0 \quad (16.23)$$

$$C_{1,0}(x,t) = f_1(x) = a_1 x exp(-a_2 x), \ C_{2,0}(x,t) = f_2(x) = 0 \quad (16.24)$$

where E_2 is an integral constant. Using Eq. (16.23), the higher-order terms can be obtained from Eq. (16.12) as follows:

$$C_{i,m}(x,t) = \chi_m C_{i,m-1}(x,t) + \left(\frac{\hbar_i}{r_i}\right) \int_0^t H_i(x,t) R_{i,m}\left(\vec{C}_{i,m-1}\right) dt + E_2, \ m = 1,2,3,... \quad (16.25)$$

where $R_{i,m}$ are given by Eqs. (16.19) and (16.20) and the constant E_2 can be determined from the initial condition for the higher-order deformation equations, which simply yields $E_2 = 0$ for all $m \geq 1$.

Now, the auxiliary functions $H_i(x,t)$ can be determined from the *rule of coefficient ergodicity*. Based on the rule of solution expression, the general form of $H_i(x,t)$ should be:

$$H_i(x,t) = x^{n_1} exp(-bn_2 x) t^{n_3} \quad (16.26)$$

where n_1, n_2, and n_3 are the integers. However, for simplicity, we can take $H_i(x,t) = 1$ (Vajravelu and Van Gorder 2013). Finally, the approximate solutions can be obtained as follows:

$$C_i(x,t) \approx C_{i,0}(x,t) + \sum_{m=1}^{M} C_{i,m}(x,t) \tag{16.27}$$

16.4   HPM-BASED SOLUTION

When applying the homotopy perturbation method (HPM), let us rewrite Eqs. (16.8) and (16.9) in the following form:

$$\mathcal{N}_i\big(C_i(x,t)\big) = f(x,t) \tag{16.28}$$

Let us construct a homotopy $\Phi_i(x,t;q)$ that satisfies:

$$(1-q)\big[\mathcal{L}_i\big(\Phi_i(x,t;q)\big) - \mathcal{L}_i\big(C_{i,0}(x,t)\big)\big]$$
$$+ q\big[\mathcal{N}_i\big[\Phi_i(x,t;q)\big] - f(x,t)\big] = 0,\ i = 1,2 \tag{16.29}$$

where q is the embedding parameter that lies on $[0,1]$, $\mathcal{L}_i$ is the linear operators and $C_{i,0}(x,t)$ is the initial approximation to the final solutions. Eq. (16.29) shows that as $q = 0$, $\Phi_i(x,t;0) = C_{i,0}(x,t)$ and as $q = 1$, $\Phi_i(x,t;1) = C_i(x,t)$. Let us now express $\Phi_i(x,t;q)$ as a series in terms of q as follows:

$$\Phi_i(x,t;q) = \Phi_{i,0} + q\Phi_{i,1} + q^2\Phi_{i,2} + q^3\Phi_{i,3} + q^4\Phi_{i,4} + \cdots \tag{16.30}$$

where $\Phi_{i,m}$ for $m \geq 1$ is the higher-order term. As $q \to 1$, Eq. (16.30) produces the final solutions as:

$$C_i(x,t) = \lim_{q \to 1} \Phi_i(x,t;q) = \sum_{k=0}^{\infty} \Phi_{i,k} \tag{16.31}$$

For simplicity, we consider the linear and nonlinear operators as follows:

$$\mathcal{L}_i\big[\Phi_i(x,t;q)\big] = r_i \frac{\partial \Phi_i(x,t;q)}{\partial t} \tag{16.32}$$

$$\mathcal{N}_1\big[\Phi_i(x,t;q)\big] = r_1 \frac{\partial \Phi_1(x,t;q)}{\partial t} - D \frac{\partial^2 \Phi_1(x,t;q)}{\partial x^2}$$
$$- \left(\frac{\partial D}{\partial x} - v\right) \frac{\partial \Phi_1(x,t;q)}{\partial x} + \left(k_1 + \frac{\partial v}{\partial x}\right) \Phi_1(x,t;q) \tag{16.33}$$

$$\mathcal{N}_2\big[\Phi_i(x,t;q)\big] = r_2 \frac{\partial \Phi_2(x,t;q)}{\partial t} - D \frac{\partial^2 \Phi_2(x,t;q)}{\partial x^2} - \left(\frac{\partial D}{\partial x} - v\right) \frac{\partial \Phi_2(x,t;q)}{\partial x}$$
$$+ \left(k_2 + \frac{\partial v}{\partial x}\right) \Phi_2(x,t;q) - k_1 \Phi_1(x,t;q) \tag{16.34}$$

Using these expressions, for $i = 1, 2$, Eq. (16.29) takes on the form:

$$(1-q)\left[r_1 \frac{\partial \Phi_1(x,t;q)}{\partial t} - r_1 \frac{\partial C_{1,0}}{\partial t} \right]$$

$$+q\left[r_1 \frac{\partial \Phi_1(x,t;q)}{\partial t} - D \frac{\partial^2 \Phi_1(x,t;q)}{\partial x^2} - \left(\frac{\partial D}{\partial x} - v \right) \frac{\partial \Phi_1(x,t;q)}{\partial x} + \left(k_1 + \frac{\partial v}{\partial x} \right) \Phi_1(x,t;q) \right] = 0$$

$$(16.35)$$

$$(1-q)\left[r_2 \frac{\partial \Phi_2(x,t;q)}{\partial t} - r_2 \frac{\partial C_{2,0}}{\partial t} \right] + q\left[r_2 \frac{\partial \Phi_2(x,t;q)}{\partial t} - D \frac{\partial^2 \Phi_2(x,t;q)}{\partial x^2} \right.$$

$$\left. - \left(\frac{\partial D}{\partial x} - v \right) \frac{\partial \Phi_2(x,t;q)}{\partial x} + \left(k_2 + \frac{\partial v}{\partial x} \right) \Phi_2(x,t;q) - k_1 \Phi_1(x,t;q) \right] = 0 \quad (16.36)$$

Using Eq. (16.30), we have:

$$(1-q)\left[r_1 \frac{\partial \Phi_1(x,t;q)}{\partial t} - r_1 \frac{\partial C_{1,0}}{\partial t} \right]$$

$$= (1-q)\left[r_1 \left(\frac{\partial \Phi_{1,0}}{\partial t} - \frac{\partial C_{1,0}}{\partial t} \right) + qr_1 \frac{\partial \Phi_{1,1}}{\partial t} + q^2 r_1 \frac{\partial \Phi_{1,2}}{\partial t} + q^3 r_1 \frac{\partial \Phi_{1,3}}{\partial t} + q^4 r_1 \frac{\partial \Phi_{1,4}}{\partial t} + \cdots \right]$$

$$= r_1 \left(\frac{\partial \Phi_{1,0}}{\partial t} - \frac{\partial C_{1,0}}{\partial t} \right) + qr_1 \left(\frac{\partial \Phi_{1,1}}{\partial t} - \left(\frac{\partial \Phi_{1,0}}{\partial t} - \frac{\partial C_{1,0}}{\partial t} \right) \right) + q^2 r_1 \left(\frac{\partial \Phi_{1,2}}{\partial t} - \frac{\partial \Phi_{1,1}}{\partial t} \right)$$

$$+ q^3 r_1 \left(\frac{\partial \Phi_{1,3}}{\partial t} - \frac{\partial \Phi_{1,2}}{\partial t} \right) + q^4 r_1 \left(\frac{\partial \Phi_{1,4}}{\partial t} - \frac{\partial \Phi_{1,3}}{\partial t} \right) + \cdots \quad (16.37)$$

$$r_1 \frac{\partial \Phi_1(x,t;q)}{\partial t} - D \frac{\partial^2 \Phi_1(x,t;q)}{\partial x^2} - \left(\frac{\partial D}{\partial x} - v \right) \frac{\partial \Phi_1(x,t;q)}{\partial x}$$

$$+ \left(k_1 + \frac{\partial v}{\partial x} \right) \Phi_1(x,t;q) = r_1 \left(\frac{\partial \Phi_{1,0}}{\partial t} + q \frac{\partial \Phi_{1,1}}{\partial t} + q^2 \frac{\partial \Phi_{1,2}}{\partial t} + q^3 \frac{\partial \Phi_{1,3}}{\partial t} + \cdots \right)$$

$$- D \left(\frac{\partial^2 \Phi_{1,0}}{\partial x^2} + q \frac{\partial^2 \Phi_{1,1}}{\partial x^2} + q^2 \frac{\partial^2 \Phi_{1,2}}{\partial x^2} + q^3 \frac{\partial^2 \Phi_{1,3}}{\partial x^2} + \cdots \right)$$

$$- \left(\frac{\partial D}{\partial x} - v \right) \left(\frac{\partial \Phi_{1,0}}{\partial x} + q \frac{\partial \Phi_{1,1}}{\partial x} + q^2 \frac{\partial \Phi_{1,2}}{\partial x} + q^3 \frac{\partial \Phi_{1,3}}{\partial x} + \cdots \right)$$

$$+ \left(k_1 + \frac{\partial v}{\partial x} \right) \left(\Phi_{1,0} + q\Phi_{1,1} + q^2 \Phi_{1,2} + q^3 \Phi_{1,3} + \cdots \right)$$

$$= \left[r_1 \frac{\partial \Phi_{1,0}}{\partial t} - D \frac{\partial^2 \Phi_{1,0}}{\partial x^2} - \left(\frac{\partial D}{\partial x} - v \right) \frac{\partial \Phi_{1,0}}{\partial x} + \left(k_1 + \frac{\partial v}{\partial x} \right) \Phi_{1,0} \right]$$

$$+ q\left[r_1 \frac{\partial \Phi_{1,1}}{\partial t} - D \frac{\partial^2 \Phi_{1,1}}{\partial x^2} - \left(\frac{\partial D}{\partial x} - v \right) \frac{\partial \Phi_{1,1}}{\partial x} + \left(k_1 + \frac{\partial v}{\partial x} \right) \Phi_{1,1} \right]$$

$$+ q^2\left[r_1 \frac{\partial \Phi_{1,2}}{\partial t} - D \frac{\partial^2 \Phi_{1,2}}{\partial x^2} - \left(\frac{\partial D}{\partial x} - v \right) \frac{\partial \Phi_{1,2}}{\partial x} + \left(k_1 + \frac{\partial v}{\partial x} \right) \Phi_{1,2} \right]$$

$$+ q^3\left[r_1 \frac{\partial \Phi_{1,3}}{\partial t} - D \frac{\partial^2 \Phi_{1,3}}{\partial x^2} - \left(\frac{\partial D}{\partial x} - v \right) \frac{\partial \Phi_{1,3}}{\partial x} + \left(k_1 + \frac{\partial v}{\partial x} \right) \Phi_{1,3} \right] + \cdots \quad (16.38)$$

$$(1-q)\left[r_2\frac{\partial\Phi_2(x,t;q)}{\partial t}-r_2\frac{\partial C_{2,0}}{\partial t}\right]$$

$$=(1-q)\left[r_2\left(\frac{\partial\Phi_{2,0}}{\partial t}-\frac{\partial C_{2,0}}{\partial t}\right)+qr_2\frac{\partial\Phi_{2,1}}{\partial t}+q^2r_2\frac{\partial\Phi_{2,2}}{\partial t}+q^3r_2\frac{\partial\Phi_{2,3}}{\partial t}+q^4r_2\frac{\partial\Phi_{2,4}}{\partial t}+\cdots\right]$$

$$=r_2\left(\frac{\partial\Phi_{2,0}}{\partial t}-\frac{\partial C_{2,0}}{\partial t}\right)+qr_2\left(\frac{\partial\Phi_{2,1}}{\partial t}-\left(\frac{\partial\Phi_{2,0}}{\partial t}-\frac{\partial C_{2,0}}{\partial t}\right)\right)+q^2r_2\left(\frac{\partial\Phi_{2,2}}{\partial t}-\frac{\partial\Phi_{2,1}}{\partial t}\right)$$

$$+q^3r_2\left(\frac{\partial\Phi_{2,3}}{\partial t}-\frac{\partial\Phi_{2,2}}{\partial t}\right)+q^4r_2\left(\frac{\partial\Phi_{2,4}}{\partial t}-\frac{\partial\Phi_{2,3}}{\partial t}\right)+\cdots \tag{16.39}$$

$$r_2\frac{\partial\Phi_2(x,t;q)}{\partial t}-D\frac{\partial^2\Phi_2(x,t;q)}{\partial x^2}-\left(\frac{\partial D}{\partial x}-v\right)\frac{\partial\Phi_2(x,t;q)}{\partial x}+\left(k_2+\frac{\partial v}{\partial x}\right)\Phi_2(x,t;q)$$

$$-k_1\Phi_1(x,t;q)=r_2\left(\frac{\partial\Phi_{2,0}}{\partial t}+q\frac{\partial\Phi_{2,1}}{\partial t}+q^2\frac{\partial\Phi_{2,2}}{\partial t}+q^3\frac{\partial\Phi_{2,3}}{\partial t}+\cdots\right)$$

$$-D\left(\frac{\partial^2\Phi_{2,0}}{\partial x^2}+q\frac{\partial^2\Phi_{2,1}}{\partial x^2}+q^2\frac{\partial^2\Phi_{2,2}}{\partial x^2}+q^3\frac{\partial^2\Phi_{2,3}}{\partial x^2}+\cdots\right)$$

$$-\left(\frac{\partial D}{\partial x}-v\right)\left(\frac{\partial\Phi_{2,0}}{\partial x}+q\frac{\partial\Phi_{2,1}}{\partial x}+q^2\frac{\partial\Phi_{2,2}}{\partial x}+q^3\frac{\partial\Phi_{2,3}}{\partial x}+\cdots\right)$$

$$+\left(k_2+\frac{\partial v}{\partial x}\right)\left(\Phi_{2,0}+q\Phi_{2,1}+q^2\Phi_{2,2}+q^3\Phi_{2,3}+\cdots\right)-k_1\left(\Phi_{1,0}+q\Phi_{1,1}+q^2\Phi_{1,2}+q^3\Phi_{1,3}+\cdots\right)$$

$$=\left[r_2\frac{\partial\Phi_{2,0}}{\partial t}-D\frac{\partial^2\Phi_{2,0}}{\partial x^2}-\left(\frac{\partial D}{\partial x}-v\right)\frac{\partial\Phi_{2,0}}{\partial x}+\left(k_2+\frac{\partial v}{\partial x}\right)\Phi_{2,0}-k_1\Phi_{1,0}\right]$$

$$+q\left[r_2\frac{\partial\Phi_{2,1}}{\partial t}-D\frac{\partial^2\Phi_{2,1}}{\partial x^2}-\left(\frac{\partial D}{\partial x}-v\right)\frac{\partial\Phi_{2,1}}{\partial x}+\left(k_2+\frac{\partial v}{\partial x}\right)\Phi_{2,1}-k_1\Phi_{1,1}\right]$$

$$+q^2\left[r_2\frac{\partial\Phi_{2,2}}{\partial t}-D\frac{\partial^2\Phi_{2,2}}{\partial x^2}-\left(\frac{\partial D}{\partial x}-v\right)\frac{\partial\Phi_{2,2}}{\partial x}+\left(k_2+\frac{\partial v}{\partial x}\right)\Phi_{2,2}-k_1\Phi_{1,2}\right]$$

$$+q^3\left[r_2\frac{\partial\Phi_{2,3}}{\partial t}-D\frac{\partial^2\Phi_{2,3}}{\partial x^2}-\left(\frac{\partial D}{\partial x}-v\right)\frac{\partial\Phi_{2,3}}{\partial x}+\left(k_2+\frac{\partial v}{\partial x}\right)\Phi_{2,3}-k_1\Phi_{1,3}\right]+\cdots \tag{16.40}$$

Using the initial and boundary conditions given by Eqs. (16.4) and (16.5), we have:

$$\Phi_{1,0}(x,0)=f_1(x),\ \Phi_{1,0}(0,t)=f_1(0),\ \lim_{x\to\infty}\Phi_{1,0}(x,t)=0 \tag{16.41}$$

$$\Phi_{1,m}(x,0)=0,\ \Phi_{1,m}(0,t)=0,\ \lim_{x\to\infty}\Phi_{1,m}(x,t)=0\ \text{for}\ m\geq1 \tag{16.42}$$

$$\Phi_{2,0}(x,0)=f_2(x),\ \Phi_{2,0}(0,t)=f_2(0),\ \lim_{x\to\infty}\Phi_{2,0}(x,t)=0 \tag{16.43}$$

$$\Phi_{2,m}(x,0)=0,\ \Phi_{2,m}(0,t)=0,\ \lim_{x\to\infty}\Phi_{2,m}(x,t)=0\ \text{for}\ m\geq1 \tag{16.44}$$

Using Eqs. (16.37)–(16.40) and equating the like powers of q in Eqs. (16.35) and (16.36), the following systems of differential equations are obtained:

$$r_1\left(\frac{\partial \Phi_{1,0}}{\partial t} - \frac{\partial C_{1,0}}{\partial t}\right) = 0 \text{ subject to } \Phi_{1,0}(x,0) = f_1(x),$$
$$\Phi_{1,0}(0,t) = f_1(0), \lim_{x \to \infty} \Phi_{1,0}(x,t) = 0 \tag{16.45}$$

$$r_2\left(\frac{\partial \Phi_{2,0}}{\partial t} - \frac{\partial C_{2,0}}{\partial t}\right) = 0 \text{ subject to } \Phi_{2,0}(x,0) = f_2(x),$$
$$\Phi_{2,0}(0,t) = f_2(0), \lim_{x \to \infty} \Phi_{2,0}(x,t) = 0 \tag{16.46}$$

$$r_1\left(\frac{\partial \Phi_{1,1}}{\partial t} - \left(\frac{\partial \Phi_{1,0}}{\partial t} - \frac{\partial C_{1,0}}{\partial t}\right)\right) + r_1\frac{\partial \Phi_{1,0}}{\partial t} - D\frac{\partial^2 \Phi_{1,0}}{\partial x^2} - \left(\frac{\partial D}{\partial x} - v\right)\frac{\partial \Phi_{1,0}}{\partial x} + \left(k_1 + \frac{\partial v}{\partial x}\right)\Phi_{1,0} = 0$$
$$\text{subject to } \Phi_{1,1}(x,0) = 0, \Phi_{1,1}(0,t) = 0, \lim_{x \to \infty} \Phi_{1,1}(x,t) = 0 \tag{16.47}$$

$$r_2\left(\frac{\partial \Phi_{2,1}}{\partial t} - \left(\frac{\partial \Phi_{2,0}}{\partial t} - \frac{\partial C_{2,0}}{\partial t}\right)\right) + r_2\frac{\partial \Phi_{2,0}}{\partial t} - D\frac{\partial^2 \Phi_{2,0}}{\partial x^2} - \left(\frac{\partial D}{\partial x} - v\right)\frac{\partial \Phi_{2,0}}{\partial x}$$
$$+ \left(k_2 + \frac{\partial v}{\partial x}\right)\Phi_{2,0} - k_1\Phi_{1,0} = 0 \text{ subject to } \Phi_{2,1}(x,0) = 0,$$
$$\Phi_{2,1}(0,t) = 0, \lim_{x \to \infty} \Phi_{2,1}(x,t) = 0 \tag{16.48}$$

$$r_1\left(\frac{\partial \Phi_{1,2}}{\partial t} - \frac{\partial \Phi_{1,1}}{\partial t}\right) + r_1\frac{\partial \Phi_{1,1}}{\partial t} - D\frac{\partial^2 \Phi_{1,1}}{\partial x^2} - \left(\frac{\partial D}{\partial x} - v\right)\frac{\partial \Phi_{1,1}}{\partial x} + \left(k_1 + \frac{\partial v}{\partial x}\right)\Phi_{1,1} = 0$$
$$\text{subject to } \Phi_{1,2}(x,0) = 0, \Phi_{1,2}(0,t) = 0, \lim_{x \to \infty} \Phi_{1,2}(x,t) = 0 \tag{16.49}$$

$$r_2\left(\frac{\partial \Phi_{2,2}}{\partial t} - \frac{\partial \Phi_{2,1}}{\partial t}\right) + r_2\frac{\partial \Phi_{2,1}}{\partial t} - D\frac{\partial^2 \Phi_{2,1}}{\partial x^2} - \left(\frac{\partial D}{\partial x} - v\right)\frac{\partial \Phi_{2,1}}{\partial x} + \left(k_2 + \frac{\partial v}{\partial x}\right)\Phi_{2,1}$$
$$- k_1\Phi_{1,1} = 0 \text{ subject to } \Phi_{2,2}(x,0) = 0, \Phi_{2,2}(0,t) = 0, \lim_{x \to \infty} \Phi_{2,2}(x,t) = 0 \tag{16.50}$$

$$r_1\left(\frac{\partial \Phi_{1,3}}{\partial t} - \frac{\partial \Phi_{1,2}}{\partial t}\right) + r_1\frac{\partial \Phi_{1,2}}{\partial t} - D\frac{\partial^2 \Phi_{1,2}}{\partial x^2} - \left(\frac{\partial D}{\partial x} - v\right)\frac{\partial \Phi_{1,2}}{\partial x} + \left(k_1 + \frac{\partial v}{\partial x}\right)\Phi_{1,2} = 0$$
$$\text{subject to } \Phi_{1,3}(x,0) = 0, \Phi_{1,3}(0,t) = 0, \lim_{x \to \infty} \Phi_{1,3}(x,t) = 0 \tag{16.51}$$

$$r_2\left(\frac{\partial \Phi_{2,3}}{\partial t} - \frac{\partial \Phi_{2,2}}{\partial t}\right) + r_2\frac{\partial \Phi_{2,2}}{\partial t} - D\frac{\partial^2 \Phi_{2,2}}{\partial x^2} - \left(\frac{\partial D}{\partial x} - v\right)\frac{\partial \Phi_{2,2}}{\partial x} + \left(k_2 + \frac{\partial v}{\partial x}\right)\Phi_{2,2}$$
$$- k_1\Phi_{1,2} = 0 \text{ subject to } \Phi_{2,3}(x,0) = 0, \Phi_{2,3}(0,t) = 0, \lim_{x \to \infty} \Phi_{2,3}(x,t) = 0 \tag{16.52}$$

Proceeding in a like manner, one can arrive at the following recurrence relation:

$$r_1\left(\frac{\partial \Phi_{1,m}}{\partial t}-\frac{\partial \Phi_{1,m-1}}{\partial t}\right)+r_1\frac{\partial \Phi_{1,m-1}}{\partial t}-D\frac{\partial^2 \Phi_{1,m-1}}{\partial x^2}-\left(\frac{\partial D}{\partial x}-v\right)\frac{\partial \Phi_{1,m-1}}{\partial x}+\left(k_1+\frac{\partial v}{\partial x}\right)\Phi_{1,m-1}$$

$$=0 \text{ subject to } \Phi_{1,m}(x,0)=0,\ \Phi_{1,m}(0,t)=0,\ \lim_{x\to\infty}\Phi_{1,m}(x,t)=0 \text{ for } m\geq 2 \qquad (16.53)$$

$$r_2\left(\frac{\partial \Phi_{2,m}}{\partial t}-\frac{\partial \Phi_{2,m-1}}{\partial t}\right)+r_2\frac{\partial \Phi_{2,m-1}}{\partial t}-D\frac{\partial^2 \Phi_{2,m-1}}{\partial x^2}-\left(\frac{\partial D}{\partial x}-v\right)\frac{\partial \Phi_{2,m-1}}{\partial x}+\left(k_2+\frac{\partial v}{\partial x}\right)\Phi_{2,m-1}$$

$$-k_1\Phi_{1,m-1}=0 \text{ subject to } \Phi_{2,m}(x,0)=0,\ \Phi_{2,m}(0,t)=0,\ \lim_{x\to\infty}\Phi_{2,m}(x,t)=0 \text{ for } m\geq 2$$

$$(16.54)$$

The initial approximation can be chosen as $\Phi_{1,0}=f_1(x)=a_1 x \exp(-a_2 x)$ and $\Phi_{2,0}=f_2(x)=0$. Using this initial approximation, we can solve the equations iteratively using symbolic software. We avoid those expressions here, as they are lengthy. Finally, the HPM-based solutions can be approximated as follows:

$$C_i(x,t)\approx \sum_{k=0}^{M}\Phi_{i,k} \qquad (16.55)$$

16.5 OHAM-BASED SOLUTION

When applying the optimal homotopy asymptotic method (OHAM), Eqs. (16.8) and (16.9) can be written in the following form:

$$\mathcal{L}_i\left[C_i(x,t)\right]+\mathcal{N}_i\left[C_i(x,t)\right]+h_i(x,t)=0 \qquad (16.56)$$

We select:

$$\mathcal{L}_i\left[C_i(x,t)\right]=r_i\frac{\partial C_i(x,t)}{\partial t} \qquad (16.57)$$

$$\mathcal{N}_1\left[C_i(x,t)\right]=-D\frac{\partial^2 C_1(x,t)}{\partial x^2}-\left(\frac{\partial D}{\partial x}-v\right)\frac{\partial C_1(x,t)}{\partial x}+\left(k_1+\frac{\partial v}{\partial x}\right)C_1(x,t) \qquad (16.58)$$

$$\mathcal{N}_2\left[C_i(x,t)\right]=-D\frac{\partial^2 C_2(x,t)}{\partial x^2}-\left(\frac{\partial D}{\partial x}-v\right)\frac{\partial C_2(x,t)}{\partial x}$$

$$+\left(k_2+\frac{\partial v}{\partial x}\right)C_2(x,t)-k_1 C_1(x,t) \qquad (16.59)$$

and:

$$h_i(x,t)=0 \qquad (16.60)$$

With these considerations, the zeroth-order problem becomes:

$$\mathcal{L}_i\left(C_{i,0}(x,t)\right)+h_i(x,t)=0 \text{ subject to } C_{i,0}(x,0)=f_i(x),$$

$$C_{i,0}(0,t)=f_i(0),\ \lim_{x\to\infty}C_{i,0}(x,t)=0 \qquad (16.61)$$

Solving Eq. (16.61), one obtains:

$$C_{i,0}(x,t) = f_i(x) \tag{16.62}$$

To obtain $\mathcal{N}_{i,0}$, $\mathcal{N}_{i,1}$, $\mathcal{N}_{i,2}$, etc., the following can be used:

$$-D\frac{\partial^2 C_1}{\partial x^2} - \left(\frac{\partial D}{\partial x} - v\right)\frac{\partial C_1}{\partial x} + \left(k_1 + \frac{\partial v}{\partial x}\right)C_1$$

$$= -D\left(\frac{\partial^2 C_{1,0}}{\partial x^2} + q\frac{\partial^2 C_{1,1}}{\partial x^2} + q^2\frac{\partial^2 C_{1,2}}{\partial x^2} + q^3\frac{\partial^2 C_{1,3}}{\partial x^2} + \cdots\right)$$

$$-\left(\frac{\partial D}{\partial x} - v\right)\left(\frac{\partial C_{1,0}}{\partial x} + q\frac{\partial C_{1,1}}{\partial x} + q^2\frac{\partial C_{1,2}}{\partial x} + q^3\frac{\partial C_{1,3}}{\partial x} + \cdots\right)$$

$$+\left(k_1 + \frac{\partial v}{\partial x}\right)\left(C_{1,0} + qC_{1,1} + q^2C_{1,2} + q^3C_{1,3} + \cdots\right)$$

$$= \left[-D\frac{\partial^2 C_{1,0}}{\partial x^2} - \left(\frac{\partial D}{\partial x} - v\right)\frac{\partial C_{1,0}}{\partial x} + \left(k_1 + \frac{\partial v}{\partial x}\right)C_{1,0}\right]$$

$$+ q\left[-D\frac{\partial^2 C_{1,1}}{\partial x^2} - \left(\frac{\partial D}{\partial x} - v\right)\frac{\partial C_{1,1}}{\partial x} + \left(k_1 + \frac{\partial v}{\partial x}\right)C_{1,1}\right]$$

$$+ q^2\left[-D\frac{\partial^2 C_{1,2}}{\partial x^2} - \left(\frac{\partial D}{\partial x} - v\right)\frac{\partial C_{1,2}}{\partial x} + \left(k_1 + \frac{\partial v}{\partial x}\right)C_{1,2}\right]$$

$$+ q^3\left[-D\frac{\partial^2 C_{1,3}}{\partial x^2} - \left(\frac{\partial D}{\partial x} - v\right)\frac{\partial C_{1,3}}{\partial x} + \left(k_1 + \frac{\partial v}{\partial x}\right)C_{1,3}\right] + \cdots \tag{16.63}$$

$$-D\frac{\partial^2 C_2}{\partial x^2} - \left(\frac{\partial D}{\partial x} - v\right)\frac{\partial C_2}{\partial x} + \left(k_2 + \frac{\partial v}{\partial x}\right)C_2 - k_1 C_1$$

$$= -D\left(\frac{\partial^2 C_{2,0}}{\partial x^2} + q\frac{\partial^2 C_{2,1}}{\partial x^2} + q^2\frac{\partial^2 C_{2,2}}{\partial x^2} + q^3\frac{\partial^2 C_{2,3}}{\partial x^2} + \cdots\right)$$

$$-\left(\frac{\partial D}{\partial x} - v\right)\left(\frac{\partial C_{2,0}}{\partial x} + q\frac{\partial C_{2,1}}{\partial x} + q^2\frac{\partial C_{2,2}}{\partial x} + q^3\frac{\partial C_{2,3}}{\partial x} + \cdots\right)$$

$$+\left(k_2 + \frac{\partial v}{\partial x}\right)\left(C_{2,0} + qC_{2,1} + q^2C_{2,2} + q^3C_{2,3} + \cdots\right) - k_1\left(C_{1,0} + qC_{1,1} + q^2C_{1,2} + q^3C_{1,3} + \cdots\right)$$

$$= \left[-D\frac{\partial^2 C_{2,0}}{\partial x^2} - \left(\frac{\partial D}{\partial x} - v\right)\frac{\partial C_{2,0}}{\partial x} + \left(k_2 + \frac{\partial v}{\partial x}\right)C_{2,0} - k_1 C_{1,0}\right]$$

$$+ q\left[-D\frac{\partial^2 C_{2,1}}{\partial x^2} - \left(\frac{\partial D}{\partial x} - v\right)\frac{\partial C_{2,1}}{\partial x} + \left(k_2 + \frac{\partial v}{\partial x}\right)C_{2,1} - k_1 C_{1,1}\right]$$

$$+ q^2\left[-D\frac{\partial^2 C_{2,2}}{\partial x^2} - \left(\frac{\partial D}{\partial x} - v\right)\frac{\partial C_{2,2}}{\partial x} + \left(k_2 + \frac{\partial v}{\partial x}\right)C_{2,2} - k_1 C_{1,2}\right]$$

$$+ q^3\left[-D\frac{\partial^2 C_{2,3}}{\partial x^2} - \left(\frac{\partial D}{\partial x} - v\right)\frac{\partial C_{2,3}}{\partial x} + \left(k_2 + \frac{\partial v}{\partial x}\right)C_{2,3} - k_1 C_{1,3}\right] + \cdots \tag{16.64}$$

Using Eqs. (16.63) and (16.64), the first-order problem reduces to:

$$r_1 \frac{\partial C_{1,1}}{\partial t} = H_{1,1}\left(x,t,D_{1,i}\right)\left\{-D\frac{\partial^2 C_{1,0}}{\partial x^2} - \left(\frac{\partial D}{\partial x} - v\right)\frac{\partial C_{1,0}}{\partial x} + \left(k_1 + \frac{\partial v}{\partial x}\right)C_{1,0}\right\}$$

subject to $C_{1,1}(x,0) = 0$, $C_{1,1}(0,t) = 0$, $\lim_{x\to\infty} C_{1,1}(x,t) = 0$ (16.65)

$$r_2 \frac{\partial C_{2,1}}{\partial t} = H_{2,1}\left(x,t,D_{2,i}\right)\left\{-D\frac{\partial^2 C_{2,0}}{\partial x^2} - \left(\frac{\partial D}{\partial x} - v\right)\frac{\partial C_{2,0}}{\partial x} + \left(k_2 + \frac{\partial v}{\partial x}\right)C_{2,0} - k_1 C_{1,0}\right\}$$

subject to $C_{2,1}(x,0) = 0$, $C_{2,1}(0,t) = 0$, $\lim_{x\to\infty} C_{2,1}(x,t) = 0$ (16.66)

The auxiliary functions can be chosen in many ways. Here, we select $H_{1,1}\left(x,t,D_{1,i}\right) = D_{1,1}$ and $H_{2,1}\left(x,t,D_{2,i}\right) = D_{2,1}$. Using $k = 2$, the second-order problem becomes:

$$\mathcal{L}_i\left[C_{i,2}\left(x,t,D_{i,i}\right) - C_{i,1}\left(x,t,D_{i,i}\right)\right] = H_{i,2}\left(x,t,D_{i,i}\right)\mathcal{N}_0\left(C_{i,0}\left(x,t\right)\right)$$
$$+ H_{i,1}\left(x,t,D_{i,i}\right)\left[\mathcal{L}_i\left[C_{i,1}\left(x,t,D_{i,i}\right)\right] + \mathcal{N}_{i,1}\left[C_{i,0}\left(x,t\right), C_{i,1}\left(x,t,D_{i,i}\right)\right]\right]$$

subject to $C_{i,2}(x,0) = 0$, $C_{i,2}(0,t) = 0$, $\lim_{x\to\infty} C_{i,2}(x,t) = 0$ (16.67)

Further, we choose $H_{1,2}\left(x,t,D_{2,i}\right) = D_{1,2}$ and $H_{2,2}\left(x,t,D_{2,i}\right) = D_{2,2}$. Therefore, Eq. (16.67) becomes:

$$r_1 \frac{\partial C_{1,2}}{\partial t} = r_1 \frac{\partial C_{1,1}}{\partial t} + D_{1,2}\left\{-D\frac{\partial^2 C_{1,0}}{\partial x^2} - \left(\frac{\partial D}{\partial x} - v\right)\frac{\partial C_{1,0}}{\partial x} + \left(k_1 + \frac{\partial v}{\partial x}\right)C_{1,0}\right\}$$
$$+ D_{1,1}\left[r_1 \frac{\partial C_{1,1}}{\partial t} - D\frac{\partial^2 C_{1,1}}{\partial x^2} - \left(\frac{\partial D}{\partial x} - v\right)\frac{\partial C_{1,1}}{\partial x} + \left(k_1 + \frac{\partial v}{\partial x}\right)C_{1,1}\right]$$

subject to $C_{1,2}(x,0) = 0$, $C_{1,2}(0,t) = 0$, $\lim_{x\to\infty} C_{1,2}(x,t) = 0$ (16.68)

$$r_2 \frac{\partial C_{2,2}}{\partial t} = r_2 \frac{\partial C_{2,1}}{\partial t} + D_{2,2}\left\{-D\frac{\partial^2 C_{2,0}}{\partial x^2} - \left(\frac{\partial D}{\partial x} - v\right)\frac{\partial C_{2,0}}{\partial x} + \left(k_2 + \frac{\partial v}{\partial x}\right)C_{2,0} - k_1 C_{1,0}\right\}$$
$$+ D_{2,1}\left[r_2 \frac{\partial C_{2,1}}{\partial t} - D\frac{\partial^2 C_{2,1}}{\partial x^2} - \left(\frac{\partial D}{\partial x} - v\right)\frac{\partial C_{2,1}}{\partial x} + \left(k_2 + \frac{\partial v}{\partial x}\right)C_{2,1} - k_1 C_{1,1}\right]$$

subject to $C_{1,2}(x,0) = 0$, $C_{1,2}(0,t) = 0$, $\lim_{x\to\infty} C_{1,2}(x,t) = 0$ (16.69)

The terms of the series can be computed using the equations developed here. One can compute these terms without any difficulty using symbolic computation software, such as MATLAB. Further, following Chapter 2, the higher-order terms can be computed in a similar manner. However, our aim is to produce an accurate solution with just two to three terms of the OHAM-based series. Therefore, we restrict our calculation up to $k = 2$. Finally, the approximate solution can be found as:

$$C_i(x,t) \approx C_{i,0}(x,t) + C_{i,1}\left(x,t,D_{i,i}\right) + C_{i,2}\left(x,t,D_{i,i}\right)$$ (16.70)

where the terms are given by Eqs. (16.62), (16.65), (16.66), (16.68), and (16.69).

16.6   CONVERGENCE THEOREMS

The convergence of the HAM- and OHAM-based solutions is proved theoretically using the following theorems.

16.6.1   CONVERGENCE THEOREM OF THE HAM-BASED SOLUTION

The convergence theorem for the HAM-based solution given by Eq. (16.27) can be proved using the following theorems.

Theorem 16.1: If the homotopy series $\Sigma_{m=0}^{\infty} C_{i,m}(x,t)$, $\Sigma_{m=0}^{\infty} \dfrac{\partial C_{i,m}(x,t)}{\partial t}$, $\Sigma_{m=0}^{\infty} \dfrac{\partial C_{i,m}(x,t)}{\partial x}$, and $\Sigma_{m=0}^{\infty} \dfrac{\partial^2 C_{i,m}(x,t)}{\partial x^2}$ converge, then $R_m\left(\vec{C}_{i,m-1}\right)$ given by Eq. (16.19), (16.20) satisfies the relation $\Sigma_{m=1}^{\infty} R_m\left(\vec{C}_{i,m-1}\right) = 0$.

Proof: The proof of this theorem follows exactly from the previous chapters. However, in place of a single equation, one needs to consider a system of equations.

Theorem 16.2: If $\hbar$ is so properly chosen that the series $\Sigma_{m=0}^{\infty} C_{i,m}(x,t)$, $\Sigma_{m=0}^{\infty} \dfrac{\partial C_{i,m}(x,t)}{\partial t}$, $\Sigma_{m=0}^{\infty} \dfrac{\partial C_{i,m}(x,t)}{\partial x}$, and $\Sigma_{m=0}^{\infty} \dfrac{\partial^2 C_{i,m}(x,t)}{\partial x^2}$ converge absolutely to $C_i(x,t)$, $\dfrac{\partial C_i(x,t)}{\partial t}$, $\dfrac{\partial C_i(x,t)}{\partial x}$, and $\dfrac{\partial^2 C_i(x,t)}{\partial x^2}$, respectively, then the homotopy series $\Sigma_{m=0}^{\infty} C_{i,m}(x,t)$ satisfies the original governing Eqs. (16.8) and (16.9).

Proof: Theorem 16.1 shows that if $\Sigma_{m=0}^{\infty} C_{i,m}(x,t)$, $\Sigma_{m=0}^{\infty} \dfrac{\partial C_{i,m}(x,t)}{\partial t}$, $\Sigma_{m=0}^{\infty} \dfrac{\partial C_{i,m}(x,t)}{\partial x}$, and $\Sigma_{m=0}^{\infty} \dfrac{\partial^2 C_{i,m}(x,t)}{\partial x^2}$ converge, then $\Sigma_{m=1}^{\infty} R_m\left(\vec{C}_{i,m-1}\right) = 0$.

Therefore, substituting the previous expressions in Eqs. (16.19) and (16.20) and simplifying further lead to:

$$r_1 \sum_{m=0}^{\infty} \frac{\partial C_{1,m}}{\partial t} - D\sum_{m=0}^{\infty} \frac{\partial^2 C_{1,m-1}}{\partial x^2} - \left(\frac{\partial D}{\partial x} - v\right) \sum_{m=0}^{\infty} \frac{\partial C_{1,m-1}}{\partial x} + \left(k_1 + \frac{\partial v}{\partial x}\right) \sum_{m=0}^{\infty} C_{1,m-1} = 0 \quad (16.71)$$

$$r_2 \sum_{m=0}^{\infty} \frac{\partial C_{2,m-1}}{\partial t} - D\sum_{m=0}^{\infty} \frac{\partial^2 C_{2,m-1}}{\partial x^2} - \left(\frac{\partial D}{\partial x} - v\right) \sum_{m=0}^{\infty} \frac{\partial C_{2,m-1}}{\partial x}$$
$$+ \left(k_2 + \frac{\partial v}{\partial x}\right) \sum_{m=0}^{\infty} C_{2,m-1} - k_1 \sum_{m=0}^{\infty} C_{1,m-1} = 0 \quad (16.72)$$

which is basically the original governing Eqs. (16.8) and (16.9). Furthermore, subject to the initial and boundary conditions $C_{i,0}(x,0) = f_i(x)$, $C_{i,0}(0,t) = f_i(0)$, $\lim_{x \to \infty} C_{i,0}(x,t) = 0$, and the conditions for the higher-order deformation equation $C_{i,m}(x,0) = 0$, $C_{i,m}(0,t) = 0$, $\lim_{x \to \infty} C_{i,m}(x,t) = 0$, for $m \geq 1$, we easily obtain

$\sum_{m=0}^{\infty} C_{i,m}(x,0) = f_i(x)$, $\sum_{m=0}^{\infty} C_{i,m}(0,t) = f_i(0)$, and $\lim_{x\to\infty} \sum_{m=0}^{\infty} C_{i,m}(x,t) = 0$. Hence, the convergence result follows.

16.6.2   CONVERGENCE THEOREM OF THE OHAM-BASED SOLUTION

Theorem 16.3: If the series $C_{i,0}(Y,T) + \sum_{j=1}^{\infty} C_{i,j}(x,t,D_{i,i})$, $i = 1,2,\ldots,s$ converges, where $C_{i,j}(x,t,D_{i,i})$ is governed by Eqs. (16.62), (16.65), (16.66), (16.68), and (16.69), then Eq. (16.70) is a solution of the original Eqs. (16.8) and (16.9).

Proof: The proof of this theorem follows exactly from Theorem 2.3 of Chapter 2.

16.7   RESULTS AND DISCUSSION

First, we discuss the expressions and the values of the parameters needed for the assessment of the developed solutions. Then, the numerical convergence of the HAM-based analytical solution is established for a specific test case, and then the solution is validated over a numerical solution. Finally, the HPM- and OHAM-based analytical solutions are validated by comparing them with the numerical solution.

16.7.1   SELECTION OF EXPRESSIONS AND PARAMETERS

For the assessment of solutions, it can be seen that the dispersion coefficient $D(x,t)$ and seepage velocity $v_0(x,t)$ are taken as given by Eq. (16.7). Using these equations, we have:

$$\frac{\partial D}{\partial x} = 2\alpha_2 D_0 (\alpha_1 + \alpha_2 x), \frac{\partial v}{\partial x} = \alpha_2 v_0 \qquad (16.73)$$

For the validation of solutions, space and time domains are considered as $0 \le x(cm) \le 3000$ and $0 \le t(hr) \le 40$. The required parameters are chosen as (Simpson and Ellery 2014; Chaudhary and Singh 2020): $a_1 = 2$, $a_2 = 0.0025$, $k_1 = 0.05 \, (/hr)$, $k_2 = 0.01 \, (/hr)$, $v_0 = 0.2 \, (cm/hr)$, and $D_0 = 0.5 \, (cm^2/hr)$. Also, the other parameters are considered as $r_1 = 2$, $r_2 = 2.5$, $\alpha_1 = 1$, and $\alpha_2 = 0.05 \, (/cm)$.

16.7.2   NUMERICAL CONVERGENCE AND VALIDATION OF THE HAM SOLUTION

As mentioned in Chapter 2, there are two ways to handle the convergence of HAM: one is based on the $\hbar$–curves and the other is by finding the squared residual errors. It can be noted that for the system of equations considered here, we have chosen $\hbar_1 = \hbar_2 = \hbar$. One can indeed choose two different auxiliary parameters; however, it is better to first consider a common $\hbar$ to see the accuracy of solutions, as this assumption simplifies the problem. Here, we calculate the so-called $\hbar$–curves to find a suitable choice for the auxiliary parameter $\hbar$, which determines the convergence of the HAM-based series solution. In this regard, for a particular order of approximation, we plot the approximate solution $C_i(x,t)$ (or its derivatives) at some point within the domain. The flatness of the $\hbar$–curve determines a suitable choice for the auxiliary parameter $\hbar$. In Figure 16.1, we plot the $\hbar$–curves for 5th- and 10th-order HAM-based approximations for the value $C_i(1,1)$. From the exact solution, those quantities can be calculated, and it

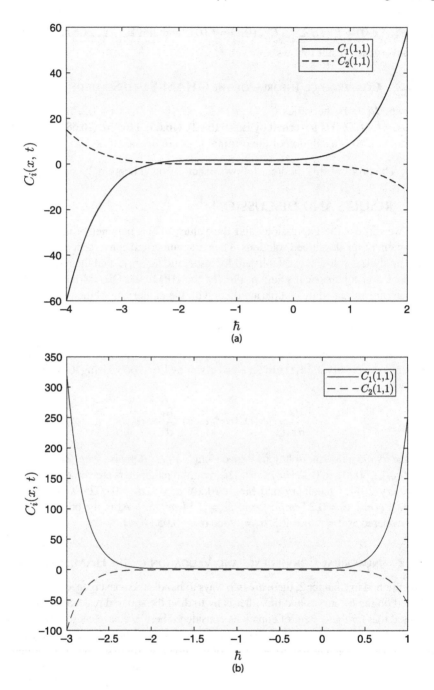

FIGURE 16.1 $\hbar$–curves for $C_i(1,1)$ of the HAM solution: (a) 5th-order approximation and (b) 10th-order approximation.

is observed from the figures that the curves exhibit the flat nature for a specific range of $\hbar$. Any choice of $\hbar$ within this range determines an optimal value for which the series solutions converge (Abbasbandy et al. 2011).

Here, we test the performance of the HAM-based analytical solutions for the present problem by comparing them with a numerical solution. The numerical solution is obtained using an efficient MATLAB tool, namely *pdepe*. This MATLAB routine uses the method of lines by discretizing a parabolic partial differential equation (PDE) (single or system) in one space direction (Skeel and Berzins 1990). For the parameters described in the previous section, the pollutant concentrations are computed using *pdepe* for $t = 0$, 20, and 40 hrs. On the other hand, HAM solutions are computed for each of the cases. The auxiliary parameter $\hbar$ is chosen as -0.87. Figure 16.2 depicts the pollutant concentration values for the selected cases. It can be seen from the figure that the third-order HAM solution agrees very well with the numerical solution. For a quantitative assessment, comparison between the HAM-based solution and numerical solution is shown in Table 16.1, where we consider some selected values of the domain. A flowchart containing the steps of the method is provided in Figure 16.3.

16.7.3 VALIDATION OF THE HPM-BASED SOLUTION

The HPM-based analytical solutions for the selected test case are validated over the numerical solution obtained using *pdepe* of MATLAB. It is seen that the five-term

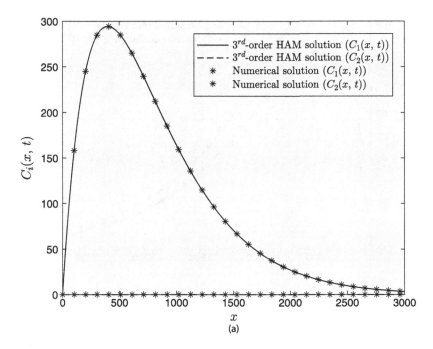

FIGURE 16.2 HAM-based pollutant concentrations versus distance for both species at (a) $t = 0$ hr, (b) $t = 20$ hr, and (c) $t = 40$ hr. (*Continued*)

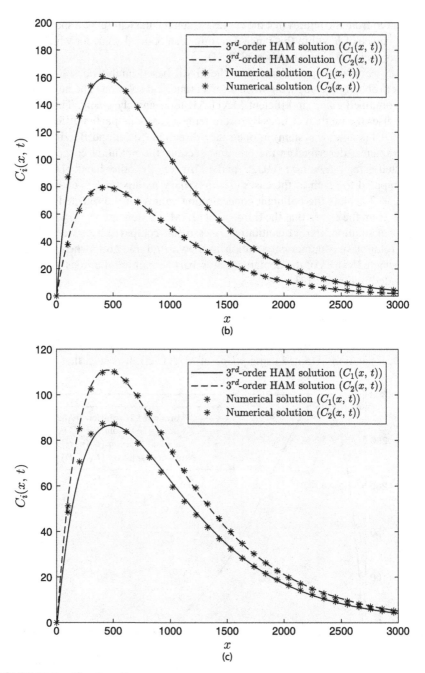

FIGURE 16.2 (*Continued*)

TABLE 16.1

Comparison between HAM-Based Approximation and Numerical Solution for the Selected Case: (a) $C_1(x,t)$ and (b) $C_2(x,t)$

(a) Species, $C_1(x,t)$

x (cm)	Numerical Solution			Third-Order HAM-based Approximation		
	$t = 0$ hr	$t = 20$ hr	$t = 40$ hr	$t = 0$ hr	$t = 20$ hr	$t = 40$ hr
0	0	0	0	0.0000	0	0
250	267.6307	140.4908	73.5227	267.6299	140.6578	72.6647
500	286.5048	157.6763	86.1393	286.5039	157.7893	86.6023
750	230.0325	133.5315	76.4081	230.0318	133.5544	77.7250
1000	164.1700	101.2569	60.9565	164.1696	101.2147	62.3186
1250	109.8423	72.5128	46.1183	109.8421	72.4397	47.0389
1500	70.5532	50.2082	33.8653	70.5531	50.1308	34.2290
1750	44.0585	34.0330	24.4288	44.0585	33.9673	24.3471
2000	26.9518	22.7495	17.4318	26.9518	22.7023	17.0942
2250	16.2295	15.0663	12.3584	16.2295	15.0383	11.9382
2500	9.6523	9.9165	8.7296	9.6523	9.9046	8.3447
2750	5.6831	6.5007	6.1556	5.6831	6.5008	5.8654
3000	3.3185	4.2511	4.3387	3.3185	4.2586	4.1579

(b) Species, $C_2(x,t)$

x (cm)	Numerical Solution			Third-Order HAM-Based Approximation		
	$t = 0$ hr	$t = 20$ hr	$t = 40$ hr	$t = 0$ hr	$t = 20$ hr	$t = 40$ hr
0	0	0	0	0	0	-1.7574
250	0	70.9251	95.3208	0.0005	70.7487	96.0586
500	0	79.2222	110.6372	0.0006	79.1049	110.1981
750	0	66.7362	97.1614	0.0005	66.7144	95.9354
1000	0	50.3063	76.6791	0.0003	50.3567	75.3736
1250	0	35.7906	57.3454	0.0002	35.8784	56.4095
1500	0	24.6059	41.5970	0.0001	24.7026	41.1632
1750	0	16.5518	29.6238	0	16.6392	29.6214
2000	0	10.9746	20.8592	0	11.0433	21.1365
2250	0	7.2061	14.5865	0	7.2536	14.9910
2500	0	4.7005	10.1591	0	4.7284	10.5758
2750	0	3.0527	7.0611	0	3.0646	7.4205
3000	0	1.9770	4.9045	0	1.9771	5.1760

HPM series produces accurate results over the selected domain. Figure 16.4 compares the numerical solutions and the HPM-based approximations. The HPM-based values and the numerical solution are also compared numerically in Table 16.2. A flowchart containing the steps of the method is given in Figure 16.5.

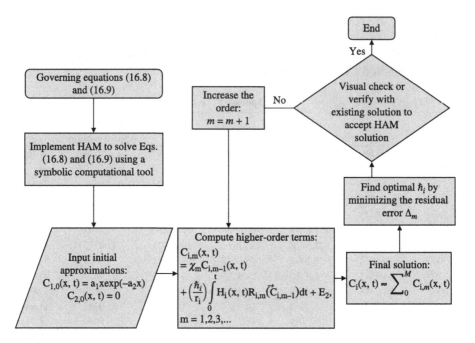

FIGURE 16.3 Flowchart for the HAM solution in Eq. (16.27).

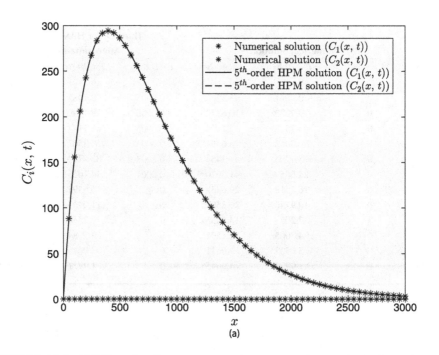

FIGURE 16.4 HPM-based pollutant concentrations versus distance for both species at (a) $t = 0$ hr, (b) $t = 20$ hr, and (c) $t = 40$ hr. (*Continued*)

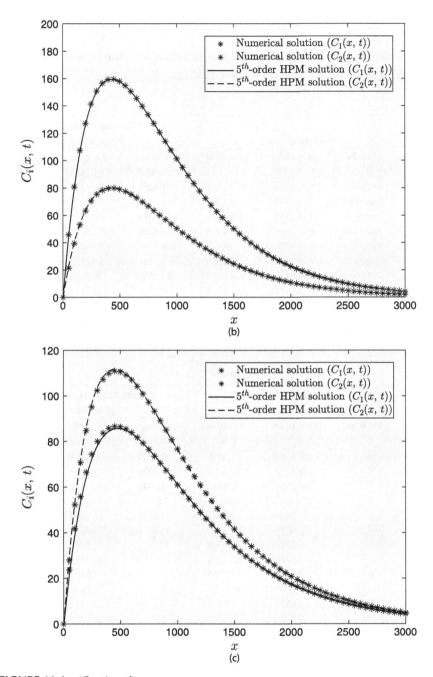

FIGURE 16.4 *(Continued)*

TABLE 16.2

Comparison between HPM-Based Approximation and Numerical Solution for the Selected Case: (a) $C_1(x,t)$ and (b) $C_2(x,t)$

(a) Species, $C_1(x,t)$

	Numerical Solution			Fifth-Order HPM-Based Approximation		
x (cm)	$t = 0$ hr	$t = 20$ hr	$t = 40$ hr	$t = 0$ hr	$t = 20$ hr	$t = 40$ hr
0	0	0	0	0.0000	−1.6183	−1.8523
250	267.6307	140.4908	73.5227	267.6299	140.4567	72.2756
500	286.5048	157.6763	86.1393	286.5039	157.6617	85.6178
750	230.0325	133.5315	76.4081	230.0318	133.5323	76.5445
1000	164.1700	101.2569	60.9565	164.1696	101.2632	61.3086
1250	109.8423	72.5128	46.1183	109.8421	72.5181	46.3814
1500	70.5532	50.2082	33.8653	70.5531	50.2103	33.9435
1750	44.0585	34.0330	24.4288	44.0585	34.0324	24.3591
2000	26.9518	22.7495	17.4318	26.9518	22.7475	17.2960
2250	16.2295	15.0663	12.3584	16.2295	15.0643	12.2277
2500	9.6523	9.9165	8.7296	9.6523	9.9151	8.6433
2750	5.6831	6.5007	6.1556	5.6831	6.5003	6.1230
3000	3.3185	4.2511	4.3387	3.3185	4.2513	4.3498

(b) Species, $C_2(x,t)$

	Numerical Solution			Fifth-Order HPM-Based Approximation		
x (cm)	$t = 0$ hr	$t = 20$ hr	$t = 40$ hr	$t = 0$ hr	$t = 20$ hr	$t = 40$ hr
0	0	0	0	0	−0.7254	−1.8424
250	0	70.9251	95.3208	0.0005	70.9525	96.4530
500	0	79.2222	110.6372	0.0006	79.2369	111.1681
750	0	66.7362	97.1614	0.0005	66.7373	97.1046
1000	0	50.3063	76.6791	0.0003	50.3017	76.3932
1250	0	35.7906	57.3454	0.0002	35.7860	57.0978
1500	0	24.6059	41.5970	0.0001	24.6035	41.4886
1750	0	16.5518	29.6238	0	16.5517	29.6432
2000	0	10.9746	20.8592	0	10.9759	20.9503
2250	0	7.2061	14.5865	0	7.2077	14.6929
2500	0	4.7005	10.1591	0	4.7018	10.2443
2750	0	3.0527	7.0611	0	3.0534	7.1103
3000	0	1.9770	4.9045	0	1.9771	4.9188

16.7.4 Validation of the OHAM-Based Solution

For the assessment of the OHAM-based analytical solution, one needs to calculate the constant $D_{i,j}$. For that purpose, we calculate the residual as follows:

$$R_i(x,t,D_{i,j}) = \mathcal{L}_i \left[C_{i,OHAM}(x,t,D_{i,j}) \right] + \mathcal{N}_i \left[C_{i,OHAM}(x,t,D_{i,j}) \right]$$
$$+ h_i(x,t), j = 1,2,\ldots,s \tag{16.74}$$

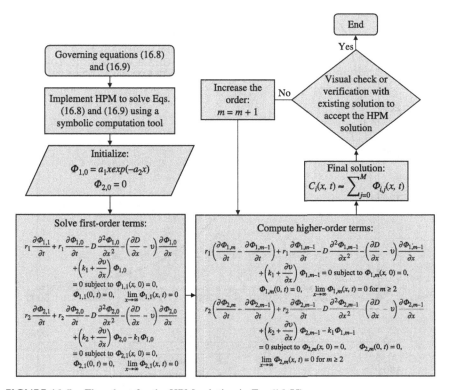

FIGURE 16.5 Flowchart for the HPM solution in Eq. (16.55).

where $C_{i,OHAM}(x,t,D_{i,j})$ is the approximate solution. When $R_i(x,t,D_{i,j}) = 0$, $C_{i,OHAM}(x,t,D_{i,j})$ becomes the exact solution to the problem. One of the ways to obtain the optimal $D_{i,j}$ for which the solution converges is the minimization of the squared residual error, i.e.,

$$J(D_{i,j}) = \int_{(x,t)\in D} \sum_{i=1}^{2} R_i^2(x,t,D_{i,j}) \, dxdt, \quad j = 1,2,\ldots,s \quad (16.75)$$

where D is the domain of the problem. The minimization of Eq. (16.75) leads to a system of algebraic equations as follows:

$$\frac{\partial J}{\partial D_{1,1}} = \frac{\partial J}{\partial D_{1,2}} = \cdots = \frac{\partial J}{\partial D_{1,s}} = 0 \quad (16.76)$$

Solving this system, one can obtain the optimal values of the parameters. We obtain the optimal values using the MATLAB routine *fminsearch*, which minimizes an unconstrained multivariate function. A three-term OHAM solution is computed and compared in Figure 16.6 with the numerical solution obtained using *pdepe* to see the effectiveness of the proposed approach. It can be seen that the OHAM-based approximation agrees well with the numerical solution for both cases. For a quantitative assessment, we also compare the numerical values of the solutions in Table 16.3. A flowchart containing the steps of the method is provided in Figure 16.7.

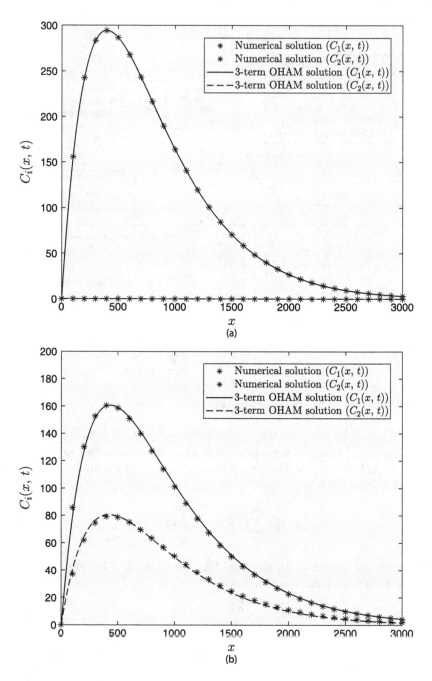

FIGURE 16.6 OHAM-based pollutant concentrations versus distance for both species at (a) $t = 0$ hr, (b) $t = 20$ hr, and (c) $t = 40$ hr. (*Continued*)

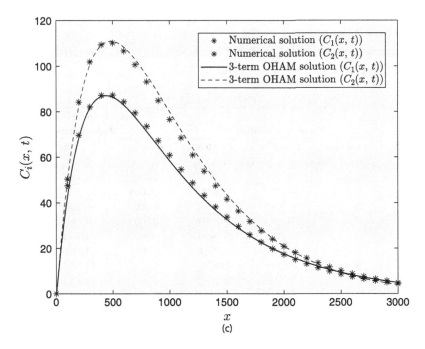

FIGURE 16.6   (*Continued*)

TABLE 16.3

Comparison between OHAM-Based Approximation and Numerical Solution for the Selected Case: (a) $C_1(x,t)$ and (b) $C_2(x,t)$

(a) Species, $C_1(x,t)$

	Numerical Solution			Three-Term OHAM-Based Approximation		
x (cm)	$t = 0$ hr	$t = 20$ hr	$t = 40$ hr	$t = 0$ hr	$t = 20$ hr	$t = 40$ hr
0	0	0	0	0.0000	0	0
250	267.6307	140.4908	73.5227	267.6299	141.3510	76.3031
500	286.5048	157.6763	86.1393	286.5040	158.9841	86.3587
750	230.0325	133.5315	76.4081	230.0319	134.9613	74.4875
1000	164.1700	101.2569	60.9565	164.1696	102.6167	58.3735
1250	109.8423	72.5128	46.1183	109.8421	73.7020	43.9499
1500	70.5532	50.2082	33.8653	70.5531	51.1863	32.5570
1750	44.0585	34.0330	24.4288	44.0585	34.7970	23.9755
2000	26.9518	22.7495	17.4318	26.9518	23.3180	17.6135
2250	16.2295	15.0663	12.3584	16.2295	15.4687	12.9082
2500	9.6523	9.9165	8.7296	9.6523	10.1857	9.4223
2750	5.6831	6.5007	6.1556	5.6831	6.6685	6.8376
3000	3.3185	4.2511	4.3387	3.3185	4.3454	4.9247

(*Continued*)

TABLE 16.3 (*Continued*)

Comparison between OHAM-Based Approximation and Numerical Solution for the Selected Case: (a) $C_1(x,t)$ and (b) $C_2(x,t)$

(b) Species, $C_2(x,t)$

x (cm)	Numerical Solution			Three-Term OHAM-Based Approximation		
	$t = 0$ hr	$t = 20$ hr	$t = 40$ hr	$t = 0$ hr	$t = 20$ hr	$t = 40$ hr
0	0	0	0	0	0	−2.4474
250	0	70.9251	95.3208	0.0005	71.4407	92.9257
500	0	79.2222	110.6372	0.0005	79.3371	110.9119
750	0	66.7362	97.1614	0.0004	66.3508	99.6570
1000	0	50.3063	76.6791	0.0003	49.5507	79.9126
1250	0	35.7906	57.3454	0.0002	34.8341	60.1912
1500	0	24.6059	41.5970	0.0001	23.5909	43.5274
1750	0	16.5518	29.6238	0	15.5772	30.5633
2000	0	10.9746	20.8592	0	10.0988	20.9755
2250	0	7.2061	14.5865	0	6.4560	14.1301
2500	0	4.7005	10.1591	0	4.0814	9.3708
2750	0	3.0527	7.0611	0	2.5566	6.1314
3000	0	1.9770	4.9045	0	1.5890	3.9647

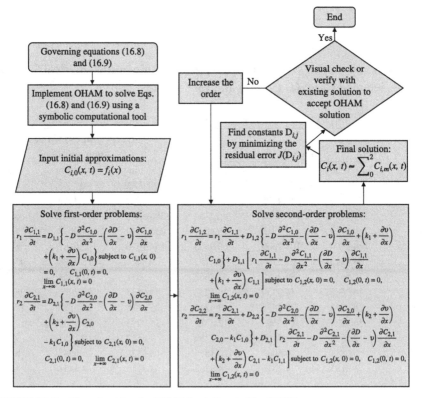

FIGURE 16.7 Flowchart for the OHAM solution in Eq. (16.70).

16.8   CONCLUDING REMARKS

Here, we derive the HAM-, HPM-, and OHAM-based analytical solutions for two-species transport equations with spatially varying dispersion coefficient and seepage velocity. The same equations were studied analytically using HAM by Chaudhary and Singh (2020). A numerical solution is also developed using the MATLAB routine *pdepe*. The proposed methods produce accurate solutions for all the cases. This work shows the potential of HAM, HPM, and OHAM in the context of obtaining a series solution for a system of parabolic PDEs. The theoretical and numerical convergences of the series solutions are provided.

REFERENCES

Abbasbandy, S., Shivanian, E., and Vajravelu, K. (2011). Mathematical properties of $\hbar$-curve in the framework of the homotopy analysis method. *Communications in Nonlinear Science and Numerical Simulation*, *16*(11), 4268–4275.

Batu V. (2005) *Applied flow and solute transport modeling in aquifers: fundamental principles and analytical and numerical methods*, 1st edn. Taylor and Francis, CRC Press, New York, NY.

Chaudhary, M., and Singh, M. K. (2020). Study of multispecies convection-dispersion transport equation with variable parameters. *Journal of Hydrology*, *591*, 125562.

Simpson, M. J., and Ellery, A. J. (2014). Exact series solutions of reactive transport models with general initial conditions. *Journal of Hydrology*, *513*, 7–12.

Skeel, R. D., and Berzins, M. (1990). A method for the spatial discretization of parabolic equations in one space variable. *SIAM Journal on Scientific and Statistical Computing*, *11*(1), 1–32.

Vajravelu, K., and Van Gorder, R. (2013). *Nonlinear flow phenomena and homotopy analysis*. Springer, Berlin.

FURTHER READING

Alkhalaf, S. (2017). Third-order approximate solution of chemical reaction-diffusion Brusselator system using optimal homotopy asymptotic method. *Advances in Mathematical Physics*, *2017*, 1–8.

Arafa, A. A. M., Rida, S. Z., and Mohamed, H. (2012). Approximate analytical solutions of Schnakenberg systems by homotopy analysis method. *Applied Mathematical Modelling*, *36*(10), 4789–4796.

Bataineh, A. S., Noorani, M. S. M., and Hashim, I. (2008). Approximate analytical solutions of systems of PDEs by homotopy analysis method. *Computers & Mathematics with Applications*, *55*(12), 2913–2923.

Bauer, P., Attinger, S., and Kinzelbach, W. (2001). Transport of a decay chain in homogenous porous media: Analytical solutions. *Journal of Contaminant Hydrology*, *49*(3–4), 217–239.

Biazar, J., and Aminikhah, H. (2009). Study of convergence of homotopy perturbation method for systems of partial differential equations. *Computers & Mathematics with Applications*, *58*(11–12), 2221–2230.

Biazar, J., Eslami, M., and Ghazvini, H. (2007). Homotopy perturbation method for systems of partial differential equations. *International Journal of Nonlinear Sciences and Numerical Simulation*, *8*(3), 413–418.

Chamkha, A. J. (2005). Modeling of multi-species contaminant transport with spatially-dependent dispersion and coupled linear/Non-linear reactions. *International Journal of Fluid Mechanics Research, 32*(1), 1–20.

Chen, J. S., Ho, Y. C., Liang, C. P., Wang, S. W., and Liu, C. W. (2018). Analytical model for coupled multispecies advective dispersive transport subject to rate-limited sorption. *Hydrology and Earth System Sciences Discussions*, 1–45. https://doi.org/10.5194/hess-2018-462

Chen, J. S., Lai, K. H., Liu, C. W., and Ni, C. F. (2012). A novel method for analytically solving multi-species advective–dispersive transport equations sequentially coupled with first-order decay reactions. *Journal of Hydrology, 420*, 191–204.

Chen, J. S., Liu, C. W., Liang, C. P., and Lai, K. H. (2012). Generalized analytical solutions to sequentially coupled multi-species advective–dispersive transport equations in a finite domain subject to an arbitrary time-dependent source boundary condition. *Journal of Hydrology, 456*, 101–109.

Chrysikopoulos, C. V., Kitanidis, P. K., and Roberts, P. V. (1990). Analysis of one-dimensional solute transport through porous media with spatially variable retardation factor. *Water Resources Research, 26*(3), 437–446.

Clement, T. P. (2001). Generalized solution to multispecies transport equations coupled with a first-order reaction network. *Water Resources Research,37*(1), 157–163.

Domenico, P. A. (1987). An analytical model for multidimensional transport of a decaying contaminant species. *Journal of Hydrology, 91*(1–2), 49–58.

Guerrero, J. S. P., Skaggs, T. H., and Van Genuchten, M. T. (2009). Analytical solution for multi-species contaminant transport subject to sequential first-order decay reactions in finite media. *Transport in Porous Media, 80*(2), 373–387.

Huo, J. X., Song, H. Z., and Wu, Z. W. (2014). Multi-component reactive transport in heterogeneous media and its decoupling solution. *Journal of Contaminant Hydrology, 166*, 11–22.

Lunn, M., Lunn, R. J., and Mackayb, R. (1996). Determining analytic solutions of multiple species contaminant transport, with sorption and decay. *Journal of Hydrology, 180*(1–4), 195–210.

Natarajan, N. (2016). Effect of distance-dependent and time-dependent dispersion on non-linearly sorbed multispecies contaminants in porous media. *ISH Journal of Hydraulic Engineering, 22*(1), 16–29.

Nawaz, R., Hussain, Z., Khattak, A., and Khan, A. (2020). Extension of optimal homotopy asymptotic method with use of Daftardar–Jeffery polynomials to coupled nonlinear-Korteweg-De-Vries system. *Complexity, 2020*, 1–6.

Quezada, C. R., Clement, T. P., and Lee, K. K. (2004). Generalized solution to multi-dimensional multi-species transport equations coupled with a first-order reaction network involving distinct retardation factors. *Advances in Water Resources, 27*(5), 507–520.

Slodička, M., and Balážová, A. (2010). Decomposition method for solving multi-species reactive transport problems coupled with first-order kinetics applicable to a chain with identical reaction rates. *Journal of Computational and Applied Mathematics, 234*(4), 1069–1077.

Suk, H. (2016). Generalized semi-analytical solutions to multispecies transport equation coupled with sequential first-order reaction network with spatially or temporally variable transport and decay coefficients. *Advances in Water Resources, 94*, 412–423.

Sun, Y., Petersen, J. N., Clement, T. P., and Skeen, R. S. (1999). Development of analytical solutions for multispecies transport with serial and parallel reactions. *Water Resources Research, 35*(1), 185–190.

Sweilam, N. H., and Khader, M. M. (2009). Exact solutions of some coupled nonlinear partial differential equations using the homotopy perturbation method. *Computers & Mathematics with Applications, 58*(11–12), 2134–2141.

Van Genuchten, M. T. (1985). Convective-dispersive transport of solutes involved in sequential first-order decay reactions. *Computers & Geosciences*, *11*(2), 129–147.

Yeh, G. T., and Tripathi, V. S. (1991). A Model for simulating transport of reactive multispecies components: Model development and demonstration. *Water Resources Research*, *27*(12), 3075–3094.

Zubair, T., Usman, M., Ali, U., and Mohyud-Din, S. T. (2012). Homotopy analysis method for system of partial differential equations. *International Journal of Modern Engineering Sciences*, *1*(2), 67–79.

Part V

Integro-Differential Equations

17 Absorption Equation in Unsaturated Soil

17.1 INTRODUCTION

During rainfall events, a large portion of rain infiltrates into the ground except for paved areas. Even for large rainfall events that cause flooding, it has been observed that the amount of rain infiltrating into the ground can be as high as 80–90%. Once infiltrated, the water moves through the pore spaces in the soil. The movement of water occurs in both vertical and horizontal (longitudinal and transverse) directions. The forces governing the movement are capillary (or suction) and gravitational. Capillary forces comprise cohesive and adhesive forces. Both gravitational and capillary forces can be transformed to corresponding energy terms which, in turn, can be expressed in terms of head. Thus, the energy head is composed of a gravitational energy head and suction head.

The movement of water in unsaturated soil can be described by the continuity equation and a flux law, which is the Darcy-Buckingham law. The flux law is a gradient law, meaning that flux is a linear function of the spatial gradient of the head. In the linear function is the parameter called hydraulic conductivity, which is function of soil moisture and in turn of the capillary head. Coupling of the gradient law with the continuity equation leads to the Richards equation. Depending on simplifications, several simplified versions of the Richards equation have been employed in hydrology. One simplification that is employed in this chapter is that the gravity head is negligible. This means the movement of water is governed by capillary forces.

17.2 GOVERNING EQUATION

It has been shown theoretically and experimentally (Richards 1931; Childs and Collis-George 1950) that Darcy's law also holds for water flow in unsaturated porous media. However, in this case, the hydraulic conductivity K becomes a function of the volumetric moisture content θ. For an unsaturated porous media, including soils, the modified Darcy law reads as follows:

$$U = -K(\theta)\nabla\Phi \tag{17.1}$$

where U is the flow velocity vector, K is the hydraulic conductivity tensor of the medium, and Φ is the total potential. For unsaturated porous media, the total potential Φ can be expressed as a sum of the moisture potential Ψ and the gravitational component z, which is the height above some datum level (Philip 1969), i.e.,

$$\Phi = \Psi(\theta) + z \tag{17.2}$$

DOI: 10.1201/9781003368984-22

427

The continuity equation can be written as:

$$\frac{\partial \theta}{\partial t} = -\nabla . U \tag{17.3}$$

where t denotes time. Combining Eqs. (17.1) and (17.3), we have:

$$\frac{\partial \theta}{\partial t} = \nabla . (K \nabla \Phi) \tag{17.4}$$

Substituting Eq. (17.2) in Eq. (17.4), we have:

$$\frac{\partial \theta}{\partial t} = \nabla . (K \nabla \Psi) + \frac{\partial K}{\partial z} \tag{17.5}$$

The moisture diffusivity D, which is a single-valued function of θ if K and Ψ are single-valued, is given as follows:

$$D = K \frac{d\Psi}{d\theta} \tag{17.6}$$

Using Eq. (17.6), Eq. (17.5) can be rewritten as:

$$\frac{\partial \theta}{\partial t} = \nabla . (D \nabla \theta) + \frac{dK}{d\theta} \frac{\partial \theta}{\partial z} \tag{17.7}$$

Eq. (17.7) is a nonlinear Fokker-Planck equation. The specific forms for Eq. (17.7) can be obtained for different physical considerations. The equation describing absorption or infiltration into horizontal systems can be obtained by neglecting the gravity term. In this case, Eq. (17.7) takes on the form of a nonlinear diffusion equation:

$$\frac{\partial \theta}{\partial t} = \nabla . (D \nabla \theta) \tag{17.8}$$

The initial condition for the governing equation can be taken as:

$$\text{At } t = 0, \ \theta = \theta_{in} \text{ for all } V \tag{17.9}$$

where V denotes the whole volume of the soil mass. Eq. (17.9) reflects a simple initial condition. In other words, infiltration into a soil mass at uniform initial moisture content is considered.

Two types of boundary conditions can be prescribed: one is the concentration condition, and the other is the flux condition. This study considers the concentration boundary condition given as follows:

$$\text{For } t \geq 0, \ \theta = \theta_{bc} \text{ at } W \tag{17.10}$$

where W denotes the surface.

In this chapter, we consider a one-dimensional absorption equation. Under such considerations, the governing equation along with the boundary and initial conditions takes on the form:

$$\frac{\partial \theta}{\partial t} = \frac{\partial}{\partial x}\left(D\frac{\partial \theta}{\partial x}\right)$$ (17.11)

subject to:

$$\theta(x,t=0) = \theta_{in} \quad for \ all \ x > 0$$ (17.12)

$$\theta(x=0,t) = \theta_{bc} \quad for \ all \ t \geq 0$$ (17.13)

In the next section, we provide the solution to Eq. (17.11) together with Eqs. (17.12) and (17.13) given by Philip (1955). Then, the homotopy analysis method (HAM), homotopy perturbation method (HPM), and optimal homotopy asymptotic method (OHAM) are used to obtain approximate solutions.

17.3 PHILIP'S SOLUTION

Eq. (17.11) can be converted into an ordinary differential equation by following the similarity transformation:

$$\varphi = \frac{x}{\sqrt{t}}$$ (17.14)

The transformation is widely called the *Boltzmann transformation*, as Boltzmann (1894) used it for the first time in solving diffusion equations. Using Eq. (17.14), the dependent variable becomes:

$$\theta(x,t) = \theta(\varphi)$$ (17.15)

Therefore, using Eqs. (17.14) and (17.15), we have:

$$\frac{\partial \theta}{\partial t} = \frac{d\theta}{d\varphi}\frac{\partial \varphi}{\partial t} = -\frac{1}{2}xt^{-3/2}\frac{d\theta}{d\varphi}$$ (17.16)

$$\frac{\partial \theta}{\partial x} = \frac{d\theta}{d\varphi}\frac{\partial \varphi}{\partial x} = t^{-1/2}\frac{d\theta}{d\varphi}$$ (17.17)

Using Eqs. (17.14)–(17.17), the governing equation and the boundary and initial conditions become:

$$\frac{d}{d\varphi}\left(D\frac{d\theta}{d\varphi}\right) + \frac{\varphi}{2}\frac{d\theta}{d\varphi} = 0$$ (17.18)

subject to:

$$\text{at } \varphi = 0, \ \theta = \theta_{bc} \tag{17.19}$$

$$\text{at } \varphi \to \infty, \ \theta = \theta_{in} \tag{17.20}$$

The solution can also be obtained in the form:

$$x(\theta,t) = \varphi(\theta)t^{1/2} \tag{17.21}$$

Philip (1954, 1957) developed an accurate method for solving Eq. (17.18). To that end, Eq. (17.18) can be rewritten in the form:

$$\frac{d}{d\theta}\left(D\frac{d\theta}{d\varphi} \right) + \frac{\varphi}{2} = 0 \tag{17.22}$$

Integration of Eq. (17.22) yields:

$$\int_{\theta_{in}}^{\theta} \varphi\, d\theta = -2D\frac{d\theta}{d\varphi} \tag{17.23}$$

subject to:

$$\text{at } \varphi = 0, \ \theta = \theta_{bc} \tag{17.24}$$

Philip used a finite difference method for Eq. (17.23) using an iterative approach. The method also used Richardson's extrapolation method to reduce the truncation error. Philip (1960) also proved that there is a large class of $D(\theta)$ functions for which $\varphi(\theta)$ can be found analytically. From Eq. (17.23), $D(\theta)$ can be written as:

$$D(\theta) = -\frac{1}{2}\frac{d\varphi}{d\theta}\int_{\theta_{in}}^{\theta} \varphi\, d\theta \tag{17.25}$$

Therefore, any functional form:

$$\varphi = F(\theta) \tag{17.26}$$

which satisfies certain conditions (Philip 1960) must be the solution of the governing Eq. (17.18) subject to the conditions in Eqs. (17.19)–(17.20). For details about the applicability and restrictions of the method, one can refer to Philip (1960).

17.4   HAM SOLUTION

Here, we use HAM to obtain a convergent series solution for the governing Eq. (17.23). Before applying HAM to the equation, we need to assume some expression for the moisture diffusivity $D(\theta)$, as it is a function of the dependent variable. To that end, we consider the diffusivity developed by Fujita (1952) given as follows:

$$D(\theta) = D_0 (1 - \mu\theta)^{-2} \tag{17.27}$$

where D_0 and μ are constants to be determined from the soil properties. Using Eq. (17.27), Eq. (17.23) can be converted to:

$$(1 - \mu\theta)^2 \frac{d\varphi}{d\theta} \int_{\theta_{in}}^{\theta} \varphi \, d\theta + 2D_0 = 0 \tag{17.28}$$

subject to the condition given by Eq. (17.24). It can be noted that Eq. (17.28) is a nonlinear first-order integro-differential equation. To solve Eq. (17.28) using HAM subject to the boundary condition given by Eqs. (17.24), the zeroth-order deformation equation can be constructed as follows:

$$(1 - q)\mathcal{L}\left[\Phi(\theta; q) - \varphi_0(\theta)\right] = q\hbar H(\theta)\mathcal{N}\left[\Phi(\theta; q)\right] \tag{17.29}$$

subject to the boundary condition:

$$\Phi(\theta_{bc}; q) = 0 \tag{17.30}$$

where q is the embedding parameter, $\Phi(\theta; q)$ is the representation of the solution across q, $\varphi_0(\theta)$ is the initial approximation, $\hbar$ is the auxiliary parameter, $H(\theta)$ is the auxiliary function, and $\mathcal{L}$ and $\mathcal{N}$ are the linear and nonlinear operators, respectively. The higher-order terms can be calculated from the higher-order deformation equation, given as follows:

$$\mathcal{L}\left[\varphi_m(\theta) - \chi_m\varphi_{m-1}(\theta)\right] = \hbar H(\theta) R_m\left(\vec{\varphi}_{m-1}\right), \; m = 1, 2, 3, \ldots$$

subject to:

$$\varphi_m(\theta_{bc}) = 0 \tag{17.31}$$

where:

$$\chi_m = \begin{cases} 0 & \text{when } m = 1, \\ 1 & \text{otherwise} \end{cases} \tag{17.32}$$

and:

$$R_m\left(\vec{\varphi}_{m-1}\right) = \frac{1}{(m-1)!} \left. \frac{\partial^{m-1} \mathcal{N}\left[\Phi(\theta; q)\right]}{\partial q^{m-1}} \right|_{q=0} \tag{17.33}$$

where φ_m for $m \geq 1$ is the higher-order term. Now, the final solution can be obtained as follows:

$$\varphi(\theta) = \varphi_0(\theta) + \sum_{m=1}^{\infty} \varphi_m(\theta) \tag{17.34}$$

The nonlinear operator for the problem is selected as:

$$N\left[\Phi(\theta;q)\right] = (1-\mu\theta)^2 \frac{\partial\Phi(\theta;q)}{\partial\theta} \int_{\theta_{in}}^{\theta} \Phi(\theta;q)d\theta + 2D_0 \tag{17.35}$$

Therefore, using Eq. (17.35), the term R_m can be calculated from Eq. (17.33) as follows:

$$R_m\left(\vec{\varphi}_{m-1}\right) = (1-\mu\theta)^2 \sum_{j=0}^{m-1}\left(\frac{d\varphi_j}{d\theta}\int_{\theta_{in}}^{\theta}\varphi_{m-j-1}d\theta\right) + 2D_0\left(1-\chi_m\right) \tag{17.36}$$

The following set of base functions is chosen to represent the solution:

$$\left\{\theta^m \mid m,n = 0,1,2,...\right\} \tag{17.37}$$

so that:

$$\varphi(\theta) = \sum_{m=0}^{\infty} a_m\theta^m \tag{17.38}$$

where a_m is the coefficient of the series. Eq. (17.38) provides us with the so-called *rule of solution expression*. Following the rule of solution expression, the linear operator and the initial approximation are chosen, respectively, as follows:

$$L\left[\Phi(\theta;q)\right] = \frac{\partial\Phi(\theta;q)}{\partial\theta} \text{ with the property } L[C_3] = 0 \tag{17.39}$$

$$\varphi_0(\theta) = \alpha\left(\frac{\theta - \theta_{bc}}{\theta_{bc} - \theta_{in}}\right) \tag{17.40}$$

where α is an additional convergence-control parameter and C_3 is an integral constant. Using Eq. (17.39), the higher-order terms can be obtained from Eq. (17.35) as follows:

$$\varphi_m(\theta) = \chi_m\varphi_{m-1}(\theta) + \hbar\int_{\theta_{in}}^{\theta} H(x)R_m\left(\vec{\varphi}_{m-1}\right)dx + C_3, \; m = 1,2,3,... \tag{17.41}$$

where R_m is given by Eq. (17.36) and the constant C_3 can be determined from the boundary condition for the higher-order deformation equations given by Eq. (17.31).

Now, the auxiliary function $H(\theta)$ can be determined from the *rule of coefficient ergodicity*. Based on the rule of solution expression in Eq. (17.38), the general form of $H(\theta)$ should be:

$$H(\theta) = \theta^{\beta} \tag{17.42}$$

where β is an integer. Substituting Eq. (17.42) into Eq. (17.41) and calculating some of φ_m, we can infer two observations. If we consider $\beta \geq 2$, then term θ^{β} will be missing in the solution expression. This does not satisfy the *rule of coefficient ergodicity*, which allows us to include all the terms of the base function for the completeness of the solution. On the other hand, if we consider $\beta \leq -1$, then the terms appearing in the solution violate the *rule of solution expression*. Therefore, we are left with two choices, $\beta = 0$ or 1, and can choose either of them. Here, we take $\beta = 0$, i.e., $H(\theta) = 1$. Finally, the approximate solution can be obtained as follows:

$$\varphi(\theta) \approx \varphi_0(\theta) + \sum_{m=1}^{M} \varphi_m(\theta) \tag{17.43}$$

17.5   HPM-BASED SOLUTION

For applying HPM, we rewrite Eq. (17.28) in the following form:

$$\mathcal{N}(\varphi) = f(\theta) \tag{17.44}$$

We construct a homotopy $\Phi(\theta; q)$ that satisfies:

$$(1-q)\left[\mathcal{L}(\Phi(\theta; q)) - \mathcal{L}(\varphi_0(\theta))\right] + q\left[\mathcal{N}(\Phi(\theta; q)) - f(\theta)\right] = 0 \tag{17.45}$$

where q is the embedding parameter that lies on $[0,1]$, $\mathcal{L}$ is the linear operator, and $\varphi_0(\theta)$ is the initial approximation to the final solution. Eq. (17.45) shows that as $q = 0$, $\Phi(\theta; 0) = \varphi_0(\theta)$ and as $q = 1$, $\Phi(\theta; 1) = \varphi(\theta)$. We now express $\Phi(\theta; q)$ as a series in terms of q as follows:

$$\Phi(\theta; q) = \Phi_0 + q\Phi_1 + q^2\Phi_2 + q^3\Phi_3 + q^4\Phi_4 \ldots \tag{17.46}$$

where Φ_m for $m \geq 1$ is the higher-order term. As $q \to 1$, Eq. (17.46) produces the final solution as:

$$\varphi(\theta) = \lim_{q \to 1} \Phi(\theta; q) = \sum_{i=0}^{\infty} \Phi_i \tag{17.47}$$

For simplicity, we consider the linear and nonlinear operators as follows:

$$\mathcal{L}(\Phi(\theta; q)) = \frac{\partial \Phi(\theta; q)}{\partial \theta} \tag{17.48}$$

$$\mathcal{N}\big(\Phi(\theta;q)\big)=(1-\mu\theta)^2\,\frac{\partial\Phi(\theta;q)}{\partial\theta}\int_{\theta_{in}}^{\theta}\Phi(\theta;q)\,d\theta+2D_0 \tag{17.49}$$

Using these expressions, we have:

$$(1-q)\left[\frac{\partial\Phi(\theta;q)}{\partial\theta}-\frac{d\varphi_0}{d\theta}\right]$$

$$=(1-q)\left[\left(\frac{d\Phi_0}{d\theta}-\frac{d\varphi_0}{d\theta}\right)+q\frac{d\Phi_1}{d\theta}+q^2\frac{d\Phi_2}{d\theta}+q^3\frac{d\Phi_3}{d\theta}+q^4\frac{d\Phi_4}{d\theta}+\cdots\right]$$

$$=\left(\frac{d\Phi_0}{d\theta}-\frac{d\varphi_0}{d\theta}\right)+q\left(\frac{d\Phi_1}{d\theta}-\left(\frac{d\Phi_0}{d\theta}-\frac{d\varphi_0}{d\theta}\right)\right)+q^2\left(\frac{d\Phi_2}{d\theta}-\frac{d\Phi_1}{d\theta}\right)$$

$$+q^3\left(\frac{d\Phi_3}{d\theta}-\frac{d\Phi_2}{d\theta}\right)+q^4\left(\frac{d\Phi_4}{d\theta}-\frac{d\Phi_3}{d\theta}\right)+\cdots \tag{17.50}$$

$$(1-\mu\theta)^2\,\frac{\partial\Phi(\theta;q)}{\partial\theta}\left[\int_{\theta_{in}}^{\theta}\Phi(\theta;q)\,d\theta\right]+2D_0$$

$$=(1-\mu\theta)^2\left(\frac{d\Phi_0}{d\theta}+q\frac{d\Phi_1}{d\theta}+q^2\frac{d\Phi_2}{d\theta}+q^3\frac{d\Phi_3}{d\theta}+\cdots\right)$$

$$\times\left(\int_{\theta_{in}}^{\theta}\Phi_0\,d\theta+q\int_{\theta_{in}}^{\theta}\Phi_1\,d\theta+q^2\int_{\theta_{in}}^{\theta}\Phi_2\,d\theta+q^3\int_{\theta_{in}}^{\theta}\Phi_3\,d\theta+\cdots\right)+2D_0$$

$$=(1-\mu\theta)^2\left[\frac{d\Phi_0}{d\theta}\int_{\theta_{in}}^{\theta}\Phi_0\,d\theta+q\left(\frac{d\Phi_1}{d\theta}\int_{\theta_{in}}^{\theta}\Phi_0\,d\theta+\frac{d\Phi_0}{d\theta}\int_{\theta_{in}}^{\theta}\Phi_1\,d\theta\right)\right.$$

$$+q^2\left(\frac{d\Phi_0}{d\theta}\int_{\theta_{in}}^{\theta}\Phi_2\,d\theta+\frac{d\Phi_1}{d\theta}\int_{\theta_{in}}^{\theta}\Phi_1\,d\theta+\frac{d\Phi_2}{d\theta}\int_{\theta_{in}}^{\theta}\Phi_0\,d\theta\right)$$

$$\left.+q^3\left(\frac{d\Phi_0}{d\theta}\int_{\theta_{in}}^{\theta}\Phi_3\,d\theta+\frac{d\Phi_3}{d\theta}\int_{\theta_{in}}^{\theta}\Phi_0\,d\theta+\frac{d\Phi_1}{d\theta}\int_{\theta_{in}}^{\theta}\Phi_2\,d\theta+\frac{d\Phi_2}{d\theta}\int_{\theta_{in}}^{\theta}\Phi_1\,d\theta\right)+\cdots\right]$$

$$+2D_0=\left[(1-\mu\theta)^2\frac{d\Phi_0}{d\theta}\int_{\theta_{in}}^{\theta}\Phi_0\,d\theta+2D_0\right]+q\left[(1-\mu\theta)^2\left(\frac{d\Phi_1}{d\theta}\int_{\theta_{in}}^{\theta}\Phi_0\,d\theta+\frac{d\Phi_0}{d\theta}\int_{\theta_{in}}^{\theta}\Phi_1\,d\theta\right)\right]$$

$$+q^2\left[(1-\mu\theta)^2\left(\frac{d\Phi_0}{d\theta}\int_{\theta_{in}}^{\theta}\Phi_2\,d\theta+\frac{d\Phi_1}{d\theta}\int_{\theta_{in}}^{\theta}\Phi_1\,d\theta+\frac{d\Phi_2}{d\theta}\int_{\theta_{in}}^{\theta}\Phi_0\,d\theta\right)\right]$$

$$+q^3\left[(1-\mu\theta)^2\left(\frac{d\Phi_0}{d\theta}\int_{\theta_{in}}^{\theta}\Phi_3\,d\theta+\frac{d\Phi_3}{d\theta}\int_{\theta_{in}}^{\theta}\Phi_0\,d\theta+\frac{d\Phi_1}{d\theta}\int_{\theta_{in}}^{\theta}\Phi_2\,d\theta+\frac{d\Phi_2}{d\theta}\int_{\theta_{in}}^{\theta}\Phi_1\,d\theta\right)\right]+\cdots$$

$$\tag{17.51}$$

Using Eqs. (17.50) and (17.51) and equating the like powers of q in Eq. (17.45), the following system of differential equations is obtained:

$$\frac{d\Phi_0}{d\theta} - \frac{d\phi_0}{d\theta} = 0 \text{ subject to } \Phi_0(\theta_{bc}) = 0 \tag{17.52}$$

$$\frac{d\Phi_1}{d\theta} - \left(\frac{d\Phi_0}{d\theta} - \frac{d\phi_0}{d\theta}\right) + (1-\mu\theta)^2 \frac{d\Phi_0}{d\theta} \int_{\theta_{in}}^{\theta} \Phi_0 d\theta + 2D_0 = 0$$

$$\text{subject to } \Phi_1(\theta_{bc}) = 0 \tag{17.53}$$

$$\frac{d\Phi_2}{d\theta} - \frac{d\Phi_1}{d\theta} + (1-\mu\theta)^2 \left(\frac{d\Phi_1}{d\theta} \int_{\theta_{in}}^{\theta} \Phi_0 d\theta + \frac{d\Phi_0}{d\theta} \int_{\theta_{in}}^{\theta} \Phi_1 d\theta\right) = 0$$

$$\text{subject to } \Phi_2(\theta_{bc}) = 0 \tag{17.54}$$

$$\frac{d\Phi_3}{d\theta} - \frac{d\Phi_2}{d\theta} + (1-\mu\theta)^2 \left(\frac{d\Phi_0}{d\theta} \int_{\theta_{in}}^{\theta} \Phi_2 d\theta + \frac{d\Phi_1}{d\theta} \int_{\theta_{in}}^{\theta} \Phi_1 d\theta + \frac{d\Phi_2}{d\theta} \int_{\theta_{in}}^{\theta} \Phi_0 d\theta\right) = 0$$

$$\text{subject to } \Phi_3(\theta_{bc}) = 0 \tag{17.55}$$

$$\frac{d\Phi_4}{d\theta} - \frac{d\Phi_3}{d\theta} + (1-\mu\theta)^2 \left(\frac{d\Phi_0}{d\theta} \int_{\theta_{in}}^{\theta} \Phi_3 d\theta + \frac{d\Phi_3}{d\theta} \int_{\theta_{in}}^{\theta} \Phi_0 d\theta + \frac{d\Phi_1}{d\theta} \int_{\theta_{in}}^{\theta} \Phi_2 d\theta + \frac{d\Phi_2}{d\theta} \int_{\theta_{in}}^{\theta} \Phi_1 d\theta\right) = 0$$

$$\text{subject to } \Phi_4(\theta_{bc}) = 0 \tag{17.56}$$

The initial approximation can be chosen as $\Phi_0 = 0.08\left(\dfrac{\theta_{bc} - \theta}{\theta_{bc} - \theta_{in}}\right)$. Using this initial approximation, we can solve the equations iteratively using symbolic software. We do not include those expressions here, as they are lengthy. Finally, the HPM-based solution can be approximated as follows:

$$\phi(\theta) \approx \sum_{i=0}^{M} \Phi_i \tag{17.57}$$

17.6   OHAM-BASED SOLUTION

When applying OHAM, Eq. (17.28) can be written in the following form:

$$\mathcal{L}[\phi(\theta)] + \mathcal{N}[\phi(\theta)] + h(\theta) = 0 \tag{17.58}$$

Here, we tackle the problem in a different way. To choose an initial approximation similar to HPM, we select $\mathcal{L}[\phi(\theta)] = (\theta_{bc} - \theta_{in})\dfrac{d\phi}{d\theta}$, $\mathcal{N}[\phi(\theta)] = \dfrac{d\phi}{d\theta} \times$

$\left[(1-\mu\theta)^2 \int_{\theta_{in}}^{\theta} \varphi d\theta - (\theta_{bc} - \theta_{in}) \right] + 2D_0 - 0.08$, and $h(\theta) = 0.08$. With these considerations, the zeroth-order problem becomes:

$$\mathcal{L}(\varphi_0(\theta)) + h(\theta) = 0 \text{ subject to } \varphi_0(\theta_{bc}) = 0 \qquad (17.59)$$

Solving Eq. (17.59), one obtains:

$$\varphi_0(\theta) = 0.08 \left(\frac{\theta_{bc} - \theta}{\theta_{bc} - \theta_{in}} \right) \qquad (17.60)$$

To obtain $\mathcal{N}_0$, $\mathcal{N}_1$, $\mathcal{N}_2$, etc., as in Eqs. (17.50) and (17.51), the following expressions are to be used:

$$\frac{d\varphi}{d\theta} \left[(1-\mu\theta)^2 \int_{\theta_{in}}^{\theta} \varphi d\theta - (\theta_{bc} - \theta_{in}) \right] + 2D_0 - 0.08$$

$$= \left(\frac{d\varphi_0}{d\theta} + q\frac{d\varphi_1}{d\theta} + q^2 \frac{d\varphi_2}{d\theta} + q^3 \frac{d\varphi_3}{d\theta} + \cdots \right)$$

$$\times \left[(1-\mu\theta)^2 \left(\int_{\theta_{in}}^{\theta} \varphi_0 d\theta + q\int_{\theta_{in}}^{\theta} \varphi_1 d\theta + q^2 \int_{\theta_{in}}^{\theta} \varphi_2 d\theta + q^3 \int_{\theta_{in}}^{\theta} \varphi_3 d\theta + \cdots \right) - (\theta_{bc} - \theta_{in}) \right]$$

$$+ 2D_0 - 0.08 = \frac{d\varphi_0}{d\theta} \left[(1-\mu\theta)^2 \int_{\theta_{in}}^{\theta} \varphi_0 d\theta - (\theta_{bc} - \theta_{in}) \right] + 2D_0 - 0.08$$

$$+ q\left[\frac{d\varphi_1}{d\theta} \left\{ (1-\mu\theta)^2 \int_{\theta_{in}}^{\theta} \varphi_0 d\theta - (\theta_{bc} - \theta_{in}) \right\} + \frac{d\varphi_0}{d\theta} \int_{\theta_{in}}^{\theta} \varphi_1 d\theta \right]$$

$$+ q^2 \left[\frac{d\varphi_2}{d\theta} \left\{ (1-\mu\theta)^2 \int_{\theta_{in}}^{\theta} \varphi_0 d\theta - (\theta_{bc} - \theta_{in}) \right\} + \frac{d\varphi_1}{d\theta} \int_{\theta_{in}}^{\theta} \varphi_1 d\theta + \frac{d\varphi_0}{d\theta} \int_{\theta_{in}}^{\theta} \varphi_2 d\theta \right] + \cdots \quad (17.61)$$

Using Eq. (17.61), the first-order problem reduces to:

$$(\theta_{bc} - \theta_{in}) \frac{d\varphi_1}{d\theta} = H_1(\theta, C_i) \left\{ \frac{d\varphi_0}{d\theta} \left[(1-\mu\theta)^2 \int_{\theta_{in}}^{\theta} \varphi_0 d\theta - (\theta_{bc} - \theta_{in}) \right] + 2D_0 - 0.08 \right\}$$

subject to $\varphi_1(\theta_{bc}) = 0$ $\qquad\qquad (17.62)$

We simply choose $H_1(\theta, C_1) = C_1$. Using $k = 2$, the second-order problem becomes:

$$\mathcal{L}\left[\varphi_2(\theta, C_i) - \varphi_1(\theta, C_i) \right] = H_2(\theta, C_i) \mathcal{N}_0(\varphi_0(\theta))$$

$$+ H_1(\theta, C_i) \left[\mathcal{L}\left[\varphi_1(\theta, C_i) \right] + \mathcal{N}_1\left[\varphi_0(\theta), \varphi_1(\theta, C_i) \right] \right] \text{ subject to } \varphi_2(\theta_{bc}) = 0 \quad (17.63)$$

Further, we choose $H_2(\theta, C_i) = C_2$. Therefore, Eq. (17.63) becomes:

$$(\theta_{bc} - \theta_{in})\frac{d\varphi_2}{d\theta} = (\theta_{bc} - \theta_{in})\frac{d\varphi_1}{d\theta} + C_2\left\{\frac{d\varphi_0}{d\theta}\left[(1-\mu\theta)^2\int_{\theta_{in}}^{\theta}\varphi_0 d\theta - 1\right] + 2D_0\right\}$$

$$+ C_1\left[(\theta_{bc} - \theta_{in})\frac{d\varphi_1}{d\theta} + \frac{d\varphi_1}{d\theta}\left\{(1-\mu\theta)^2\int_{\theta_{in}}^{\theta}\varphi_0 d\theta - (\theta_{bc} - \theta_{in})\right\} + \frac{d\varphi_0}{d\theta}\int_{\theta_{in}}^{\theta}\varphi_1 d\theta\right]$$

subject to $\varphi_2(\theta_{bc}) = 0$ \hfill (17.64)

The terms of the series can be computed using the equations developed here. One can compute these terms without any difficulty using symbolic computation software, such as MATLAB. Further, following Chapter 2, the higher-order terms can be computed in a similar manner. However, our aim is to produce an accurate solution with just two to three terms of the OHAM-based series. Therefore, we restrict our calculation up to $k = 2$. Finally, the approximate solution can be found as:

$$\varphi(\theta) \approx \varphi_0(\theta) + \varphi_1(\theta, C_1) + \varphi_2(\theta, C_1, C_2) \tag{17.65}$$

where the terms are given by Eqs. (17.60), (17.62), and (17.64).

17.7 CONVERGENCE THEOREMS

The convergence of the HAM-based and OHAM-based solutions are proved theoretically using the following theorems.

17.7.1 CONVERGENCE THEOREM OF THE HAM-BASED SOLUTION

The convergence theorems for the HAM-based solution given by Eq. (17.36) can be proved using the following theorems.

Theorem 17.1: If the homotopy series $\Sigma_{m=0}^{\infty}\varphi_m(\theta)$ and $\Sigma_{m=0}^{\infty}\varphi_m{}'(\theta)$ converge, then $R_m(\vec{\varphi}_{m-1})$ given by Eq. (17.38) satisfies the relation $\Sigma_{m=1}^{\infty}R_m(\vec{\varphi}_{m-1}) = 0$. [Here "'" denotes the first derivative with respect to θ.]

Proof: The proof of this theorem follows exactly from the previous chapters.

Theorem 17.2: If $\hbar$ is so properly chosen that the series $\Sigma_{m=0}^{\infty}\varphi_m(\theta)$ and $\Sigma_{m=0}^{\infty}\varphi_m{}'(\theta)$ converge absolutely to $\varphi(\theta)$ and $\varphi'(\theta)$, respectively, then the homotopy series $\Sigma_{m=0}^{\infty}\varphi_m(\theta)$ satisfies the original governing Eq. (17.28).

Proof: Using the Cauchy product defined in Theorem 2.2 of Chapter 2 in relation to Eq. (17.36), we have:

$$\sum_{m=1}^{\infty}\sum_{j=0}^{m-1}\left(\frac{d\varphi_j}{d\theta}\int_{\theta_{in}}^{\theta}\varphi_{m-j-1}\,d\theta\right) = \left(\sum_{m=0}^{\infty}\varphi_m{}'\right)\left(\int_{\theta_{in}}^{\theta}\sum_{m=0}^{\infty}\varphi_m\,d\theta\right) \tag{17.66}$$

Theorem 17.1 shows that if $\sum_{m=0}^{\infty} \varphi_m(\theta)$ and $\sum_{m=0}^{\infty} \varphi_m'(\theta)$ converge, then $\sum_{m=1}^{\infty} R_m(\vec{\varphi}_{m-1}) = 0$.

Therefore, substituting the previous expressions in Eq. (17.36) and simplifying further lead to:

$$(1-\mu\theta)^2 \left(\sum_{m=0}^{\infty} \varphi_m' \right) \left(\int_{\theta_{in}}^{\theta} \sum_{m=0}^{\infty} \varphi_m \, d\theta \right) + 2D_0 \sum_{m=1}^{\infty} (1-\chi_m) = 0 \qquad (17.67)$$

which is basically the original governing equation in Eq. (17.28). Furthermore, subject to the boundary condition $\varphi_0(\theta_{bc}) = 0$ and the condition for the higher-order deformation equations $\varphi_m(\theta_{bc}) = 0$, for $m \geq 1$, we easily obtain $\sum_{m=0}^{\infty} \varphi_m(\theta_{bc}) = 0$. Hence, the convergence result follows.

17.7.2 CONVERGENCE THEOREM OF THE OHAM-BASED SOLUTION

Theorem 17.3: If the series $\varphi_0(\theta) + \sum_{j=1}^{\infty} \varphi_j(\theta, C_i)$, $i = 1, 2, \ldots, s$ converges, where $\varphi_j(\theta, C_i)$ is governed by Eqs. (17.60), (17.62), and (17.64), then Eq. (17.65) is a solution of the original Eq. (17.28).

Proof: The proof of this theorem can be followed exactly from Theorem 2.3 of Chapter 2.

17.8 RESULTS AND DISCUSSION

First, the numerical convergence of the HAM-based analytical solution is established for a specific test case, and then the solution is validated over the existing quasi-analytical/numerical solution proposed by Philip (1960). Finally, the HPM- and OHAM-based analytical solutions are validated by comparing them with the solution of Philip (1960).

17.8.1 NUMERICAL CONVERGENCE AND VALIDATION OF THE HAM-BASED SOLUTION

A suitable choice for the auxiliary parameters α and $\hbar$ determines the convergence of the HAM-based series solution. For that purpose, the squared residual error regarding Eq. (17.28) can be calculated as follows:

$$\Delta_m = \int_{\theta \in \Omega} \left(\mathcal{N}[\varphi(\theta)] \right)^2 d\theta \qquad (17.68)$$

where Ω is the domain of the problem. The HAM-based series solution leads to the exact solution of the problem when Δ_m tends to zero. Therefore, for a particular order of approximation m, it is sufficient to minimize the corresponding Δ_m, which yields optimal values of the parameters α and $\hbar$. Here, we consider a test case, where parameters

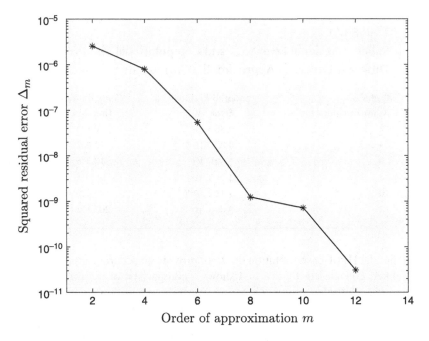

FIGURE 17.1 Squared residual error (Δ_m) versus different orders of approximation (m) of the HAM-based solution for the selected case.

are chosen as $D_0 = 1.67 \times 10^{-5}$ cm^{3}s^{-1}, $\mu = 1.97$, $\theta_{in} = 0.2376$, and $\theta_{bc} = 0.4950$. It may be noted that HAM can indeed provide a convergent solution for the whole domain of θ. Using these parameters, the HAM solution is assessed, and the squared residual errors are calculated. Figure 17.1 shows the squared residual errors for a different order of approximation for the selected case. It can be seen from the figure that as the order of approximation increases, the corresponding residual error decreases. Thus, the adequacy of selecting the linear operator, initial approximation, and auxiliary function is established. Also, for a quantitative assessment, numerical results are reported in Table 17.1 along with the computational time taken by the computer to produce the corresponding order of approximation. For the selected case, comparison between Philip's solution and the 15th-order HAM-based approximate solution is shown in Figure 17.2. A favorable agreement between the computed and observed values is found from the figure. To get a clear comparative idea, the numerical values are calculated for both solutions and shown in Table 17.2. All computations are performed using the BVPh 2.0 package developed by Zhao and Liao (2002). A flowchart containing the steps of the method is provided in Figure 17.3.

17.8.2 VALIDATION OF THE HPM-BASED SOLUTION

The HPM-based analytical solution for the selected test case is validated over the analytical solution given by Philip (1960). After assessing the solution, it was

TABLE 17.1

Squared Residual Error (Δ_m) and Computational Time versus Different Orders of Approximation (m) for the Selected Case

Order of Approximations (m)	Squared Residual Error (Δ_m)	Computational Time (sec)
2	2.56×10^{-6}	0.895
4	8.02×10^{-7}	5.032
6	5.40×10^{-8}	12.345
8	1.24×10^{-9}	30.058
10	7.15×10^{-10}	89.478
12	3.09×10^{-11}	203.112

found that the HPM-based solution does not provide an accurate approximation for the problem considered. Figure 17.4 shows a comparison of the three-term HPM solution with Philip's analytical solution. It can be noted that one can indeed try some other initial approximations, which might work in producing more accurate solutions to the problem. The HPM-based values and the existing analytical solution are compared in Table 17.3. A flowchart containing the steps of HPM is given in Figure 17.5.

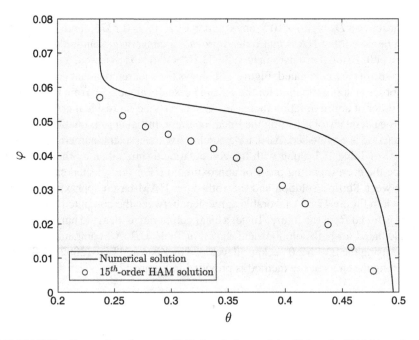

FIGURE 17.2 Comparison between Philip's solution and the 15th-order HAM-based solution for the selected case.

TABLE 17.2

Comparison between HAM-Based Approximation and Philip's Solution for the Selected Case

θ	Philip's Solution	HAM-Based Approximation 15th Order
0.2376	0.08000	0.05699
0.2576	0.05867	0.05162
0.2776	0.05691	0.04844
0.2976	0.05579	0.04622
0.3176	0.05467	0.04425
0.3376	0.05371	0.04204
0.3576	0.05258	0.03926
0.3776	0.05130	0.03569
0.3976	0.04986	0.03122
0.4176	0.04810	0.02587
0.4376	0.04537	0.01977
0.4576	0.04088	0.01312
0.4776	0.03142	0.00616

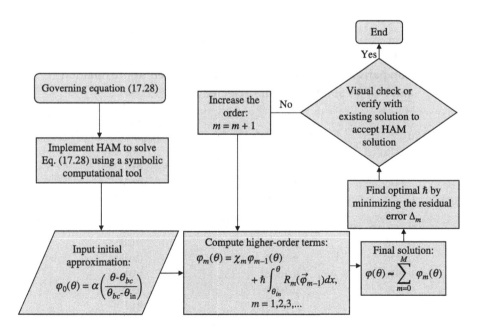

FIGURE 17.3 Flowchart for the HAM solution in Eq. (17.43).

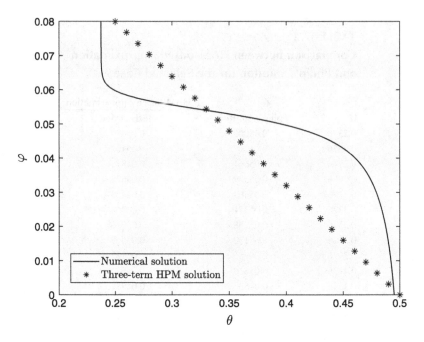

FIGURE 17.4 Comparison between Philip's solution and the three-term HPM-based approximation for the selected case.

TABLE 17.3
Comparison between Three-Term HPM-Based Approximation and Philip's Solution for the Selected Case

θ	Philip's Solution	Three-Term HPM-Based Approximation
0.2376	0.08000	0.08393
0.2576	0.05867	0.07753
0.2776	0.05691	0.07113
0.2976	0.05579	0.06473
0.3176	0.05467	0.05834
0.3376	0.05371	0.05195
0.3576	0.05258	0.04555
0.3776	0.05130	0.03916
0.3976	0.04986	0.03276
0.4176	0.04810	0.02637
0.4376	0.04537	0.01997
0.4576	0.04088	0.01357
0.4776	0.03142	0.00717

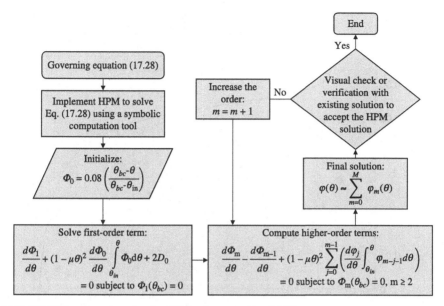

FIGURE 17.5 Flowchart for the HPM solution in Eq. (17.57).

17.8.3 VALIDATION OF THE OHAM-BASED SOLUTION

For the assessment of the OHAM-based analytical solution, one needs to calculate the constant C_i. For that purpose, we calculate the residual as follows:

$$R(\theta, C_i) = \mathcal{L}\left[\varphi_{OHAM}(\theta, C_i)\right] + \mathcal{N}\left[\varphi_{OHAM}(\theta, C_i)\right] + h(\theta), \; i = 1, 2, \ldots, s \quad (17.69)$$

where $\varphi_{OHAM}(\theta, C_i)$ is the approximate solution. When $R(\theta, C_i) = 0$, $\varphi_{OHAM}(\theta, C_i)$ becomes the exact solution to the problem. One of the ways to obtain the optimal C_i for which the solution converges is the minimization of the squared residual error, i.e.,

$$J(C_i) = \int_{\theta \in D} R^2(\theta, C_i) d\theta, \; i = 1, 2, \ldots, s \quad (17.70)$$

where $D = [0, 0.08]$ is the domain of the problem. The minimization of Eq. (17.70) leads to a system of algebraic equations as follows:

$$\frac{\partial J}{\partial C_1} = \frac{\partial J}{\partial C_2} = \cdots = \frac{\partial J}{\partial C_s} = 0 \quad (17.71)$$

Solving this system, one can obtain the optimal values of the parameters. We obtain the optimal values using the MATLAB routine *fminsearch*, which minimizes an unconstrained multivariable function. It is found that a three-term solution produces favorable results. Therefore, the three-term OHAM solution is computed and is compared in Figure 17.6 with the existing analytical solution of Philip (1960) to see the effectiveness of the proposed approach. For a quantitative assessment, we also compare the numerical values of the solutions in Table 17.4. A flowchart containing the steps of OHAM is provided in Figure 17.7.

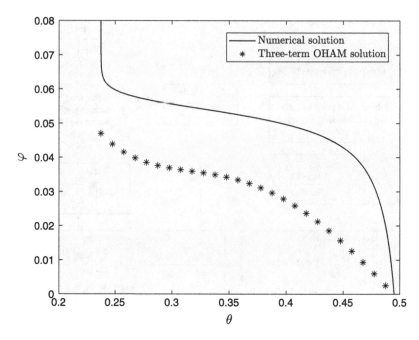

FIGURE 17.6 Comparison between Philip's solution and the three-term OHAM-based solution for the selected case.

TABLE 17.4

Comparison between the Three-Term OHAM-Based Approximation and Philip's Solution for the Selected Case

θ	Philip's Solution	Three-Term OHAM-Based Approximation
0.2376	0.08000	0.04703
0.2576	0.05867	0.04156
0.2776	0.05691	0.03854
0.2976	0.05579	0.03694
0.3176	0.05467	0.03594,
0.3376	0.05371	0.03491
0.3576	0.05258	0.03341
0.3776	0.05130	0.03113
0.3976	0.04986	0.02791
0.4176	0.04810	0.02370
0.4376	0.04537	0.01855
0.4576	0.04088	0.01260
0.4776	0.03142	0.00603

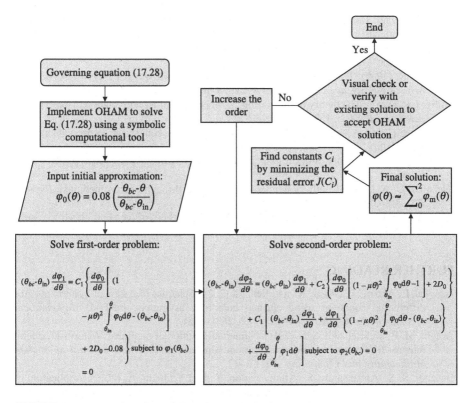

FIGURE 17.7 Flowchart for the OHAM solution in Eq. (17.57).

17.9 CONCLUDING REMARKS

The HAM, HPM, and OHAM are applied to the governing equation of absorption in unsaturated soils. Following the work of Philip, the original partial differential equation is first converted to a second-order boundary value problem using the Boltzmann transformation and then to an integro-differential equation form. Philip provided both quasi-analytical and numerical solutions by solving the equivalent integro-differential equation. This chapter summarizes the work of Philip and develops three kinds of series solutions for the equation. The HAM- and OHAM-based solutions seem to provide favorable results for a selected test case. It is also found that the HPM cannot produce an accurate solution for the problem. The theoretical and numerical convergences of the series solutions are provided.

REFERENCES

Boltzmann, L. (1894). About the integration of the diffusion equation in the case of variable diffusion coefficients. *Annalen der Physik und Chemie*, 53, 959–964.

Childs, E. C., and Collis-George, N. (1950). The permeability of porous materials. *Proceedings of the Royal Society of London. Series A. Mathematical and Physical Sciences*, 201(1066), 392–405.

Fujita, H. (1952). The exact pattern of a concentration-dependent diffusion in a semi-infinite medium, part I. *Textile Research Journal, 22*(11), 757–760.

Philip, J. R. (1954). An infiltration equation with physical significance. *Soil Science, 77*(2), 153–158.

Philip, J. R. (1955). The concept of diffusion applied to soil water. *Proceedings National Academy of Science India Section A, 24*, 93–104.

Philip, J. R. (1957). The theory of infiltration: 3. Moisture profiles and relation to experiment. *Soil Science, 84*(2), 163–178.

Philip, J. R. (1960). A very general class of exact solutions in concentration-dependent diffusion. *Nature, 185*(4708), 233–233.

Philip, J. R. (1969). Theory of infiltration. In *Advances in hydroscience* (Vol. 5, pp. 215–296). Elsevier.

Richards, L. A. (1931). Capillary conduction of liquids through porous mediums. *Physics, 1*(5), 318–333.

Zhao, Y., and Liao, S. (2002). User guide to BVPh 2.0. School of naval architecture, ocean and civil engineering, Shanghai, *40*.

FURTHER READING

Agarwal, P., Akbar, M., Nawaz, R., and Jleli, M. (2021). Solutions of system of Volterra integro-differential equations using optimal homotopy asymptotic method. *Mathematical Methods in the Applied Sciences, 44*(3), 2671–2681.

Araghi, M. F., and Behzadi, S. S. (2011). Numerical solution of nonlinear Volterra-Fredholm integro-differential equations using homotopy analysis method. *Journal of Applied Mathematics and Computing, 37*(1), 1–12.

Behzadi, S. S., Abbasbandy, S., Allahviranloo, T., and Yildirim, A. (2012). Application of homotopy analysis method for solving a class of nonlinear Volterra-Fredholm integro-differential equations. *Journal of Applied Analysis and Computation, 2*(2), 127–136.

Biazar, J., Ghazvini, H., and Eslami, M. (2009). He's homotopy perturbation method for systems of integro-differential equations. *Chaos, Solitons & Fractals, 39*(3), 1253–1258.

Du Han, Y., and Yun, J. H. (2013). Optimal homotopy asymptotic method for solving integro-differential equations. *International Journal of Applied Mathematics, 43*(3), 120–126.

El-Shahed, M. (2005). Application of He's homotopy perturbation method to Volterra's integro-differential equation. *International Journal of Nonlinear Sciences and Numerical Simulation, 6*(2), 163–168.

Ghasemi, M., Kajani, M. T., and Babolian, E. (2007). Application of He's homotopy perturbation method to nonlinear integro-differential equations. *Applied Mathematics and Computation, 188*(1), 538–548.

Ghoreishi, M., Ismail, A. M., and Alomari, A. K. (2011). Comparison between homotopy analysis method and optimal homotopy asymptotic method for nth-order integro-differential equation. *Mathematical Methods in the Applied Sciences, 34*(15), 1833–1842.

Parlange, J. Y. (1971). Theory of water-movement in soils: I. One-dimensional absorption. *Soil Science, 111*(2), 134–137.

Philip, J. R. (1957). The theory of infiltration: 2. The profile of infinity. *Soil Science, 83*(6), 435–448.

Philip, J. R. (1960). General method of exact solution of the concentration-dependent diffusion equation. *Australian Journal of Physics, 13*, 1.

Philip, J. R. (1990). Inverse solution for one-dimensional infiltration, and the ratio A/K1. *Water Resources Research, 26*(9), 2023–2027.

Philip, J. R. (1992). Exact solutions for redistribution by nonlinear convection-diffusion. *The ANZIAM Journal, 33*(3), 363–383.

Philip, J. R., and Farrell, D. A. (1964). General solution of the infiltration-advance problem in irrigation hydraulics. *Journal of Geophysical Research*, *69*(4), 621–631.

Saberi-Nadjafi, J., and Ghorbani, A. (2009). He's homotopy perturbation method: An effective tool for solving nonlinear integral and integro-differential equations. *Computers & Mathematics with Applications*, *58*(11–12), 2379–2390.

Yıldırım, A. (2008). Solution of BVPs for fourth-order integro-differential equations by using homotopy perturbation method. *Computers & Mathematics with Applications*, *56*(12), 3175–3180.

Index

Printed in the United States
by Baker & Taylor Publisher Services